INTRODUCTION TO FORESTS AND RENEWABLE RESOURCES

SEVENTH EDITION

Grant W. Sharpe
University of Washington

John C. Hendee
University of Idaho

Wenonah F. Sharpe
Editor; Co-Author Fifth, Sixth, and Seventh editions

 **Higher Education**

Boston   Burr Ridge, IL   Dubuque, IA   Madison, WI   New York   San Francisco   St. Louis
Bangkok   Bogotá   Caracas   Kuala Lumpur   Lisbon   London   Madrid   Mexico City
Milan   Montreal   New Delhi   Santiago   Seoul   Singapore   Sydney   Taipei   Toronto

McGraw-Hill Higher Education

A Division of The **McGraw-Hill** Companies

INTRODUCTION TO FORESTS AND RENEWABLE RESOURCES
SEVENTH EDITION

This book is printed on recycled, acid-free paper containing 10% postconsumer waste.

3 4 5 6 7 8 9 0 QPD / QPD 0 9 8 7

ISBN 978-0-07-366172-8
MHID 0-07-366172-4

Publisher: *Margaret J. Kemp*
Sponsoring editor: *Thomas C. Lyon*
Freelance developmental editor: *Scott Spoolman*
Executive marketing manager: *Lisa L. Gottschalk*
Project manager: *Joyce Watters*
Production supervisor: *Sherry L. Kane*
Lead media technology producer: *Judi David*
Designer: *David W. Hash*
Cover photo: *©Corbis Stock Market*
Compositor: *Shepherd-Imagineering Media Services, Inc.*
Typeface: *10.5/12 Times Roman*
Printer: *Quebecor World Dubuque, IA*

Library of Congress Cataloging-in-Publication Data

Sharpe, Grant William.
    Introduction to forests and renewable resources / Grant W. Sharpe, John C. Hendee, Wenonah F. Sharpe.
        p. cm.
    Rev. ed. of: Introduction to forests and renewable resources / Grant W. Sharpe . . . [et al.]. 6th ed. (c) 1995.
    Includes bibliographical references and index.
    ISBN 0-07-366172-4
    1. Forests and forestry. 2. Forests and forestry—United States. I. Hendee, John C. II. Sharpe, Wenonah. III. Introduction to forests and renewable resources. IV. Title.

SD373 .I585 2003
634.9—dc21

200267098
CIP

www.mhhe.com

McGraw-Hill Series in Forest Resources

Avery and Burkhart: Forest Measurements
Dana and Fairfax: Forest and Range Policy
Daniel, Helms, and Baker: Principles of Silviculture
Davis and Johnson: Forest Management
Duerr: Introduction to Forest Resource Economics
Ellefson: Forest Resources Policy: Process, Participants, and Programs
Harlow, Harrar, Hardin, and White: Textbook of Dendrology
Knight and Heikkenen: Principles of Forest Entomology
Laarman and Sedjo: Global Forests: Issues for Six Billion People
Panshin and De Zeeuw: Textbook of Wood Technology
Sharpe, Hendee, and Sharpe: Introduction to Forests and Renewable Resources
Sinclair: Forest Products Marketing
Stoddart, Smith, and Box: Range Management

Walter Mulford was Consulting Editor of this series from its inception in 1931 until January 1, 1952.

Henry J. Vaux was Consulting Editor of this series from January 1, 1952 until July 1, 1976.

Paul V. Ellefson, University of Minnesota is currently our Consulting Editor.

ABOUT THE AUTHORS

GRANT W. SHARPE received his B.S.F. and M.F. from the University of Washington in 1951, and his Ph.D. there in 1956. He taught from 1956 to 1967 in the School of Natural Resources, University of Michigan, followed by 24 years in the College of Forest Resources, University of Washington. He retired in 1990. He was visiting scientist lecturer (SAF/NSF) (1964–1966), and Sigma Xi National Lecturer (1982–1984). He served on the Pacific Crest National Scenic Trail Advisory Council (1978–1981), and on the editorial boards of the *Journal of Forestry* and the *Journal of Leisure Research.* He is an elected Fellow in the Association of Interpretive Naturalists, and was the 1990 University of Washington Foresters' Alumni Association Honored Alumnus. In 2001, the Society of American Foresters awarded him the Golden Membership Award in appreciation for a half-century of advancing science, technology, education, and practice of professional forestry in America. He has been associated with *Introduction to Forestry* since the third edition. His other major publications include *A Comprehensive Introduction to Park Management* (Sagamore Publishing, 1994) and *Interpreting the Environment* (John Wiley, 1982).

WENONAH FINCH SHARPE writes on forestry and natural history and is a free lance editor with a B.A. degree, *magna cum laude,* in English from the University of Washington, where, upon graduating, she was elected to Phi Beta Kappa. Together with her husband, Grant W. Sharpe, and Clare Hendee, she co-authored the fifth edition of *Introduction to Forestry* and the sixth edition when John C. Hendee joined the co-authors. She also co-authored two editions of *A Comprehensive Introduction to Park Management* with Grant Sharpe and Charles

Odegaard, Pacific Northwest Regional Director of the National Park Service. From 1978 to 1994 she was an instructor for University of Washington Distance Learning classes in Forest Management.

JOHN C. HENDEE, now Professor Emeritus, was dean of the College of Natural Resources at the University of Idaho from 1985 to 1994, and then Professor and Director of the UI Wilderness Research Center from 1994 to his retirement in September 2002. He earned a B.S. degree in forestry from Michigan State University, a masters degree in forestry from Oregon State University, and a Ph.D. in forestry from the University of Washington. Prior to joining the University of Idaho in 1985, he worked for 25 years with the U.S. Forest Service in field forestry, research, and legislative affairs positions. He has authored or co-authored more than 140 articles, monographs, and scientific papers primarily on recreation, wilderness, wildlife, and human behavior aspects of forests and renewable resources. He is senior co-author of the textbook, *Wilderness Management*, now in its third edition, co-author of the book *Wildlife Management in Wilderness*, and founding managing editor in 1995 and now editor-in-chief of the *International Journal of Wilderness*. His professional work since 1994 has focused on the use of wilderness for personal growth and therapy and he assists his wife, Marilyn Riley, in leading trips with their nonprofit company, Wilderness Transitions, Inc. With his father, Clare Hendee (a co-author of the fourth, fifth, and sixth editions of this book), he developed the Hendee Tree Farm, which was named a Michigan Regional Tree Farm in the year 2000.

In Memory of Clare W. Hendee,
1908–2001.
Formerly, Deputy Chief of the
U.S. Forest Service and forestry
instructor at the University of
Maryland; co-author of the fourth,
fifth, and sixth editions of this book;
wonderful father, grandfather, and friend.
"Do the right thing anyway."

(Courtesy of USDA Forest Service)

CONTENTS

ix

PART V

Organization for Forest Resources Management

APPENDICES

PREFACE

It is said that the only constant in the world is change, and certainly policies and practices in the conservation and management of forests and other renewable resources are proof of this observation. This seventh edition of *Introduction to Forests and Renewable Resources* embraces these changes in the United States and beyond.

In Part I, the centrality of forests to the human enterprise is evident in Historical Uses, as well as in the story of the United States's awakening to the necessity of forest conservation and environmental protection. Policy is shown to be shaped by the pressures of events and ideas, underscoring its susceptibility to change. Policy today must respect major environmental protection laws and incorporate public input.

Part II deals with the distribution, management, and protection of forests, beginning with an overview of North American Forest Regions. Forest Ecology follows, wherein the focus shifts to the natural processes operating on trees to influence their growth in forest ecosystems.

The central management tool of forestry—silviculture and its several varied applications—and its evolution in meeting the emerging mandates for biodiversity and ecosystem management, comes next. A section on restoring and maintaining forest health reminds us that past practices have not always had good long-term results. Closely allied with this issue, the next three chapters in Part II address other forest health issues: Damage from Insects and Mammals, Disease and the Elements, and finally, the complex issues of Fire Management, an increasing focus in the United States under a new 10-year national fire plan.

Renewable resources associated with forests are examined in Part III. Gerry Wright's absorbing chapter on Wildlife Conservation and Management, and Mike Falter's discussion of Watersheds and Streams, are followed by Kendall Johnson's chapter on Rangeland Resources. These three chapters give full coverage to important forest and natural resource topics, thereby improving our understanding of important current issues.

Part III concludes with Outdoor Recreation and Wilderness Management, in which Steve Hollenharst joins the text's authors in addressing outdoor recreation benefits, providers, and management techniques. Next, the beginning and growth of the wilderness movement and principles of wilderness management are discussed.

Moving out of the forest and into the marketplace in Part IV, we start off with Harvesting Trees and associated considerations, going next, with contributions from Tom Gorman, to a complete coverage of Forest Products, stressing the role of efficiency and innovation in both these areas. A chapter on Economics illustrates principles of economic analysis applied to forests and renewable resources and how federal and state assistance programs seek to stimulate private forest productivity and conservation. Finishing up Part IV, Mike Vasievich covers traditional and new methods of Measuring and Analyzing Forests, including evolving computerized techniques and internet applications now affecting management.

Part V describes how all the activities are applied by federal, then state governments, and on private lands, including the timber industry and those private but nonindustrial owners who control a major portion of U.S. timberlands. These chapters present vital information for those confused over who owns what and how they manage it, and the emerging influence of green certification of forests and forest products. Also in Part V, Mike Bowman describes Urban Forestry, including the rewards and problems of trees, and in some instances, forest ecosystems existing side by side with high-density populations. In the final chapter of this section and of the book, we discuss the importance of International Forestry, including the organizations involved, global warming, and the advent of global markets and their effect on forests and renewable resources today.

This text, through its six previous editions, has been likened to an encyclopedia in that it has always had not only a broad coverage of the forestry enterprise, but also a comprehensive index and useful appendices. These features remain. The list of resource legislation, and the metric conversion factors, tree mammal and bird names, professional societies for natural resource personnel and influential environmental groups, still serve the inquiring student and general public, as does the glossary. New to this edition are numerous Internet references to information sources, a list of literature cited in each chapter, and a bibliography of suggested additional readings. At the request of instructors, study questions are included.

We are confident this seventh edition will introduce tomorrow's natural resource managers to the expanded scope of their responsibilities. Furthermore, this edition will assist those in related disciplines, as well as the general public, in understanding the complexities of forest and renewable resource conservation and management. We trust this edition will advance the stewardship of forests and renewable resources wherever it is used.

ACKNOWLEDGMENTS

A book on such a comprehensive field as forests and renewable resource management requires a good deal of help from our colleagues. As in the sixth edition, we called on several individuals to draft chapters on topics beyond our expertise, and they are listed here and in the particular chapters to which they contributed. Other colleagues contributed specialized material to supplement our efforts. Nevertheless, having reworked every chapter many times, we accept full responsibility for the technical accuracy of this seventh edition.

For contributing individual chapters we thank the following colleagues: Chapter 9, Wildlife Conservation and Management, Dr. Gerald Wright, National Park Service, Cooperative Park Studies Unit and Professor of Wildlife, University of Idaho; Chapter 10, Watersheds and Streams, Dr. Mike Falter, Professor of Fisheries, University of Idaho; Chapter 11, Conservation and Management of Rangeland Resources, Dr. Kendall Johnson, Professor and Head, Department of Range Resources, University of Idaho; the first half of Chapter 12 on Outdoor Recreation and Wilderness Management, Dr. Steve Hollenhorst, Professor and Head of Resource Recreation and Tourism Dept., University of Idaho; Chapter 14, Forest Products, Dr. Tom Gorman, Professor and Head, Department of Forest Products, University of Idaho; Chapter 16, Measuring and Analyzing Forests and Renewable Resources, Dr. Mike Vasievich, USDA Forest Service, Human Dimensions Unit, USDA Forest Service, East Lansing, Michigan; and Chapter 20, Urban Forestry, Mr. Michael Bowman, USDA Forest Service, retired and presently Urban Forester, Lewiston, Idaho.

For substantive contributions and material for individual chapters, we gratefully acknowledge material on: global warming, Dr. Peter Morrisette; international environmental education, Dr. Sam Ham, Professor of Resource Recreation and Tourism, University of Idaho; nonindustrial private forest landowner, Dr. Brett Butler, USDA Forest Service, NE Forest Expt. Stations.

For review of past editions of the textbook, we acknowledge Frank Armstrong, University of Vermont; William R. Bentley, State University New York College of Environmental Science & Forestry; Ronald W. Boldenow, Central Oregon Community College; John M. Edgington, University of Illinois; Paul Ellefson, University of Minnesota; John H. Harris, California Polytechnic State University, San Luis Obispo; Ray R. Hicks, Jr., West Virginia University; Maureen McDonnaugh, Michigan State University; Charles Newlon, Delaware State University; Richard G.

Oderwald, Virginia Polytechnic Institute; Doug Piirto, California Polytechnic University; Charles L. Shilling, Louisiana State University; Frank B. Shockley, Stephen F. Austin State University; and Carl Vogt, University of Minnesota.

For technical review of individual chapters we thank: Chapter 1, Dr. Dennis Roth, Historian, U.S. Department of Agriculture, Economic Research Service. Chapter 2, Mr. Terry West, USDA Forest Service, History Section, Washington, DC; Dr. Sally Fairfax, University of California, Berkeley, California. Chapter 4, John Marshall, Department of Forest Resources, University of Idaho. Chapter 5, Dr. Dave Adams, Professor of Forest Resources, University of Idaho; Dr. Ralph D. Nyland, Professor of Silviculture, State University of New York at Syracuse. Chapter 6, Dr. Robert Gara, Professor of Forest Entomology and Dr. Stephen West, Associate Professor of Wildlife Management, University of Washington; Dr. Dave Adams, Professor of Forest Resources, University of Idaho; Dr. Nancy Lee, Assistant Professor of Forestry-Fire Science, Central Oregon Community College. Chapter 7, Dr. Robert Edmonds, Professor of Forest Pathology, University of Washington. Chapter 8, Dr. Leon Neuenschwander, Associate Dean and Professor of Forest Resources, University of Idaho; Mr. Steve Robinson, Assistant Director for Fire and Law Enforcement, U.S. Bureau of Land Management, Boise, Idaho; Mr. Gardner Ferry, BLM National Fire Program Leader. Chapter 11, Dr. Charles Driver, Professor Emeritus of Forest Resources, University of Washington; Dr. Bill Laycock, Professor and Head, Department of Range Management, University of Wyoming; Mr. Charles B. Rumberg, Executive Vice President, Society for Range Management. Chapter 12, Dr. Perry Brown, Professor and Associate Dean of Forestry, Oregon State University; Dr. Jim Fazio and Dr. Ed Krumpe, Department of Resource Recreation and Tourism, University of Idaho; Dr. Alan Watson, Research Social Scientist, Wilderness Management Research Unit, USDA Forest Service, Missoula, Montana; Dr. Chad Dawson, Assistant Professor of Forest Recreation; State University of New York, Syracuse. Chapter 13, Dr. Leonard Johnson, Professor and Head, Department of Forest Products, and Dr. Harry Lee, Associate Professor of Forest Products, University of Idaho; Dr. Penn Peters, Logging Engineering Research Unit Leader, USDA Forest Service, Morgantown, W.Va.; Rick Niederhoff, forestry faculty member at Central Oregon College, Bend, Oregon. Chapter 14, Dr. Fran Wagner, then associate professor of Forest Products, Mississippi State University; Dr. John Erickson, Director, and Dr. Ted Wegner, Dr. Erv Schaffer and Dr. John Zerbe, Assistant Directors, USDA Forest Service, Forest Products Laboratory, Madison, Wisconsin. Chapter 15, Dr. John Fedkiw, Chief Forest Economist, U.S. Department of Agriculture; Dr. George Dutrow, Professor of Forest Economics, Duke University; Dr. David Jackson, Professor of Forest Economics, University of Montana; Art Benefiel, Associate Professor of Forestry-Fire Science, Central Oregon Community College. Chapter 16, Mr. G. Robinson Barker, Consultant, Bob Barker and Associates, Athens, Georgia; Dr. Harold Burkhart, Professor of Forestry, Virginia Polytechnic Institute and State University. Chapter 17, Ms. Gladys Daines and Mr. Gordon Small,

USDA Forest Service, Washington, DC.; Mr. Henry Kipp, USDI Bureau of Indian Affairs, Washington, DC.; Mr. Larry Biles, USDA Extension Service, Washington, DC.; Mr. Frank Snell and Mr. Keith Corrigall, U.S. Bureau of Land Management, Washington, DC. Chapter 18, Mr. Ron Abraham, Chief, Ohio Division of Forestry; Mr. John Mixon, Director, Georgia Forestry Commission; Mr. Bruce Miles, Texas State Forester; Mr. David Stere, Director of Planning, Oregon Dept. of Forestry; Mr. Richard A. Wilson, Director, California Dept. of Forestry and Fire Protection; Mr. Al Schacht, Director, Northeastern Area State and Private Forestry; and Mr. Bill Cannon, Asst. Director of Cooperative Forestry (ret.), USDA Forest Service, Washington, DC. Chapter 19, Dr. Con Schallau, Chief Forest Economist, American Forest and Paper Association; Dr. Jay O'Laughlin, Director, Idaho Forest, Wildlife and Range Policy Analysis Group; Dr. Keith Argow, President, National Woodland Owners Association; Dr. Steve Brunsfeld and Dr. John Marshall, Assistant Professors of Forest Resources, University of Idaho. Chapter 20, Dr. James Fazio, Professor of Resource, Recreation and Tourism, University of Idaho; Dr. Michael Kuhns, Extension Forester, Utah State University; Dr. Rita S. Schoeneman, Community Tree Specialist, USDA Forest Service, Washington, DC.; Mr. Gary Moll, Vice President for Urban Forestry, American Forests; Dr. J. Alan Wagar, Professor of Forest Resources, University of Washington; Dr. James Kielbaso, Professor of Forestry, Michigan State University, and President, International Society of Arboriculture; Ms. Michelle Mazzola, Director, Kansas Urban Forestry Council, Manhattan, Kansas; Dr. Leonard E. Phillips, Landscape Architect and author, West Peabody, Maine. Chapter 21, Dr. Thomas R. Waggener, Professor of Forest Resources, University of Washington; Dr. Jan Laarman, Professor of Forest Resources, North Carolina State University; and Mr. Douglas Kneeland, Mr. William Helin, Mr. Walter Dunn, and Dr. Gary Wetterberg of the USDA Forest Service International Forestry Staff, Washington, D.C. Beyond these contributions to this seventh edition, we know that this edition continues to benefit from contributed material and reviews by scores of colleagues in prior editions, particularly those mentioned in the preface of the sixth edition.

We also thank the many individuals, agencies and companies who provided photos for this and prior editions. Recognition is given with each photo. Finally, we thank Mrs. Susan Goetz for administrative secretarial assistance.

Grant W. Sharpe
John C. Hendee
Wenonah Finch Sharpe

Forests:

Historical Uses and Future Values

INTRODUCTION

Even as forests recede in size, they move closer to center stage in matters of global concern. Demand for wood products is higher than ever and continues to increase. Furthermore, attention is being focused on forest resources in new and unprecedented ways. Scientific research is developing new uses for wood as well as discovering the importance of forests and forest ecosystem processes to the global climate and the atmosphere. The understanding that forests are more than trees—that they are ecosystems whose health is reflected in the quantity and quality of water emerging from them and the biodiversity they support—has made some headway against exploitive attitudes in many parts of the world, our own included. We are also rediscovering the connection among forests, outdoor experiences, and the spiritual and emotional well-being of urbanized populations. Amid all this we must remind ourselves that millions of people still burn wood to cook their food and to keep warm, and that in many developing countries forests are still slashed and burned to provide land on which to grow food.

HUMAN USE OF THE FOREST

Even though we might think of our ancestors primarily as cave dwellers, the long prehistory of humans is deeply involved with forests and the diverse riches to be found there. Not only does it seem likely that we descended from tree-dwelling primates, but when the move to the ground was accomplished, the forest and its edge continued to function as a place of safety and as a source of food, fuel, clothing, and materials for shelter. Primitive dwellings with pole frames, as shown in Fig. 1-1, often had a thatch of grasses or bark or the skins of animals stretched over them. Later, logs and timbers were used. The forest has always provided wood for human use.

Nuts, berries, fruits, buds, and roots, as well as fish and game animals, were to be found in, or in association with, forested areas. The first tools, as well as the first weapons, were possibly splinters or branches. The discovery of how to keep and use fire, the idea of the wheel, and the concept of the lever are among the concepts that grew from our use of wood. From the stone-ax handle and the branch that served as a crude spit, through the period of wooden ships to the innumerable forest products of the present, hu-

FIGURE 1-1

A wigwam of poles and ash bark built in 1899 by Chippewas, Lac Courte Oreille Reservation, Wisconsin. *(Courtesy of the Smithsonian Institution)*

mans have coupled their inventive genius with the materials from the forest to make life both easier and more complicated.

Water represents another, and even more immediate, need of humans, but without the covering forest, watersheds cease to produce a regulated flow of clear water, and navigable channels may become silted in. Forests serve in this way wherever they exist.

In this new century and millennium, wood remains highly valued for its beauty, warmth, and reassuring "naturalness." Although many substitutes exist, wood itself appears indispensable in some situations. As it has yielded up its secrets to science, use of the chemical and structural components of wood fiber has undergone a rapid evolution in recent decades, and the results are pervasive, although often unrecognized. Besides paper in all its complex and vital forms, residues and chemicals from wood and various treatments of wood provide components for items ranging from photographic film to chewing gum and vitamins. Lignin, a basic component of wood, may some day yield food value for humans, if we can unlock the secrets that allow some animals to convert it to energy through their digestive processes.

Despite this highly technological use, part of the world continues to depend on wood for fuel and shelter; and shifting cultivation, which can deplete both forests and soils, is still found in developing countries, where it is intensified by population growth (Fig. 1-2).

FIGURE 1-2
Wood is the most accessible and cheapest source of energy for rural populations in developing countries. These rural Pretoria South Africans are bringing in water, and *acacia* branches for fuel. *(Photo by R. J. Poynton; courtesy of National Academy of Sciences)*

Given the productivity of the forest for domestic use, it is not unreasonable to believe that it must have been equally important in early defense and aggression. Weapons, fortifications, concealment—these and other uses are apparent in historical accounts, but it is the role of timber in naval activities that we see most clearly. Until the advent of the iron ships in the mid-nineteenth century, a supply of naval timber was an absolute necessity for nations seeking world trade and power. Concern over the scarcity of high-quality naval timbers in northern Europe in the sixteenth and seventeenth centuries prompted the planting and protection of desirable species. When one considers the amount of timber needed to build and maintain merchant vessels, let alone the huge, fragile naval fleets that rotted even as they were being built, or burned and sunk in battle and broke up as a result of storm or navigational damage, the political and strategic significance of certain forests becomes evident.

Trees and their associated plants and animals (assemblages that we call *forests*) from ancient times have figured prominently in animism and other religious expressions. The size, strength, and longevity of trees, added to their beauty and their bounty, inspire respect and even affection. Yet fear has tinged human attitudes toward the forests, too; at various times they were seen as the haunt of pagan gods or of Christian devils as well as savage beasts. In the new world, vengeful natives were added to this list. Forests serve as hiding places for illicit activities and for fugitives even today, as well as sanctuaries for those seeking privacy and solitude for more conventional reasons.

Emerging Uses and Values

The word *forest* originated in the idea of an area where the common people might not go. *Silva forestis* meant woods lying outside those for common use, or wooded areas reserved to protect the beasts of the forest and their habitat. At first this was done to protect the hunting grounds of Norman kings and nobles; later the reason was financial returns from use taxes and "exits," as forest products were called (Young, 1979). For Europeans, one of the great advantages of getting out from under feudalism and its lingering effects may have been the right to fish and hunt in the forests of the new world. This privilege is so much a part of our heritage that we take it for granted. Out of this use of the forest, the phenomenon of recreational use emerged in the twentieth century, and seems likely to continue and grow in the twenty-first. In the United States and

Canada, millions of people annually spend time in a forest setting and enjoy forest-associated outdoor activities. Since the 1950s, forest recreation has grown to the point that many forests are dedicated to such use, and some areas have even been altered by overuse for recreational activities.

The back-to-the-woods movement, on the other hand, may express and help advance a new spirit of reverence toward forests and other manifestations of nature, as evidenced in the environmental movement. Today a majority of Americans proclaim themselves to be environmentalists. Even during the recession-plagued 1980s, national environmental groups doubled their membership (Hendee & Pitstick, 1992) and have become a major influence on forest and natural resource policy. Today, following several years of prosperity, there are 8,000 environmental organizations, but the 20 largest groups took in 29 percent of the contributions reported to the Internal Revenue Service (IRS). The top 10 were listed in the *Chronicle of Philanthropy*'s list of America's wealthiest charities (Knudson, 2001) (see Appendix H). Millions of Americans have moved from cities to ever-more-distant suburbs, thus giving rise to the term *urban–wildland interface,* and creating a new set of problems for wildlife and pet owners, as well as for fire management in subdivisions or solitary homes that back up against forested areas. The move to protect forests, whether to save endangered species from extinction, such as the spotted owl and marbled murrelet in the Pacific Northwest and the red-cockaded woodpecker in the South, or to save ancient forests for their inherent values and scarcity, has grown to be a major political issue, as has the reintroduction of native predator species in remote areas. Homeowners in the wooded suburbs also struggle to keep the surrounding area untouched to protect their view and the value of their property.

More and more we see that Americans are in love with their forests. The rise of urban and community forests, much of it based on volunteer and citizen initiative, and the surging interest in the national forests, national parks, and wilderness as parts of our national heritage, reflect today's era of forest appreciation. But it was not always so—in earlier years a young nation turned to its forests as one of its major sources of wealth and material assets to build the country.

Forests for Growing People as Well as Trees

The American view of the environment has evolved to include concern for the conservation and rehabilitation of forests, as well as appreciation for the virtues of outdoor activity and work as a healthy endeavor for body and spirit. This is reflected in the popularity of outdoor recreation activities and volunteerism in conservation activities by many individuals and organizations. Forests are now also regarded as proper places for the inspiration and renewal of people, and the simplicity and rigors of outdoor life and work faced by early settlers have acquired a romantic quality in retrospect.

During the 1930s, as the nation faced serious economic recession and contemplated enormous needs for restoration of natural resources depleted by fire, logging, farming, drought, and erosion, Roosevelt's "new deal" for national rehabilitation included the Civilian Conservation Corps (CCC). From 1933 to 1943, the CCC played an important role in helping the country weather the depression by paying CCC enrollees to plant billions of trees, reclaim millions of acres of eroded land, and build thousands of parks, bridges, ranger stations, erosion control dams, and fire trails (Cohen, 1980). The CCC took a total of about 4 million unemployed young men off the streets, gave them hope, work, and money to send home, and restored their self-respect and dignity. Many learned to read and write, obtained diplomas, and perhaps learned a trade.

The CCC was among the first and most successful programs that used forests as "growing places for people as well as trees," to provide for human renewal, education, training, and inspiration (Fig. 1-3, upper). Such programs also included federal public works programs such as the WPA (Works Progress Administration), NIRA (National Industrial Recovery Act), and APW (Accelerated Public Works) during the depression and national recovery era of the 1930s through the 1950s. During the environmental awareness era of the 1960s and 1970s, federal youth programs emerged such as the Job Corps, YACC (Young Adult Conservation Corps), and YCC (Youth Conservation Corps), all of them seeking to provide environmental education as well as healthy outdoor work. This same era saw the emergence of many other conservation–human development programs such as the widespread volunteer efforts that included all age groups from youth to mid-career

FIGURE 1-3
Helping both individuals and the nation through conservation work is a time-honored tradition in the United States. *Upper:* CCC enrollees thinning lodgepole pine in the Medicine Bow National Forest, Wyoming, 1933. *(Courtesy of USDA Forest Service) Lower:* A modern Youth Conservation Corps crew works on a stream improvement project in Blue Knob State Park, Pennsylvania. YCC was the first program to require an equal number of males and females in the work crew, between 15 and 18 years of age. *(Courtesy of Pennsylvania Department of Environmental Resources)*

teachers to senior citizens working summers in parks or forests; many of them paid subsistence wages or expenses through various agencies.

From the 1980s to today, new and more diverse programs have evolved at state and local levels and in the private sector aimed at using forest-based experiences to develop human potential and positive qualities (Fig. 1-3, lower). Youth and adult state-sponsored conservation corps programs, such as the California, Pennsylvania, and Arizona Conservation

Corps, and community-based programs such as the Los Angeles Conservation Corps, are now widespread. The urban-based programs do traditional conservation work including trail construction and maintenance, but are also a resource for help following traumatic events such as riots, tornados, hurricanes, or earthquakes.

A related and striking development in the past three decades has been the emergence of outdoor experience and adventure-based programs, sponsored by private, nonprofit education enterprises and used by the public, government, and business corporations for personal growth, therapeutic adjustment, and leadership development of participants. The best known of such programs are Outward Bound and the National Outdoor Leadership School (NOLS), but there are hundreds of similar programs aimed at developing self-confidence, mental strength, physical skills, and inspiration in people. These programs make up a growing wilderness- and wildland-experience industry (Friese et al., 1998). They aim at youth development as well as therapeutic purposes such as strengthening recovery for victims of alcoholism, drug addiction, grief, and mental illness, and the adjustment of persons struggling with problems of poverty, delinquency, or other life changes (Russell & Hendee, 2000). "Outdoor adventure training," much of it forest- or outdoor experience-based, is now widely used in private business for the development of leadership and competitiveness among employees (Petrini, 1990).

Thus, forests in today's world can be a place where people retreat to renew their humanity through simplified living and contact with the primal forces of nature in respite from the hectic pace of modern civilization.

BUILDING THE COUNTRY WITH FORESTS

It is no exaggeration to say that without forests, there would have been no United States of America as we know it, yet it must be remembered that in building the nation, forests were both in the way and in demand.

Where natural resources are abundant, they are logically used to meet the most pressing demands at that time and place, often leaving much of the resource underutilized or wasted by later standards. Then, when basic necessities such as food and shelter are met, more complete and refined use of the re-

source is possible to meet other demands. This was the case in colonial times as the clearing of forests for farming and use of wood for homes, fences, and fuel slowly expanded into crude extracting, manufacturing, and wood-based exporting industries. Ship masts, boards, staves for casks, and potash from wood ashes, tar, pitch, and turpentine were the principal items of trade. This expansion was inevitable. Together with fish and tobacco, forest products could be bartered or converted into money with which to buy the commodities the colonists could not produce. Ambitious, shrewd, and thrifty individuals wanted to make as good a living as possible, and the still-vague concept of land and timber as property at this time offered unusual freedom for logging and other forest businesses.

Some historians have interpreted the history of forest practices as a positive part of the building of the country rather than as pure exploitation. The following example is taken from Nash (1968).

> Most accounts of American conservation contain at some point an elegiac description of the "unspoiled" continent: once the country was beautiful and rich in resources, but then came the "greedy exploiters" who "raped" the "virgin" land. Such a representation unjustly uses the emotions of the present to describe the actions of the past. It fails to employ historical sympathy, to understand the past in its own terms. Neither the pioneers nor most subsequent resource developers considered themselves unthinking spoilers or were regarded as such by their contemporaries. Instead they acted in a manner consistent with their environmental circumstances and intellectual heritage. When the forest seemed limitless, cut-out-and-get-out was an appropriate response. Certainly early Americans made mistakes in using the land, but they became such only in the opinion of later generations. Rather than shaking moralistic fingers, conservation historians would do well to attempt to understand why men acted as they did toward their environment.

As is pointed out in Chapter 15 on Economics, forests worldwide have gone, and continue to go, through a familiar pattern of exploitation, recovery, management, and finally, appreciation. Economic and social conditions of the time dictate how forests are viewed and treated.

Forest Industries in Colonial Times

The first sawmill in America was in York, Maine, in 1623, and a second in South Berwick, Maine, in

1634 (Howard, 1985). The circular saw was still to come, and the crude sash saws, driven by water power and slicing away vertically, could be tended by a crew of two or three hands and might produce 500 board feet a day. The boards that were exported in 1631 from New Hampshire to England were possibly rived or hewed. At any rate, they found a ready market, and eastern white pine, a soft, clean, fragrant, and reasonably strong wood, entered a market it was to dominate for 250 years.

Meanwhile, the rulers of England were counting on the forests of these new colonies to supply the mast timber for the Royal Navy. Regular traffic in ship masts began in 1645 (Howard, 1985). By about 1695, the surveyors-general of William and Mary and their deputies went dutifully about, picking out the best and straightest trees and reserving them with the king's broad-arrow mark (Fig. 1-4). This was done in accordance with the new charter of 1691 issued to the Province of Massachusetts Bay in New England, which set aside as property of the Crown and for the use of the Royal Navy all trees more than 24 in (60 cm) in diameter and 1 ft (30 cm) from the ground. The decree was in force throughout areas comprising present-day New England, Nova Scotia, New York, and New Jersey, and penalties were provided for violation, though they were not quite as harsh as those in medieval England, where, for a time, castration was the punishment for disobeying forest laws (Young, 1979).

The reaction of colonial lumbermen to the broad arrow announced a new world attitude, which declared in effect, "We're a long way from the old country; her needs may be important, but the wilderness we live in is for us—let the job go on." And with that, the axes continued to ring in the forest, and the whips cracked over the oxen. This was in spite of the fact that England's navy was fighting to maintain its mastery over the French and to defend its commerce with the American colonies, all of which constituted service to those colonies.

The desire of the new country for a good international reputation for its forest products is reflected in various early regulations of exports imposed by the colonial legislatures. Masts must be without defect, staves of good sound wood, and those who would set fires to a forest that could produce these merchantable goods were to be punished. Trade was something to keep alive. Customers must be kept

FIGURE 1-4
The king's broad arrow. The trees so marked were to be reserved for masts for the English navy during colonial days. Such a restriction was one of the irritations that led to the Revolutionary War.

supplied, whether Portuguese, Spanish, or British. Other laws had to do with abating thievery and keeping the wharves free of piled-up timber waiting for export—which had by now acquired the name *lumber.* The idea of conservation as wise use of resources was not to be found in these early laws and regulations, and yet their flavor of self-interest indicates some respect for the forest as a source of valuable products.

Building a Wooden Navy Following a period of growing resentment against the broad-arrow policy and of frayed nerves over other family quarrels with England came the hard-fought war of the American Revolution, and out of that emerged a young nation. One of the first surprising pieces of legislation was the reenactment in 1783 by the Massachusetts Assembly of a "broad-arrow" policy of its own, which later, in modified form, became the policy of the nation in the Forest Reserve Act of 1891.* The United States had a commerce to protect from pirates, and wood to build ships must come from the same forests that England had sought to reserve for her navy (Fig. 1-5).

---

*A list and brief description of legislation cited in this textbook is found in Appendix A.

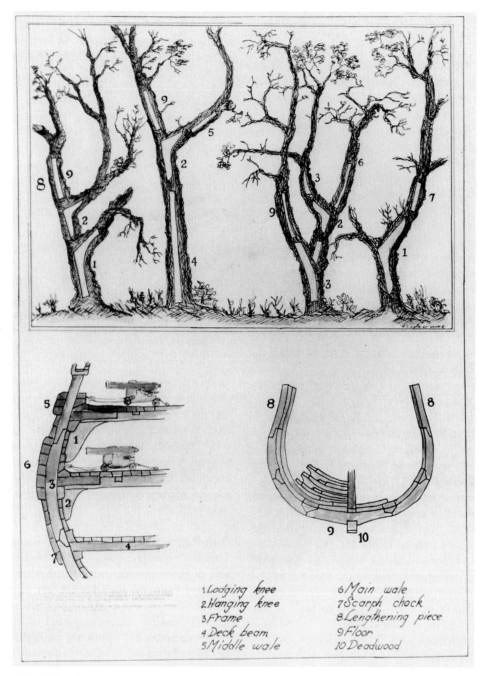

1. Lodging knee
2. Hanging knee
3. Frame
4. Deck beam
5. Middle wale
6. Main wale
7. Scarph chock
8. Lengthening piece
9. Floor
10. Deadwood

FIGURE 1-5

This drawing appeared in a book for naval architects in 1789. *Upper:* Shows shipbuilders where on the oak tree they would find pieces of the desired shape. *Lower:* Shows how the pieces were to be fitted together in the construction of the ship's hull. *(Courtesy of U.S. Naval Historical Center)*

FIGURE 1-6
U.S. Navy ships of the Mediterranean Squadron departing the Balearic Islands, off Spain, January 26, 1825. From left to right, ship of the line *North Carolina,* frigates *Constitution* and *Brandywine,* and sloops *Erie* and *Ontario.* American sea power was highly dependent on forests for ship timbers. *(Courtesy of U.S. Naval Historical Center)*

Following the establishment of the United States Navy in 1798 and up to 1833, several laws were enacted that provided for reserving lands on the Florida Gulf coast as live oak forests for the use of the Navy, purchasing actual ship timber and storing it for naval use, policing the federal live-oak reserves, and even authorizing the purchase of land for growing live oak. Although these purchases and experiments failed to achieve the objective of a continuous and satisfactory supply of naval construction material, they do reflect the fact that forests and sea power in those days were intimately related (Fig. 1-6). Even then, the sustainability of forest productivity was a concern.

Forests and the Early Villages It is of particular interest that more than the material products of forests were recognized early on as valuable forest dividends. One of these "extras" was the power of forest cover to hold soil in place. For example, as a result of severe grazing of domestic animals and some very heavy timber cutting near the town of Truro on Cape Cod peninsula, the sand was beginning to invade town lands. In 1709, the Massachusetts legislature attempted to regulate such practices to forestall the encroachment of sand on the village. The productive character of soil enriched by the forest, and the necessity to protect it from injury by forest fires, were emphasized by fire laws in Massachusetts in 1743 and in North Carolina in 1777.

Also of significance in the literature of colonial days is the mention of local shortages of wood. It was, of course, natural to use first the forest that was at hand, as long-distance or rapid transport had not yet been developed. There was a lack of wood for particular purposes in certain localities after the American Revolution, and as early as 1720 the French in Canada had been forced by local shortages to enact severe timber trespass laws (Cameron, 1928). But even before this, some communities in

New Hampshire had limited the number of trees a citizen might cut at any one time, and in one village a fine was imposed for every tree taken other than for home use.

An interesting comment quoted by Cameron (1928) illustrates the importance of wood as a fuel during colonial days in this country. After the Revolutionary War, a surgeon who had been with the Hessian troops revisited the new nation and worried in his writings lest farmers and landowners should not "be taught how to manage their forests so as to leave for their grandchildren a bit of wood over which to hang the teakettle."

Industrial Revolution Finds the Lumber Business

As in other instances where established trade relations are broken by war, the interruption of trade with England during the American Revolution and the War of 1812 promoted in America what is now known as *economic nationalism*. The new nation sought to make itself industrially as well as politically independent of the old world. The factory system, which had made a weak start in Rhode Island after the Federal Tariff Act of July 4, 1789, finally became well-established by 1815. Although cotton mills were the first manifestation of the jump of the industrial revolution across the Atlantic, there is little doubt that boom times and the stimulation of invention affected the embryonic forest industries. Local wood demand had not yet expanded appreciably, but population was moving westward. The first steam sawmill started at New Orleans in 1803, and river rafting of logs was by then a well-established technique (Fig. 1-7). At about this time also, the wood-burning locomotive, which at first traveled on wooden rails, made its appearance, thus heralding acceleration in timber use as well as the dawn of widespread rapid transportation. The first logging locomotive, however, was not to arrive until 1852. Another steam sawmill was built in 1830 near Pontiac, Michigan. But the center of lumber production was still in southeastern Pennsylvania. The United States had entered upon that important period between 1820 and 1870, when population quadrupled. There was an accompanying demand for forest products, and the home market became more important than foreign shipments of masts, boards, and staves.

Subsidizing Farms and Transportation with Land and Forests

By the mid-nineteenth century agricultural expansion had pushed beyond the central hardwood forest, and the prairie communities of Illinois, Iowa, and even Kansas were looking to the day when they would be states instead of territories. The sod-house farmer of the prairie had wood problems unknown to the log-cabin dwellers of West Virginia or Ohio. To the latter, after the cabin, barn, fences, and woodpile were accounted for, the forest was literally in the way. For the prairie farmers, lumber for their farm buildings, the towns where they must trade and market their crops, and the railroad ties on which these crops were to be shipped all had to come from the forests of the states bordering the Great Lakes (the Lake states) (Fig. 1-8).

During the mid-nineteenth century, the United States had rounded out its boundaries by the annexation of Texas, and by other cessions, purchases, and settlements. Two important groups of land laws enacted between 1841 and 1871 speeded up the development of widespread rapid transportation, the settlement of the country, and the exploitation of forests. The first group of laws included the Preemption Act of 1841, the Homestead Act of 1862, and others of less importance. The Preemption Act enabled heads of families with certain qualifications to cheaply purchase 160 acres of public land for their own settlement, use, and benefit. The Homestead Act was similar but broader in scope. It provided free patent after five years of residence and cultivation, and the claimant could elect to "commute" or pay up and secure patent at any time six months after the time of filing. These two laws induced settlement but also made forest exploitation the standard way to operate.

The second group of laws had to do with grants of public land to subsidize internal improvements, such as post roads, canals, and railroads, and grants to states and territories in aid of education. The builders of a railroad, for example, were to have as an incentive, in addition to an adequate right-of-way, a gift of land including each alternate square mile or "section" in a belt 6 miles (9.6 km) wide on each side of the road. They were also to have the right to select lands elsewhere in lieu of those due them but already granted to other owners. The remaining government sections were to be sold at $2.50 per acre,

FIGURE 1-7
Waterways were the highways for timber and log transport in America before the coming of the railroads. *Upper:* Log driving on the Ausable River in New York State in the late 1800s. Here log drivers keep loose logs moving. *(From Bureau of Forestry Bulletin no. 34, 1902. Courtesy of Forest History Society, Inc.) Lower:* Rafting logs on the Mississippi River about 1895. The steamer in the foreground is pushing; the bow boat at the other end of the raft gives the raft its direction by moving forward or reverse. *(From A Raft Pilot's Log, the Arthur H. Clark Company)*

FIGURE 1-8

Two of the thousands of logging crews in the states bordering the Great Lakes, supplying the needs of the settlers as they moved westward to the treeless prairies. An interesting comparison between winter and summer log hauling is shown in these photographs. *Upper:* This sleigh of logs, purported to be the largest ever hauled by one team, was hauled from Minnesota to the Chicago World's Fair in 1892. Winter logging made use of a wide bunk sleigh. Because of the almost frictionless action of the weighted runners on the ice, large loads could be hauled great distances. *(Courtesy of St. Louis County Historical Society, Duluth, Minn.) Lower:* Summer logging near Newaygo, Mich., about 1875. Trucks (wheeled bunks) were narrower than sleighs and were loaded with fewer logs, as horses were able to haul far less by wheeled vehicles than by sleigh in winter. For this reason, areas near rivers were logged in summer; the areas further away were left for winter. *(Courtesy of The University of Michigan)* and the railroad company was to sell its acquired lands at a low fixed price to actual settlers. The railroad was then obligated to transport troops, supplies, and mail free of charge or at equitable rates fixed by the Congress.

The first of these railroad land-grant acts, passed in 1850, subsidized construction of the Illinois Central Railroad. The transcontinental railway grants were made from 1862 on, and by the close of President Ulysses Grant's administration in 1877, more than 155 million acres (62.7 million ha) of land had been allotted by the United States in place of money subsidies for internal improvement (Donaldson, 1884).

Most of this land was unfit for agriculture, but much of it was heavily timbered. No attempt was made to inventory soil, cover, or minerals before the grants were made. As might be expected, therefore, title to vast areas of good timberland passed to private owners and, from them, to large timber and mining corporations. This was particularly true in the non-agricultural sections of the Lake states and in the west.

The development of transportation brought settlement and farms on the better soils, inspired the establishment of towns and cities, and generally invited the population to spread out over the middle west and prairie states (Fig. 1-9). Thus, a new and great demand for building material was created, and by 1870, close to 270,000 lumber manufacturing plants were in operation.

FUTURE VALUES

There is no question that forests are increasingly valued worldwide, and for new uses that may directly compete with traditional uses.

In rural areas of developing nations, wood for fuel and heating is a pressing need. Here too, forests may be slashed and burned to create land for growing food in a destructive cycle of forest removal, soil depletion, and abandonment (Peters & Neuenschwander, 1988). The same nations may be granting, in one form or another, timber harvest rights to domestic, foreign, or multinational companies in order to meet their countries' needs for wood products and other goods and to realize capital from such exports. These practices are depleting forests in developing nations but they reflect the typical first stage in the historical approach to forest management as nations develop. "Exploitation," hopefully followed by "recovery,"

"management," and finally "appreciation," has been the pattern in the United States and other mature economies. We logically hope for the same pattern in developing nations.

At the same time that forest waste and destruction may continue in developing nations, new uses for wood and its elemental fiber and chemicals continue to evolve in developed nations, leading to increasingly strict utilization standards and to innovative uses for previously discarded wood. Demand for wood and wood products, driven by expanding populations and consumer-oriented lifestyles, continues to grow, but at the same time appreciation of forest resources has also grown. Forest policy battles in the United States now focus on how much of the remaining old-growth forests and roadless areas are to be preserved. Stewardship of all forest and range habitat through "ecosystem management" to protect endangered species and ecological services such as clean air, water, soil, and biodiversity, has become recognized and accepted as a desirable goal.

So how will these extremes affect the future use and values of forests? We see hopeful signs and trends. Increased speed of communication and access to information over the Internet and satellite-transmitted media have made the world smaller. More people than ever before are aware of environmental conditions and practices around the world. While developed nations account for most of the wood consumption in the world, their citizens are also most appreciative of forests and other natural values, and most concerned about alarming trends such as global warming and species extinction. This awareness has given rise to developments that could not have been predicted even a decade ago. A "greening" of the wood products industry is occurring in response to consumer environmental concerns. Many major wood products companies, including multinational corporations, now certify that their raw materials come from well managed forests. Ecotourism has become a huge industry worldwide, thereby providing incentives for developing nations to preserve their forests and wildlife to attract tourists from developed nations. Tourism revenues will be far more sustainable than will exploitive practices.

Another valuable attribute of forests lies in their ability to absorb emissions of carbon dioxide, the main heat-trapping gas produced by human activities. Climate change is now an international concern

FIGURE 1-9

These Currier and Ives lithographs idealize homesteading and prairie settlement but allow us a glimpse of the conditions in each situation. *Upper:* "The Home of the American Pioneer." The clearing for cabin and cornpatch in the deciduous forest, and the bountiful game borne home by the buckskin-clad father, show the best of frontier cabin life. *Lower:* "The Railroad Spans the Continent." Note the heavy use of wood in the frontier village. The land-grant railroad spread settlement to the treeless prairies and created a new demand for wood. By 1900 there were nearly 200,000 miles of railway lines, resting on wooden ties and accompanied by wooden telegraph poles.

and priority, with protection and establishment of forests worldwide seen as important to mediating global warming and extreme weather phenomena linked with it. A "debt swap for conservation" movement encourages developing nations to protect resources and adopt conservation practices in return for having their foreign debt forgiven or reduced.

Environmental concerns among the fortunate people living in developed nations, expressed through their actions as consumers demanding green products, as ecotourists, and as citizens encouraging and supporting forest environmental protection at home and abroad, are making a difference. Forests are growing in value as objects of appreciation as well as for consumptive use, and also as a means to help combat adverse global weather changes. The mandate for sustainability in forest management and use has never been so clear or widespread. It gives us hope. But can it withstand the counterforces of population growth and associated exploitive pressures for survival? Will this mandate be heard above the internal and external armed conflicts in which control of natural resources may be at stake but that nevertheless damages them? The outcome is uncertain.

SUMMARY

In the closing decades of the twentieth century, recreational and therapeutic values of forests, other wildlands, and wilderness areas, as well as appreciation for their beauty and ecological complexity, have become recognized in the more developed countries. These trends will no doubt continue.

Though we see exploitation and misuse arising from past practices, in judging the results of land distribution policies, we must remember that land, and also forests (because they occupied so much of the land), were used in place of money—an asset the young nation did not have. Yet, the vast, empty lands had to be settled, or they might be lost to settlement by rival nations. Further, by subsidizing internal improvement, the government opened up a market for land and timber. The fact that these natural resources did not bring a good price and that fraud was practiced in many transactions is unfortunate, but the land laws did generate much-needed revenue for the treasury and assisted in putting a good deal of land into the ownership of homesteading families.

Beyond the current use of forests and other natural resources, which is the content of this text, we see changing future values that may derive from forests. The paradox of prodigious consumption by developed nations, coupled with our widespread appreciation of the forest environment, gives us pause, but also hope. As the dangers of global warming and its connection to deforestation become well known, consumer pressures are being applied to initiate sustainable forest practices in all countries that supply forest products. A growing desire to see natural areas protected everywhere, and increasing worldwide ecotourism further encourage developing nations to protect at least some of their forests. But war, greedy manipulation of resources, and overpopulation temper our hopes for the future of forests and renewable resources.

LITERATURE CITED

Cameron, Jenks. 1928. *Development of Governmental Forest Control in the United States,* Johns Hopkins Press, Baltimore, Md.

Cohen, Stan. 1980. *The Tree Army: A Pictorial History of the Civilian Conservation Corps,* Pictorial Histories Publishing Co., Missoula, Mont.

Donaldson, Thomas Corwin. 1884. *The Public Domain,* Washington, D.C.

Friese, G., J. C. Hendee, and M. Kinziger. 1998. The Wilderness Experience Program Industry in the United States: Characteristics & Dynamics. *J. Experiential Education* **21**(1):40–45.

Hendee, John C., and Randall Pitstick. 1992. "The Growth of Environmental and Conservation-Related Organizations 1980–1991," *Renewable Natural Resources Journal* **10**(2):6–19.

Howard, Theodore E. 1985. The Lore of Eastern White Pine. From Eastern White Pine: Today and Tomorrow. Symposium Proceedings. U.S. Department of Agriculture, Forest Service General Technical Report WO-51.

Knudson, Tom. 2001. Fat of the Land: (Environmental) Movement's Prosperity Comes at a High Price, *Sacramento Bee* Special Report. April 22, 2001.

Nash, Roderick (ed.). 1968. *The American Environment: Readings in the History of Conservation,* Addison-Wesley Publishing, Reading, Mass.

Peters, William J., and Leon Neuenschwander. 1988. *Slash and Burn: Farming in the Third World.* University of Idaho Press, Moscow.

Petrini, Catherine M. 1990. "Over the River and Through the Woods." *Training and Development Journal,* May 1990, pp. 25–36.

Russell, K. C., and John C. Hendee. 2000. *Outdoor Behavioral Healthcare: Definitions, Common Practice, Expected Outcomes, and a National Survey of Programs.* Tech. Report 26. Forest, Wildlife & Range Expt. Sta. Moscow, Id. 87 pp.

Young, Charles R. 1979. *The Royal Forests of Medieval England,* University of Pennsylvania Press, Philadelphia.

ADDITIONAL READINGS

Alexander, Thomas G. 1989. "Timber Management, Traditional Forestry, and Multiple-Use Stewardship: The Case of the Intermountain Region 1950-1985," *J. Forest History* **33**(1):21–34.

Beazley, Mitchell (ed.). 1981. *The International Book of the Forest,* Simon and Schuster, New York.

Dana, Samuel Trask, and Sally Fairfax. 1980. *Forest and Range Policy: Its Development in the United States* (2nd ed.), McGraw-Hill Book Co., New York.

Davis, Richard (ed.). 1983. *The Encyclopedia of Forest and Conservation History,* Macmillan, New York.

DiCerbo, G. C. 1988. "Legislative History of the Youth Conservation Corps," *J. Forest History.* **32**(1):22–31.

Fernow, B. E. 1911. *History of Forestry,* University Press, Toronto.

Frederick, Kenneth D., and Roger A. Sedjo (eds.). 1991. *America's Renewable Resources: Historical Trends and Current Challenges.* Resources For the Future, Washington, D.C. 296 pp.

Hotchkiss, George W. 1889. *History of the Lumber and Forest Industry of the Northwest,* George W. Hotchkiss & Co., Chicago, Ill.

Ise, John. 1920. *United States Forest Policy,* Yale University Press, New Haven, Conn.

Miller, Char. 2000, Spring. "Back to the Garden," *Forest History Today,* 16–23.

Miller, Char, and Rebecca Staebler. 1999. *The Greatest Good: 100 Years of Forestry in America.* Soc. American Foresters, Bethesda, Md. 125 pp.

Pyle, Charlotte, and Michael P. Schafale. 1988. "Land Use History of Three Spruce-Fir Forest Sites in Southern Appalachia," *J. Forest History* **32**(1):4–21.

Reynolds, R. V., and A. H. Pierson. 1925. "Tracking the Sawmill Westward," *Amer. Forests and Forest Life,* **31**:646.

Steen, Harold K. (ed.). 1999. "Forest and Wildlife Science in America: A History," Forest History Society, Durham, N.C.

Whitney, Gordon G., and William C. Davis. 1986. "From Primitive Woods to Cultivated Woodlots: Thoreau and the Forest History of Concord, Massachusetts," *J. Forest History* **30**(2):70–81.

Williams, Michael. 1987. "Industrial Impacts of the Forests of the United States 1860–1920," *J. Forest History* **31**(3):108–121.

Wright, Jeffery A., Anthony DiNicola, and Eduardo Gaitan. 2000. "Latin American Forest Plantations: Opportunities for Carbon Sequestration, Economic Development, and Financial Returns." *J. Forestry* **98**(8):20–23.

WEBSITE

www.foresthistory.org

STUDY QUESTIONS

1. Discuss some historical uses of forests and new uses that are evolving.
2. Discuss why people engaged in forest exploitation at various periods in our nation's history and how this was important at the time.
3. What role did forest resources play in the discontent of the American colonists?
4. In what sense are forests important for growing people as well as trees?
5. How might our appreciation of forest values change forest management in other countries around the world?

Forest and Renewable Resource Policy:

Historical Development and Current Applications

INTRODUCTION

Chapter 2 provides definitions and explanations pertinent to forest and renewable resource policy and the groups involved in its formation and implementation. The evolution of such policy is reviewed, covering the historical era, as well as the post-World War II-to-current era, with an emphasis on the new directions of the federal natural resource agencies as they respond to environmental protection legislation. Public sentiment, public involvement and the environmental movement are emphasized as major influences. The application of current natural resource policy is illustrated with two case studies and examples. Federal forest policy receives the most attention because of its formative influence on the management of state and private lands. More specific information on forest management by the states is found in Chapter 18, on private lands in Chapter 19, and in relation to forest economics, in Chapter 15.

Key legislation shaping forest policies is found in Appendix A, professional societies influencing policy are listed in Appendix F, and major environmental organizations influencing policy appear in Appendix H.

THE RELATIONSHIP OF POLICY AND LAW

It is useful to distinguish between *policy* and *law.* For our purposes here, policies are governing principles, plans, and courses of action that guide the management of forests and renewable resources (Cubbage et al., 1992). Policies are often broad and diffuse—an expression of goals together with procedures for achieving them. They give direction to how resources will be managed. The mission of an agency may be defined here, along with the philosophical reasons for its existence and actions.

Law, both legislative and administrative, is the encoding of regulations to achieve the objectives of policy. *Legislative law,* embodied in legislation, bills, or legal codes, is usually broad and general in its guidelines for action, although it may contain specific provisions or prohibitions. Federal law emanates from the U.S. Congress. State law emanates from state legislatures. *Administrative law* deals with specifics and is comprised of the regulations agencies formulate in order to comply with the mandates of legislative law.

The ambiguous nature of policy is evident when we realize that, although policy might be presumed to represent a settled or definite course or direction to be followed by a governmental organization or individual, it is only settled and definite in a relative sense. Policy is not, and cannot be, immutable. Time inevitably brings new conditions and thus, revisions in directions for management, so policies are in a constant state of flux. For example, in a broad sense the policy of the U.S. government toward federal lands has evolved over the past 150 years from one of disposal in order to get them in private hands; to policies aimed at stopping the disposal, waste, and destruction of resources; to support for acquisition and scientific management of these lands and other natural resources; to concern today for protection and restoration of the environment along with sustainable uses.

Policy regarding the management of resources—including forest and range policy—is both a major influence on, and an expression of, the social, economic, and political structure—the culture—of the United States (Dana & Fairfax, 1980). This is also true in other countries. Ideas regarding the appropriate approach to forests and renewable resources evolve as nations change and develop.

Actors in Policy Formulation

Policies must be broad enough to provide latitude for responsive decision making yet be specific enough to enable the decision makers to accurately reflect intent (Sharpe et al., 1994). The formulation of policy results from the interaction of several groups. The actors in federal policy formulation, both executive and legislative, are categorized here in four groups.

The Bureaucracy Public agency employees, and the politically appointed agency officials of the executive branches of government and their supporting staffs, comprise the bureaucracy. In the federal government, the departments most involved in forest and renewable resource issues are the Departments of Agriculture and Interior, their respective resource agencies, and the Environmental Protection Agency (EPA), as well as the Office of Management and Budget (OMB). For example, the Forest Service in the Department of Agriculture is answerable to the secretary of agriculture. They are constrained in their policies and programs by environmental regulations administered by EPA, and budgets and fiscal regulations administered by OMB. The Forest Service is directed by a chief, several deputy chiefs in functional areas, and their staffs. Below these are regional foresters, national forest supervisors, and district rangers and their staffs, who implement policy. This work force, with whom final implementation of policy rests, consists of personnel trained in many disciplines, such as foresters, fish and wildlife biologists, range conservationists, and water quality, recreation, and cultural resource specialists.

Most policies emanate from federal or state legislation. With the growing complexity of government, agencies have taken on a bigger role in formulating regulations to implement broad policies set by legislation or by the agency itself.

Society   This includes the social force of individual voters who express themselves through election of federal, state, and local leadership, and of organized voting blocks of people who adhere to certain directional principles (e.g., Democrats or Republicans). Societal forces have become a stronger influence on policy formation during the past three decades because of the legal requirements that the public be involved in developing natural resource policy and in specific decisions where environmental impacts may occur. Citizen environmental organizations (see Appendix H) have grown tremendously as an influence on natural resource policy, as they now have memberships of millions, large budgets, technical staffs, and networks of communication through magazines, newsletters, e-mail list servers, and Websites (Hendee & Pitstick, 1992; Knudson, 2001).

Consumers are also becoming a major influence on policy through what they choose to buy. This is demonstrated by the demand for so-called "green forest products"— certified as coming from forests meeting high standards of management. This has led to recent adoption of certified wood products by major home-building outlets (e.g., Home Depot)— thereby also pressuring wood products producers to meet forest management standards qualifying for certification (Forest Stewardship Council, www.fscus.org).

Industry   Many industries use natural resources. An obvious example is the forest products industry, which harvests timber as a raw material for making lumber and paper. Other examples are the livestock industry, which grazes rangelands; the tourism industry, using wildlands for recreation; and the many industrial manufacturers who use water for cooling, power, or waste disposal. These industries not only have an economic interest in resource policy, but technical expertise in many matters. Industries organize and express their vested interests through trade associations such as the American Forest and Paper Association and the American Tree Farm System, as well as cattle associations, woolgrowers, associations of recreation interests, and many others. Labor unions may represent worker interests in natural resources utilization.

Natural Resource Professions   There are many professional groups, societies, and organizations to which persons with education or expertise in particular fields belong, such as the Society of American Foresters, Wildlife Society, American Fisheries Society, Society for Range Management, and others. Members of these professional societies and groups usually make their living in a natural resource field, and their professional society or organization keeps them abreast of issues through its meetings and journals. A few organizations, such as American Forests (formerly the American Forestry Association), have both public and professional members. Descriptions of the major natural resource professional societies and organizations are found in Appendix F.

All these categories of groups—the bureaucracy; society, including political parties and citizens; environmental groups; industry; and the natural resource professions—may advocate their various points of view to the U.S. Congress, its separate committees, or agency decision makers; and likewise, to legislatures agencies at the state level. They seek to formulate, shape, support, or oppose legislation and policy according to their special interests and views on resource use.

POLICY DEVELOPMENT: THE HISTORICAL CONTEXT 1841–1945

During the past 160 years, renewable resource policy has continuously evolved. We are currently caught up in vigorous public debate over forest practices and what constitutes sustainable use and management of natural resources. This debate is not limited to North America, for the world is becoming more crowded and our natural resources grow more precious as we perceive their limits. Natural resources have become major political issues locally, nationally, and internationally. Our focus is national, in the United States. To try to understand these complex issues, we need to explore the origins of U.S. conservation in the nineteenth century, and its development through World War II.

Disposal of the Public Domain

As mentioned in Chapter 1, the period from 1841 to 1891 is significant as a time characterized by three ideas: (1) the national sovereignty would be bolstered through settlement; (2) the government should subsidize internal improvements in education and agriculture with land in lieu of money, since it did not have the necessary funds; and (3) natural resources should be transferred to private ownership and be managed without governmental intervention. These three ideas, right or wrong, brought about the disposal of more than two-thirds of the public domain, an empire of 1,500 million acres (607 million ha) (exclusive of Alaska), and four-fifths of the timberland contained therein, to individuals, corporations, and new states.

Under stipulations of the Morrill Acts of 1862 and 1890, grants of public land were made to the states for the establishment of agricultural and mechanical arts colleges. The states also received other "school lands" reserved at the rate of first one, then two, and then four sections per township, as well as swamp and overflow lands granted to them by acts of 1849 and 1850. But the states showed little interest in anything that approached management for these other "school lands" until the modern era, when the value of revenues for education from use of these lands became recognized.

By 1891 the country was largely settled and homesteaded, and the government had generously subsidized internal improvements in education and agriculture. But natural resources had not gravitated into entirely satisfactory ownership and management under the disposal policy that had prevailed. Many large tracts were in the hands of a comparatively few owners, a trend that persisted and increased. Most of the choice and accessible timber had passed to those with sufficient capital, and wholesale cutting practices prevailed on much of it.

Forest fires, often originating in slash from the poor logging practices, began to assume unaccept-

able proportions. In 1871, the same day as the great Chicago fire, the Peshtigo forest fire (named after a local river) swept rural Wisconsin, killing 1,200 people and blackening 2,400 square miles. A few thoughtful people were beginning to worry constructively about the government's forest policies. From 1870 on, there was growing protest, and profound changes would soon be suggested for the care and use of the public forests that remained. In 1872 Congress set aside 2 million acres as Yellowstone National Park, and the first Arbor Day was celebrated in Nebraska to encourage tree planting in arid lands (West, 1992).

Building the Federal Forests

The decades from 1871 to 1891 witnessed the emergence of forestry and conservation into public policy. At the American Association for the Advancement of Science (AAAS) annual meeting held at Portland, Maine, in 1873, a paper entitled "On the Duty of Government for the Preservation of Forests" was presented by a physician from New York. Dr. Franklin B. Hough urged the retention of forests on the public lands. A key focus was the importance of forests, when supplemented by reservoirs, for regulating the flow of streams. The next day the AAAS passed a resolution "on the importance of promoting the cultivation of timber and the preservation of forests."

Eventually, in 1876, a Division of Forestry was set up in the Department of Agriculture (USDA), and in 1881 Dr. Hough was named its chief. Thus, Dr. Franklin B. Hough (Fig. 2-1), a well-informed physician, became the chief pioneer in forestry in the United States and is credited with the nation's first book on the subject, *Elements of Forestry,* published in 1882 (Steen, 1976).

Out of the new USDA Division of Forestry, the present U.S. Forest Service eventually emerged. However, at the time, the Department of the Interior, whose General Land Office already had a Bureau of Forestry, considered itself the logical agency to manage the nation's forestlands when they were reserved. From these origins grew embittered agency relationships that evolved into the invigorating competition that remains today between the U.S. Forest Service in the Department of Agriculture and the Bureau of Land Management, National Park Service, and Fish and Wildlife Service in the Department of the Interior.

FIGURE 2-1
Dr. Franklin B. Hough, first head of the Division of Forestry, U.S. Department of Agriculture, in 1876. *(Courtesy of American Forestry Association)*

A subsequent service of the AAAS that illustrates the influence of science on policy at that time was a recommendation to President Benjamin Harrison in 1890 urging investigation of the preservation of forest areas on public land for maintenance of favorable water conditions. This recommendation also urged that, pending the investigation and report, sales of public timberlands should be suspended until they could be put under adequate administration and protection for the supplying of local wood and lumber needs.

The next year the Forest Reserve Act of 1891 was passed, so called because of its Section 24 on Timber Reserves, which added to the national parks and national monuments already in existence. This law marked a turn in public land policy from disposal to retention, since it repealed the Preemption Act of 1841 and the Timber Culture Act of 1873 and slowed sales of public land. President Harrison proclaimed the first "Forest Reserves," now known as National Forests, in the same year, 1891. Thus, policy changed after 20 years of discussions. General revision of certain land laws had been under consideration by Congress, and the American Forestry Association (organized in 1875), among others, had been urging the formation of forest reserves for many years (Steen, 1991). The Audubon Society of New York was formed in 1876, and the Boone and Crockett Club in 1887. These earliest of the environmental groups formed around influential leaders who supported enlightened ideas on conservation and provided a means for advocacy through their magazines and meetings. The slaughter of wildlife for trophies and for food by market hunters gave added impetus to the forest conservation movement as railroads entered more remote regions (Trefethen, 1975).

Upon signing the law that gave him the power to reserve public forestlands, President Harrison promptly proclaimed some 2.5 million acres (1 million ha) in Wyoming and Colorado as forest reserves. The combined total of lands affected by his proclamations reached more than 13 million acres (5.3 million ha).

These forest reserves were impressive areas, and the new policy of establishing public forests in the Department of the Interior was well launched, except for the necessary authority to administer and protect them and to make their products available. Dr. Bernhard Fernow (Fig. 2-2), then head of the Division of Forestry in the Department of Agriculture and alert to the need for putting forests to use, was probably behind a request to enlist the National Academy of Sciences in a study of the forestry situation so that it might advise the government on how to provide for rational handling of the forestlands of the country.

The resultant commission urged upon President Grover Cleveland the proclamation of a new group of forest reserves involving more than 21 million acres (8.5 million ha). This the president did with a suddenness that created much western opposition. The whole project appeared to some western groups as a drag on development in their part of the country and as unwarranted interference by the eastern states. But the final outcome, following this opposition, was a law that cleared the way for real progress in public forest policy. Known as the Forest Reserve Organic

FIGURE 2-2
Dr. Bernhard E. Fernow, who became chief of the Division of Forestry in 1886. *(Courtesy of Department of Manuscripts and University Archives, Cornell University)*

Act of 1897, it provided, among other things, the following policy guidelines and directives.

- The making of rules and regulations by the secretary of agriculture for the use and occupancy of the reserves, which should have the force of law
- Appropriations for surveying the existing and projected reserves
- Modification of reserve boundaries by the President
- Creation of the reserves for purposes of producing timber and protecting water supply only
- Protection of the reserves from fire and trespass
- Sale of mature timber at market value
- Free use of timber by miners and settlers under permit
- Selection by landholders of other public lands in lieu of patented lands or valid claims within the reserves (this opportunity was abused and was repealed in 1906)
- Various other concessions to local users and to the jurisdiction of local courts

This legislation, with modifications by the National Forest Management Act of 1976, provides the major direction under which the national forests are administered today by the U.S. Department of Agriculture's Forest Service.

Setting the Forest Reserves in Order

The Forest Reserve Organic Act of 1897 by no means marked the end of all withdrawals of public lands and presidential proclamation of them as forest reserves. That procedure was to continue for at least another 15 years, but the law did start to set in order millions of acres of forestlands. Dr. Fernow found himself with a Division of Forestry in one department (Agriculture), while the forest reserves and parks were in another (Interior). He retired in 1898 with the observation that the time had come for the Division of Forestry to have charge of the public forest reservations, which were in need of systematic management. This did not come about until 1905.

Meanwhile Dr. Fernow had turned over the division to his successor, Gifford Pinchot (Fig. 2-3), who, at Fernow's suggestion some years before, had studied forestry in France and had acquired varied resource experience in the United States upon his return. He had managed the great Vanderbilt forest estate at Biltmore, North Carolina, had served on the special Forest Commission of the National Academy of Sciences in 1896, and, for a year previous to following Fernow at the Division of Forestry, had traveled over the forest reserves of the west as a special agent of the Department of the Interior.

Pinchot continued forestry work in what had become a bureau in 1901 in the Department of Agriculture and which had a public to inform but no forests under its control. During this time Pinchot and his organization made studies and plans for management of privately owned timberlands, and, although comparatively few of these were adopted by owners, the educational value was significant.

The Pinchot Years

In 1905, the forest reserves, administered until then by the Department of the Interior, were transferred to the Department of Agriculture. This transfer was assisted when Interior Secretary Hitchcock noted that *"the presence of properly trained foresters in the Agriculture Department, as well as the nature of the subject itself, makes the ultimate transfer, if found*

FIGURE 2-3
Gifford Pinchot, who served from 1898 to 1910 as chief of the Division of Forestry, the Bureau of Forestry, and the Forest Service, all in the Department of Agriculture.

practicable, of the administration of the forest reserves to the Department essential to the best interest, both of the reserves and the people who use them" (Dana & Fairfax, 1980). With its own secretary supporting the transfer, the General Land Office lost its Forestry Division and its forests. In the same year, the name of Agriculture's Bureau of Forestry was changed to the Forest Service, with Gifford Pinchot as its chief, and, in 1907, the name *forest reserve* was changed to *national forest.*

The act of February 1, 1905, that had effected the transfer from the Department of the Interior to the Department of Agriculture also provided that all receipts for 5 years from the then forest reserves should be earmarked for expenditure by the Forest Service; that supervisors of national forests and forest rangers should be picked, where practicable, from

the states in which the forests were situated; and that forest officers should have the authority to arrest without warrant. Thus, the legal foundation for managing the federal forests was strengthened; there was unity of command, and some hope of enforcement of protective regulations in the west.

By now a crop of technically trained graduates was appearing each year from Biltmore, where Dr. Carl Schenck, Pinchot's successor there, had organized a "master school" (of forestry) in 1898, patterned after German institutions. Other graduates were available from forestry schools being organized in universities, the New York State College of Forestry, Yale University, the University of Michigan, and others. The Society of American Foresters was organized in 1900 and by 1905 had begun to publish its proceedings, and *The Forestry Quarterly* began publishing technical articles and news in 1902. Forestry graduates pooled their talents with those of western forest supervisors and rangers. Sawmill owners and workers, ranchers, cowmen, sheepmen, settlers, miners, and land agents began slowly, and sometimes unwillingly, to adjust themselves to increasing controls on the national forests. President Theodore Roosevelt was keenly interested in the national forests and gave the movement his aggressive support, having brought the total area of forest reserves to 100 million acres (40.5 million ha), with a disposition to add additional lands when examination indicated their suitability.

Certain policy changes were also adopted in this period, including the opening to settlement and entry of agricultural lands that were situated within the forest reserves (thus creating an administrative problem for the Forest Service); the granting of 10 percent of the receipts from the reserves for the support of schools and roads to the states in which they were located as compensation for loss of tax revenues (in 1908 the grant was raised to 25 percent and made a permanent arrangement); the charging of a fee per head for grazing stock on the reserves; the limiting of authority for additional presidential withdrawals of land for reserves (President Roosevelt had before him a report on some 16 million acres [6.5 million ha] of proposed reserves that he neatly proclaimed before signing the Agricultural Appropriation Bill that carried a rider limiting his authority to do so); the changing of names to *Forest Service* and *national forests;* and the provision of a generous appropriation for administrative operations. In 1908, the Forest Service was organized into field districts (now called regions) with 15 to 20 national forests each to administer, and the first of its Forest Experiment Stations was organized near Flagstaff, Arizona, a forerunner of the eight regional forest experiment stations and huge Forest Products Laboratory now operated by the Forest Service. This decentralized organization responded to regional and local outbursts of dissatisfaction with Forest Service policies that could not be addressed solely from Washington, D.C.

The Role of Women

Women's clubs provided early support for conservation, and many women leaders in conservation were also active in the suffrage movement (West, 1992). Mary Eno Pinchot, mother of Gifford Pinchot, who was chief of the Forest Service in 1905, headed a 100-member conservation committee for the 77,000-member Daughters of the American Revolution. The American Forest Congress in 1905 was addressed by Lydia Phillips, chair of the Conservation Committee of the 800,000-member General Federation of Women's Clubs. But the 1910 dispute over Hetch Hetchy Dam, and subsequent split between the wise use conservation policy espoused by the Forest Service and the preservation philosophy favored by the Sierra Club, saw many women favoring the preservation side of the argument. By 1915 more than half the members of the Audubon Society were women, and by 1929 more than half the members of the National Parks Association were female. Women were further alienated from early resource management policies as the professions became more technical and women were excluded (Ranney & Gumaer, 1990).

Only within the past two decades have serious efforts begun to remove barriers to women joining the renewable natural resource professions. Today about one in four students in natural resource programs is a woman, but female representation in the professional work force, and especially in leadership positions, still lags below that level. Aggressive affirmative action in the natural resource agencies seeks to change this ratio. The quarterly journal *Women in Natural Resources* documents the growing importance and contributions of women to the field, while uniting and empowering them with information, job listings and web links (winr&uidaho.edu). A similar effort has been mounted to increase minority employment in the natural resource field but without as much success as with the inclusion of women.

The White House Conference of 1908

The business of conserving natural resources is characterized by interlocking problems. This fact had impressed itself so deeply upon the members of President Theodore Roosevelt's Inland Waterways Commission (Fig. 2-4) that they suggested to him the advisability of calling the governors of the states together to confer on the broad natural resource problems of the country. The conference is significant, not only for what it accomplished, but also because it was the first assembly of this kind ever held. The invitation list had grown from the names of the governors to those of the Supreme Court, the cabinet officers, representatives of various national societies, and various groups of special guests. Significantly, John Muir and other preservationist leaders were not invited to this utilitarian-oriented conference (Nash, 1982). Disputes over dams and logging in national parks had alienated these two major forces in forest policy making. However, politicians, scientists, and members of the business community exchanged ideas under the leadership of the president and adopted a "Declaration of Principles," which said the following about forestry.

> We urge the continuation and extension of forest policies adapted to secure the husbanding and renewal of our diminishing timber supply, the prevention of soil erosion, the protection of the headwaters, and the maintenance of the purity and navigability of our streams. We recognize that the private ownership of forestland entails responsibilities in the interests of all the people and we favor the enactment of laws looking to the protection and replacement of privately owned forests.

The governors went home, and within a year and a half 41 of them had appointed conservation commissions for their own states (see Chapter 18). The president, shortly after the conference, appointed a National Conservation Commission of scientists, business people, and politicians in equal proportions, with Chief Forester Gifford Pinchot as chairman. They met, undertook a natural resource inventory with the help of the federal bureaus, and reported to the president and the second meeting of governors in less than a year. The commission gave the conservation movement, including forestry, a tremendous start toward an adequate program.

FIGURE 2-4
President Theodore Roosevelt and Chief Forester Gifford Pinchot on the river steamer
Mississippi in 1907, on a trip promoting interest in the development of the inland waterways.
(Courtesy of USDA Forest Service)

William Howard Taft became President in 1909, and Pinchot soon had a falling out with the new Secretary of the Interior, Richard Ballinger, and was dismissed. Pinchot remained active in forest policy matters, however. He and his intensely loyal lieutenants, many of them still in the Forest Service, took part in the decades-long struggle to ensure good forestry practices on public and private lands. Although direct regulation has not come about, some state forest

practice acts do, to a degree, regulate action on private land, as do several federal environmental and wildlife protection laws.

Establishment of the National Park Service

The setting aside of public domain land for forest reserves to protect watersheds and timber resources was preceded by the designation of certain more limited areas of the public domain for aesthetic and recreational purposes. National pride in the scenic beauty of these areas, coupled with assurances of their economic "worthlessness," were also important factors (Runte, 1997).

The Yosemite Grant was the first area of national significance to be set aside for these purposes. It included two separate areas, the Yosemite Valley with its spectacular granite walls and waterfalls, and the nearby Mariposa Grove of Big Trees. The federal government at that time was unwilling to assume administrative authority and gave this responsibility to the state of California in 1864. The Yosemite Grant thus became the first state park in the United States (Sharpe et al., 1994). Twenty-six years later, the two units were added to the new Yosemite National Park.

By 1872 legislation aiming at preservation of the Yellowstone region was successfully urged by "a comparatively few outdoorsmen, a handful of ideal-ists . . . a few men of vision in Congress" and their supporters in the press and conservation groups (Everhart, 1972). Rather than a reserve, Yellowstone was made "a public park or pleasuring ground for the benefit and enjoyment of the people with regulations . . . for the preservation from injury or spoilation, of all timber, mineral deposits, natural curiosities, or wonders within said park and their retention in their natural condition."

No development or management funds were appropriated for several years; in fact, this was a condition for passage of the act that created Yellowstone National Park (Haines, 1977). Eventually the preservation policy, which excluded many accustomed uses such as grazing, hunting, and mineral entry, had to be enforced in certain areas by the United States Cavalry (Fig. 2-5) and continued from 1883 until 1917. During these years the commanding officers of the

FIGURE 2-5

Well before the establishment of the National Park Service, the U.S. Cavalry administered both Yosemite and Yellowstone National Parks. Here, in 1899, F Troop, posed on and in front of the Fallen Monarch in Mariposa Grove. *(Courtesy of Yosemite National Park Research Library)*

assigned units served as acting superintendents. In other areas, where parklands were carved out of existing national forests or were adjacent to them, the forest supervisor also served as the principal administrative officer of the national park or monument (Brockman & Merriam, 1979). Not surprisingly, the Forest Service was opposed to the establishment of a separate bureau for parks, believing that, with minor adjustments to its operating procedures, it could handle the recreational and inspirational needs of the American public, fitting them in with the somewhat more specialized requirements of the mining, grazing, lumbering, irrigation, hunting, and other interests of the multiple-use fraternity. But parks were being defined differently as public pleasuring grounds, while providing public cabin sites on its attractive and accessible lakes and rivers was the early Forest Service recreational emphasis. The debate over the merits of using versus preserving natural resources continued to develop.

A classic confrontation between the utilitarian conservation versus aesthetic preservation took place in Yosemite National Park. San Francisco wanted Yosemite's Hetch Hetchy valley in the high Sierra for a dam site to supply city water. The intensity of the demand increased after the 1906 earthquake and fire. Access to water for fire control was seen as vital. A protracted struggle ensued. Although San Francisco eventually got the Hetch Hetchy site, the preservationists, headed by the Sierra Club and John Muir, came out of the losing battle determined to prevent further use of park land for utilitarian purposes. They had their victory in 1916 with the establishment of the National Park Service, whose mandate precluded dams in parks. Eventually they would also propose preservation or use restrictions on certain areas of public lands under other jurisdictions.

In an effort to encourage a policy of greater protection, as well as to foster the establishment of new areas, supporting groups such as the Appalachian Mountain Club and the Sierra Club began to enlarge upon the parks' commercial value as tourist attractions. This emphasis, in conjunction with scenic nationalism and the desire of the railroad companies to assure their passengers unimpaired vistas, was helpful in the final establishment of a parks bureau in the Department of the Interior (Runte, 1997).

The willingness of Stephen T. Mather (Fig. 2-6) to take on the formidable task of heading what was to become the National Park Service, and his personal and social qualifications for the job, were fortunate occurrences for the cause of the preservationists, and for all Americans interested in the public park concept. By this time there were 31 national parks and monuments, the latter classification having been formulated under the Antiquities Act of 1906. This act, originally inspired by the problems that unregulated access was causing to Native American dwellings and artifacts in the southwest, provided the president (by means of a presidential proclamation) with a quick administrative way to set aside areas containing these objects of cultural or scientific interest. It also expanded the concept of preservation to these cultural resources.

Mather worked with dedication and skill for the remainder of his life, setting the Park Service on a firm and respected foundation. He fought off serious

FIGURE 2-6
Stephen T. Mather, first director of the National Park Service, 1917 to 1929. Much of the present-day policy in the NPS was established by Mather. *(Courtesy of National Park Service)*

threats to national park integrity, not only from utilitarian development interests, but from states advocating national park status for scenic areas that were not of national significance. To solve this dilemma, he worked with the states, fostering the establishment of state park systems to protect and administer their lands having considerable aesthetic and recreational values but which fell short of being nationally significant.

The Weeks Act of 1911

This act paved the way for closer cooperation between the federal and state governments by making possible an extension of the fire-control program on the national forests to many states. An approved fire plan was necessary in order to gain the federal assistance, a provision that put the Forest Service in a position of leadership on an issue in which all landholders could agree. The Weeks Act also made provision for the creation of national forests east of the Mississippi. Most public lands available for national forests or national parks were situated in the western half of the continent. To obtain cut-over and abandoned timberland or tax-delinquent farmlands available in the eastern United States, special provisions had to be enacted, as these lands were not in the public domain but had to be purchased.

Forest Policy in the 1920s

Agitation for control of forest devastation on private timberlands led to hearings by a Senate Select Committee on Reforestation and eventually the Clarke-McNary Act of 1924. Yet this act was actually much more significant in its provisions affecting forestry on public lands than in control of destructive practices of private lands. Congressional support for a far-reaching policy of enforced federal forest regulation on private lands would not materialize, then or later. Not until the Endangered Species Act, in a prime example of the power of unintended effects, would federal regulations apply to private lands, and then only in certain situations.

Other legislation of the 1920s sought to revitalize the program of purchasing available lands for national forests, particularly in the South and Lake states, as well as to ensure financial support for forestry research and timber surveys. State and private cooperation with federal forestry expanded under Chief Forester W. B. Greeley (Fig. 2-7), who

FIGURE 2-7
William B. Greeley, who served as chief forester of the Forest Service from 1920 to 1928.

headed the Forest Service from 1920 to 1928, and reforestation efforts expanded. Although it did not come until after his time in office, legislation providing for the use of a portion of the stumpage price received from sales of national forest timber to cover the cost of replanting or otherwise silviculturally improving the timber-sale area was the result of this long advocacy. He also strongly supported the emerging policy of setting aside primitive areas within certain national forests, and tried to obtain congressional appropriations to deal with the growing recreation demand on forest service lands (Dana & Fairfax, 1980). It is not difficult to imagine the political complications of these funding requests, as the National Park Service felt it should be the primary administrator of recreation lands. While the debate continued, automobiles were bringing more and more citizens to camping sites in all jurisdictions. Recreational use of the forest was underway.

FIGURE 2-8
Young men of the Civilian Conservation Corps performing timber-stand improvement work on a forest plantation in Montana in 1935. The Corps provided forestry jobs for thousands of unemployed men during the Great Depression. *(USDA Forest Service photo by K.D. Swan)*

The Depression Years

Forest policy was affected by depressed economic conditions in the 1930s in that conservation activities were chosen by the Democratic administration as one means of implementing social and economic policies aimed at recovery.

One of the first important decisions of President Franklin Roosevelt upon assuming office in 1933 was to attack the problem of unemployment, which had been growing since the 1929 financial crash. President Roosevelt, a lawyer by training and former governor of New York, also claimed many times that he was a forester (Armstrong & Oates, 1992). He called for the establishment of a Civilian Conserva-

tion Corps (CCC) of unemployed young men organized for the purpose of conserving natural resources, with a focus on forestry work such as reforestation, fire control, erosion and flood control, and construction of trails in national forests and national parks (Fig. 2-8). Authority was granted the president to extend the work to state, county, and private lands as long as it was of a nature in which the federal government was legally authorized to cooperate.

The huge success of the CCC and its works expanded public forestry and recreation and provided a still-admired example of conservation of human resources. It was not the only natural resource program of the depression years, but it was the most important in its far-reaching influence. During that era, refor-estation and windbreak forests expanded in the Great Plains through cost share programs with private landowners, and state forestry programs emerged. The country entered the war and postwar years of the 1940s with renewed awareness of its natural resource heritage and opportunities. During the postwar years, many returning young men pursued natural resource college educations, some stimulated by experience with, or the stories about, the CCC.

Thus, the CCC had a profound effect on most of the enrollees, the communities, the land resources of the United States, and the subsequent policies of the federal and state agencies. The sudden availability of funds and manpower through this and other New Deal programs strengthened fire control and timber management; expanded roads, bridges, housing, and other facilities for administering forests and parks; and enhanced public enjoyment through camp-grounds and visitor-use facilities (Cohen, 1980).

POST-WORLD WAR II TO CURRENT ERA

The period 1950 to 1990 brought dramatic change to forest and renewable resource management. By the 1950s the issue of regulation of forest practices on private lands began to be dealt with by individual states; however, other major policy issues remained unresolved. Control and disposition of the residual public domain remained a concern. Ultimately this issue was resolved by the Federal Land Policy and Management Act of 1976 (FLPMA), which empow-ered the stewardship and retention of those lands under the Bureau of Land Management (BLM).

Industry participation in the forest policy debate increased greatly in the 1950s because wood prod-ucts were in demand, and research, both governmen-tal and private, opened up new wood and fiber uses. The Forest Service and other government agencies were managers of timber to which these burgeoning industries now needed access. Timber sales on the national forests and public land administered by the BLM became important sources of raw material for the forest industry, while providing revenues for local schools and roads, financing the construction of roads into the forests, and stimulating growth of rural towns with influx of workers hired by local sawmills and loggers.

Recreation

An expanding system of forest roads made possible the dispersal of a variety of recreation uses such as hunting, fishing, hiking, and camping throughout the forest. Ultimately this led to intense debate over allo-cation of land use between recreation and other uses, and questions regarding the kinds of recreational use to be emphasized. Lloyd and Fisher (1972) summed it up in an address at the seventh World Forestry Congress.

> In the United States we are involved in a forest policy struggle over allocation of scenic and recreational op-portunities between dispersed and concentrated types of uses. This, of course, is taking place within the broader political struggle over allocation of forest land among its many possible multiple-use combinations.

Much of the federal policy response to such post-war pressures emerged from the Outdoor Recreation Resources Review Commission (ORRRC) reports in 1961. These reports were important for the recogni-tion they gave to the need for mass recreation oppor-tunities, as well as for wilderness set-asides, and for the impetus they gave to recreation research and planning. The commission advocated cooperation among all levels of government and made state plan-ning essential to receiving federal funding for out-door recreation.

Wilderness

After nine years of debate, a consensus favoring preservation of the nation's most pristine areas in their roadless natural condition emerged. A legisla-tive landmark, the Wilderness Act of 1964, estab-lished a National Wilderness Preservation System (NWPS).

The Wilderness Act was more important than the 9 million acres of wilderness it instantly created; the process it mandated for study and review of other roadless lands that might be designated as wilder-ness, and especially the role it created for the public in that process, forever changed forest and renewable resource management. By early 2001, the National Wilderness Preservation System had grown to nearly 106 million acres. More than 130 laws had been passed adding new areas to the system (Hendee & Dawson, 2002), and the ultimate size of the Wilder-ness System seems destined to reach 120 to 140 mil-

FIGURE 2-9
These trail riders were among the first to enjoy entry into the National Wilderness Preservation System. Upper Jean Lake, above 11,000 feet (3350 m), Bridger Wilderness Area, Wyoming.
(Courtesy of USDA Forest Service)

lion acres once the remaining roadless areas in the national forests, parks, and fish and wildlife refuges, and on the BLM national resource lands, are studied and considered by Congress for their wilderness potential (Fig. 2-9). Once the issue of how much land will be designated as wilderness is resolved, wilderness values will be retained only by serious attention to wilderness management (Hendee & Dawson, 2002). See Chapter 12, "Outdoor Recreation and Wilderness Management," for a more detailed discussion of wilderness issues.

Multiple-Use

The Multiple-Use, Sustained-Yield (MUSY) Act of 1960 legally applied only to the national forests, but the public embraced the concept as a part of the answer in stretching all forest and renewable resources. The law stated it to be the policy of the Congress that the national forests should be administered for out-door recreation, range, timber, watershed, and wildlife and fish habitat purposes, supplemental to, but not in derogation of, the original intent for which the national forests were established as set forth in the Organic Act of 1897. The national forests would be lands of many uses, with harvest of timber limited to what could be perpetually sustained.

Despite substantial commitment to multiple-use and sustained-yield management in the 1960s and early 1970s, the proper emphasis and priority of allocations for each use were not easy to determine. Each situation and combination of uses was different from region to region and forest to forest. Although multiple-use management by definition included all uses, recreational use was unquestionably the most delicate issue (Steen, 1976). Some kinds of recreation, such as sightseeing, driving for pleasure, and hiking, seemed to conflict with timber harvest, especially where clearcutting was practiced. To complicate matters further, sufficient funding for recreational operations did not materialize. The Forest Service, however, was still obligated to placate diverse groups of users, none of whom would admit they had been allocated sufficient amounts of land or project funds.

The Environmental Movement and the Conservation Eras

A number of societies and organizations representing a broad spectrum of conservation concerns were formed before the turn of the twentieth century. For example, the formation of the American Ornithologists Union (1873), American Forestry Association (1875), American Fisheries Society (1884), Audubon Society (1886), and Sierra Club (1892) documented early concern and interest of citizens and natural resource professionals in the natural environment. In this first era of U.S. conservation, reform to stop wasteful activities and liquidation of natural resources was an outgrowth of their concern, as has been described. During the 1930s, the economic depression, dust storms, and catastrophic floods brought a second conservation era, a temporary revival of the mood and tempo of turn-of-the-century conservation concerns. In this era, dedicated professionals and conservation groups provided mechanisms for putting management and use of land, water, forest, rangeland, and wildlife on at least a partially scientific basis. The country had moved gradually from a purely custodial mode toward a managerial approach to renewable resources—an undramatic, but highly significant change.

Shortly after the end of World War II the United States entered its third conservation era, which extended from the late '60s, into the '70s and '80s, and until today. This era has had much wider social and ecological implications than the first two and has become part of a worldwide reaction to an increasing decline in the quality of life. The earlier two conservation eras were impelled primarily by wholesale dissipation and deterioration of natural resources as raw materials, and they sought to reduce waste and establish wise use as the prevailing philosophy. In contrast, the third era was motivated by real and perceived threats to people's personal health and welfare presented by disruption of essential life-support systems. While we included the early 1990s as part of the third era of conservation, some argue that we

then entered a fourth conservation era characterized by heightened concern over global change and environmental deterioration, and of international programs directed at their mitigation (Armstrong & Oates, 1992).

During the 1950s and '60s, attention focused on pollution of air and water caused by industrial wastes. The immediate effects of this long-accepted practice could be seen, felt, tasted, and smelled. The hidden or delayed public health implications also became the object of intense research. Similarly there was growing anxiety about widespread use of new chemicals, particularly synthetic organic materials, because so little was known about their biological consequences. Rachel Carson (1962), in her seminal book *Silent Spring,* vividly dramatized the potential dangers of pesticides, keynoting the popular reaction against injury to wildlife as a result of the introduction of toxic chemicals into their food chains (Fig. 2-10). Finally

FIGURE 2-10

Rachel Carson, an American biologist well known for writings on environmental pollution and marine biology, created a worldwide awareness of the dangers of introducing toxic chemicals into the wildlife food chain. *(Photo by Brooks Studio. Courtesy of Shirley Briggs, Rachel Carson History Project)*

there were growing and insistent demands that massive physical assaults on air, water, and land be halted until the full nature and extent of their effects could be determined. A memorable manifestation of these demands was the first Earth Day in April 1970. Thus, various developments such as the supersonic transport, offshore oil drilling, strip mining, construction of the Alaska pipeline, and clearcutting of timber were stopped, delayed, or subjected to restrictive regulations. As the twentieth century drew to a close, these and new environmental issues aroused public concern: aquaculture and its effects on native fish stocks, the fish-stock-depleting practice of driftnet fishing, and world deforestation and its effects on global warming are three such issues.

Much of the impetus and effectiveness of the environmental movement came from a proliferation of concerned citizens' organizations, some of long standing, most more recent, and many almost spontaneously created to deal with a specific issue. Greatly expanded funding for environmental research became available from federal agencies, universities, foundations, and other sources. In a short time a great deal was learned, about both the true nature of problems and the best strategies for halting or changing environmentally damaging practices. The nature of people's impacts on the operation of the biosphere was elucidated and proclaimed, as was the close interrelationship and interdependence of all living organisms. The opening sentence of Barry Commoner's book *The Closing Circle* put it very well: "The environment has just been rediscovered by the people who live in it" (Commoner, 1971).

Resource Managers
and the Environmental Movement

The environmental movement profoundly affected resource managers and their work. In the time-honored tradition of Pinchot, resource managers' self-image embodied all that was good in resource protection and management. They were the people who stopped the wholesale, exploitative destruction of the forest, the ones who put out fires, planted trees, and took care of wildlife and fish. In the decades since Teddy Roosevelt, they had progressed from custodians of natural resources to managers. They had accepted as natural and proper the roles of pioneers, leaders, and molders of public attitudes in conservation matters. Wildlife managers were sure

that hunting was the way to contain wildlife populations, and that some predators should be controlled. If any group understood cause-and-effect relationships in nature, the natural resource professionals were convinced that surely they did. They had been taught that conservation questions require technical answers. Their competence and versatility had been tested and proved in the management and control of forest regeneration, deer herds, insects, fire, fish, stream flow, livestock grazing, and activities of logging contractors and campers. Through their dedication and professionalism, resource managers felt they were largely responsible for keeping conservation alive during long periods of public apathy.

But as the environmental movement gathered momentum, many conventional resource management practices came under fire. Suddenly resource managers found themselves accused of indiscriminate use of pesticides, destructive cutting practices, tolerance of pollution, disregard of aesthetics and public sentiment, neglect of wildlife, and opposition to wilderness. The resource management professions stood indicted for default of principles, insensitivity to national needs and public attitudes, and defiance of the will of the people.

Resource managers then generally came from rural backgrounds, were predominately white males, and espoused a conservative philosophy. Their contacts were often limited to people of similar orientation. They were professional in outlook but better trained in consumptive as opposed to appreciative uses. Too often they assumed a paternalistic, self-righteous posture that assumed, as resource professionals, they knew best how America's natural resources should be used and managed. Unfortunately, their influence had too long been confined to technical problems and technical solutions. They recognized shifting patterns of natural resource use but failed to sense or understand the profound changes in society's goals and values. Nor did they comprehend the degree to which "their" dominion of forest and rangeland, fish, wildlife, and water had become part of a much larger and more complex environmental dominion. Their traditional role was not just being challenged, it was being ignored by the new generation of self-proclaimed ecologists who had forgotten, never knew, or did not care about past history, and who had suddenly and dramatically captured the imagination and support of large segments of the public. Technical competence was no longer enough. Without a sensitive and responsive environmental and social conscience, professional resource managers could not sustain public trust and support.

On the other hand, the environmental movement had become a vast but also fragmented crusade. It included many special-interest groups that tended to define environmental problems and solutions in the narrow terms of their own particular concerns. Many were strongly polarized and showed little sympathy for compromise, which they regarded as a sellout. When agreements were forged with agencies, other environmental groups would frequently repudiate them. This led to unreasonable demands and often irreconcilable conflicts between equally dedicated groups. Many were impatient and wanted instant solutions, not understanding how long it takes to develop safe and workable alternatives. They repeatedly advanced the unanswerable argument that, if a practice could not be conclusively shown to be free from all environmental hazards, then it was unacceptable and should be discontinued.

So pervasive became environmentalism that, by 1990, more than three-fourths of all Americans proclaimed themselves to be environmentalists. The largest 42 national environmental organizations claimed nearly 15 million memberships, after doubling during the decade of the 1980s; employed 3,500 persons; and had collective annual budgets totalling more than $500 million. Internal Revenue Service data showed that the more than 3,500 environmental, animal-related, conservation, and natural resource groups combined reported $2.47 billion in expenditures in 1990 and reported assets of $6.7 billion (Hendee & Pitstick, 1992). Clearly, environmentalism was a big enterprise. Today it is much larger.

A major study of the environmental movement at the turn of the twenty-first century revealed that there are now more than 8,000 environmental organizations. According to the National Center for Charitable Statistics, in 1999 environmental groups received $3.5 billion in contributions from individuals, companies, and foundations, with about three-fourths coming from an estimated 8 to 17 million individuals. According to IRS tax records, the nation's 20 largest groups received 29 percent of these contributions, with the top 10 listed among America's wealthiest charities.

As with major businesses, the leaders of the 10 largest organizations are paid well, earning an average salary of $235,918.00 per year. Fig. 2-11 show public contributions, total revenue, spending, and top executive salaries for the 20 environmental organizations with the largest total contributions in 1999 (www.sacbee.com; Knudson, 2001). Some of the major national environmental organizations and their Websites are shown in Appendix H.

Resource management, now constrained by investigations, appeals, injunctions, petitions, protests, restrictive legislation, and lawsuits, must adapt or lose even more influence. However, grudging compliance with court orders has not, and will not, be sufficient. What was and is required is a new approach in theory and practice of resource management that will reflect contemporary values and priorities, as well as the needs of the resources. Biodiversity must be maintained. Environmental pollution and waste must be controlled and limited. Aesthetics and public access are important. Sustainability in some measure is being demanded. This will not be easy, since, despite growing environmental concern, demands for all natural resource uses, consumptive as well as appreciative, are higher than ever before and are still growing as our population increases.

Through their professional societies and organizations, the natural resource professions are responding and changing. This is evident in the content of their professional journals, the topics addressed by professional task forces and committees, the campaign messages by candidates for professional society and organization offices, and the growing number of interdisciplinary conferences and programs for professional audiences. (Several major professional organizations to which natural resource professionals belong are listed in Appendix F, along with their Websites).

Public Involvement

Reform in natural resource management, embodied in a change from "we professionals know best" to greater sharing of responsibility with the public, began with new requirements for public involvement in natural resource decisions. First, the Wilderness Act of 1964 required public hearings during wilderness allocation processes; then other environmental legislation and agency policies began to require public involvement. Today this activity requires consider-

	Group	Public contributions	Total revenue*	Spending	Top Executive Salary
1	The Nature Conservancy	$403.4	$704.0	$359.4	$210,151
2	Trust for Public Land	$94.9	$105.7	$51.4	$157,868
3	Conservation International	$76.7	$83.5	$26.2	$203,049
4	World Wildlife Fund	$68.4	$111.3	$89.7	$241,638
5	Ducks Unlimited	$63.4	$108.6	$109.1	$346,882
6	Natural Resources Defense Council	$32.6	$36.1	$30.6	$238,964
7	Conservation Fund	$32.5	$41.9	$27.7	$211,048
8	National Wildlife Federation	$31.2	$88.1	$85.9	$247,081
9	National Audubon Society	$30.7	$64.7	$53.6	$239,670
10	Environmental Defense	$28.4	$32.0	$26.3	$262,798
11	Sierra Club	$19.1	$56.5	$54.3	$199,577
12	Rocky Mountain Elk Foundation	$17.5	$36.3	$34.9	$186,369
13	The Wilderness Society	$17.4	$18.8	$14.3	$204,591
14	Sierra Club Foundation**	$16.4	$17.8	$12.8	$100,000
15	National Parks Conservation Association	$14.6	$18.3	$16.6	$172,879
16	Earthjustice Legal Defense Fund	$12.2	$16.1	$13.3	$157,583
17	Defenders of Wildlife	$10.3	$14.9	$13.3	$201,337
18	Greenpeace Inc.	$9.9	$14.0	$11.1	$54,033
19	Save the Redwoods League	$9.8	$11.4	$8.9	$165,110
20	Center for Marine Conservation	$8.6	$9.9	$8.7	$135,806

*Includes public contributions and government grants, etc. **The Sierra Club Foundation is the tax-deductible fund-raising arm of the Sierra Club.

Source: Bee Research

FIGURE 2-11
The greening of the environmental movement. This figure shows public contributions, total revenue, spending, and top executive salaries, of 20 environmental groups with the largest contributions. Reprinted with permission from the Sacramento Bee. (Knudson, 2001)

able time and effort for resource managers. Input to natural resource proposals comes from public meetings, listening sessions, workshops, focus groups, surveys, and other forms of public involvement including letters, phone calls, e-mails, and Website hits (Fig. 2-12). Under the National Environmental Policy Act of 1969, possible environmental impacts are identified and assessed, safeguards proposed, and total costs (social, environmental, and economic) and benefits analyzed. Safe, effective alternatives to unacceptable practices must be developed and tested. Performance standards and controls must be strengthened to assure proper protection of soil, water, wildlife, scenic beauty, and other nonmarket values. A major result of all this public involvement is that intangible resource values and public opinion figure more prominently in management objectives and decisions.

With public involvement now the accepted way of doing natural resource business, the resource managers who are good at working with the public stand out, succeeding and moving to positions of responsibility. Thus, a good natural resource education today includes coursework and class exercises in public involvement methods and group process skills.

Applying Renewable Resource Policy Today

Today, on public lands at least, resource managers operate in a highly social and political environment where their decisions and processes may be tightly circumscribed by laws and challenged by the public, including special-interest groups. Their influence on natural resource decisions may now depend on the clarity of their presentation to the public of their analysis of proposed alternatives (even the best analysis will be wasted if the public can't understand it), and their skill in forging a consensus for action among opposing interest groups. Even on private lands, managers are accountable to stockholders, owners, neighbors, applicable federal, state, and local regulations, and a public concerned about the land,

FIGURE 2-12
Two different approaches to public meetings where citizens make their views known. *Upper:* A citizen expresses his viewpoint to a panel, while those in the audience listen and wait their turn to speak. *(Photo courtesy of National Park Service) Lower:* Representatives of interest groups work through issues seeking a resource management decision. Workshop approaches currently seem more effective than traditional hearings and public meetings. *(Photo by Ed Krumpe)*

forests, water, and wildlife, regardless of who owns them. When a course of action emerges, it must run a gauntlet of legal and procedural requirements and may be tested or challenged with editorials, letters, demonstrations, administrative appeals, and lawsuits.

A FRAMEWORK OF LAWS

Several major laws affect resource management on public (and often on private) lands. The most important of these laws are the National Environmental Policy Act, the Endangered Species Act, National Forest Management Act, Air and Water Quality Acts, the National Historic Preservation Act, and the Forest and Rangeland Renewable Resources Planning Act of 1974 (RPA). These laws, and the case studies illustrating their application, show how resource management has been complicated by the necessity of meeting the procedural and substantive requirements of these laws.

The National Environmental Policy Act of 1969 (NEPA)

NEPA requires federal agencies to utilize a systematic, interdisciplinary approach that will ensure the integrated use of the natural and social sciences in planning programs that may have an impact on the human environment. Broadly construed, this means the environment of all living organisms that affect humans. Agencies must make certain that presently unquantified environmental amenities and values are given appropriate consideration. Under NEPA, proposals for major federal actions that may significantly affect the quality of the human environment are to include a detailed statement on the potential environmental impacts of the proposed action. This statement, called an *environmental impact statement* (EIS) must consider any adverse environmental and social and economic effects that cannot be avoided; alternatives to the proposed action; the relationship between local, short-term uses of the human environment and the maintenance and enhancement of long-term productivity; and any irreversible and irretrievable commitments of resources (Fig. 2-13).

If the need for an EIS is uncertain, the proposing agency prepares an *environmental assessment* (EA). If the EA determines that the proposed federal action will *not* result in significant impacts, a finding of *no significant impact* is issued. But if a proposed federal action may have a significant effect on the human environment, the responsible agency must prepare an EIS. During its preparation, the agency must inform the public of the proposed action and its possible consequences, and give interested and affected people an opportunity to express their views. Other concerned agencies and Indian tribes must be consulted and provide formal reviews of a *draft environmental impact statement* (DEIS). Feedback from all individuals and groups must be considered, and used where appropriate in the final EIS. This final statement and a record of decision is filed with the Environmental Protection Agency and made available to the public.

Consideration is given, not only to visible, immediate, and local changes, but also to what may be expected in terms of obscure, long-range, cumulative, and off-site effects. Thus, environmental statements are prepared for projects that will cause extensive ecosystem disturbance; that may result in delayed chain reactions in birds, fish, or other animals (including humans); or that could impair unique values of parks, potential wilderness, archeological sites, or other special-use areas.

The NEPA has had important effects on the conduct of federal operations and, to a lesser degree, on other public and private activities, especially since most states now have comparable state laws. Further, because of NEPA, the environmental consequences of all kinds of development and management practices are becoming better understood. Preparation of environmental impact analyses has obliged planners and managers to dig deeply into available knowledge about ecosystems and how they are altered by various types of stresses induced by natural resource uses. Decision makers now must be more sensitive and responsive to public attitudes and needs. But NEPA has greatly increased workloads of public resource managers, slowing decisions and permit issuance. Finally, compliance with NEPA has helped to bring resource managers face to face with public concerns, social climate, and cultural values, thus reducing their isolation, broadening their environmental sensitivities, and perhaps instilling in them a measure of humility. It has enabled them to make better decisions resulting in more environmentally sensitive management activities that better serve the American people. And activists themselves, as well as other interested parties, have had to learn how local and

FIGURE 2-13
Recovery methods in the 1987 Silver Fire on the Siskiyou National Forest, Oregon, provide an excellent illustration of public involvement at work. A management plan was chosen for this lightning-caused burn only after extensive on-the-ground data collection, analysis of nine alternatives, and numerous public hearings. Through the NEPA process, a solution that stood up against legal challenge was reached in a timely fashion. *(Courtesy of USDA Forest Service)*

regional authorities administer their areas of interest. Political awareness has thus been raised in the general public.

National Forest Management Act (NFMA)

The National Forest Management Act of 1976 (NFMA) amended the Forest and Rangeland Renewable Resources Planning Act of 1974. In this amendment Congress states its policy that all forested lands in the national forest system be maintained in appropriate forest cover with species of trees, degree of stocking, rate of growth, and stand conditions designed to secure the maximum benefits of multiple-use sustained-yield management in accordance with land management plans. The Act went on to require the secretary of agriculture to develop, maintain, and, as appropriate, revise land and resource management plans for units of the national forest system, coordinated with the land and resource management planning processes of state and local governments and other federal agencies.

The Act also required the secretary of agriculture to issue regulations that, among other things:

1. specified procedures to ensure that land management plans were prepared in accordance with the National Environmental Policy Act of 1969, including, but not limited to, direction on when and for what plans environmental impact statements would be required;
2. specified guidelines that require the identification of the suitability of lands for resource management;
3. specified a number of guidelines for land management plans.

One of the most important guidelines required forest plans to provide for diversity of plant and animal communities based on the suitability and capability of the specific land area in order to meet overall multiple-use objectives, and to the degree practicable, to provide for steps to be taken to preserve the diversity of tree species similar to that existing in the region controlled by the plan.

Another important guideline required forest plans to ensure that clearcutting, seed tree cutting, and other timber harvest methods designed to regenerate even-aged stands of timber would be used on national forest lands only where they were determined to be the optimum method to meet the objectives and requirements of the relevant land management plans. Forest plans also had to ensure that timber would be harvested from national forest systems lands only where the agency could assure that such lands could be adequately restocked within five years after harvest.

Using the planning process required by this Act, the Forest Service prepared comprehensive forest plans for each of its 123 national forests. They were developed through the NEPA process with full public involvement and guide the multiple-use management of the natural resources on 191 million acres of land in the National Forest System (USDA Forest Service, 1993).

Endangered Species Act (ESA)

The Endangered Species Act of 1973 (ESA) was designed to conserve the ecosystems or habitat upon which endangered or threatened plant and animal species depend. It requires all federal agencies to use their authorities to achieve the purposes of the Act. Section seven of the Act requires federal agencies to consult with the Fish and Wildlife Service on any proposed action that might adversely affect a threatened or endangered species or its critical habitat. If the Fish and Wildlife Service finds that the action will jeopardize or adversely affect a species, it must provide terms and conditions that the agency must meet to prevent or mitigate the anticipated adverse affects.

This Act has had a profound effect on the management of federal lands. Management requirements for some threatened and endangered species, such as the northern spotted owl, the marbled murrelet, and several Pacific salmon runs, literally dictate management of millions of acres of federal forest and range lands in the Pacific Northwest, and also on private lands determined to contain critical habitat for the endangered species (O'Laughlin, 1992).

Clean Water Act

The objective of the Clean Water Act was to restore and maintain the chemical, physical, and biological integrity of the nation's waters through the control of both point (specific) and nonpoint (widespread and diffuse) sources of pollution.

In order to implement the nonpoint source management program, the Act required the governors of each state, after notice and opportunity for public comment, to prepare and submit to the administrator of the Environmental Protection Agency a plan that identified those navigable waters within the state that, without additional actions to control nonpoint sources of pollution, cannot reasonably be expected to obtain or maintain applicable water quality standards or the goals and requirements of the Act. It also identified those categories and subcategories of nonpoint sources that add significant pollution to these waters and described the process, including governmental coordination and public participation, for identifying best management practices and measures to control each category and subcategory of nonpoint sources. Nonpoint sources included such things as timber harvest, implementation of silvicultural practices, and similar management activities. The idea was that, by implementing the best management practices developed by the state, land managers could meet state water quality requirements.

Clean Air Act

This legislation, passed in 1965, with important amendments in 1977, sought to improve, strengthen, and accelerate progress for the prevention and abatement of air pollution. The following objectives were stated:

1. to protect and enhance the quality of the nation's air resources to promote the public health and welfare and the productive capacity of its population;
2. to initiate and accelerate a national research and development program to achieve the prevention and control of air pollution;
3. to provide technical and financial assistance to state and local governments;
4. to encourage and assist development and operations of regional air pollution prevention and control programs.

As in the Clean Water Act, this Act required states to prepare a plan that provided for implementation, maintenance, and enforcement of each primary and secondary standard in each air-quality control region (or portion thereof) within the state.

One part of the Act of special interest to forest land managers dealt with the prevention of significant deterioration in air quality. One of the purposes of this part was to protect the air quality in national parks, wilderness areas, national monuments, national seashores, and other areas of special national value. Federal land managers were given the affirmative responsibility to protect the air-quality related values (including visibility) of lands within these areas. In situations where the federal land manager demonstrated to the satisfaction of the state that emission from a proposed major emitting facility (such as a power plant) would have an adverse impact on the air-quality-related values of such lands, the state would not allow the facility to be constructed.

Federal and nonfederal forest managers must also meet state air-quality standards. This is especially important where the use of prescribed burning, including slash burning, is proposed. Normally, forest managers work directly with the state forestry organization to coordinate their plans with the plans of others and to obtain the necessary permits. In some cases, detailed smoke management plans may be developed. The increase in wildfires in recent years poses a special concern, as many nearby communities may be plagued for weeks with smoky conditions that do not meet ambient air-quality standards.

Air Quality in Wilderness Areas

The following example from Washington State, and examples from California and Virginia, illustrate how wilderness areas and parks play a role in air-quality management and monitoring (Stokes, 1999). The Stokes article reports:

> With passage of the Clean Air Act (CAA) amendments in 1977, the U.S. Congress designated 88 wilderness areas managed by the USFS as Class I federal areas for air quality, thus mandating special protection and monitoring for visibility. For wildernesses not designated as Class I areas, protection of their naturalness is still mandated under The Wilderness Act (TWA), although monitoring is not required by the CAA. Along with negatively impacting the experience of wilderness users, reduced visibility is an indicator of particulate matter and other pollutants with potential for adverse effects on vegetation, soils, and water quality. The existence of visibility problems in relatively remote, unoccupied locations, such as the Class I federal areas and other

wildernesses, is a further indicator of the breadth and growth of air quality problems throughout the country.

Many wilderness areas are being adversely impacted by air quality deterioration. We have identified the following Air Quality Related Values (AQRVs) in wilderness that may be affected:

- visibility, which also serves as an indicator of alternate adverse effects on other AQRVs
- soils, which may be impacted by air quality with long-term ecosystem damage
- vegetation, impacted directly, leading to potential change in plant communities and naturalness
- water quality, stream and lake chemistry, and dependent fish and wildlife

Air-quality impacts on wilderness are occurring nationwide, and the Alpine Lakes and Goat Rocks Wildernesses in Washington State are a key example. In 1996 the USFS notified the state of Washington that visibility in these two wildernesses was being adversely affected by pollution from a coal-fueled power plant in Centralia, with supporting information also documenting impacts on water quality. The polluting source, then the largest in the western United States, emitted up to 75,000 tons per year of sulfur dioxide, more than half of the sulfur emitted in the state. In this case, the USFS, in collaboration with the NPS, worked with the plant owners and regulatory agencies to achieve a mediated settlement that will provide a 90% reduction in sulfur emissions from that plant by the year 2002.

More difficult to resolve are the following examples, also from Stokes:

> The San Gorgonio Wilderness near Los Angeles sustains major air-quality impacts from surrounding urbanization, with 30 years of research documenting ozone damage to vegetation. Visibility in San Gorgonio is so severely diminished on the most polluted days that visibility-monitoring cameras produce only black pictures (there is nothing to photograph). The James River Face and St. Mary's Wildernesses in Virginia are being impacted by pollutants generated by industrial centers in the Ohio Valley of the Upper Midwest, which are transported to Virginia by prevailing air currents. The St. Mary's River within the wilderness, a state-recognized blue ribbon trout stream, is now so acid from pollution that the USFS is proposing to add lime to the stream to reduce the acidity and to preserve its natural character.

National Historic Preservation Act

This Act, passed in 1966, established a national policy that the federal government, in cooperation with others, would use measures, including financial and technical assistance, to foster conditions under which modern society and prehistoric and historic resources can exist in productive harmony. Among other things, the Act authorized the secretary of the interior to expand and maintain a National Register of Historic Places composed of districts, sites, buildings, structures, and objects significant in American history, architecture, archaeology, engineering, and culture.

The effects of this Act on natural resource management are far-reaching. Before taking any management action, forest land managers must make a reasonable effort to determine whether or not prehistoric and historic resources may be affected. If these resources are present, the managers must consult with the state historic preservation officer and take whatever measures are necessary to protect the resource or to mitigate possible adverse affects.

The Resources Planning Act (RPA)

The Forest and Rangeland Renewable Resources Planning Act of 1974 (RPA) requires the secretary of agriculture to conduct an assessment of the nation's renewable resources every 10 years. Only the essential framework is described here, but RPA details are available in publication or online (USDA, 2001a; www.fs.fed.us/pl/rpa/what.htm).

The RPA is important because it provides a national database and analysis every 10 years (with 5-year updates) as a basis for renewable resource policies. Thus, renewable resource policy and programs can evolve in response to changing field conditions, supply, and demand. The data and analysis, prepared by the U.S. Forest Service, cover all land ownerships, and thus guide programs designed to enhance productivity of all lands, not just the national forests. For example, RPA data showing the substantial forest resources owned by nonindustrial private forest landowners (NIPFs) has been an important guide to federal and state programs of technical and financial assistance to improve management of NIPF lands. The RPA assessments include many technical supporting documents, which can be accessed at the Forest Service Website, www.fs.fed.us/pl/rpa/publications. These reports are a source of the best, current data about renewable resources nationwide. A good overview, with references to supporting documents, is the U.S. Forest Service (2001b) Renewable Resource Planning Act 2000 Assessment Summary Report.

CASE STUDIES IN THE APPLICATION OF LAW AND POLICY

Federal agencies must integrate the requirements of NEPA into their planning and decision-making process. This integrated process is frequently referred to as the "NEPA process." Regulations also require agencies to preclude delay by integrating NEPA requirements with other environmental review and consultation requirements, such as those discussed for the National Forest Management Act and Endangered Species Act. Thus, the NEPA process becomes a very practical way for considering and meeting the requirements of other environmental laws and their implementing regulations, policy, and procedures as a matter of routine for any proposed resource management action. The NEPA process is now the primary framework for federal agency decision making, and for many state agencies that have similar legislation.

The following case studies describe federal agency actions that used the NEPA process on very complex proposals. These complied fully with all environmental laws, and they were able to be implemented within a reasonable period of time.

CASE STUDY 1

Control of Douglas-fir Tussock Moth in the Pacific Northwest

The Douglas-fir tussock moth (DFTM) is a native tree defoliator that in the larval (caterpillar) stage lives by eating needles of live trees. It attacks Douglas-fir *Psuedotsuga menziesii* and true firs: grand fir *Abies grandis,* subalpine fir *Abies lasiocarpa,* and white fir *Abies concolor.* Tussock moth populations are cyclic, with an epidemic every 7 to 13 years. Each outbreak lasts two to four years and ends with a sudden crash. The insect can go from sub-outbreak to destructive outbreak populations in one year; with substantial damage occurring before land managers can take action. In the 1970s, an outbreak defoliated 700,000 acres. As a result of research in response to prior outbreaks, the Forest Service developed a survey technique, the "Douglas-fir tussock moth early warning system," which indicated that a serious outbreak was likely to occur on national forests in eastern Washington and Oregon in 2000–2002.

The imminent outbreak created a need to protect specific areas of concern where the tussock moth defoliation would create conditions jeopardizing *endangered species habitat* (e.g., adjacent salmon spawning streams; spotted owl nesting, roosting and foraging habitat), *areas for health and safety reasons* (e.g., campgrounds, municipal watersheds, residential areas), and *areas where the Forest Service had made substantial investment* (e.g., scenic foregrounds, seed orchards, and previously protected areas such as old-growth forests). Specific "areas of concern" comprising approximately 250,000 acres were identified in an analysis of 4 million on nine national forests in eastern Washington and Oregon.

To protect the areas of concern, the Forest Service proposed to use two biological pesticides. *Bacillus thuringiensis* var. *kurstaki* [B.t.k.] is a bacterium that occurs naturally in the soil and kills only butterflies and moths. TM-BioControl, an insecticide made of the natural virus specific to only the tussock moth, is the primary cause of its population collapse under natural conditions. While the ensuing NEPA process required that many alternatives and adjustments be considered to address public concerns about environmental impacts, the potential use of these biological pesticides was the preferred alternative of the Forest Service.

A 'Notice of Intent' (NOI) to prepare an EIS was published in the Federal Register on June 18, 1999. Public "scoping" comments to identify concerns were received until August 20, 1999. Then issues of concern and alternatives for treating the tussock moth outbreak were identified based on the public comments. A Draft Environmental Impact Statement (DEIS) was published and public comments were invited on January 11, 2000. Comments were accepted until February 29, 2000. The availability of the final EIS (FEIS) was announced in the Federal Register on April 21, 2000, and a Record of Decision (ROD) was issued 30 days after release of the FEIS. Here it gets a little complicated because the ROD was appealed on technicalities and timing to the Chief of the Forest Service, with the appeal subsequently denied, but this led to the issuance of a second ROD for three of the national forests on November 7, 2000. This process and the dates are important because they illustrate the requirements of NEPA and the importance of strict compliance.

Issues

The DEIS and FEIS analyzed the proposed and additional action alternatives for their effects on issues identified from public comments and analysis by an interdisciplinary team of Forest Service specialists. A total of 11 issues were identified, including effects from treating and not treating moth infestations on such things as human health; protection of timber values; risk of killing nontarget *Lepidoptera,* thus leading to additional direct and indirect effects; maintaining healthy forests; fuel buildup and fire risk from trees killed by tussock moths; impact on fish and wildlife; water quality; economic impact of reduced tourism; and a few other issues of less significance.

Alternatives

Four alternatives were considered in the final analysis: [1] no action; [2] the proposed action alternative—to treat 628,000 acres in "areas of concern" with both TM-BioControl and B.t.k.; [3] expanded protection alternative, which was developed in response to public scoping and concerns for maintaining a healthy forest and protection of timber and dispersed recreation; and [4] TM-BioControl-only alternative, which was developed in response to public comments to the DEIS.

It is important to understand that dozens of other federal and state agencies and all Native American tribes in the target areas also received copies of the Notice of Intent and DEIS, with requests for comment. Thus, the DEIS, FEIS, and ROD had the benefit of expertise beyond that of the Forest Service, in addition to public comment. This input helped shape the decision in important ways. For example, while the proposed alternative was selected, it would be implemented first by spraying only TM-BioControl until the supply was exhausted—enough was available to treat about 210,000 acres. The areas for immediate treatment were also reduced to 240,000 acres of specific areas of concern, with cocoon/egg mass surveys in the fall and spring used to identify any additional areas having sub-outbreak population levels that clearly needed treatment. Some of these areas would be treated with B.t.k if TM-BioControl supplies were not available. Furthermore, proposed treatment was refined and adjusted with special considerations addressing the identified issues and particularly sensitive areas.

Public Involvement

Public involvement has been important to decisions with potential environmental impacts since passage of NEPA. In this case, public involvement included comments in response to the original Notice of Intent seeking input to help identify the issues of concern, and response to the DEIS and FEIS. The ROD listed 75 specific comment categories received in response to the FEIS (where the issue was addressed in the FEIS [and sometimes the DEIS]) and summarized the responses to the comments.

In conclusion, this case study illustrates the NEPA decision process and steps that have become a way of life for decisions impacting the environment (i.e., Notice of Intent for an action to elicit scoping comments identifying issues; DEIS analyzing the impact of various alternatives, including the agency's preferred alternative; FEIS containing refined analysis with the benefit of public comment and other agency input; and ROD explaining the decision and its rationale).

Interested readers can find further information on this case at www.fs.fed.us.r6/nr/fid/eisweb/dftmeis.htm.

CASE STUDY 2

Plantation Lakes Vegetation Management Projects in Michigan

The Plantation Lakes area of the Ottawa National Forest in Michigan's Upper Peninsula has extensive jack pine and red pine plantations that were planted in the 1930s, second-growth northern hardwood and hemlock stands, paper birch, aspen stands of several age classes, and lowland conifer in riparian areas. Many of the stands are over-mature, with dead and dying trees, susceptible to disease and repeated insect attacks. The area is managed primarily for natural dispersed recreation, including upland bird, deer, and bear hunting, and viewing scenery and wildlife from forest roads. The Ottawa Forest Plan identified the need to maintain and improve habitat conditions to promote ecosystem health, including reducing the potential for insect and disease infesta-

tions and long-term fire hazard. Within this framework the Plantation Lakes Vegetation Management Project was conceived to assess some stands for old-growth classification; convert some stands to species better adapted to site capabilities; and enhance natural diversity and bring insect and disease potential below epidemic levels. Timber harvests were the primary means by which these objectives could be accomplished.

May 26, 2000, the Ontonagon and Kenton Ranger Districts of the Ottawa National Forest in Houghton County, Michigan, issued a Notice of Intent to harvest timber, classify stands as old growth, do watershed and wildlife habitat improvements, improve dispersed recreation opportunities, and provide the transportation system to serve these projects within its 17,700 acres. Then in early 2001, incorporating public input to the NOI, a Draft EIS was issued in two volumes, with maps and appendices that identified and analyzed six alternatives and identified an agency-preferred alternative (USDA Forest Service Ottawa National Forest, 2001a). Finally, on June 11, 2001, a Record of Decision (ROD) and FEIS were signed, responding to comments on the DEIS and documenting rationale for the selected alternative [USDA Forest Service National Forest, 2001b).

The FEIS, already shaped by analysis and public input to the DEIS, further analyzed the six alternatives including (1) no action; (2) the modified proposed action; (3) early successional species emphasis; (4) late successional species emphasis; (5) no road construction; and (6) temporary openings of 40 acres or less. It is apparent that the alternatives contrasted with some of the choices made in the preferred alternative, such as the extent of timber harvesting, the size of openings to be created by harvests, and the extent of road construction and restrictions on access. The ROD explained why the preferred alternative was selected for implementation, relating it to public and agency concerns and the Forest Plan, and why each of the other alternatives was not selected. The ROD also explains how the FEIS decision conforms to requirements of law, regulations, and agency policy (e.g., the National Forest Management Act and its associated regulations and policy, the Clean Water Act and state water quality standards, Endangered Species Act, National Historic Preservation Act, and Wild and Scenic

Rivers Act). To meet the vegetation management objectives, a total of 7,010 acres was proposed for harvest, utilizing modified clearcut, shelterwood, salvage sanitation, commercial thinning, and selection harvest methods.

The two-volume DEIS, plus the ROD and FEIS in one spiral-bound document, totaled 2 full inches of text, illustrating the documentation, analysis, input from the public and other agencies, and legal compliance that is involved in modern natural resource management. Yet this complex process was completed within 13 months from start to finish, although the project implemented actions rooted in a Forest Plan prepared much earlier. The time required might have been longer but the Forest Service, like other agencies, is now experienced and skilled at executing the legal, process-oriented decision now required.

FOREST AND RENEWABLE RESOURCE POLICY IN THE FUTURE

Policy is dynamic, responding to issues of the day and the energy invested by interested groups seeking change. Several major issues are emerging as potential targets for policy revision at the turn of the twenty-first century.

From the late 1980s to the early 1990s, timber harvest practices, especially on public lands, became a policy hot spot fueled by perceived environmental impacts on threatened and endangered species such as the spotted owl, marbled murrelet, red cockaded woodpecker, and several stocks of salmon. Driven by these concerns and pressure from the environmental community, forest management responded with proposed new approaches that seem to have merged under the heading of *ecosystem management* (see Chapter 5, Silviculture). While this approach is still in the process of definition as to what it means on the ground, it seeks forest practices that better simulate natural processes, protect biodiversity, focus on the landscape level as well as forest stands, and retain fully functioning forest ecosystems. This concern with protecting the natural and aesthetic qualities of forests on public lands will be a major issue of public natural resource policy into the next century.

Other key issues are related. The role of public forests and rangelands in producing goods and services will continue to be an issue as public timber harvests are reduced, grazing fees are increased, and critical habitat for endangered wildlife is protected. These actions will inevitably affect private lands also. For example, the protective regulations and area set-asides on public land put more pressure on private lands to produce timber and range products. So incentives to intensify management of private forests and rangelands are a likely focus. The public's ever-growing appreciation for nature, perhaps intensified by deteriorating social and environmental conditions in urban areas where most people live, will be reflected in continuing demand for more parks, wilderness, and other protected areas. More people are likely to be involved with natural resources through public involvement and a variety of human resource programs in which conservation work is accomplished, such as the Job Corps, conservation corps, and volunteer programs for both youth and senior groups.

While we struggle in the United States to address such issues, on a worldwide scale the pressures of uncontrolled population growth represent the next assault on natural resources (Gardner-Outlaw & Engelman, 1999). What policies must we devise as world citizens and leaders to protect, regulate, and conserve on this scale? As the world shrinks in the face of accelerating means of communication, transportation, and access to markets, natural resource policy in the United States must increasingly address the effects of global natural resource demands.

SUMMARY

Policies are governing principles, plans, and courses of action that guide management. They are usually broad and general in direction but may contain specific provisions or prohibitions. Four major groups are influential in shaping laws and policy (i.e., the bureaucracy of the executive agencies of state or federal government; society, through the force of individual voters as well as special-interest groups; organized industry; and the natural resource professions).

Forest and renewable resource policy has evolved over 150 years from disposal of two-thirds of an empire of 1,500 million acres in the mid 1800s, to the setting aside of forests reserves and park lands, to today's heightened concern over the management and protection of public lands. Four conservation eras are recognized: First, the move to stop waste, destruction, and disposal of public lands near the turn of the twentieth century; second, the move, beginning in the 1930s, to bring public lands under scientific management; third, an era of environmental concern beginning in the 1960s and continuing today; and fourth, an era of global environmental concern coupled with a realization of population pressures on all resources worldwide.

The environmental movement resulted in several new laws to protect natural resources, as well as the initiation of public involvement in decisions as required under the National Environmental Policy Act of 1969. The National Forest Management Act, Endangered Species Act, clean air and clean water acts, and the National Historic Preservation Act are key laws circumscribing actions and prescribing resource management process. Consequently, forest and renewable resource management must wait on analysis to see if provisions of these laws will be violated by any proposed project. Mitigation of anticipated impacts must be accomplished. While management is more complex now because of these requirements, expedient decisions are possible through expert application of the NEPA process and public involvement, as the two case studies demonstrate.

LITERATURE CITED

Armstrong, Frank H., and Marguerite Oates. 1992. Window Seat: An Aerial Perspective of America's Forests with General Enlightenment for America's Civic Leaders. Bull Run of Vermont.

Brockman, C. Frank, and Lawrence C. Merriam, Jr. 1979. *Recreational Use of Wildlands,* McGraw-Hill Book Co., New York.

Carson, Rachel. 1962. *Silent Spring,* Houghton Mifflin, Boston, Mass.

Cohen, Stan. 1980. *The Tree Army: A Pictorial History of the Civilian Conservation Corps.* Pictorial Histories Publishing Co., Missoula, Mont.

Commoner, Barry. 1971. *The Closing Circle,* Alfred A. Knopf, New York.

Cubbage, Frederick W., Jay O'Laughlin, and Charles S. Bullock, III. 1992. *Forest Resource Policy.* John Wiley & Sons, New York.

Dana, Samuel T., and Sally K. Fairfax. 1980. *Forest and Range Policy: Its Development in the United States* (2nd ed.), McGraw-Hill Book Co., New York.

Everhart, William C. 1972. *The National Park Service,* Praeger, New York.

Gardner-Outlaw, Tom, and Robert Engelman. 1999. Forest Futures Population, Consumption and Wood Resources. Population Action International, Washington, D.C. www.populationaction.org.

Haines, Aubrey L. 1977. *The Yellowstone Story,* Vol. I, Yellowstone Library and Museum Association in cooperation with Colorado Associated University Press.

Hendee, John C., and Randall Pitstick. 1992. "The Growth of Environmental and Conservation-Related Organizations 1980–1991," *Renewable Natural Resources Journal,* **10**(2):6–19.

Hendee, John C., and Chad Dawson. 2002. *Wilderness Management: Stewardship and Protection of Resources and Values.* North American Press, Fulcrum Publishing, Golden, Colo.

Knudson, Tom. 2001. Fat of the Land: Environmental Movement's Prosperity Comes at a High Price. *The Sacramento Bee,* Special Report. April 22, 2001 pp.A,1, A10–12.

Lloyd, R. Duane, and V. L. Fisher. 1972. "Dispersed versus Concentrated Recreation as a Forest Policy," Proceedings of the Seventh World Forestry Congress, Buenos Aires, Argentina.

Nash, Roderick. 1982. *Wilderness and the American Mind* (3rd ed.). Yale University Press, New Haven, Conn.

O'Laughlin, Jay. 1992. "The Endangered Species Act: What the Law Is and What It Might Become," *J. Forestry* **90**(8):6–12.

Ranney, Sally, and Ann Gumaer. 1990. "Women and the History of American Conservation," *Women in Natural Resources* **11**(3):44–50. University of Idaho, Moscow.

Runte, Alfred. 1997. *National Parks: The American Experience.* University of Nebraska Press, Lincoln.

Sharpe, Grant W., Charles Odegaard, and Wenonah F. Sharpe. 1994. *A Comprehensive Introduction to Park Management.* Sagamore Publishing, Champaign, Ill.

Steen, Harold K. 1976. *The U.S. Forest Service. A History.* University of Washington Press, Seattle.

Steen, Harold K. 1991. *The Beginning of the National Forest System.* USDA Forest Service, FS–488.

Stokes, Jerry. 1999, April. "Wilderness Management Priorities in a Changing Political Environment." *Int'l. Jour. Wilderness* **5**(1):4–8.

Trefethen, James B. 1975. *An American Crusade for Wildlife.* Winchester Press and Boone and Crockett Club. New York.

U.S. Department of Agriculture, Forest Service. 1993. *The Principal Laws Relating to Forest Service Activities.* U.S. Govt. Printing Office, Washington, D.C. ISBN 0-16-041927-1.

U.S. Department of Agriculture, Forest Service, Ottawa National Forest. 2001a. *Draft Environmental Impact Statement: Plantation Lakes Vegetative Management*

Projects, Volumes 1 and 2. Ontonagon and Kenton Ranger Districts, Ontonagon, Mich.

U.S. Department of Agriculture, Forest Service, Ottawa National Forest. 2001b. *Record of Decision and Final Environmental Impact Statement: Plantation Lakes Vegetative Management Projects.* Ontonagon and Kenton Ranger Districts, Ontonagon, Mich.

U.S. Department of Agriculture, Forest Service. 2001a. *The RPA Assessment: Past, Present and Future.* Washington, D.C. www.fs.fed.us/pl/rpa/what.htm.

U.S. Department of Agriculture, Forest Service. 2001b. *Renewable Resources Planning Act 2000 Assessment Summary Report.* Washington, D.C. www.fs.fed.us/pl/rpa/publications

West, Terry. 1992. *The National Forests: Centennial Mini-Histories, Part One 1864–1891.* USDA Forest Service, Washington, D.C.

Frederick, Kenneth, and Roger Sedjo. (ed.). 1991. "America's Renewable Resources: Historical Trends and Current Challenges. Resources for the Future." Washington, D.C.

Freemuth, John, and R. McGreggor Cawley. 1993. "Ecosystem Management: The Relationship Among Science, Land Managers, and the Public," *The George Wright Forum* **10**(2):26–32.

Lewis, James G. 2000. "Raphael Zon & Forestry's First School of Hard Knocks." *J. Forestry* **98**(11):13–17.

Mayberry, B. D. 1991. *A Century of Agriculture in the 1890 Land Grant Institutions and Tuskegee University—1890–1990.*

McCleary, Douglas W. 1992. *American Forests: A History of Resiliency and Recovery.* Forest History Society. Durham, N.C.

Miller, Char. 2000. "The Pivotal Decade. American Forestry in the 1870s." *J. Forestry* **98**(11):6–10.

ADDITIONAL READINGS

Adams, David A. 1993. *Renewable Resource Policy: The Legal-Institutional Foundations.* Island Press. Covelo, CA.

Aplet, Gregory, H. Nels Johnson, Jeffrey T. Olson, and Sample Alaric. 1993. *Defining Sustainable Forestry.* Island Press. Covelo, CA.

Ellefson, Paul V. 1992. *Forest Resources Policy; Process, Participants and Programs.* McGraw-Hill, Inc. New York.

Ellefson, Paul V. 2000. Has Gifford Pinchot's Regulatory Vision Been Realized?" *J. Forestry* **98**(5):15–22.

Ellefson, Paul, and Fred H. Kaiser. 1991. "Forest Policy Development and Administration: Potential Research Directions for the Future," *J. World Forest Resource Management,* **5**:85–100.

Fedkiw, John. 1989. *The Evolving Use and Management of the Nation's Forests, Grasslands, Croplands and Related Resources.* U.S. Department of Agriculture, Forest Service. General Technical Report RM–175.

Floyd, Donald W., Sarah L. Vonhof, and Heather E. Seyfang. 2001. "Forest Sustainability. A Discussion Guide for Professional Resource Managers." *J. Forestry* **99**(2):8–27.

STUDY QUESTIONS

1. What is the relationship between laws and policies?
2. Define four groups involved in the policy process.
3. Describe three eras of conservation. What would be a suitable name for a fourth era?
4. Describe the role of women in early conservation policy. What about today?
5. Contrast the methods and aims of Gifford Pinchot and John Muir. Was the conflict between these philosophies inevitable? Can you see any parallels today within either the resource professionals or the environmental movement itself? Describe these briefly, if so.
6. What were the effects of public involvement on forest and renewable resource policy?
7. Explain the meaning of these acronyms: NEPA, EIS, DEIS. Name a project, present or past, that has had to undergo the NEPA process.
8. How is the work of resource managers different today than it was until the 1960s?
9. What does the Endangered Species Act seek to do?
10. Why is closely following the NEPA process important in proposing a resource management action?

CHAPTER 3

Distribution of North American Forests

Chapter Outline

INTRODUCTION

FOREST CLASSIFICATION

In Chapter 4 we will discuss the various parts of a tree and look at associations of trees in terms of biotic communities called *forests.* To understand better where the forests of North America are located, the classification of these larger units will be considered here.

Throughout the world, plant ecologists have mapped the locations of tree and other plant species and classified land into vegetation *cover types.* The classifications may be very broad, such as *evergreen* forests or *deciduous* forests. The forest may be named for a region such as the southern latitudes, where *tropical* forests flourish, or the north country, which in some parts of the world is called the *boreal* forest, and in others the *taiga* forest. The name may be more specific and denote a mixture of species, possibly containing both evergreen and deciduous trees. For instance, in the United States we recognize the oak–pine forest of the South and the birch–beech–maple–hemlock forest, or *type,* of the Northeast. The word *type* or *region* may be used instead of *forest.*

Classifications are sometimes based on the original forest, as seen by early surveyors, before extensive land-use conversion or strict wildfire prevention measures altered the cover type. Where forests are abundant, people make use of them, and thus forests ultimately may become the basis of a wood-utilizing industry. In that situation the name of the forest may reflect the economic importance of certain commercial species, such as the *Douglas-fir region,* or the *Southern pine region.* Of course, the commercial importance of species fluctuates with time and markets. Land clearing for agriculture, and for the expansion of towns and cities, has modified the original forest cover. Catastrophic tree diseases also impact forests and skew classifications. For example, as we see in the section on United States Forests, the once important oak–chestnut type of the Appalachian Mountains no longer exists because the chestnut blight disease virtually eliminated chestnut trees. Further complicating forest cover typing is the fact that each country, in North America at least, has its own classification system, based on different criteria.

One can easily see why different classifications developed within the three countries of North America. The forests of the eastern part of the continent differ greatly from those of the west, and are separated by areas where rainfall is less than about 16 in (40 cm). These areas are practically treeless. The eastern forests have both hardwoods and softwoods of commercial value. The western forests have few important hardwoods, although the variety and commercial values of softwoods are notable. These two forest groups or regions meet, not in the United States, but in Canada. There, as a transcontinental cover type, this union of species assumes a special importance. The dry western interior forests of the United States are also represented in northern Mexico, where its species mingle with those of Mexico's northern states. The Tropic of Cancer crosses Mexico at midpoint, placing the southern half of the country in the tropical zone. However, Mexico's climate is orographically modified by mountain ranges of considerable height. Resultant variations in rainfall and other factors create a wide range of climates and a diversity of ecological zones, producing tree species not found elsewhere in North America.

Even the demands placed on forests vary from country to country, as well as from region to region. Factors such as landownership patterns, forest and economic policy, available energy resources, and social structure affect the use and, thus, the extent of forests. For these reasons, we treat the forests of Canada, Mexico, and the United States individually.

United States Forests

There are six major *natural forest types* in the conterminous United States (Fig. 3-1). Four are in the eastern half of the country: the Northern Forest, the Central Forest, the Southern Forest, and the Bottomland Forest. West of the prairie states lie the Western Interior Forest and the West Coast Forest.* Since

---

*The location of these six general forest types comes from the extensive work undertaken by A. W. Küchler in his *Potential Natural Vegetation of the Conterminous United States,* 1964, published by the American Geographical Society. The map in Fig. 3-1 represents the present knowledge of geographical distribution of vegetation in conterminous United States, and makes no attempt to predict what influence human activity will have in the future. If anything, the map posits the vegetative cover that would continue if the human species were suddenly removed from the scene. Students interested in the larger scale map, and in his timely approach to vegetation classification, are encouraged to read Küchler's book.

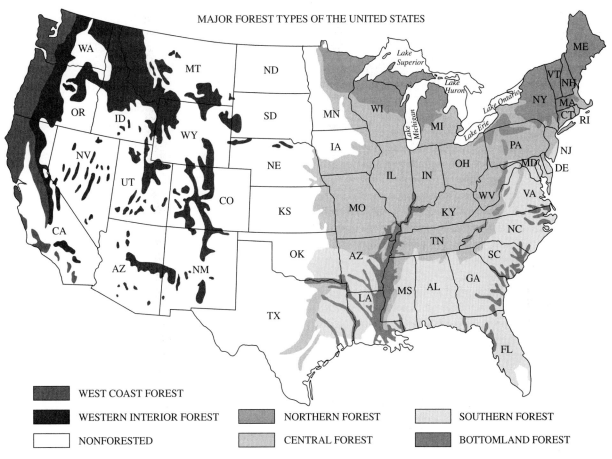

FIGURE 3-1
The major forest types of the contiguous 48 United States. (Adopted by the authors from A. W. Küchler's *Potential Natural Vegetation of the Conterminous United States*)

within these regions the tree species have determined the character of the forest industries to a large extent, the most common uses of major species will also be cited.

Following this section is a discussion of the forests of Alaska, Hawaii, and Puerto Rico. All tree names are in accord with Little's (1979) *Checklist of United States Trees (Native and Naturalized)*. The scientific names for all common tree species are found in Appendix C.

The Northern Forest

Location   This region covers almost all New England, New York, and the upper parts of the three Lake states—Michigan, Wisconsin, and Minnesota. Also, it extends in a narrow interrupted strip from

northern and central Pennsylvania southwest through the Appalachian Mountains.

Principal Tree Species   Of the five trees most characteristic of the Northern Forest, four are conifers. These are balsam fir, northern white-cedar, eastern white pine, and eastern hemlock. The characteristic hardwood is the yellow birch. Sugar maple is a common tree also, but it is even more typical of the Central Forest.

Within the Northern Forest (Fig. 3-2) are found several local forest types of importance. In the northern portion of Minnesota, Wisconsin, and Michigan are eastern white pine and red pine (Fig. 3-2). Jack pine, the principal pine marketed in the Lake states today, now occupies great areas in these three states

FIGURE 3-2
The Northern Forest. *Left:* A red pine stand in northern Minnesota. *Right:* Paper birch in northern Wisconsin. *(Courtesy of USDA Forest Service)*

where white and red pine were once abundant. After logging and repeated uncontrolled fires in the late 1880s and early 1900s destroyed the humus layer of the soil and the seed source of red and white pine, jack pine flourished. In moist or swampy areas, northern white-cedar, black spruce, and tamarack are common associates. Balsam fir requires moderate moisture, and in wet places in the northern part of the region, it may occur in pure stands. On drier sites of the north, it occurs with white spruce; in the northeast with white and red spruce; and in the Appalachians with red spruce.

A common hardwood type throughout the Northern Forest is a mixture of American beech with yellow birch and sugar maple. Eastern hemlock is commonly associated with these three hardwoods. Aspen (quaking and bigtooth), which sprouts root suckers after harvesting or burning, and paper birch

(Fig. 3-2), a prolific seeder, are both common over large, previously burned areas of the Lake states where moist mineral soil has been exposed. A dozen more hardwoods are found to some extent, and these include walnuts, hickories, elms, ashes, and American basswood.

In the southern extension of the Northern Forest (Appalachian Mountains), in addition to red spruce and balsam fir, we find the shortleaf, eastern white, pitch, and Virginia pines; eastern hemlock; white, northern red, and scarlet oaks; yellow-poplar; magnolia; black tupelo; and other hardwoods common in the northern part. American chestnut was at one time the most abundant tree in the Appalachians. The chestnut blight, a virulent disease accidentally introduced into the United States from Europe in 1903, has obliterated this species from the Northern Forest.

Historical and Commercial Importance   Most of the soils of this region (or forest) can be classified as poor agriculturally, although the forest has given way to many farms. Some of the poorer ones are surrounded with rock fences built through years of clearing. For more than two centuries, they have yielded a living only through application of intense human effort. The Northern Forest presents a vast problem in planning and in reconciling its use for mining, agriculture, recreation, and manufacturing, and for regulation and utilization of rainfall runoff.

It is difficult to think of the entire Northern Forest region as having much in common socially, industrially, or in the matter of population. The earliest establishment in this country of sawmills, pulp mills, wood-distillation plants, and shipyards occurred in the eastern section. Forest industries moved westward to the Lake states as the old-growth forests of the Atlantic seaboard were consumed, but the eastern section has now come back into the forest production picture. In Massachusetts forests were cleared for farms, eliminating deer habitat to the point where, for a 100-year period, no deer hunting was allowed. Now Massachusetts has an extensive forest that supports a thriving deer herd and modest forest industry.

About the middle of the nineteenth century, New York, Pennsylvania, and Maine led in lumber production. By 1870 Michigan, which had for 10 years been close to these, became the scene of greatest activity, and continued so until 1890. Minnesota and Wisconsin also produced heavily from 1860 to 1890. Shortly after 1899 the Northern Forest lost its lumber production lead to the Southern Forest.

It was the Northern Forest that furnished a large share of the lumber used in this country for the first 250 years of settlement. The bulk of this lumber was eastern white pine. Today it is still considered a desirable northern lumber species, followed closely by red pine and white and red spruce. A lower-quality softwood lumber is manufactured from jack pine, eastern hemlock, tamarack, and balsam fir. The Northern Forest used to be the leader in the production of pulpwood. The three species of spruce and balsam fir are the most desirable pulp species for paper making, although all northern conifers except tamarack and northern white-cedar are used successfully today in one pulp process or another. Aspen,

once considered a forest weed, and the most widespread species in North America, accounts for significant pulpwood harvest in the Lake states. Aspen is also an important source of fiber in the wafer-board industry. The main uses for the very durable northern white-cedar are for poles, posts, and shingles.

The region's hardwoods have kept thousands of woodenware plants supplied. Handles, flooring, furniture, ladder rungs, novelties, and athletic equipment such as bats and bowling pins come from the denser woods such as sugar maple, yellow birch, black cherry, white ash, and American beech. Yellow birch supplies a highly prized veneer for kitchen cabinets. Visit any northwoods gift shop and you will see a variety of novelties made from the bark of paper birch. Tons of charcoal are produced from the lower-grade, dense hardwoods and some from wood residue. Basswood lumber provides cooperage and woodenware; as veneer it is made into cabinet linings, mirror backings, and drawing boards, and is utilized in other places where a fine-grained, softer wood is desirable. Basswood flowers are favored by some bee keepers for the tart honey they yield. Millions of Christmas trees are cut annually in the Northern Forest from both native and nonnative evergreens grown on hundreds of plantations.

Finally, the Northern Forest, with its relatively small human population, provides a source of pleasure for millions of visitors who go north from crowded cities, particularly in the summer, to enjoy a northwoods vacation. Three national parks—Acadia, Isle Royale, and Voyageurs—and an elaborate system of state parks and forests, plus hundreds of private resorts, provide access to the region's many lakes, streams, and other forest amenities. The million-acre-plus Adirondack Forest Preserve in New York, with its mandate to be protected "forever wild," is the largest state area and the oldest area in the nation legally protected and managed for wilderness values.

The Central Forest
Location   This vast area of broad-leaved trees stretches from Cape Cod almost to the Rio Grande in Texas and northwest to Minnesota. Its western boundary is very irregular where it meets the essentially treeless midcontinental prairies. This forest is invaded by the Northern Forest along the Appalachian Mountains

and by the Bottomland Forest up the Mississippi Valley. The Central Forest is scattered over a huge area involving about 30 states.

Principal Tree Species   More than 20 genera and perhaps 75 important species of trees are represented here. Because of the great diversity of climate and soils, however, few species grow throughout the entire region. White oak, the region's most important and widely distributed tree, is one exception. The northern red; the southern red; the black, bur, and chestnut; and, to a lesser extent, the post, over-cup, swamp white, and chinkapin comprise the rest of the oaks found here. Most of these trees lose their identity and become either "white" or "red" oaks when manufactured into wood products. American beech, sugar maple, American and slippery elm, black wil-

low, eastern cottonwood, black walnut, shagbark and bitternut hickory, hackberry, yellow-poplar, American sycamore, black cherry, honeylocust, and white and green ash, although enjoying differing degrees of importance and abundance, are all found throughout the Central Forest (Fig. 3-3). Other, less widespread species of hardwoods include butternut, several more oaks and hickories, sweet and river birch, cucumbertree, sassafras, sweetgum, and black locust; black, red, and silver maple; and Ohio and yellow buckeye, American basswood, black tupelo, black and blue ash, osage-orange, and northern catalpa. American chestnut, once making up 25 percent of the Eastern Hardwood Forest, has been reduced by chestnut blight to suckers on the stumps of formerly noble specimens, its niche now taken up by oaks and yellow-poplar (Ronderos, 2000).

FIGURE 3-3
An example of the Central Forest, showing an all-aged deciduous stand in Ohio. Note the abundant reproduction of sugar maple in the foreground. The person is standing next to a yellow-poplar. *(Courtesy of Ohio Agricultural Experiment Station, Wooster, Ohio)*

The conifers are confined to shortleaf, pitch, and Virginia pine, eastern redcedar, and, occasionally, eastern hemlock.

Historical and Commercial Importance   In numbers and utility of species, the region compares well with others, but because much of the forest originally occupied deep, rich soils, thousands of productive farms have appeared, displacing the original cover.

With such a variety of tree species, it is understandable that the Central Forest produces a large assortment of forest products. Among the more important are products from the white oaks—railroad ties, tight cooperage, flooring, furniture, cabinets, and ship and boat timbers. From the harder maples come flooring, fancy veneer, furniture, and a wealth of sporting equipment such as bowling pins and croquet mallets.

Handles of various shapes and sizes come from the hickories, oaks, and ashes. Fine-quality furniture, cabinets, veneer, and gun stocks are produced from black walnut. Other cabinet woods are black cherry and yellow-poplar. Hardwood lumber is cut from sweetgum, elm, and sycamore. Other products, such as mine timbers, posts, poles, piling, light millwork, and a host of lesser items, come from the hardwoods of this forest.

The nutlike fruit of such trees as American beech and the walnuts, oaks, and hickories, known as *mast,* serves wildlife as a food.

From eastern redcedar, a highly aromatic and colorful wood with moth-repelling properties, comes material for chests and closets. Wood pulp for paper products is derived, not only from the region's limited conifers, but also from an ever-increasing tonnage of hardwood species, such as cottonwood, oak, beech, aspen, and maple, to name a few. Much of the charcoal available on the market comes from hardwoods of this forest.

Over the past 40 to 50 years, several national forests have been established in this forest region. There are also two major national parks, Shenandoah and Great Smoky Mountains, at its eastern edge, several other units of the National Park System, and several small but important state parks and forests. The Tennessee Valley Authority has jurisdiction over a significant portion of this area's forested lands. Generally, however, the majority of the private forestland lies in small ownerships. Because there are large population centers here, the forests of this region also serve as major sites for outdoor recreation. Most states also have extensive county and state park systems that attempt to supply this need.

The vast extent of this forest precludes social or economic unity, making it the most diverse of the forest regions.

The Southern Forest

Location   This forest extends along the Atlantic coastal plains from southern New Jersey south to all but the lower end of Florida, and west along the Gulf plains to east Texas, and northward to Oklahoma. The Bottomland Forest bisects it up the Mississippi Valley and penetrates it along the river bottoms of the Atlantic coastal states.

Principal Tree Species   Four important pines characterize the Southern Forest—shortleaf, longleaf, slash, and loblolly (Fig. 3-4). Two less significant

FIGURE 3-4
Loblolly and shortleaf pine in the Southern Forest. *(Courtesy of International Paper Company)*

pines are the pitch and pond pines. In moist areas we find baldcypress, water tupelo, sweetgum, black tupelo, and numerous oaks, such as Nuttall, willow, and swamp chestnut oak, and in other moist areas, American elm, red maple, and eastern cottonwood. Less common species are sugarberry and common persimmon. The southern magnolia and live oak, the latter festooned with "Spanish moss," are the picturesque southern trees of drier sites. Other upland trees include cucumbertree, yellow-poplar, pecan, mockernut hickory, American beech, and several more species of oak.

Historical and Commercial Importance This wide southeastern border of the United States was once the scene of intense timber production, capturing the lumber production lead from the Lake states group around 1900 and holding it until 1920, when the lead shifted to the West Coast. The once great cotton industry of the South moved westward also; thus thousands of acres of plantations on sandy soil, ill adapted to agriculture, reverted to the various species of southern pine. Because of a mild climate, an abundant rainfall, and a long growing season, in a comparatively few years these second-growth forests produced huge volumes of wood, which because of its structure, was more suited to high-grade wood pulp than lumber. The northern pulp industry, looking for places with raw material in which to expand, moved south. Today approximately 70 percent of the U.S. pulpwood harvest comes from the Southern Forest. Its mills, favorably located with respect to markets, generally have an ample supply of raw materials because of the large land area and rapid tree growth. Excellent transportation facilities, and an abundant supply of relatively cheap labor, water, chemicals, and power complete the picture. Once again the southern pines have assumed a position of great commercial importance. No species or group of species in the United States equals the four southern pines in their annual pulpwood cut.*

Lumber milling from the southern pines still represents an important employer in many Southern Forest communities. Timbers for heavy construction, such as bridges and large buildings, come mostly

from longleaf pine. Slash pine produces some heavy construction timbers also. From loblolly and shortleaf come material for lighter construction.

Longleaf and slash pine also produce gum or crude resin, the raw material of a small naval-stores industry (see page 322). This region once supplied more than 60 percent of the world's natural production of turpentine and rosin. Much of the naval-stores production today, however, comes as a by-product of the southern kraft-paper pulping process.

The hardwoods of the Southern Forest serve the same economic purposes as the hardwoods of the Central Forest. However, because of the heat and humidity, and perhaps because of snakes, ticks, and chiggers, less recreational use takes place in the forest, although hunting remains a popular forest-based activity in the South.

The Bottomland Forest

Location The Bottomland Forest, as noted in Fig. 3-1, is located primarily on the brown-loam bluffs along the Mississippi River system, south of Illinois and Indiana, down to Louisiana. Smaller isolated units flourish along swamps and streams throughout the Atlantic coastal plain and the Gulf coast.

Principal Tree Species Baldcypress, a needleleaf tree that annually sheds its needles, predominates (Fig. 3-5). It forms pure stands in permanent swamps. Major associates are water tupelo in the alluvial floodplains, and swamp tupelo in swamps and estuaries. Other important deciduous associates, on slightly higher ground that invites more competition, include numerous oaks such as cherrybark, water, and swamp chestnut, and also American elm, red maple, green ash, river birch, eastern cottonwood, and American sycamore. Less common species are swamp poplar, waterelm (planertree), water hickory, and sweetgum. Two conifer associates grow here too: Atlantic white-cedar and pond pine.

Historical and Commercial Importance The durability of baldcypress has made it a favorite species of wood for over four centuries, although this swamp species is relatively inaccessible. Parts of the Bottomland Forest comprise a remnant wilderness—the refuge of both real and fictional isolates, hermits, escapees and other desperados. On the other hand, the unusual qualities of old-growth baldcypress, particu-

FIGURE 3-5
Baldcypress in the Bottomland Forest. Note the fluted bases and the conical root outgrowths called *knees*. A typical tree of permanent swamps, it sheds its needles each year. Berkeley County, South Carolina. *(Courtesy of Westvaco)*

larly its resistance to decay, make it a sought-after species rivaling western redcedar from the Northwest. Major uses include wood for construction timbers, caskets, stadium seats, boxes, cooperage, railroad ties, and boat building.

Other species in this forest have uses similar to those in adjacent forest types, including hardwood lumber, veneer paneling, furniture stock, and pulpwood.

The Western Interior Forest

Location   A glance at Fig. 3-1 reveals the irregular pattern of the Western Interior Forest extending from Canada on the north, where it is generally classified as the Montane Region, to Mexico on the south, where it continues as the Temperate Forest of Mexico's mountain ranges. From the prairies it runs up the east slopes of the Cascades and Sierra Nevadas. Moreover, the treeless areas between these patches of forest are mainly mountainous outcrops, not rich agricultural lands. Natural prairie and other grasslands account for some of them, however, and it is said that the early Spanish ranchers and their descendants, desiring pasture for sheep and goats, burned the mountains regularly to extend the open stretches.

Native Americans also fired the forests and grasslands frequently. This centuries-old history of interaction among climate, fire, and cultural practices has left its imprint on the Western Interior Forest.

In terms of ownership, roughly 75 percent of this region's forestland is owned by federal public agencies, with 60 percent of it in national forests managed by the U.S. Forest Service. The next largest ownership is state forestlands in scattered tracts, often two sections per township, the result of the original disposition of federal grant lands in the 1800s. In Utah and Nevada the Bureau of Land Management is the chief land manager for the vast, arid federal lands, although some of the higher mountains are in national forests.

Principal Tree Species   If a vanload of visitors, for example, were to set out to see the Western Interior Forest, their wanderings would not only be over great distances horizontally, but to see all the types, the travelers would have to climb to higher elevations than those common in the East. Starting at 5,000 ft (1,500 m) above sea level in the drier sites of the central Rockies of New Mexico or Colorado, they could see two or three species of juniper, several pinions or nut pines, and some shrubby species of oak. Willows would appear if water were nearby. At about 6,000 ft (1,800 m), they would notice the beginning of the ponderosa pine forest characteristic of all western states (Fig. 3-6). Should their day's journey carry them upward even further, they would see more and more quaking aspen, Douglas-fir, and white fir mixing with the now-disappearing ponderosa pine, until at their camping place at night, there would be al-most a pure forest of Douglas-fir and white fir. Above the upper fringes of the ponderosa pine, they might also see patches of lodgepole pine in old burns. The next morning on their upward climb, they would travel through almost pure forests of blue and then Engelmann spruce. At 9,000 ft (2,700 m), they would run into a forest of Engelmann spruce, sub-alpine fir, and limber pine, which would continue until they reached timberline at 11,000 ft (3,300 m) (Fig. 3-6).

FIGURE 3-6
The Western Interior Forest. *Upper:* Ponderosa pine in the Black Hills of South Dakota. Note the wide spacing of trees in this dry interior forest. *Lower:* Engelmann spruce and subalpine fir in the mountains of Colorado. *(Courtesy of USDA Forest Service)*

Tired of climbing, our travelers might drive from Denver northwest to Missoula, Montana, and the "Inland Empire," a region comprised of eastern Washington and Oregon, northern Idaho, and the west slope of the Montana Rockies. Much of what they would see would be ponderosa pine and heavy patches of lodgepole pine, a species that grows in thick, pure stands soon after a fire. Upon reaching Missoula they could travel northwest to Idaho, through fine stands of ponderosa pine, into the heart of the Inland Empire, where they would also find magnificent forests of western white pine, western larch, and occasional mixtures of western hemlock, grand fir, Douglas-fir, and western redcedar. At higher elevations this far north would be Engelmann spruce and subalpine fir, again the timberline species. This close to the Canadian border, timberline is around 7,000 ft (2,100 m) above sea level. Travelers to the Black Hills of South Dakota would find forests largely of ponderosa pine (Fig. 3-6, *upper*), with white spruce in the valleys and some scattered bur oak and hophornbeam. In eastern California they would find ponderosa pine mixed with Jeffrey, lodgepole, western white, and sugar pines; white and California red fir; and Douglas-fir.

The region has several pines of lesser importance; however, two varieties of bristlecone pine are found in the southern half of the Western Interior Forest. The Rocky Mountain bristlecone pine at high elevations in Colorado, New Mexico, and Arizona, and the great basin bristlecone pine in Utah, Nevada, and central California, are noted for their longevity (Harlow et al., 1991). One tree in east central California, growing at an elevation of 10,000 ft (3,000 m), is known to have been growing for over 4,500 years, making it the oldest known living tree.

Historical and Commercial Importance The principal commercial timbers of the Western Interior Forest as a whole are ponderosa pine, western white pine, Douglas-fir, lodgepole pine, and Engelmann spruce. The rather spotty distribution of the forest makes large-industry utilization less practical and so encourages smaller forest industries. Major operations are confined largely to the heavily timbered northwestern portion and, to a lesser extent, to the southeastern section.

The southern Western Interior Forest lies in Arizona, Colorado, Nevada, New Mexico, Utah, and eastern California, with smaller units in the mountains of southern California. Most of the commercial forest is in national forests. Industry is limited primarily to sawmills that prepare lumber for local use, although there have been some integrated operations in recent years; that is, a forest products company may have a mill in the area from which a variety of products result, some for local remanufacture and others for shipment to processing plants elsewhere.

The northern Western Interior Forest includes eastern Washington and Oregon, Idaho, Montana, Wyoming, and western South Dakota. Sawmills, some pulp mills, and increasingly, chipboard plants, represent the forest products industry.

The Western Interior Forest contains a total of 141 million acres (56 million ha) of forestland. However, less than half of this—57.8 million acres (23.4 million ha)—is commercial timberland. As a result, while important locally, this forest plays a minor role in the production of timber nationally, accounting for only about 6 percent of the roundwood output of all regions. A majority of the wood products produced is used in the manufacture of lumber. Most of the remaining volume is used in the production of plywood and chipboard.

In recent years, the area classified as commercial timberland has declined on public ownerships, chiefly on national forests. Most of the reduction resulted from the withdrawal of commercial timberland for wilderness consideration and because of other environmental sensitivities, including water quality and protection of habitat for endangered species such as salmon and grizzly bear. Wildfires are also consuming large areas of forest in the region, the result of recent drought, as well as decades of fire control allowing fuel buildup.

For the forest as a whole, high mountain recreation, forage for sheep and cattle, and water production for domestic and municipal purposes as well as for irrigation are its most important uses. Here lie eight of the country's best-known national parks (including Glacier, Grand Canyon, Grand Teton, Rocky Mountain, and Yellowstone), over 50 national forests, vast areas of public lands under Bureau of Land Management jurisdiction, and many wilderness areas. There are also many large reservoirs in this region. The lumber industry, which is centered mostly around ponderosa pine, has never been in serious competition with the Pacific coast, the Lake states, or the South. It has had a significant local importance, however, and also furnishes lumber, poles, and railroad ties, in considerable quantity, to the Great Plains region.

The West Coast Forest

Location The forests of this region extend southward from British Columbia's Coast Forest Region through western Washington and Oregon, and into California to the San Francisco Bay area. The entire forest lies west of the crest of the Cascade and Sierra Nevada ranges.

Principal Tree Species In the northern part of the range, Douglas-fir is the most important and characteristic tree (Fig. 3-7). Along the coast its principal associates are Sitka spruce, western hemlock, western redcedar, and Pacific silver and grand fir. At higher elevations these species give way to subalpine fir and whitebark pine. Further inland but still west of the Cascades, Douglas-fir is found with western white pine and noble fir. The most important hardwoods of the Northwest are red alder, bigleaf maple, black cottonwood, Pacific madrone, and Oregon ash.

In the north coastal California section of the West Coast Forest, redwood is an important tree. Its common coniferous associates include Douglas-fir, Sitka spruce, western hemlock, and grand fir. Port-Orford-cedar is limited to a small coastal area of Oregon and California. The hardwood associates of redwood include red alder, Pacific madrone, bigleaf maple, tanoak, and California-laurel. As one moves inland, away from the 50-mile-wide (80-km-wide) fog belt, most coastal species disappear, although Douglas-fir and bigleaf maple are still encountered. Incense-cedar, western white and sugar pine, Pacific madrone, Oregon white oak, California black oak, and giant chinkapin become more common. At somewhat higher elevations, Jeffrey and ponderosa pine and giant sequoia appear with sugar pine and incense-cedar. At about 6,000 ft (1,800 m) white, California red, and Shasta red fir appear. Further up the mountain slopes grow mountain hemlock and lodgepole pine. Several lesser-known pines are found in this region. One of these is the Monterey pine. Although of no commercial value within its range, it has been widely planted in countries in the southern hemisphere where it grows rapidly and becomes a commercial, viable product, providing a notable ex-

FIGURE 3-7
The West Coast Forest. A densely growing stand of 40-year-old Douglas-fir in western Oregon. Compare with the more widely spaced ponderosa pine in Fig. 3-6. *(Courtesy of American Forest Products Industries)*

ception to the rule that species do not thrive outside their provenance.

Historical and Commercial Importance The lead in lumber production shifted from the South to the West Coast Forest in 1920. Washington was the nation's number one producer of lumber from 1905 until 1938, when the lead went to Oregon—which has held it ever since. From the three Pacific coast states with their valuable forests of Douglas-fir comes a high grade of softwood lumber, timbers, and veneer for plywood, which is exported throughout the world. Substantial quantities also go into pulp, poles, and piling, and a variety of other items. Douglas-fir is considered to be the most widely used commercial species in the world. Western hemlock, Sitka spruce, and the several species of western true firs are the region's important pulp

FIGURE 3-8
Old-growth western redcedar was once very abundant in western Washington. This historical photograph, taken in 1898, shows two loggers posing on spring boards and holding their 9-foot falling saw. *(Courtesy of University of Washington Special Collections Library)*

species. Western redcedar, because of its durability, has many uses, but is best known as the nation's leading wood for shingles and increasingly for decks and decorative fencing (Fig. 3-8). Redwood, because of its durability and strength, is used for many items, ranging from decks and fences to boxes and coffins to siding and items for building construction. Redwood burls are highly prized for a variety of products. Incense-cedar is the principal wood used in making pencils. Port-Orford-cedar is known to the archer as the ideal wood for arrows. The region's hardwoods are limited. However, red alder, long considered an infe-

rior species and for years cleverly disguised as other woods by the local furniture industry, has come into its rightful place as the Northwest's leading furniture wood. Red alder is also an important tree for pulpwood and fuel. Bigleaf maple is cut for flooring and furniture, and black cottonwood is used for pulp, excelsior, and veneer for boxes and crates.

Although the smallest of the natural forest regions in the United States, the West Coast Forest produces over one-third of the country's total timber products.

The entire West Coast Forest produces trees of such remarkable size that it is no wonder this forest

FIGURE 3-9
The giant sequoia is perhaps the most famous tree species in the world. Part of the Congress Group, Sequoia National Park, California. *(Courtesy of the National Park Service; photo by George Grant)*

is one of the most famous in the world. Trees of such enormity are found nowhere else. In the rain forests of Olympic National Park, for example, on the Olympic Peninsula in the state of Washington, where the annual rainfall is over 150 in (380 cm), four of the Northwest's most important trees reach their maximum sizes. Farther south, along the fog belt of coastal California, are the magnificent redwoods with average diameters of 8 to 12 ft (2.4 to 3.6 m). Even more impressive are their heights, which average 200 to 300 ft (60 to 90 m); the loftiest is 368 ft (110 m) (see frontispiece). Further inland, in groves at middle elevations in the Sierra Nevada, stand the most famous of all trees, giant sequoias (Fig. 3-9). What these 270-ft (80-m) trees lack in height, they make up for in diameter. The General Sherman tree in Sequoia National Park is nearly 30 ft (9 m) in diameter at breast height. Diameters of 12 to 14 ft (3.6 to 4.2 m) are relatively common in the giant sequoia groves. Most of the giant sequoia, however, as well as the record-sized trees of the other species, are under some form of protection. They contribute to the economy of the region through their aesthetic value and recreation attraction rather than through their use as lumber. A multitude of tourists annually spends millions of dollars just to look at or camp near these dignified western giants.

The forests of the region's vast high country are also important for recreational use, as indicated by the eight mountainous national parks (Mount Rainier, Olympic, North Cascades, Crater Lake, Lassen Volcanic, Yosemite, Kings Canyon, Sequoia), and by Redwood at a lower elevation, plus several national forests that contain many wilderness areas.

The May 18, 1980 eruption of Mount St. Helens in Washington State focused interest on the region. Most of the salvable timber in the blast zone has been removed, and the area, now largely reforested, has become an important tourist attraction (Figs. 7-10 and 7-11, pages 149 and 150). Part of the site was established as the Mount St. Helens National Volcanic Monument in 1982.

The growing population of the West Coast is also dependent on these forests for water production. The very size of the timber, its location in rugged sites, and its distance from the greatest consuming centers of the country have required a different approach to forest practices. In no other region are logging and milling of the product more costly.

The region's timber potential, its ruggedness, vulnerability to wildfires, and importance in hydroelectric and irrigation projects, present one set of imperatives. Invasion of some prime growing sites by vacation homes as well as urban sprawl, plus demand for recreation and wilderness designations, present another set of imperatives. The multidimensional conflicts produce social and planning dilemmas. In 1990 the northern spotted owl was declared endangered, and permanent protection for several million acres of old-growth forests that contain nesting habitat for the owl has been mandated. In 1992 the marbled murrelet, another old-growth dependent species, was classified as threatened. Other forest-dependent species may be added. This, plus proposed protection for remaining roadless areas, will result in substantial reduction of the timber harvest volume in Washington, Oregon, and northern California.

The Alaskan Forest

Location The forests of this huge state are naturally divided into two forest regions; the coastal and the interior (Fig. 3-10). The coastal forest extends 800 mi (1,300 km) from the panhandle of southeastern Alaska, north along the coast through the Kenai Peninsula to Kodiak Island, from sea level to approximately 2,500 ft (750 m) in the south to 1,500 ft (450 m) in the north. The greater part of this region, largely encompassed by the Tongass National Forest (Fig. 3-11), lies adjacent to northwestern British Columbia.

Principal Tree Species The tree species in this area are chiefly western hemlock and Sitka spruce, and in some areas western redcedar, Alaska-cedar, and mountain hemlock. In the muskegs, lodgepole pine becomes a component of the vegetation; the hardwoods are red alder and black cottonwood.

Commercial Importance Most of the 7 million acres (2.8 million ha) of this coastal commercial timber lies within a few miles of tidewater. Western hemlock makes up about 74 percent of the timber volume and Sitka spruce about 20 percent. These two trees, considered as pulpwood species, cover nearly 4.3 million acres (1.7 million ha) of commercial forest, the vastness of which gives some indication of the pulp potential of Alaska's coastal forest. The pulp industry has become well-established in southeastern Alaska.

Alaska's interior forests of commercial value cover 4.1 million acres (1.67 million ha). Although yields here are for the most part much inferior to those of the coastal forest, some of the local stands take on added value because of the excessive cost of transporting wood products either in or out. Insects and diseases also take a large annual toll, and nearly all the interior forest has been burned at one time or another. Limited rainfall, low humidity, high winds, high temperatures, and long hours of summer sunshine combine to create one of the most dangerous fire conditions in any North American forest. An expanding road system, increasing population, and greater visitation by tourists may aggravate the fire situation, as it is estimated that 75 percent of the interior fires are caused by people.

In spite of this, there are commercial-grade forests of white spruce, quaking aspen, balsam poplar, black cottonwood, and paper birch in the Alaskan interior, especially along stream courses, lake shores, flood plains, and other well-drained lowlands. Black spruce is a common muskeg type and, depending on the locality, may be associated with tamarack. These

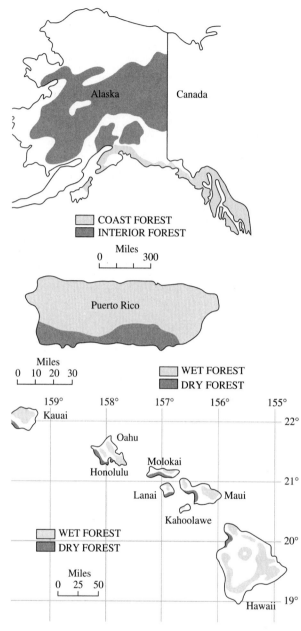

FIGURE 3-10
Maps showing the forest areas of Alaska, Puerto Rico, and Hawaii.
(Courtesy of USDA Forest Service)

FIGURE 3-11
The Alaskan Forest. An aerial view of the tree-covered islands of southeast Alaska near Admiralty Island. Tongass National Forest. *(Courtesy of USDA Forest Service)*

muskeg forests are similar to those found in Canada and the northeast United States and Lake states. Several species of willows and alders occur in the interior but seldom grow to marketable sizes.

Alaska's interior forest is going through the same phase of use and misuse that the forests of other states witnessed 50 to 250 years ago. The interior forests of Alaska have been interdependent with fire for eons and this will no doubt continue to be their most important source of renewal, despite the best fire prevention and control efforts. Greater utilization of the interior forest and an expansion of markets in the state are possible, since all but logs and rough materials for house construction are at present brought in from the "outside."

The Alaska National Interest Lands Conservation Act of 1980 (PL 96-487), often called ANILCA, made major changes in the administration of federal lands by placing 104 million acres (42 million ha) of land in federal conservation systems such as national parks, wildlife refuges, wild and scenic rivers, and wilderness, providing for scientific management of fish and wildlife in these systems and providing for subsistence uses by rural residents. A struggle continues between elements of the national citizenry wanting to conserve Alaska's natural resources and many of the state's residents wanting to develop and use them.

The Hawaiian Forest
Location Almost 2 million acres (808,500 ha) of assorted wet and dry tropical forest cover about one-half of the total area of the seven main islands in the Hawaiian chain (Fig. 3-10). The islands are noted for their diverse ecological environments that support a wide range of vegetative conditions from rain

FIGURE 3-12
The Hawaiian Forest. The koa, at left, is one of Hawaii's largest trees. At right is the ohia lehua, one of the first trees to invade new lava flows. The tree is widespread throughout Hawaii. *(Photo by Grant W. Sharpe)*

forests to desert to high mountain scrub forests. Annual rainfall ranges from less than 20 in to 500 in (50 cm to 12.7). Of the approximately 2,200 native plants, nearly 97 percent are endemic (found only in Hawaii).

Principal Tree Species The predominant forest tree in Hawaii is the ohia lehua, which occurs as a pioneer on lava flows, as a climax canopy tree reaching heights in excess of 80 ft (24 m) in the rain forest, and as a dwarf shrub in the wet montane bogs. The most valuable native commercial tree is the koa (Fig. 3-12), which is similar in wood quality to black walnut.

Commercial Importance The wood of the koa, hard, dark, and fine-grained, is used locally in the manufacture of paneling, furniture, and cabinets. The everpresent ohia lehua serves as flooring, pallets, poles and posts, pile-driver cushions, and wharf fenders.

Mimani, which occurs as a component of a montane scrub forest, is a durable native wood used locally for fence posts and fuel. Kiawe, a species introduced in 1860, grows well in the dry lowlands and is valuable for fuel, forage, fence posts, and honey.

Many trees have been introduced to Hawaii, particularly eucalyptus species from Australia. Their growth rate can be extremely rapid, with increments of 10 ft (3 m) in height per year commonly observed. There are approximately 46,000 acres (18,500 ha) of commercial forest plantations in the state.

A commercially valuable nut tree, the macadamia, continues to increase in importance. Many sugarcane fields have been converted to macadamia orchards, and the nut is touted as the finest edible nut in the world.

Water is the principal product derived from the Hawaii forests. The original act of the territorial legislature, which created the Division of Forestry in 1903, stressed this activity and directed that ways and means be devised for "protecting, extending, increasing, and utilizing the forests and forest reserves, more particularly for protecting and developing the springs, streams, sources of water supply so as to increase and make such water available for use." The Hawaii Division of Forestry predates the USDA Forest Service.

Today about 808,500 acres (327,000 ha) of state-owned forestland are included in 54 forest reserves. In addition there are many acres of privately owned forestland that are strictly zoned for protection.

Two national parks, Hawaii Volcanoes on the island of Hawaii and Haleakala on the island of Maui, have extensive areas of forestland including designated wilderness, all managed for the protection of the native Hawaiian flora and fauna, for recreation and for wilderness values.

The Tropical Forest

United States   Covering an area of not more than 400,000 acres (162,000 ha) the Tropical Forest within the United States lies at the southern tips of Florida and Texas. More subtropical than tropical, this commercially unimportant forest is composed largely of palms and mangroves and, in Texas, of mesquite.

Puerto Rico   Mixed forests more representative of the tropics are found in the Commonwealth of Puerto Rico, a 100-mile- (116-km-) long rectangular island in the eastern Caribbean (Fig. 3-10). Formerly completely forested, the island still has a variety of coastal and mountain vegetation ranging from dry to rain forest. The more than 500 native tree species include Spanish cedar and lignumvitae, but mahogany and pine are absent. The moist evergreen forests on the north side of the east-west mountain chain called the Cordillera Central contain as many as 40 tree species per acre, none of which are widely distributed in the continental United States (Fig. 3-13).

Only about 15,000 acres (6,000 ha) are still unmodified forest, but the recent decline of agriculture in the interior has led to rapid extension of second-growth forests, which now cover nearly 40 percent of the island. These new forests contain many valuable and fast-growing hardwoods with potential for management and use. Mangrove forests cover estuaries and other protected coastal lands. The island, self-sufficient only in fence posts, imports almost all of its lumber, plywood, and other forest products.

The Forest Service of the U.S. Department of Agriculture has administered the 28,000-acre (11,300-ha) Caribbean National Forest in the eastern mountains since its proclamation in 1903. Silvicultural research by the International Institute of Tropical Forestry, operated also as the Luquillo Experimental Forest, is an important activity. The Caribbean National Forest includes a natural area containing four types of unmodified forests, recreation areas visited by nearly 1 million persons each year, and many forest experimental and demonstration areas. The institute is also heavily involved in re-

FIGURE 3-13

A plantation of a veneer species, *Anthocephalus chinesis,* in Puerto Rico. Growth in the Tropical Forest is rapid; the trees here are only five years old. *(Courtesy of USDA Forest Service)*

search and forestry technology transfer throughout the American tropics.

Ownership of United States Forests

According to the USDA Forest Service (2001), there are 2.3 billion acres of land in the United States, one-third of which are forested (Smith & Sheffield, 2000; see Tables 3-1, 3-2, and 3-3). This nationwide analysis of ownership divides the nation into four regions (North, South, Rocky Mountains, and Pacific Coast), including the subregions as shown in Fig. 3-14.

Forest Land Base   Forestland is defined as land having 10 percent or more tree cover. Of the 2.3 billion-acre total in the United States, 33 percent (747 million acres) is forestland and 67 percent (1,516 million acres) is nonforestland (Table 3-1), Forests are found in abundance in every region, ranging from sparse scrub forests of the arid, interior west to the highly productive forests of the Pacific Coast and the South.

TABLE 3-1

LAND AREA IN THE UNITED STATES BY TYPE OF LAND AND REGION, 1997

Type of land	Total	North	South	Rocky Mountains	Pacific Coast
	(million acres and hectares)				
Timberland	504 (204)	159 (64.4)	201 (81.4)	71 (29)	72 (29)
Forestland, reserved	52 (21)	8 (3.2)	4 (1.6)	18 (7.3)	22 (8.9)
Other forestland	191 (77.3)	3 (1.2)	9 (3.6)	54 (22)	125 (50.6)
All forest	747 (302.5)	170 (69)	214 (86.7)	143 (58)	219 (88.7)
Other [nonforest land]	1,516 (614)	243 (98.4)	321 (130)	598 (242.2)	354 (143.4)
Total, all land	2,263 (916.5)	413 (167.3)	535 (217)	742 (300.5)	573 (232)

TABLE 3-2

AREA OF FOREST LAND IN THE UNITED STATES BY REGION AND OWNERSHIP, 1997

Region	Total	National forest	Other federal	State and local	Private
	(million acres and hectares)				
North	170 (68.8)	12 (4.9)	2 (.8)	27 (10.9)	129 (52.2)
South	214 (86.7)	12 (4.9)	7 (2.8)	8 (2.4)	189 (76.5)
Rocky Mountains	143 (57.9)	72 (29)	26 (10.5)	6 (2.4)	38 (15.4)
Pacific Coast	219 (88.7)	50 (20)	64 (25.9)	30 (12.1)	74 (29.9)
Total, all regions	747 (302.5)	147 (59.5)	100 (40.5)	70 (28.3)	430 (174)

TABLE 3-3

AREA OF FOREST LAND IN THE UNITED STATES BY REGION AND PRODUCTIVITY CLASS, 1997

Region		Productivity class[1]			
	Total	High (85+)	Medium (20–85)	Low (<20)	Reserved[2]
	(million acres and hectares)				
North	170 (68.9)	34 (13.8)	126 (51)	3 (1.2)	8 (3.2)
South	214 (86.7)	91 (36.9)	110 (44.6)	9 (3.6)	4 (1.6)
Rocky Mountains	143 (57.9)	12 (4.9)	59 (23.9)	54 (21.9)	18 (7.3)
Pacific Coast	219 (88.7)	44 (17.8)	29 (11.7)	125 (50.6)	22 (8.9)
Total, all regions	747 (302.5)	181 (73.3)	323 (130.8)	191 (77.4)	52 (21)

[1]A measure of mean annual growth in cubic feet per acre per year obtainable in fully stocked natural stands
[2]Reserved forestlands are not classified for timber productivity.

About two-thirds (504 million acres) of the nation's forests are classed as *timberland*—defined as productive forests capable of producing at least 20 cubic ft per acre of industrial wood annually and not legally reserved from timber harvest. An additional 52 million acres of forest, reserved for nontim- ber uses, are managed by public agencies as parks or wilderness areas (Table 3-1). There are also 191 million acres of "other forestlands" not capable of producing 20 cubic ft per acre of industrial wood annually, but of major importance for watershed protection, wildlife habitat, domestic livestock graz-

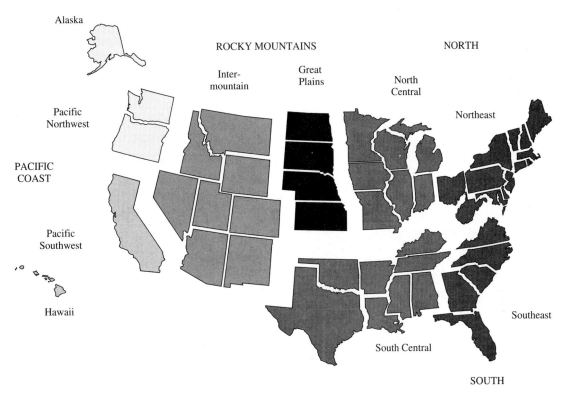

FIGURE 3-14
The four forest regions of the United States and their subregions. *(Courtesy of Brad W. Smith and Raymond M. Scheffield, 2000. A Brief Overview of the Forest Resources of the United States, 1997. USDA Forest Service)*

ing, and other uses, including production of tree products. Almost all of the other forests are in the South and West, and over half are in Alaska. For example, fuelwood is a primary product in many areas that have nontimber forests, such as the pinyon-juniper forests of the Southwest.

Ownership In 1997 about one-third of the forestland (247 million acres) was owned by the federal government, another 9 percent (70 million acres) by state, county, and municipal agencies, and 58 percent (430 million acres) by the forest industry and other private individuals (Table 3-2).

The federal lands are administered primarily by five organizations: the Forest Service, Bureau of Land Management, National Park Service, Fish and Wildlife Service, and Defense/Energy departments. Nonfederal forestlands are concentrated in the East, and national forest and other public lands are concentrated in the West (Table 3-2).

There are 9.9 million private forest landowners in the nation, the vast majority of whom (94 percent) do not identify timber production as either their primary or secondary reasons for owning forestland. However, the 6 percent of forest landowners who claim timber production as a primary or secondary ownership reason own 38 percent of the land area. The vast majority of the noncommercial forest landowners cite reasons such as "part of residence or farm" (39 percent), "aesthetic enjoyment" (14 percent), "estate" (10 percent), "land investment" (9 percent), and "recreation" (9 percent) as primary reasons for owning forestland (USDA Forest Service, 2001)

Productivity Productivity of forestland (Table 3-3) is defined as the amount of wood per acre that can be produced in fully stocked natural stands. The "natural potential" has been used because measures of the potential are available for most regions of the United States, and these provide a uniform means of describing and comparing productivity. This estimate of biological productivity is also a useful indicator of forest capacity for "other uses" since it is based principally on soil, climate, and topography, which are factors that affect all uses.

Most of the nation's "high-" productivity forestlands are located in the South (91 million acres, 36.85 ha) and in the Pacific Coast region (44 million acres, 17.82 ha). In the East (north and south), the highly productive sites are found in the loblolly-shortleaf pine and oak-gum-cypress ecosystems of the lower Mississippi drainage and the Atlantic coastal plain. In the West, the largest areas in the high-productivity class are in the coastal Douglas-fir and hemlock-sitka spruce forest types.

The East (north and south) also has the most (about 73 percent) of the medium-productivity forestland (Table 3-3). The Pacific Coast has the greatest area of low-productivity lands—including Alaska, which has 105 million acres of interior forests with the potential of producing less than 20 cubic ft per year of industrial wood. The Rocky Mountains also have large areas of low-productivity forests dominated by pinyon-juniper trees.

In 1997, some 52 million acres of forestland were classified as reserved (unavailable for timber harvest) and include wilderness areas on federal and state lands and national and state parks. The reserved forestland area in 1997 amounted to 7 percent of the total forestland areas—more than double the area classified as reserved in 1953. This estimate does not include any allowance for private lands in conservation easements or held by private groups such as the Nature Conservancy, Ducks Unlimited, and other convervation organizations.

Canada's Forests

The Forest Regions   Canada recognizes eight distinct forest regions (Fig. 3-15). While some of them can be seen as analogous to U.S. regions, others are more closely delineated formations, bearing only partial relationship to other U.S. and Canadian regions.

Largest in area is the Boreal, which covers about three-quarters of Canada's productive forestland. This forest runs in a wide swath from the Yukon Territory across the northeastern corner of British Columbia, through the northern to central reaches of the prairie provinces, centrally through Ontario and Que-

bec and out to the island of Newfoundland. The region embraces a wide belt of white and black spruce, tamarack, jack pine, subalpine fir, and lodgepole pine. Paper birch, quaking aspen, and balsam poplar are also common.

The Coast Region is contiguous with the West Coast Forest of the United States, and consists principally of western redcedar and western hemlock, with Douglas-fir common in the south and Sitka spruce in the northern part (Fig. 3-16). Black cottonwood, red alder, bigleaf maple, Pacific madrone (arbutus), and Oregon white oak (Garry oak) have a limited distribution.

Lying mostly in British Columbia, the Montane Region can be seen as a northern extension of the Rocky Mountain region of the United States, with ponderosa pine and interior Douglas-fir in the southern and central areas, and lodgepole pine and quaking aspen more abundant in the northern reaches. Spruce, subalpine fir, and western paper birch are also present.

Canada identifies a third and fourth region still largely within the borders of British Columbia, although one of these, the Subalpine Region, spills over the mountains into Alberta on the Rockies' east slopes. The characteristic species of the Subalpine Region are Engelmann spruce, subalpine fir, and lodgepole pine. It is clearly related to the Boreal Forest and has some characteristic species intruding from there, as well as Douglas-fir from the Montane Forest, Pacific silver (amabilis) fir and western redcedar from the Coast Region, and western hemlock from the Columbia Forest, the fourth region. This Columbia Forest Region, which comprises the interior wet belt, resembles the Coast Region. Western redcedar, western hemlock, interior Douglas-fir, some western white pine, western larch, grand fir, and Pacific yew, as well as some Engelmann spruce, are found in the Columbia Region.

The Deciduous Region, the most southern portion of the province of Ontario, comprises an extension of the Central Forest of the United States. There are few conifers; those present include eastern white pine, tamarack, eastern redcedar, and eastern hemlock.

Canada's Great Lakes-St. Lawrence Region, like the Northern Forest region of its neighbor, presents a mosaic of forest types. The characteristic trees are eastern white pine, eastern hemlock, and yellow

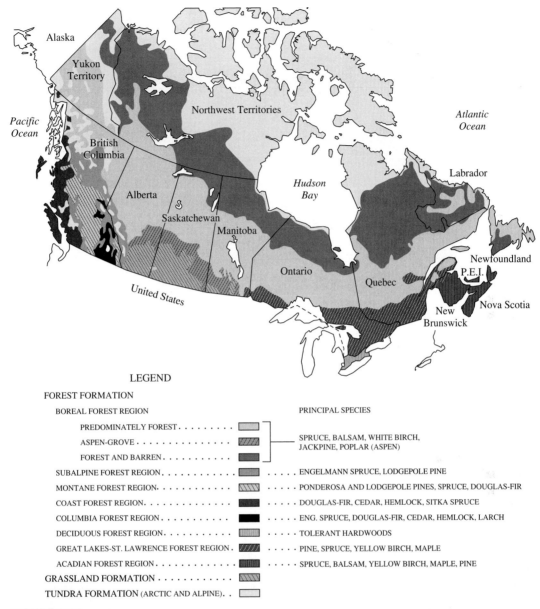

LEGEND

FOREST FORMATION

BOREAL FOREST REGION   PRINCIPAL SPECIES

     PREDOMINATELY FOREST . . . .

     ASPEN-GROVE . . . .   SPRUCE, BALSAM, WHITE BIRCH,
                                     JACKPINE, POPLAR (ASPEN)

     FOREST AND BARREN . . . .

     SUBALPINE FOREST REGION . . . . ENGELMANN SPRUCE, LODGEPOLE PINE

     MONTANE FOREST REGION· . . . . PONDEROSA AND LODGEPOLE PINES, SPRUCE, DOUGLAS-FIR

     COAST FOREST REGION . . . . DOUGLAS-FIR, CEDAR, HEMLOCK, SITKA SPRUCE

     COLUMBIA FOREST REGION . . . . ENG. SPRUCE, DOUGLAS-FIR, CEDAR, HEMLOCK, LARCH

     DECIDUOUS FOREST REGION . . . . TOLERANT HARDWOODS

     GREAT LAKES-ST. LAWRENCE FOREST REGION .   . . . . PINE, SPRUCE, YELLOW BIRCH, MAPLE

     ACADIAN FOREST REGION . . . . SPRUCE, BALSAM, YELLOW BIRCH, MAPLE, PINE

GRASSLAND FORMATION . . . .

TUNDRA FORMATION (ARCTIC AND ALPINE) . .

FIGURE 3-15
The forests of Canada. *(From Timber Resources for America's Future)*

birch, with red spruce abundant in the eastern and some central portions as well.

Broadleaved species common to the Deciduous Region are here too, such as the maple, northern red oak, basswood, and elm. Some elements of the Boreal Region, such as the spruces, balsam fir, jack pine, aspens, balsam poplar, and paper birch, are present.

The eighth region is the Acadian, growing over most of the eastern maritime provinces. It most resembles the Great Lakes-St. Lawrence (Central) Region, although it has some elements of the Boreal

FIGURE 3-16
Sitka spruce and western hemlock in the coast forest region of British Columbia.
(Courtesy of USDA Forest Service)

(or Northern) Forest also. Red spruce, balsam fir, yellow birch, and sugar maple are the most common species. However, white spruce has become important, as it has invaded abandoned farmlands.

Forest Ownership   Canada has a total forest land base of 1,119 million acres (453.3 million ha) of which 602 million acres (243.7 million ha) have been classified as suitable for regular harvest (Table 3-4). Approximately 90 percent of Canada's forestland is publicly owned, and each of the 10 provincial governments is responsible for the public lands within its boundaries. The federal government administers forests in the Yukon and Northwest Territories, as well as areas in the provinces such as First Nation and military reserves, national parks and wildlife preserves, wilderness, and forest research areas.

Less than 10 percent of Canada's commercial forestland resides in private ownership, compared with Mexico, where 40 percent is in private ownership, and the United States, where 72 percent is privately owned (see Fig. 19-1, page 426).

TABLE 3-4

FOREST LAND IN NORTH AMERICA BY TYPE

	(Million Hectares)		
Type of area	**Canada**	**United States**	**Total area**
Productive	243.7	209.6	453.3
Unproductive	150.8	86.4	237.2
Unspecified	3.5	—	3.5
Uninventoried	55.4	—	55.4
Total Area	453.4	296.0	749.3

Source: United Nations *Timber Trends,* 1990.

Although most of Canada's production (commercial) forests are publicly owned, these crown lands are harvested by private industry, which obtains cutting rights to public lands under long-term leasing and other arrangements with the provincial government. Substantial numbers of Canadians work in logging and wood-using industries, which export large volumes of wood products to the United States and worldwide.

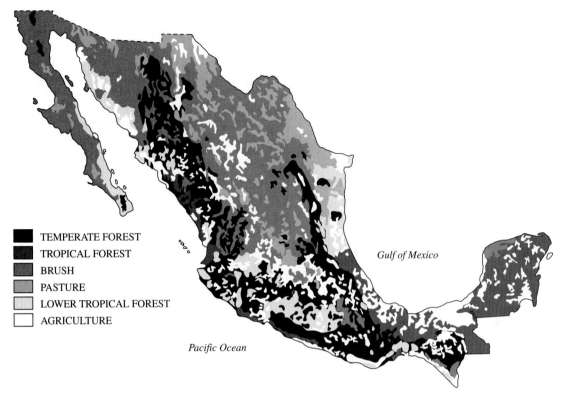

FIGURE 3-17
Present vegetative types in the Republic of Mexico. *(Courtesy of Instituto Nacional de Investigaciones Forestales)*

Mexico's Forests

Topography and Climate Mexico is crossed by the Tropic of Cancer at about midpoint and so has both a tropical and a semitropical region. However, Mexico is a very mountainous country presenting a wide range of climatic and ecological zones, from humid tropical in the south to cold temperate in the high mountains, including snow-capped peaks to 18,701 ft (5,700 m). A great deal of land also lies in arid and semiarid zones.

There are two major mountain ranges. One is the Sierra Madre Occidental, a continuation of the western mountain ranges that extend southward from Alaska through Canada and the United States to Patagonia at the tip of South America. The other range, the Sierra Madre Oriental, lies near the Gulf of Mexico on the east coast and adjoins the easternmost ranges of the Rocky Mountains and the Big Bend area of Texas. These ranges average 8,000 to 9,000 ft

(2,440 to 2,740 m) in altitude, with peaks rising to 12,000 ft (3,600 m) in some instances.

Between the two ranges lies a high plateau, 5,000 to 8,000 ft (1,500 to 2,400 m), rising to the south where it terminates in a third mountain range, the Volcanic Axis. This runs east–west to join both the Sierra Madre Occidental and the Sierra Madre Oriental.

Secondary ranges, such as the Sierra Madre del Sur, Sierra Madre de Chiapas, and the Maya Mountains, among others in the south, and the Sierra de San Pedro Martir in Baja California, also exert local climatic effects.

The Forest Regions As seen on the map of Mexico's vegetation types in Fig. 3-17, this topography is broadly outlined by the temperate forest regions. Mexico classifies any land that is not an urban area, or devoted to agriculture, as "forestland." Approximately 73 percent of Mexico falls into this category

(349 million acres or 141.5 million ha). However, the area actually supporting trees, and that could be seen as forested land, is estimated to be about 123.3 million acres (49.6 million ha). These forested areas can be categorized as:

Temperate forests	63.8 m ac	25.5 m ha
Tropical high- and medium-tree forests	21.5 m ac	8.7 m ha
Low-tree forests	38.0 m ac	15.4 m ha
Total	123.3 m ac	49.6 m ha

Tropical high-forest refers to trees more than 100 ft (30 m) in height; medium-tree means trees 50 to 100 ft (15 to 30 m), and low-tree indicates trees less than 50 ft (30 m) in height (Inventario Nacional Forestal de Gran Vision, 1991).

Temperate forests contain both dense broadleaved stands and coniferous stands. The broadleaved forests include evergreen oak forests of pure or mixed stands with other species such as American sweetgum, beech, dogwood, waxmyrtle, and broadleaved oaks. Mexico is rich in pines; 55 species are currently recognized (Fig. 3-18). Some of these species have their major distribution in Central America or the United States, with only a few stands or perhaps just a few trees in Mexico. Other species are found exclusively in Mexican forests (Perry, 1991). Also of importance in the temperate forests are the true firs and cedars, which occur in varying concentrations.

Tropical forests, found from sea level to around 1,700 ft (500 m), are characterized by large numbers of species per unit of area. Typically, these would in-

FIGURE 3-18

The ocote pine, one of the 12 pine species native to the mountains in the state of Michoacan, in southwestern Mexico. Note the piles of fuelwood. *(Courtesy of USDA Forest Service; photo by J. L. Whitmore)*

clude mahogany, redcedar *(Cedrela odorata),* and the gum tree. The coastal lowlands flanking the mountain ranges are tropical; however, the major tropical forests are found in the southeastern region of the country, particularly the Yucatan Peninsula.

Another of Mexico's diverse habitats is mangrove swamp, found mostly on the Pacific Coast. These forests and associated biota form important natural ecosystems. Tannin bark, used for tanning leather, is obtained from several species growing there. Other habitats include mesquite stands, which are scattered throughout several regions. Palms and pinions grow in arid and semiarid zones (Caballero et al., 1977).

Forest Ownership   Estimates place 2 percent of Mexico's forested lands in state or federal ownership, 58 percent in communal farm and Indian communities *(ejidos),* and 40 percent as private property. However, more than 80 percent of all commercial quality timberland is in *ejido* ownership. Government restrictions on the sale of timber from these lands, although meant to protect the communities, discourage investment by not allowing contracts for more than one year at a time. Private timber companies have been unwilling to incur the costs of upgrading forest management under these terms; consequently, yields have been low. As a part of modernization of the rural sector by the Mexican government, these laws are now under revision, so a boost in forestry development is expected.

Commercial Importance   Mexico could have a much stronger forest industry sector. At present, its production of wood is about one-third of estimated capacity and is concentrated in primary transformation (Table 3-5). With the exception of pulp and paper and resin plants, forest industry plant capacity is underutilized. Other forest products not included in Table 3-5, but of economic importance, are waxes, gums, fibers, rhizomes, and fuelwood.

Even though lumber is seldom used for homes or commercial construction in Mexico, the internal demand for wood and wood products is not satisfied by the Mexican forest industry. Distribution is a problem, with fine woods such as mahogany sometimes used for concrete forming, rather than for furniture. Mexico both imports and exports forest products; major trading partners are the United States and Canada, with some trade involving Europe and South America.

Eighty percent of Mexico's forest products come from pine species, 3 percent from true firs *(Abies species),* 4 percent from oaks, and 13 percent from other hardwoods including fine tropical woods.

Land Use and Shifting Agriculture   Along with other Latin American countries, Mexico has a disturbingly high rate of deforestation. Between 25 and 30 percent of the forest cover in Latin America was lost between 1850 and 1985. Half of that loss has occurred since 1960. Reduction in the forest base during those years was due to expansion of pastures (44 percent), conversion to cropland (25 percent), loss to erosion and other problems (20 percent), and, through shifting agriculture, 10 percent (Houghton et al., 1991).

A cycle of shifting agriculture in the tropical forest involves cutting down the larger trees, slashing the small trees and shrubs, burning the slash, cultivating the cleared land for a few years, and, as crop

---

TABLE 3-5

MEXICAN FOREST INDUSTRY

Industry type	Number of plants	Capacity used (%)	Number of people employed
Sawmills	1,543	44.6	18,516
Wood containers	1,144	n/a	8,008
Impregnation	14	37.6	280
Pulp and paper	74	77	35,361
Veneer/plywood	49	47	11,433
Resin	13	83	9,000

Source: Camara Nacional de la Industria Forestal. 1992. Memoria Economica 1991–1992.

productivity declines due to soil depletion of nutrients, moving to another location to repeat the cycle. The abandoned area is left fallow for a variable period of years, depending on the region or on site conditions. After the fallow years, the farmer returns to the same place to start a new cycle (Hernandez-Xolocotzi et al., 1987; Peters & Neuenschwander, 1988).

These farmers in tropical Mexico are practicing their own kind of silviculture, and there is a diversity of profits involved that contributes to their subsistence. Farmers generally prefer secondary forest over primary forests. When cutting, they leave some useful trees standing for a variety of uses, such as fruit, seed, medicine, ornaments, or shade. Usually, when farmers slash, they leave the stumps of several trees in order to accelerate fallow succession, or to provide support for cultivated vine beans. Additionally, part of these stumps become firewood once the stump sprouts from the base (Gomez-Pompa & Kaus, 1990).

Since there are an estimated 12 million acres (5 million ha) under the shifting agricultural system in the Mexican humid tropics, both the accelerated depletion of the forested land base and a reduction in the length of the growing cycles have caused concern.

Some argue that shifting agriculture conflicts with forestry and even that it caused the decline of the Maya culture (Peters & Neuenschwander, 1988). Others have documented the existence of high population densities that were historically integrated with forest ecosystems, suggesting that the extensive areas of tropical forest that have been cut in the last 50 years were not untouched environments at all, but the result of the last cycle of abandonment in the shifting agriculture system (Gomez-Pompa & Kaus, 1990). Mayan areas contain the richest mahogany stands of the hemisphere. These are thought to be a product of Mayan abandonment of cleared lands where mahogany regenerates easily.

Other authors postulate that the dominance of one or two economically important tree species in many tropical forests is the result of the past forest management practices of ancient Mexican cultures, such as the Olmec and the Maya, who protected these trees. The effect of past civilizations on the structure and composition of today's forests is more than just an intriguing question—it is important in determining those practices used by past civilizations to maintain tropical biodiversity (Gomez-Pompa & Kaus, 1990).

SUMMARY

Forest types are a biotic response to climatic and soil-related conditions. The varieties of tree species present in a given area, their various associations, and the vigor of their growth are indicators of latitude, longitude, elevation, and other geographic dimensions, as well as past cultural practices and events. In an area as extensive as North America, it is difficult to remember the details of all cover types, but familiarity with the concept of forest types as a response to climate and soil, and an acquaintance with the species and associations within one's own area, add to the interested citizen's understanding of the out-of-doors. Such an understanding will prepare the student of forestry for more intensive study of the subject.

LITERATURE CITED

Caballero, Miguel, Victor Sosa, and Juana Marin. 1977. "Forestry in Mexico," *J Forestry,* **78**(8):473–477.

Camara Nacional de la Industria Forestal. 1992. Memoria Economica 1991–1992.

Gomez-Pompa, Arturo, and Andreas Kaus. 1990. Traditional Management of Tropical Forests in Mexico. In Anthony B. Anderson (ed.) 1990. *Alternatives to Deforestation: Steps Toward Sustainable Use of the Amazon Rain Forest,* Columbia University Press, New York.

Harlow, W. M., E. S. Harrar, J. W. Hardin, and F. M. White. 1991. *Textbook of Dendrology* (7th ed.). McGraw-Hill, New York.

Hernandez-Xolocotzi, Efram, Samual Levi, and Luis Arias-R. 1987. Hacia una Evaluation de los Recursos Naturales Renovables Bajo el Sistema Roza-tumba-quema en Mexico. In *Proceedings of the International Conference and Workshop on Land and Resource Evaluation for National Planning in the Tropics.* Chetmaul, Mexico, January 25–31, 1987. USDA Forest Service Technical Report WO-39. Washington, D.C.

Houghton, R. A., D. S. Lefkowitz, and D. L. Skole. 1991. Changes in the Landscape of Latin America between 1850 and 1985. In Progressive Loss of Forests. *Forest Ecology and Management* **38**(1991):143–172.

Inventario Nacional Forestal de Gran Vision, 1991. SARH-Subsecretaria Forestal Y de Fauna Silvestre. Inventario Nacional Forestal 1961–1985.

Küchler, A.W. 1964. *Potential Natural Vegetation of the Conterminous United States.* American Geographical Society, New York.

Little, Elbert L. 1979. *The Checklist of U.S. Trees (Native and Naturalized).* Agriculture Handbook No. 541. U.S. Department of Agriculture Forest Service, Washington, D.C.

Perry, Jesse P., Jr. 1991. *The Pines of Mexico and Central America.* Timber Press, Portland, Ore.

Peters, William J., and Leon F. Neuenschwander. 1988. *Slash and Burn: Farming in the Third World Forest.* University of Idaho Press, Moscow.

Ronderos, Ana. 2000. Where Giants Once Stood: The Demise of the American Chestnut. *J. Forestry* **98**(2): 10–11.

Smith, Brad W., and Sheffield, Raymond M. 2000, July. A Brief Overview of the Forest Resources of the United States, 1997. U.S. Department of Agriculture Forest Service, Washington D.C.

U.S. Department of Agriculture Forest Service. 2001. *The 2000 Resource Planning Act Assignment Summary Report.* Washington, D.C. American Chestnut Foundation www.acf.org

ADDITIONAL READINGS

Bailey, Robert G. 1980. Descriptions of the Ecoregions of the United States. US Dept. Ag. Misc. Publ. 1391 Washington, D.C.

Hall, Fred C. 1983. Forest Lands of the Continental United States, *Using Our Natural Resources,* 1983 Yearbook of Agriculture (Jack Hayes, ed.). U.S. Government Printing Office, Washington, D.C.

Laarman, Jan G., and Roger A. Sedjo. 1991. *Global Forests: Issues for Six Billion People.* New York, McGraw-Hill.

United Nations. 1990. Timber Trends and Prospects for North America. Prepared by Forestry Canada and the USDA Forest Service. United Nations, New York.

Walker, Laurence C. 1998. *The North American Forests: Geography, Ecology and Silviculture.* CRC Press, Boca Raton, Fla.

Walker, Laurence, and Brian Oswald. 2000. *The Southern Forest: Geography, Ecology, Silviculture.* CRC Press, Boca Raton Fl.

STUDY QUESTIONS

1. Name the six major forest types or regions of the United States. In which one do you reside? Name its major tree species. If you live in Canada or Mexico, do the same.
2. How did extensive and unregulated logging change the cover type in certain areas of the Northern Forest?
3. What tree species grow throughout the huge and diverse Central Forest of the United States?
4. What factors have led to the predominance of southern pine in the pulpwood industry?
5. What is the nutlike fruit of hardwoods such as hickories, oaks, and beech called?
6. Besides supplying local forest products, what contributions do the Western Interior Forests make to the region?
7. What national forest encompasses most of Alaska's Coastal Forest Region?
8. Name five factors that make Alaska's Interior Forest Region vulnerable to wildfire.
9. Why does Mexico have such a wide diversity of climatic and ecological zones?
10. Outline the controversy surrounding shifting agriculture in Mexico.
11. Which entity owns most of the commercial forestland in the United States? Can you think of a NIPF personally known to you? What does this person do with the land?

Forest Ecology

Chapter Outline

INTRODUCTION

The relationship of forest managers to forest ecosystem processes can be likened to the relationship of sailors to the wind. Good sailors know they cannot control the wind; rather they seek to understand it. Their craft must be adapted to its power and variations, and they must learn to harness the wind's strength to their own best advantage. Similarly, forest management does not try to fight natural biological and ecological processes; rather, it tries to understand these processes and their subtle variations. With this knowledge forest management attempts to influence forest growth toward patterns desired by the owners.

Forest management does not consider a tree, a tree species, or any part of their surroundings as isolated units for study. Interactions occurring between members of the forest community, from the smallest to the largest, or between members of a community and their environment, are the focus of a discipline called *forest ecology*. All members of the forest community are seen to affect the others by their presence, their vigor, or their removal. Of course, the strength of these effects ranges from almost negligible to very strong.

Forest ecology concerns itself with how forest systems grow and respond. The study of individual organisms in relationship to their environment is called *autecology*. The consideration of communities of organisms as related to their environment is termed *synecology*. The consideration of organisms together with their environment yields the concept of the *ecosystem*.

Chapter 4 first describes the components of the environment—the soils, climate, and other external influences; then it examines the growth of trees and the interactions within plant communities. Finally, it outlines various classifications of trees and forests, including such things as their common division into hardwoods and softwoods, the system of scientific names, the composition of forests, and the position of trees with respect to one another in a forest.

Soils are discussed first, but keep in mind that, although soils supply physical support and important nutrients analogous to certain vitamins and minerals in the human diet, *light, carbon dioxide,* and *moisture* are the food and drink of trees; without these they cannot grow on any soil.

COMPONENTS OF THE FOREST HABITAT

Forest Soils

Forest trees germinate in, are physically supported by, and depend upon stored water and nutrients that are found in the soil. Soil is a relatively thin layer of weathered mineral and organic material. It varies according to its parent material, its age, and the vegetation and environment in which it developed. There are many different types of soils, and their classification is complex. Soils found in a particular forest have a direct effect on tree species composition and forest productivity. Forest soils are characterized by their associated *litter,* which is composed of fallen branches, twigs, leaves, flowers, and fruits, as well as animal residues, all in various stages of decomposition. This blanket of organic material retards runoff, cushions the impact of raindrops on the soil surface, holds water, and lessens temperature and moisture extremes that could have deleterious effects on regeneration. It also provides mineral nutrients as it is decomposed. Other characteristics of forest soils are the interlacing of somewhat stable root systems and the presence of a rich biota above, on, and in the soil.

Soil Profile   In a forest habitat, as in any other, the "parent" material of the soil is a major influence on its characteristics. The parent materials, be they rock or organic, will have been exposed to weathering for varying periods of time. Other soil-forming and accumulating activities may well have been, or may still be, at work on the land. These include mass soil movements such as soil creep or landslides, flooding, and associated wind transport of soil particles, and deposit of tephra (ash) from volcanic activity. In northern and north temperate latitudes, soil displacement by glaciation is a pervasive and complicating factor in soil formation.

The amount of rainfall and the range of temperatures influence weathering properties of soil, and hence its ability to support tree growth. In undisturbed soil, there are individual horizontal layers called *horizons* formed by chemical reactions and leaching. Together the soil horizons make up a *soil profile*. In North America, soil scientists have classified over 75,000 individual kinds of soils, based on their depth, size and arrangement of particles; mineral composition; and water-holding capacity.

The uppermost layer of a soil profile is largely undecomposed litter. In forest soils where litter is an important element, this O (*organic*) layer has been further categorized into L (*litter*), F (*fermentation*), and H (*humus*) layers (Fig. 4-1). In the uppermost or L (litter) layer, decomposition is negligible. The F (fermentation) layer denotes the area where litter breakdown is occurring. Plant structures may still be recognized here, even though partial decomposition has occurred. In the H (humus) layer, the litter has been reduced to an amorphous black or dark brown substance. Depending on the development of the soil, this H layer may or may not be incorporated into the A horizon.

The topsoil, or next layer of the profile, also contains considerable amounts of organic matter, as well as mineral matter. This upper layer, usually gray or black in color, is called the A horizon (Fig. 4-1). Here most of the roots from the surface vegetation are found. *Leaching,* the process by which water percolates down through the topsoil, carrying dissolved material along with it, takes place first in this layer. Dissolved materials are moved downward by soil water from the A horizon into the B horizon, the layer immediately below. This horizon contains less organic matter. In most climates the B horizon contains an accumulation of materials leached from the A horizon above. The C horizon consists of decayed or disintegrated parent rock, grading downward into unweathered parent material. The boundaries of the various layers of the soil profile may be gradual or well defined.

Soil Properties   Several interdependent factors affect the rate at which trees grow by their influence on the roots' ability to take up essential chemicals, oxygen, and water. Water availability (for trees) is determined by the amount of precipitation and the *soil water-holding capacity.* Soil water-holding capacity is determined by the size of the pores in the soil, which is in turn determined by the soil parent material and weathering history. The size of pores is also related to the *soil texture* (the size of individual soil particles) and *soil structure* (the way these particles are arranged and the resultant pore spaces between them). For example, in a typical loam topsoil, about 45 percent of the volume will be mineral particles. (*Loam* is a mixture of sand, silt, and clay in varying

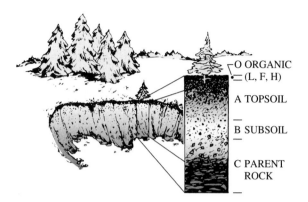

FIGURE 4-1

The principal layers of a forest soil profile. The O (organic) layer may be further subdivided into L (litter), F (fermentation), and H (humus) layers. The A (topsoil) horizon contains both organic and mineral matter. The B (subsoil) layer contains less organic matter and more mineral matter, and the C (parent rock) horizon grades down into bedrock. The A, B, and C layers also exhibit numerous subhorizons, depending on local conditions. *(Illustration by Gale Gruza)*

proportions.) Large pore spaces, caused by the presence of large particles such as sand, will allow oxygen in the air to get to the roots easily, but will allow water to flow too rapidly through the soil and beyond the root zone. Soils such as clays, which often have very small pores, will not allow oxygen to penetrate rapidly and may also hold water too tightly for it to be available for optimum tree growth. Slowly, root growth can create large pores in fine-textured (clay) soils, which will increase their potential to support rapid tree growth.

A classification of soil particle sizes (soil texture), as well as of gravel and peat, is presented in Table 4-1. Soils can be various mixtures of different particle sizes, which strongly influences their water-holding ability. The particle sizes are also important to forest managers for engineering purposes, such as for construction of logging or recreation roads, trails, or buildings, or in developing land-use plans. For example, activity on fine-textured soils should be avoided, especially when they are wet, since the resulting compaction may harm the soil's ability to grow trees.

Soil *organic matter,* the accumulation of decomposed plant and animal residues (litter), is constantly being consumed by soil organisms such as earthworms, insects, rodents, fungi, and bacteria, and

TABLE 4-1

SOIL CLASSIFICATION OF PARTICLE SIZES

1. *Gravel* is coarse, with particles over 2 mm in diameter. It is well-drained, stable material that will support heavy loads.
2. *Sand* is 0.05 to 2 mm in diameter and is gritty to the touch. It also drains well and holds up under heavy loads. When dry, sand lacks cohesion and must be confined. Fine sand is less stable when saturated and tends to flow if not confined.
3. *Silt* is 0.002 to 0.05 mm in diameter, is barely visible, and feels smooth to the touch. When dry it lacks cohesion, remains stable, but compresses under loads. When wet it becomes very unstable, and it heaves when frozen. In a disturbed state it is subject to wind erosion.
4. *Clay* is under 0.002 mm in diameter, and individual grains are invisible to the naked eye. When dry, it is cohesive and supports heavy loads. When wet, clay becomes plastic, slippery, and impervious. Porous soils lying over clay are subject to slippage on slopes greater than 10 percent. Clay is also subject to wind erosion if powdered, such as on trails or roads.
5. *Peat* soils are of organic origin, fibrous to the touch, dark in color, and lack cohesion. They are unstable and will not normally support development. They are usually associated with high water tables, either at present, or at sometime in the past.

From: Sharpe et al., 1994

therefore must be constantly replaced. As the organic matter decomposes, mineral elements previously incorporated into the tissue before it died are released to be recycled. Approximately 5 percent of the total volume of a silt loam topsoil is organic matter.

About 50 percent of the soil volume is pore space, divided roughly in half between air and water. As one increases, the other decreases. For optimum plant growth, the balance between air and water space must be carefully maintained.

Soil fertility refers to the capability of the soil to supply chemicals essential for the growth of plants in proper balance for plant requirements. This capability depends on a number of factors, such as the quantity and forms of certain key elements, primarily nitrogen and phosphorus, as well as certain chemical, physical, and biological factors that influence the rate at which they are released from mineral and organic matter into forms available for uptake by the plants. Nitrogen comes from the atmosphere but is first incorporated into the soil by microorganisms before it can be used by trees. Carbon, hydrogen, and oxygen, of course, are taken directly from the air and water.

The other essential elements generally come from weathered rock or fertilizers, or are recycled from decomposing organic matter. These include four major elements—sulphur, calcium, potassium, and magnesium. The nine minor elements, needed in smaller amounts, are iron, manganese, copper, zinc, boron, cobalt, sodium, chlorine, and molybdenum. Each of these minerals, taken up from the soil through the roots, is necessary for proper functioning of the forest. For example, phosphorus and sulfur build protein, iron and magnesium become part of the chlorophyll molecule, and potassium aids in cell division.

The fertility of soils ranks as a prime concern to forest managers seeking to grow timber crops as quickly and economically as possible. Deep soils store more water and release more nutrients and thus favor tree growth. Fertilizers or other soil amendments may be applied to increase soil fertility on certain sites.

Productivity and Site Index The ability of trees to absorb from the soil essential water, chemicals, and oxygen (for the roots) influences the rate at which they grow. Each soil has certain properties that affect tree growth rates. The "growth ability" of a soil is referred to as its *soil productivity potential*—often expressed as a *site index* (SI). This refers to the height dominant trees have attained by a certain age in that soil. (Height is used because this factor is relatively free of the influence of crowding or single-year weather fluctuations, although it is affected by genetics and other above-ground conditions.) As an example, a very good soil for Douglas-fir in western Washington would have an SI of 150, 50-year index, meaning the dominant trees will be 150 ft tall at age 50. A poorer soil would be one having an SI of 85 at 50 years. Oak trees on upland soils in the southeastern United States can have an SI of 55 to 75, 50-year index.

Climatic Factors

Just as trees have their roots in the soil, their trunks and crowns extend into the atmosphere, and thus are exposed to the climate that prevails on their *site,* which is what foresters call *habitat.* Climatic factors limit where trees will grow—not enough rainfall or permanently frozen soil are major deterrents. Within the areas where trees can exist, climatic and soil factors determine species and ranges and affect the vigor of specimens within these ranges.

Of course, geographic location affects climate. Our concern here is with the ecosystems of North America, an area encompassing a wide range of local and continental influences on climate. Chapter 7, Disease and the Elements, discusses climatic factors from the forest manager's point of view.

Light and Temperature   Light and carbon dioxide are the primary requirement for tree growth. While some trees demand relatively more light than others at certain periods of their growth, the process of photosynthesis is absolutely dependent on light. Further, given adequate moisture, nutrients, and temperature, there is a direct correlation between the amount of light or solar energy a tree receives and its rate of growth.

Not all *insolation,* or incoming solar energy, is available for photosynthesis; much of it is absorbed or dissipated in other ways. But in the flow of energy through the ecosystem, light striking a tree leaf is most readily seen as the initial input.

Extremes of temperature present difficulties for trees, although trees have the ability to modify their temperature through *transpiration,* given sufficient moisture. Transpiration is discussed under "Water Requirements."

Geographic location affects the number of hours of daylight available to trees. The net productivity of a plant community strongly depends on the number of days when light of suitable quality occurs in a temperature range that sustains the net growth of plants. *Provenance studies,* which deal with genetic variations within species populations, suggest that trees adapt through genetic variation to environmental factors such as light, temperature, and length of growing season.

Slope and Aspect   Variations in topography result in temperature and moisture differences. Trees on steep, south-facing slopes will experience a different microclimate than will those trees growing in low-lying "frost pockets" or those on the north slope of a mountain. Slope and aspect will combine with other environmental factors to limit or encourage the regeneration of various species of trees on a particular site. These phenomena are further discussed under "Forest Competition."

Precipitation   As with other components of the forest habitat, precipitation has different effects in different situations. Although rainfall and fog drip have some direct results, water is made available to trees mainly through the soil. Thus, the quantity and quality of the soil, especially with regard to its water-retention capabilities, are important. A forest does not materially affect the amount of precipitation that falls in an area, but it is critical in the distribution of this moisture. Depending on their numbers and species, trees retard runoff through their contribution to litter and soil moisture retention, and by shading snow and thus delaying snow melt.

Other Abiotic Factors

Disturbances of natural or anthropogenic (human) origin, often termed *perturbations,* affect tree and forest growth. Such perturbations include fire, erosion, wind, avalanche, flood, drought, and volcanic eruption. Human disturbances, such as timber harvest or prescribed fire, mimic the results of natural disturbances, but the sequence and timing are different. Most of these topics are dealt with at some length in later chapters. They are mentioned here so that they can be recognized as components of the forest habitat.

Even though a type of erosion (geological weathering) is one of the activities that creates soil, soils are, nonetheless, extremely vulnerable to the erosional force of wind and rain. Removal of vegetative cover by fire, overgrazing, poor agricultural practices, or, in some instances, clearcut harvesting, also causes accelerated soil erosion. Unchecked, this means irrevocable loss of the resource. Soil is carried away, sometimes in devastating sluice-outs, creating further problems as it silts streams, damages fish habitat, clogs reservoirs, and possibly impedes navigation by filling channels. Worldwide, abuses of cropland, pasture, and rangeland expose extensive areas of soil to erosion and soil movement year after year.

Careful timber harvesting activities, under modern forest management practices required on federal lands or under state forest practice laws, will generally prevent severe erosion. Knowledge of underlying soil types; care in timing and method of road building, harvesting, and slash disposal; and prompt reforestation will help keep erosion in well-managed forests to a minimum.

Biotic Factors

There are several major biotic factors in forest ecosystems that concern forest management.

Trees   Just as soils are diverse and complex, so too are the "magnificent vegetables" called trees. The plant kingdom is so varied and adaptive that the best method of classifying it is still a matter for active debate. Even in defining a tree, it is necessary to allow for departures from any rigid rule. Trees can be defined as perennial woody plants that usually grow upright with single stems and have their roots well anchored in the soil. Yet the line of demarcation between trees and shrubs is unclear. A given species may be shrubby near the extremities of its range, or near timberline, and still attain large proportions elsewhere. Some authorities require the woody plants in question to reach at least 20 ft (6 m) or more in height, and stipulate that the single trunk should be unbranched for at least several feet above the ground and that the specimen have a recognizable crown in order to be classified as a tree. Other taxonomists pose stem diameters as an additional requirement, stipulating a diameter of 2 to several inches.

Trunk   This supports the *crown*—the leaves and living branches. It has many internal parts. The innermost is the *pith,* an area of dead cells around which the first woody growth forms. Next comes the *heartwood,* an inactive area, usually darker than the wood outside it. Surrounding this is the *sapwood* or *xylem,* where water and dissolved minerals pass upward from the roots to the leaves. As new sapwood is added, the old becomes heartwood (Fig. 4-2). Growth takes place in the microscopic layer of cells called the *cambium.* As these cells under the bark divide, the inner form sapwood or xylem; the outer form *phloem* or *inner bark,* a layer of soft moist tissue that carries carbohydrates from the leaves down to the branches, trunk, and roots. Gradually the inner bark changes to *outer bark.* The trunk increases in diameter each year as a layer of wood is added and an *annual ring* is formed. Early in the growing season, a layer of large, thin-walled, light-colored cells is formed *(springwood);* a darker layer of smaller, thick-walled cells *(summerwood)* forms later. Side by side, these appear as a single annual ring. The number of rings tell the age of the tree; their width indi-

cates how fast it is growing and, thus, something of its growing conditions.

Roots   These structures are responsible for absorption and conduction of water and mineral salts, for anchorage of the tree in the soil, and, in some species, for reproduction by sprouting. Fig. 4-3 shows a diagrammatic representation of different kinds of tree roots.

The roots of herbaceous plants, shrubs, and trees, and the associated organisms or soil biota, constitute the forest underground. Roots are seldom harvested, although tree, shrub, and plant roots are used to some extent in medicine and the arts, such as for paint. Decaying roots make a significant contribution to the organic content of the soil, thus providing nourishment to living plants.

Water and dissolved mineral salts enter the tree by absorption. This occurs primarily in the tiny, thin-walled, growing root tips. Even the oldest trees must be continuously growing new roots. Trees generally depend on cooperative growth with a fungus—in an association called *mycorrhiza*—to increase the nutrient absorption capacity of the roots. The fungus produces a sheath around the roots from which grow microscopic, fingerlike *mycelia* that extend into the soil, helping the tree absorb minerals and moisture. Once the solution of water and minerals enters the roots, it is carried up the trunk, through the xylem of the tree, to the leaves.

The close, mutualistic interaction between higher plants and mycorrhizal fungi is a key factor in tree growth and forest health. Forest managers now recognize the importance of maintaining soil conditions that benefit microorganisms such as mycorrhizae.

Leaves   Leaves might be called the "solar-powered chemical laboratories" of the tree. They contain small green bodies called *chloroplasts.* These bodies contain *chlorophyll,* a substance that gives leaves their green color. In the presence of sunlight, these chloroplasts absorb light and use the energy to start a chemical reaction in which the raw materials, carbon dioxide (taken from the air entering the leaves through millions of tiny openings called *stomata*) and water (which comes up from the roots), are involved. This reaction breaks down the molecules of carbon dioxide and the molecules of water. Called

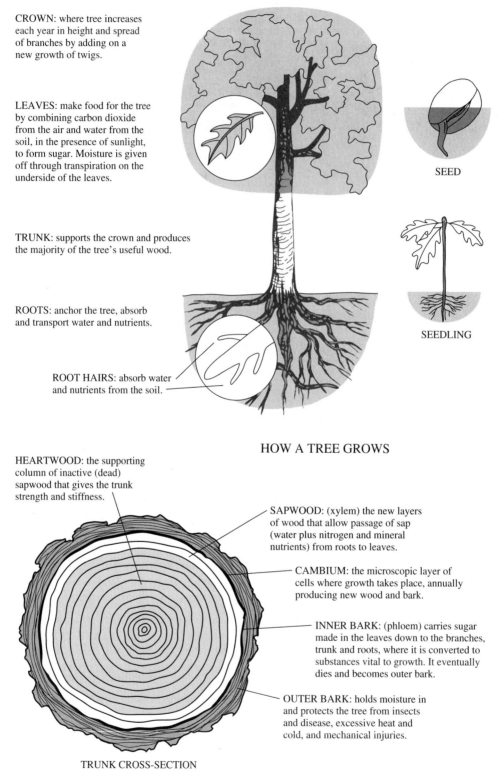

CROWN: where tree increases each year in height and spread of branches by adding on a new growth of twigs.

LEAVES: make food for the tree by combining carbon dioxide from the air and water from the soil, in the presence of sunlight, to form sugar. Moisture is given off through transpiration on the underside of the leaves.

TRUNK: supports the crown and produces the majority of the tree's useful wood.

ROOTS: anchor the tree, absorb and transport water and nutrients.

ROOT HAIRS: absorb water and nutrients from the soil.

SEED

SEEDLING

HOW A TREE GROWS

HEARTWOOD: the supporting column of inactive (dead) sapwood that gives the trunk strength and stiffness.

SAPWOOD: (xylem) the new layers of wood that allow passage of sap (water plus nitrogen and mineral nutrients) from roots to leaves.

CAMBIUM: the microscopic layer of cells where growth takes place, annually producing new wood and bark.

INNER BARK: (phloem) carries sugar made in the leaves down to the branches, trunk and roots, where it is converted to substances vital to growth. It eventually dies and becomes outer bark.

OUTER BARK: holds moisture in and protects the tree from insects and disease, excessive heat and cold, and mechanical injuries.

TRUNK CROSS-SECTION

FIGURE 4-2
How a tree grows. *(Modified from USDA Forest Service)*

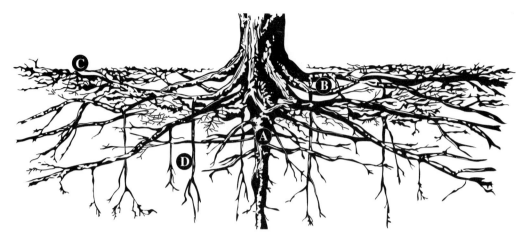

FIGURE 4-3
A root system. *Tap root (A):* Provides main support for tree and anchors it firmly in the ground. (Not all tree species have one.) *Lateral roots (B):* Help support and anchor trunk. May extend far beyond crown spread. *Fibrous roots (C):* Masses of fine feeding roots close to ground surface. *Deeply descending roots (sinkers) (D):* Grown downward from lateral roots. *(Courtesy of USDA Forest Service and Soil Conservation Service)*

photosynthesis, it produces a new combination of atoms of carbon, oxygen, and hydrogen, and may be expressed in the following equation:

$$6CO_2 + 6H_2O + \text{Radiant energy} \rightarrow$$
$$\text{Carbon} \quad \text{Water} \quad \text{(expended)}$$
$$\text{dioxide}$$

$$C_6H_{12}O_6 + 6O_2$$
$$\text{Glucose} \quad \text{Oxygen}$$

This equation indicates that six molecules of carbon dioxide are combined with six molecules of water to form one molecule of glucose (a carbohydrate) and six of oxygen. The oxygen, a by-product, is released to the atmosphere, and the carbohydrate (or *photosynthate*), which may be converted into glucose and later into either cellulose or starch, is transported to all living parts of the tree. *Cellulose* is the main constituent of wood, making up approximately 60 percent of the tree. Starch is a storage material; it can be transformed back into sugars and used by the trees for energy or for making other chemical compounds. *Lignin,* also composed of carbon, oxygen, and hydrogen, but of a different chemical structure than cellulose, makes up most of the remainder of the tree.

Another process, called *respiration,* takes place in all living parts of the tree. The chemical reaction of respiration is the opposite of the reaction of photosynthesis; it may be presented as follows:

$$C_6H_{12}O_6 + 6O_2 \rightarrow$$
$$\text{Glucose} \quad \text{Oxygen}$$

$$+ 6CO_2 + 6H_2O + \text{Energy}$$
$$\text{Carbon} \quad \text{Water} \quad \text{(released)}$$
$$\text{dioxide}$$

In this instance, some of the energy is retained by the tree and is used to do much of the internal work needed for its growth. Excessive respiration could be fatal to the tree. *Note that, in one process (photosynthesis), oxygen is released, and that, in the other (respiration), carbon dioxide is released.* Carbon dioxide for photosynthesis comes from the air. Respiration, which uses up food (the glucose), goes on 24 hours a day and takes place throughout the living parts of the tree, whereas photosynthesis, which manufactures food, takes place only during the day and only in the leaves. Therefore, trees can respire themselves to death if they are shaded by other trees and cannot get enough sunlight. Where exposed to full sunlight, photosynthesis can proceed during the day at a rate 10 times that of respiration, and from this excess comes the food necessary for growth.

In most natural, mature forests, the dead organic material is "eaten" (respired by microorganisms) in the process of decomposition. Absorption of oxygen during this respiration process may be equal to the production of oxygen during the photosynthesis that produced the wood and, thus, the net production of oxygen from such a forest may be zero. Vigorously growing young forests produce more oxygen and consume more carbon dioxide, and in those respects are considered important to retarding climate change.

Water Requirements It has already been noted that water is needed to carry dissolved minerals from the soil to the leaves. Water is also necessary in the chemical processes of photosynthesis, where its hydrogen molecules combine with those of oxygen and carbon dioxide to make a sugar called glucose. We have seen how water moves this food to all parts of the tree. But water has other functions as well. Water pressure, or *turgor pressure* as it is called, builds up inside the growing cells of the tree to "inflate" those cells and give them rigidity. Soft fruits of trees, such as apples and peaches, are 85 percent water. Water is the major constituent (75–90 percent) of *cytoplasm,* the living material of cells themselves.

As vital as water is to trees and their growth, most of the water entering trees simply passes through. The water exits through stomata, small pores in the surface of the leaves, as water vapor. This process is called *transpiration.* The leaves must open their stomata to allow carbon dioxide to enter if photosynthesis is to take place. However, when the stomata are open, the tree loses its water by transpiration. The more the tree evaporates, or transpires, the more water is taken up from the soil.

The amount of transpiration varies among species and with the time of day, and is often greatest on hot, dry summer days. The drying action of the wind tends to increase transpiration. Transpiration is also influenced by the availability of moisture in the soil. As soil dries, the water molecules are more tightly held to soil particles. If the amount of water being transpired begins to exceed the amount being taken up, water stress develops and the cells around the leaf stomata, called *guard cells,* lose *turgor pressure,* closing the stomata and thus shutting off or reducing transpiration. Although this conserves water, the closed stomata stop the intake of carbon dioxide and the process of photosynthesis, thus stopping growth of the tree.

One acre of pulpwood-sized southern pine (about 100 trees, 10 in [25.4 cm] in diameter and 80 ft [240 m] tall) transpires 750,000 gallons (2.85 million liters) of water per year. One sawlog-size tree in the south, about 20 in (51 cm) in diameter and 110 ft (335 m) tall, can transpire 80 gallons (300 liters) of water per day during a hot summer. However, trees will maintain maximum transpiration rates only as long as the supply of soil moisture is bountiful.

Tree Form and Growth As with all living entities, trees require certain conditions and nutrients in order to survive and grow. Their growth rate and shape can be affected by variations in these conditions. To help trees grow in a way most desirable for the owner is a major focus of forest management. Life in a forest means competing with other plants for the essentials of light, water, and nutrient minerals. Survival is threatened, not only by lack of these essentials, but also by mechanical damage and predation by other forest biota such as damaging insects and disease-producing fungi.

Each tree, according to its genetic endowment, has its own timetable for growth and its own growth pattern, or form. Plant hormones and other factors stimulate growth in certain areas of the developing tree and inhibit it in others. This action helps produce the distinctions in tree form so readily noted in temperate forests.

Some trees, especially conifers, have unforked, upright, central stems, and often a cone-shaped crown. Others, usually deciduous trees, can have a less defined, central stem, large branches, and a broad, flatter crown. These shapes, as well as tree growth rates, are strongly influenced by genetic, soil, and weather conditions, but can be modified by forest management activities, such as spacing and pruning.

The true "food" of a tree—its energy source—consists of sunlight and CO_2. These are absorbed through the leaves. The leaves feed their own branches, and then feed the trunk and root system. The accumulating effect of excess photosynthates feeding their own branch cambium, and then the tree-stem cambium, causes tree growth. When trees grow close together, the lower branches are shaded and so do not receive as much sunlight. If leaves cannot produce enough photosynthates at least to feed the cambium of their own branch, the branch will die. The dead branches fall off and, as a result, the tree will have a columnlike stem with a clean bole. Because the photosynthetic (leaf) area of a tree growing closely with others is less than in a tree growing in a more widely spaced situation, these trees generally do not grow as rapidly in diameter. Trees that become too shaded cannot produce enough photosynthates for their own respiration and maintenance requirements— that is, new leaves, roots, xylem, and phloem—or for the mechanisms of resistance to insects and diseases. Thus, they may be more easily infected, killed by insects or diseases, or die of "starvation."

In a typical deciduous hardwood tree, the main stem is not as dominant as in typical conifers. The result is that hardwoods' branches often assume dimensions as great as the main stem and a rounded crown is formed, whereas in conifers the main stem is dominant and a conical shape emerges. Strong, evolutionary *selective* pressures of snow, ice, and wind have apparently produced this form, just as competition for sunlight on the larger leaf surfaces may have influenced the spreading habit of the broad-leaved trees.

A seedling that escapes the various hazards and is adequately supplied with its needs eventually will grow to its reproductive, or flowering, phase. Again, inner controls or balances regulate the process. A tree is considered to be *mature* during the period of years in which it can both grow and put out seed.

Understory Plants   The understory is familiar to us; this is what we encounter when we walk through a forest, and, in most instances, it affords much pleasure and beauty. This portion of the forest habitat is vital to mammals and birds, providing them with food and cover. These plants typically recycle nutrients at a faster pace than do trees.

Seedlings of certain tree species may have a difficult time competing with understory vegetation, especially in their first years (Fig. 4-4). Only a few will survive to rise up eventually through the understory to add their crowns to the canopy (Stubblefield, 1978). It must be remembered, however, that young or suppressed trees are also a component of understory vegetation.

Trees may dominate in size, and supply the greatest weight to biomass calculation, but the shrubs and herbaceous vegetation growing under the tree canopy represent a greater number of plants as well as of species. These plants exhibit complex and changing relationships with all components of the forest ecosystem. They help provide forests with *biodiversity*.

Animals   Forests are home to many species of wildlife that find shelter in the diversity of habitat.

FIGURE 4-4
A forest floor covered with mosses, herbaceous plants, and conifer seedlings. These seedlings must compete with the understory vegetation in their first few years, and with each other in later years. Eventually only one seedling will have enough room to grow. *(Photo by Grant W. Sharpe)*

The survival of animals is well understood to be intimately involved with ecosystem functioning. For example, the bird, Clark's nutcracker, and ground squirrels collect and bury seeds from whitebark pine (*Pinus albicaulis*) in the high elevation forests of the northern Rocky Mountains. Grizzly bear eagerly seek these caches of whitebark nuts as they are an important source of food with which to build fat for hibernation—up to 40 percent of their diet in good seed years (Mattson et al., 1991). Together the Clark's nutcracker, squirrels and bear are thus interdependent with the whitebark pine, depending on it for food and in turn distributing the seeds, which regenerate the tree. Over 110 wildlife species have been found to utilize whitebark pine seed crops (Hutchins & Lanner, 1982).

This example, involving just one tree species, whitebark pine, is illustrative of the complex, interdependent relationships between forests and wildlife.

Disturbances, whether natural or humanly induced, affect all forest life to some degree. The forest furnishes shelter from weather extremes, cover for protection from predators, and food in the form of plants or other animals. Although trees themselves are important for birds and some mammals and insects, the shrubby and herbaceous vegetation, as well as the forest litter, are all vital resources for wildlife. Later chapters consider questions of forest management objectives in relation to wildlife.

Soil Organisms   Soil organisms such as fungi, bacteria, insects, and earthworms reduce the leaves, branches, bark, berries, and other litterfall components to minute particles. These organisms perform a second helpful function in mixing the fragments with the mineral soil. The presence of the reduced organic litter particles, or humus, also makes the physical structure of the soil more congenial to all users, thus further enriching it by increasing all the contributing activities of the whole biota.

Pathogenic Fungi   Certain fungi are destructive to trees as they reduce the rate of tree growth and induce rot. Their activities cause loss of vigor, leaving their hosts prey to further depredations by insects or other destructive agencies, and thus may lead to the death of the tree. Chapter 7 on Disease and the Elements expands on this subject.

THE FOREST ECOSYSTEM

Forests are biotic communities characterized by woody plants of greater than human height. Here trees grow in complex ecological relationships, including associations of tree species, shrubs, and herbaceous plants, with a complement of biota that can range from large mammals to minute soil organisms.

The forest community is interactive; that is, each component is directly or indirectly related to all others within the association. What affects one part affects all other parts. For instance, a thinning operation that takes out certain trees temporarily increases light, lessens transpiration, lessens competition, increases growth of the remaining specimens, increases available growing space on the forest floor, and affects the activities of associated soil biota, as well as those of insects, birds, and mammals.

Forest Competition

One of the things foresters are particularly interested in is why there are so many tree species, and why they grow where they do. *Competition* for the necessities of life is regarded as a significant factor in this diversity.

All trees compete for the same basic requirements of life—light, water, essential elements, oxygen, and other necessities. They will grow until their roots have saturated the soil or their foliage has saturated the area, and the lack of one or more of these limits expansion. Trees that are able to monopolize these factors will generally dominate the vegetation. Light is so frequently the limiting factor that evolution has produced trees that will grow rapidly in height and vigorously extend crowns to gain a competitive advantage. In a given situation, the competitive advantage depends on the particular soils, disturbances, and other conditions. Some species, especially the broadleafed, are often able to outcompete conifers on soils of high growth potential. While conifers are not able to grow as well on poor sites as on good sites, they can do better than hardwoods on poor sites—that is, they *outcompete* hardwoods on poor sites. As a result, conifers are often found on poor sites and hardwoods on good sites. Only a few species are found where their own growth is maximized; most tree species are found where they *compete* successfully.

On a given soil, often a particular type of disturbance can give one or another species a competitive advantage. After a fire, sprouting species often are able to grow rapidly from surviving root systems and thus dominate an area. Alternatively, if the same site is flooded and the root systems that otherwise might sprout are killed, after the waters recede the competitive advantage may be with the species that rapidly invades through the advantage of readily airborne seeds. A good understanding of forest ecology allows forest managers to manipulate conditions to give a preferred species a competitive advantage.

Northern red oak (*Quercus rubra*), a highly valued hardwood species in the southern Appalachians, is difficult to regenerate and provides an example. Intermediate in shade tolerance, it is routine to find 1,000 or more red oak seedlings per acre under mature mixed hardwood stands, especially following bumper crops of acorns that are produced once or twice a decade. There will also be regeneration from stump and root sprouts. Most of these oak seedlings and sprouts will gradually die if the overstory is not cut to allow more sunlight to reach them. But if all the overstory is cut, other, more aggressive, shade intolerant species such as yellow-poplar will overtop the oak. So the silvicultural prescription developed to provide a red oak component in the future forest is to harvest part of the overstory (60–70 percent of the basal area) so as to provide enough light to encourage growth of red oak seedlings in the understory, and to chemically treat competing seedlings of other species, if necessary. The sheltering overstory is removed in about 10 years when the advanced regeneration is large enough to ensure its survival in the new forest. This shelterwood approach to regenerating oak on productive sites in the southern Appalachians is based on more than two decades of forest research and now provides a means of greatly increasing the value of forests by increasing their red oak component (Loftus, 1988, 1990).

As mentioned under "Slope and Aspect," we sometimes find a north-facing slope covered by one species and a south-facing slope covered by another. Differing environmental conditions exist on the two slopes. South slopes receive more sunlight and hence are warmer and dryer than north slopes (in the northern hemisphere). So, the tree species on south slopes are usually better adapted to warmer, dryer conditions, whereas those on the north slope have been able to survive in the cooler, moister conditions present there. Thus, in the northern Rocky Mountains one might find the more drought-tolerant ponderosa pine on a south slope but on the north slope of the same ridge find species adapted to cooler and moister conditions, such as Douglas-fir, grand fir, and western redcedar.

Forest Succession and Development Patterns

Not only are plants in a forest community affected by their environment, but they in turn exert modifying influences on this environment. Thus, the forest environment may be changed to the extent that some species no longer may compete successfully. Therefore, the composition and interactions of the forest community are modified by tree species themselves as they increase or decline in importance.

Understanding the forest succession principles that control the development patterns and structures of forests (sometimes referred to as *forest-stand dynamics*) is fundamental to the management of the stand in order to promote desired objectives. Forest development patterns vary with soils, climates, disturbance types, species characteristics, and time, but seem to follow general trends. Historically this has been referred to as *forest succession,* since in some cases one group of species modifies the environment in ways that allow another group to outcompete it, and thus succeed it.

There are two types of succession: *primary succession,* which involves establishment of plants on bare soil material, and *secondary succession,* which deals with plant invasion on well-developed soils already covered with vegetation. Here the previous vegetation was removed or altered by natural or human actions. Foresters are now recognizing that the invasion patterns can vary greatly, depending on the kind of disturbance or *perturbation* that occurs. There are differences in the disruptions caused by felling or removal of a part of the forest by insect damage, by fire, by wind, by erosion, or by the various timber-harvesting methods (Fig. 4-5).

In understanding forest development patterns, it is often convenient to begin with typical invasion patterns after a perturbation, although in natural forests large disturbances may occur only at very infrequent

FIGURE 4-5
Upper: A clearcut area photographed after logging. *Lower:* The same area 10 years later, showing the invasion of pioneer species. St. Joe National Forest, Idaho. *(Courtesy of USDA Forest Service)*

intervals, perhaps even centuries apart. Species generally continue to invade a disturbed area until these potential invaders are excluded by competition from existing vegetation. Also, if the disturbance is small, there will be serious competition from large trees surrounding the disturbed area. These will expand their limbs and roots as quickly as possible into the newly available light and soil growing space.

Often one species will begin to dominate the others in height. The remaining trees either die, or, if they can live beneath the shade of the dominant ones, form lower *canopy strata.* Subtleties of spacing and early growth potential can give certain species a competitive advantage (Fig. 4-6). Individuals of the same species within a given stratum may also compete, with the result that some trees die. Forest managers

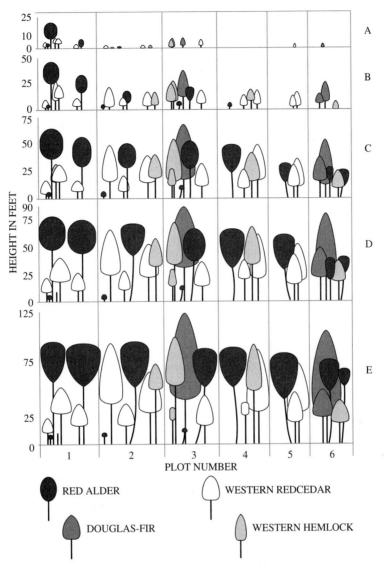

FIGURE 4-6

The reconstructed development of a mixed forest in western Washington following a clearcut and fire in 1927. The sequence illustrates how spacing and early growth give certain species a competitive advantage. *A:* Four years after the fire. All trees start from seed. Alder appears only in plot 1. *B:* Fourteen years after the fire. Red alder has height advantage only in plot 1. Douglas-fir dominates plots 3 and 6. *C:* Twenty-four years after the fire. Red alder has height advantage in all but plots 3 and 6, where Douglas-fir outgrows the alder. The hemlock and redcedar survive, but grow slowly in the shade of alder. *D:* Thirty-four years after the fire. The Douglas-fir in plot 1 has died, but this species continues to outgrow alder in plots 3 and 6. Alder dominates in all other plots except plot 4, where hemlock was growing widely spaced from alder and was not overtopped. *E:* Forty-nine years after the fire. Only those species that had an early competitive advantage and later rapid growth were not overtopped by red alder. *(Courtesy of George W. Stubblefield)* can often influence the growth of stands by anticipating which trees will die and removing them, or by spacing individuals to obtain desirable stem forms and sizes.

Sometimes small perturbations, the death of large individual trees, or climatic fluctuations make light, soil moisture, and soil nutrients available so that new stems invade a stand, creating a new age class. Certain species are successful when small perturbations occur; to favor such species foresters mimic these small perturbations through silvicultural manipulations.

Trees have a life span, a range of ages, that their species commonly attains before passing beyond maturity to old age. As trees get older, and thus larger, they become more susceptible to windthrow and lightning strikes. The amount of regeneration tissue needed (cambium, sapwood, and phloem) constantly increases with tree size, and yet photosynthetic tissues (leaves) do not necessarily increase proportionally. When a tree no longer is able to photosynthesize fast enough to meet its own root and leaf production needs, or its xylem and phloem maintenance, the ability to live and resist insects and diseases declines. Time and circumstance possibly have brought this veteran tree other problems as well, such as wounds from fire, recreationists' activities, or animals, wind breakage, or damage from logging operations. Of course, such traumas could allow disease or insect entry at any point in the tree's life, but now, because of its declining vigor, it will be less successful in fighting off invasions. A common fate is for the disease- or insect-weakened tree to be blown over in a storm. But it may die and discolor where it stands, eventually becoming a skeleton, a snag, a rotting trunk—possibly serving as a den tree for wildlife— or, riddled with holes, as a source of food for foraging mammals, birds, and insects. Once on the ground, in moist coniferous forests, it may become a nurse log, covered with herbs and tree seedlings, providing habitat on the forest floor for rodents, insects, fungi, and other micro-organisms. Eventually it is reduced to humus. Fire plays an important role in this recycling of nutrients.

The life span of a tree is usually much longer than a human life span, so that often a forest will appear "stable," especially if it is an old forest. In forests where replacement of dead overstory trees by younger individuals of the same species occurs, an equilibrium, or "climax forest," is sometimes assumed to exist. Where a species is apparently being "replaced" by another species, the forests have been referred to as "subclimax" forests. However, increasingly it is recognized that forests are dynamic and in a constant state of change, so less currency is now given to the idea of forests in an equilibrium or climax state. "The bottom-line message to those who must manage natural ecosystems, is that the world is considerably less tidy than we thought. Furthermore, this untidiness may be an integral part of maintenance of many ecosystems" (Christensen, 1988).

Societal Input

Some environmental groups feel that certain forestry practices are undesirable, for a variety of reasons. They speak of virgin forests and ancient forests, and often object to interference with natural cycles. Technically speaking, the term *virgin forest* could refer to any mature forest that has not been modified by human activity. Such undisturbed areas are now hard to find, because exclusion even of fire from a forest is a form of human modification, further blurring the meaning of "virgin" as applied to old-growth forests. In any event the trees comprising these ancient forests will become overmature, and they will be replaced through natural processes. Certain forests are set aside for preservation in their natural state, such as in national parks, wilderness areas, state parks, and dedicated natural areas where natural processes of growth and decay are not interrupted, because the continuance of unmodified processes (naturalness) is a management objective (Fig. 4-7). Here, various damages from insects and diseases are accepted as part of the natural ecological processes of nutrient recycling, and usually no attempt is made to control these processes. Sometimes this can be a hindrance to the adjacent landowner with different objectives. However, in forests managed to produce wood for harvesting, trees are not left to become overmature. The great argument comes over which public lands should be protected from harvest and to what degree, and which lands should be used for harvest and to what degree, considering both environmental and social requirements.

There are many ways in which forest management can modify forest processes through silvicultural practices in order to achieve specific objectives. Understanding competitive processes and forest successional patterns becomes essential. The forest ecosystem may be managed to produce primarily wood

FIGURE 4-7
A stand of western redcedar under protection in the Ross Creek Recreation Area.
This 100-acre (40-ha) stand is the recreation area's major visitor attraction.
Kootenai National Forest, Montana. *(Courtesy of USDA Forest Service)*

products, or to produce habitats suitable to other objectives such as aesthetics, recreational use, wildlife, or water quality, or some mix of all of these may be sought. Such possibilities are limited by the extent of knowledge, by the innate character of forests, and by the social forces at work to modify and guide the management direction of forests. These topics are further considered in "Silviculture and Forest Ecosystem Management."

TREE AND FOREST CLASSIFICATION SYSTEMS

Despite the realization that all parts of the forest ecosystem are linked by ecological processes, forest managers must still deal with individual tree species and separate forest stands. Thus, learning the classification and nomenclature for trees, forests, and their habitats is an essential part of a forestry education. The same principle holds for range, wildlife, and fisheries resource managers, who must work with individual species as well as the basic principles of ecology applicable to their broader field.

Dendrological Classification

A basic classification in dendrology (the taxonomy of trees) is that of *angiosperms* (generally hardwoods or deciduous trees) and *gymnosperms* (generally coniferous or evergreens). Angiosperms are further divided into *monocotyledons* and *dicotyledons*. Fig. 4-8 may be helpful in understanding these divisions, but one

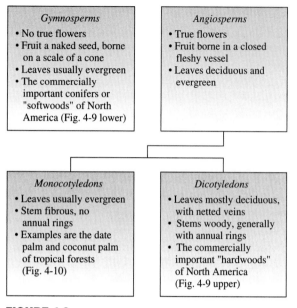

Gymnosperms	*Angiosperms*
• No true flowers • Fruit a naked seed, borne on a scale of a cone • Leaves usually evergreen • The commercially important conifers or "softwoods" of North America (Fig. 4-9 lower)	• True flowers • Fruit borne in a closed fleshy vessel • Leaves deciduous and evergreen

Monocotyledons	*Dicotyledons*
• Leaves usually evergreen • Stem fibrous, no annual rings • Examples are the date palm and coconut palm of tropical forests (Fig. 4-10)	• Leaves mostly deciduous, with netted veins • Stems woody, generally with annual rings • The commercially important "hardwoods" of North America (Fig. 4-9 upper)

FIGURE 4-8

A generalized diagram showing the relationship among gymnosperms, angiosperms, monocotyledons, and dicotyledons.

FIGURE 4-9

Upper: The white oak, a hardwood in the lumber trade, is an example of an angiosperm dicotyledon. Note the net-veined leaves and fleshy fruit. *Lower:* The Virginia pine, a softwood, and more precisely an example of a gymnosperm. Note the needlelike foliage. The naked seeds are borne on the scale of the closed cone. The squares indicate one inch. *(Photos by C. Frank Brockman)*

must realize there are exceptions to this generalization. Figs. 4-9 and 4-10 illustrate these divisions.

Another common division, not botanical, but probably originating from early-day sawmills, is that of *hardwoods* and *softwoods*. When these terms came into usage, the harder-textured species of broadleaved trees such as maples and oaks (Fig. 4-9, upper), and the softer textured conifers such as pines (Fig. 4-9, lower), were the commonly milled species. As might be expected, much confusion exists within this imprecise classification, since the wood of some conifers is "harder" than that of some broadleaved species. Yet this usage persists, chiefly in the wood products industry, where conifers are often referred to as "softwoods" and broadleaved species as "hardwoods." In order to distinguish the major tree species relevant to forestry concerns, a common division in professional terminology is that of broadleaved species and conifers (which have needles or scalelike leaves).

Common Names of Trees

Many people, including foresters, landscape architects, botanists, wildlife biologists, naturalists, and tree growers, are interested in the names and classification of trees. Certain members of the general public are also interested in tree nomenclature as a way to increase their enjoyment of the out-of-doors. Others, such as mill employees and woods workers, have a vocational interest in trees, as do the people who market wood products.

Sometimes commercial marketing techniques invite confusion in order to make a product seem more durable or more prestigious. Thus, Douglas-fir was once called Oregon pine. The common names for trees and lumber may vary depending on who is referring to them, and the same tree species may have

FIGURE 4-10
This coconut palm is an example of an angiosperm monocotyledon. The leaves are evergreen with parallel veins; the fibrous trunk has no annual rings. Eniwetok, in the Marshall Islands, South Pacific. *(Courtesy of Stanley P. Gessel, University of Washington)*

several common names throughout its region or marketing area. For instance, Douglas-fir in the northern Rocky Mountains is commonly called "red-fir," mature lodgepole pine is commonly called "black pine," and ponderosa pine is commonly called "yellow pine."

Scientific Names of Trees

Because of confusion arising from the use of common names, it is necessary to have some universal system that can be used, not only within a single country, but worldwide. Such a system was developed through the utilization of Latin names, many of which were first applied when Latin was used by scientists of all countries. Binomial scientific names have been in use since the middle of the eighteenth century, when the great Swedish botanist Linnaeus published his monumental work, *Species Plantarum* (1753). His methods have led to the practice of assigning a generic name, followed by a species name, to each plant, with this combination of genera and species designation referred to as the *Latin binomial.* For example, the Latin name for ponderosa pine is *Pinus ponderosa,* white oak has the Latin name *Quercus alba,* and longleaf pine has the Latin name *Pinus longifolia.*

Appendix C contains a list of the common and scientific names of trees cited in this book from the official list in Little's (1979) *Checklist of United States Trees, Native and Naturalized.*

Forest Classification

To facilitate management, forests have been arbitrarily classified in several ways. One deals with age and species. Forests with trees that are essentially the same age, such as those developing after a devastating fire, are said to be *even aged.* Forests with trees ranging from small *seedlings* to *poles* to large *veterans** are said to be *all aged* or *uneven aged.* Foresters speak of a *pure forest* or monoculture if it is composed mainly of one species, and of a *mixed forest* if it contains several species. Botanists and foresters refer to *shade-tolerant* or *shade-intolerant* trees, meaning those whose seedlings can, or cannot, grow in the shade of other trees.

Where trees grow closely spaced and the crowns unite in a continuous canopy, the area is classified as a *forest* (Fig. 4-11), whereas when the trees are more openly spaced and the canopy is discontinuous, the area is called a *woodland* or *savannah.*

Forest managers use another type of terminology when they want to know the *stand density* of a tract of timber. This tells whether the stand is *well-stocked, medium stocked* or *understocked.* A *medium-stocked* stand has 40 to 70 percent of its potential optimum stocking; a *well-stocked* stand more than 70 percent, and an *understocked* stand less than 40 percent.

---

*Seedling—up to 3 ft (0.9 m) tall; sapling—a young tree less than 4 in (10 cm) DBH (diameter breast height) and from 3 to 10 ft (0.9 to 3 m) tall; pole—from 4 to 12 in (10 to 30 cm) DBH; standard—from 1 to 2 ft (0.3 to 0.6 m) DBH; veteran—over 2 ft (0.6 m) DBH.

FIGURE 4-11
The forest is a complex community of organisms. This one has been altered many times by the elements, and, most notably, by human beings. Yet the forest persists, eventually covering all scars. Pisgah National Forest, North Carolina. *(Courtesy of USDA Forest Service)*

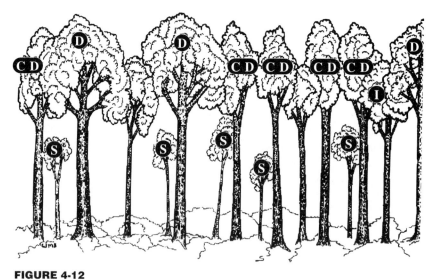

FIGURE 4-12
A crown classification showing dominant (D), codominant (C), intermediate (I), and suppressed (S). *(Illustration by Wenonah F. Sharpe)*

Crown Classification

Another common forestry-related classification of trees is based on the relative position of their crowns within the forest canopy (Fig. 4-12). Here is a common crown classification system.

Dominant—trees with large crowns that extend above the general level of the forest canopy and receive full light from above and partial light from the sides.

Codominant—trees whose crowns form the general level of the forest canopy. Their crowns receive full light from above but relatively little from the sides; crowns are smaller than those of the dominants.

Intermediate—trees with small crowns crowded into the general level of the forest canopy, receiving some light from above but none from the sides.

Suppressed—trees with small crowns that are entirely below the level of the canopy, receiving no direct light from above or from the sides. They are also called *overtopped*.

SUMMARY

Forest ecology is concerned with the interactions among all members of the forest community, from the smallest to the largest, and the environment. This includes soil; climatic factors such as light, temperature, and precipitation; trees and their physiological processes and responses; understory plants; animals; and microorganisms. These relationships, and the ecological processes that govern them, are basic to understanding the cultural practices applied in order to manage forests for the various objectives desired by their owners.

Competition pervades the interactive forest community, influencing forest development patterns as various species struggle for a share of light, water, and nutrients. In attempting to understand forests and their components, accurate terminology is necessary; hence, ecological parlance must include several different kinds of classifications, as well as the time-honored system of Latin binomials.

LITERATURE CITED

Christensen, Norman L. 1988. "Succession and Natural Disturbances: Paradigms, Problems, and Preservation of Natural Ecosystems." In James K. Agee and Darryll R. Johnson (eds.) *Ecosystem Management in Parks and Wilderness.* University of Washington Press, Seattle.

Hutchins, H. E., and R. M. Lanner. 1982. The Central Role of Clark's Nutcrackers in the Dispersal & Establishment of Whitebark Pine. *Oecologia* 55:192–201.

Linnaeus, C. 1753. *Species Plantarum,* Stockholm, Sweden.

Little, E. L., Jr. 1979. *Checklist of United States Trees (Native and Naturalized).* Agriculture Handbook 541, U.S. Department of Agriculture Forest Service, Washington, D.C.

Loftus, David. 1988. "Regenerating oaks on high quality sites: An update. In Clay Smith, Arlyn W. Perkey, and William E. Kidd, Jr. (eds.) *Guides for Regenerating Appalachian Hardwood Stands.* Society of Am. Foresters, publ. 88–03, pp. 199–209, Bethesda, Md.

Loftus, David. 1990. "A Shelterwood Method for Regenerating Red Oak in the Southern Appalachians." *For. Sci.* **36**(4):917–929.

Mattson, D. J., Blanchard, B. M., Knight, R. R. 1991. Food Habits of Yellowstone Grizzly Bears, 1977–1987. Canadian Jour. of Zoology 9:1619–1629.

Sharpe, Grant W., Charles Odegaard, and Wenonah F. Sharpe. 1994. *A Comprehensive Introduction to Park Management* (2nd ed.). Sagamore Publishing, Champaign, Ill.

Stubblefield, George W. 1978. Reconstruction of a Red Alder/Douglas-fir/Western Hemlock/Western Redcedar Mixed Stand and Its Biological and Silviculture Implications, Master's Thesis, University of Washington, Seattle.

ADDITIONAL READINGS

Aber, J. D., and J. M. Metillo. 1991. *Terrestrial Ecosystems.* Saunders College Publishing, Philadelphia.

Kimmins, J. P. 1987. *Forest Ecology,* Macmillan, New York.

Pregitzer, Kurt S., P. Charles Goebel, and T. Bently Wigley. 2000 "Evaluating Forestland Classification Schemes as Tools for Managing Biodiversity." *J. Forestry* **99**(2):33–40.

Wilson, Edward O. 1992. *The Diversity of Life.* Harvard University Press, Cambridge, Mass.

STUDY QUESTIONS

1. With what does forest ecology concern itself? Why is an understanding of forest ecology essential to forest management?
2. Name five factors in soil formation.
3. Explain what is meant by *site index*.
4. What are the primary requirements for tree growth? How do climatic factors affect this?
5. Explain why leaves might be called the solar-powered chemical laboratories of the tree.
6. Explain the following terms: *soil organic matter; transpiration; chlorophyll; mixed forest; stand density; co-dominant trees.*
7. Do deciduous trees or conifers do better on poor sites?
8. Briefly discuss forest competition.
9. Explain what is meant by *primary succession and secondary succession, perturbations, mature forest.*
10. Why are Latin binomials, introduced in the eighteenth century, still used?
11. Why are the terms *climax* and *subclimax forest* less used today?

CHAPTER 5

Silviculture and Forest Ecosystem Management

This chapter was prepared for the 5th edition by David R. M. Scott, professor of Silviculture, College of Forest Resources, University of Washington, and revised and expanded for the 6th and 7th editions by the authors.

Chapter Outline

INTRODUCTION

There are a number of ways to define *silviculture,* but in the broad sense, this aspect of forestry includes all the management operations that go into the development and maintenance of a socially determined form of forest stands. Usually these practices include site preparation, planting, pruning, fertilization, and thinning immature forest plant communities, as well as choice of harvest method for the mature stand and consequent reproduction practices. Also, deliberately stopping or not stopping a wildfire or insect epidemic in a forest, park, or wilderness can be considered a silvicultural practice, as would igniting a carefully planned, prescribed burn. Certainly, silvicultural practices interact with much of the subject matter of this book. Thus, silviculture is basic to renewable resource management in the broadest sense, whether for producing wood products, enhancing wildlife habitat, or managing a forest for multiple values that include water, soil protection, outdoor recreation, and visual qualities.

Forest ecology deals with how the forest (or forest ecosystem) interacts, while silviculture is the applied science of reproducing and managing forest stands to meet landowner objectives—drawing of course on ecological principles. Chapter 5 presents the classical forest reproduction and vegetation manipulation (management) methods used in replacing one forest stand with another, covers the tending of existing forest communities so they will develop in a desired manner, and reviews the emerging and broader ecosystem management approaches.

Responding to public questions over aesthetic, wildlife, and environmental impacts of forest practices, as well as forest health issues associated with fire exclusion, drought, and suspected global warming, new forest management systems are being developed under the headings of *new forestry, adaptive forestry, environmental forestry,* and perhaps other designations. These new approaches are all related in their search for a broader concept, and the term *ecosystem management* may best capture their goals. The new systems draw upon classical silvicultural techniques but focus on managing diversity across the landscape and on opportunities for management throughout the life of individual stands for biodiversity and other purposes. Silviculture, then, can be seen as a current frontier of forestry, where new and broader forest ecosystem management goals are being addressed.

REPRODUCTION METHODS

Replacing one age class of trees with another, or starting a forest in an area dominated by a nonforest plant community, calls for three broad categories of activities: (1) altering the existing forest or plant community to provide ecological space for the desired tree reproduction, (2) treating the surface of the soil (site preparation) to provide conditions necessary for the desired reproduction, and (3) providing sources of reproduction for the new stand or "community" of trees.

Altering the Existing Stand By Removing Trees

Harvesting methods offer a wide range of possibilities. At one extreme the existing forest stand is entirely removed in order to create a new community. This is called *clearcutting* and creates for the subsequent reproduction an environment that is not influenced by remaining members of the previous stand of trees. At the other extreme, only certain individual trees of the existing stand are removed to provide space for reproduction. This partial cutting is called the *selection* system, and it creates an environment where seedlings are strongly influenced by the proximity of residual trees from the existing stand.

The reader will note that the silvicultural methods for altering the stand by removing trees are referred to interchangeably as harvest methods and as forest regeneration methods. While *harvest* suggests the commodity to be hauled away, and *regeneration* suggests the land and life source to be renewed, both terms refer to the various means by which trees are removed and a certain kind of environment is created for reproducing the next forest. These different environments favor different species of trees based on the species' ability to tolerate shade.

Clearcutting makes the entire area available for reproduction and results in an *even-aged stand,* one in which most of the trees interact for most of their lives with trees that are not ecologically different by reason of age (Fig. 5-1). The selection system makes only part of the area available for reproduction and creates an uneven-aged stand, one in which the trees interact for most of their lives with trees that are ecologically different because of age (either younger or older) (Fig. 5-2). The clearcutting method is appropriate only for tree species that can reproduce in direct sunlight. These species would normally reproduce following severe natural disturbances. Such species usually are

FIGURE 5-1
A clearcutting method of reproduction. Seeding came from the uncut trees above. *(Courtesy of USDA Forest Service)*

FIGURE 5-2
A selection method of reproduction in a mixed hardwood stand in northern Michigan. Note the seedlings and younger trees in the background. *(Courtesy of USDA Forest Service)*

FIGURE 5-3
A shelterwood cutting of ponderosa pine in eastern Oregon. The old forest is serving as a shelter for the new one. Once reproduction becomes established, the remaining stand will be removed. *(Courtesy of USDA Forest Service)*

termed *intolerant,* because they are "intolerant of shade" and grow best at high light levels.

The selection system is appropriate for obtaining reproduction of tree species that naturally reproduce in the partial shade of small openings in the forest canopy. Such openings develop as mature trees of the preceding stand are reduced in numbers, and openings occur as large trees die from competition, insects, disease, and old age. Species that can reproduce in a standing forest are termed *shade-tolerant,* because they can tolerate at least some shading. Thus, they naturally become the dominant species as the forest matures. This is because tolerant species become established in the shade of the sun-loving intolerant species that first invade a disturbed site, and then continue to reproduce under the forest canopy. Only a major disturbance, such as fire or wind throw that opens up the site, will create conditions for shade-intolerant trees to replace the shade-tolerant ones.

Intermediate between clearcutting and selection is the *shelterwood* method of providing space for reproduction. This method removes part of the existing stand to make an area available for reproduction (Fig. 5-3). The rest of the stand remains to provide seed and shade, modifying the environment of the new generation for the first few years. Once a new community has become established, the rest of the original stand is removed, and only a new even-aged, young stand remains.

Partially shaded environments are also created near the edges of areas where all the trees are cut

(Fig. 5-4). In fact, side light from the cut area brightens the understory for a short distance into the adjacent uncut area. The conditions there also become suitable for new trees to become established as advance regeneration underneath the uncut mature stand. Likewise, along borders of the cut area there exists a zone where young tree reproduction is partly sheltered by the border trees in the adjacent uncut, older stand. This is a shelterwood type of environment. Only in the middle of a clearcut, beyond two or three mature tree heights from the cut edge, is there a true clearcut seedling environment. Note that clearcut areas of similar total size but differing configurations can present quite different seedling environments (Figs. 5-4 and 5-5). Changing a clearcut square to a long, narrow shape of similar total area creates a partial shaded environment because of bordering tall trees.

The shelterwood method can be used to produce a wider range of seedling environments than either the clearcutting or selection system. Leaving a high stocking of residual trees creates an environment similar to that of the selection method, since considerable shade is present. Conversely, a shelterwood cutting leaving more widely scattered trees gives seedling conditions that more closely resemble clearcutting. Therefore, variations of the shelterwood method are applicable to a wide spectrum of ecological requirements in tree reproduction.

Quite possibly, most successful natural reproduction in forests occurs in what is essentially a shelterwood environment. Natural disturbances rarely result in as complete a removal of the existing stand as does clearcutting. Even a forest of dead trees, following a catastrophic fire or insect epidemic, has some sheltering effect for ensuing reproduction.

Site Preparation

Site preparation, as commonly practiced, may aim to (1) control nontree plant species that both compete with and modify the environment for future forest reproduction, (2) reduce debris or slash left from the previous stand after the harvesting operation has removed the material of economic value, and (3) prepare the soil, such as by scarifying the surface to disturb the organic layer so seedlings can become established. Sometimes all of these site preparation objectives are accomplished in a single operation, such as by using prescribed fire, herbicides, or a

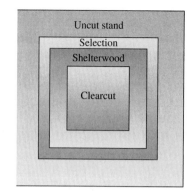

FIGURE 5-4
A schematic of a hypothetical 20-acre (8-ha) square cut showing the various types of environments for reproduction.

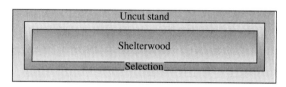

FIGURE 5-5
A schematic showing how altering the shape of a cutting (with size remaining the same) changes the proportion of environments for reproduction (see Fig. 5-4).

heavy machine that breaks down living and dead organic material and tills the surface soil. All three site preparation efforts may be essential to successful establishment of tree reproduction. Ecologically speaking, control of competing vegetation is a very important part of altering the existing stand to promote regeneration of a new age class.

Many stands, particularly those older ones that have not previously been under a management regime (except possibly the exclusion of fire), may have well-established vegetation beneath the main canopy. Reducing the density of the tree canopy may not provide sufficient site resources for reproduction. With removal of the overstory, understory plants may respond vigorously to the increased sunlight and provide serious competition for the desired tree reproduction. Of course, such vegetation may be beneficial to wildlife, or in modifying the microclimate on certain sites.

Logging debris or slash may offer a physical impediment to tree reproduction or to the operations

FIGURE 5-6
A duplex brush cutter used in site preparation prior to planting in Florida.
(Courtesy of USDA Forest Service)

necessary in tending a new stand. It may also be a fire hazard and can harbor the buildup of insects and small animals that might harm reproduction. On the other hand, this same material may serve to shade the new trees, reducing undesirable high surface temperatures on some sites, and provide cover for desired wildlife.

Soils in forested ecosystems have surface layers of organic material in various stages of decomposition. This litter layer forms from the annual deposit of foliage, dead tree parts, and individual trees that have died in the normal course of events. These organic layers may fluctuate in moisture content and temperature, and may not serve as a good rooting media for tree seedlings, especially when the tree canopy, with its ability to modify the microclimate, is removed. Mineral soil, or mixture of mineral and organic materials, is often much more favorable. However, a significant portion of the nutrient supply may be in the humus layers of the soil, and these same layers may also protect the mineral soil from erosion and keep its properties fairly stable.

Generally, site preparation methods must be matched to the ecological situation. A treatment that is too severe may be a deterrent to reproduction, while measures that are too conservative may result in poor stocking and growth rates of new trees.

Prescribed fire is commonly used as site preparation in many forest areas. Historically, natural fires have been the most common disturbance creating conditions for the regeneration of new forests; thus, prescribed fire imitates nature. Under some conditions fire is effective and economical to use. However, in other forest systems fire does not result in the site preparation required; it may be too dangerous or have unacceptable side effects, such as high soil surface temperature and loss of soil fertility. Under these conditions mechanical site preparation is an alternative.

The logging operation itself may result in a satisfactory site preparation on some areas, since the skidding equipment and the moving logs crush vegetation, break down slash, and scarify the soil. This does not always occur, such as in winter when snow covers the ground. For this reason equipment has been developed specifically for site preparation. This includes disks, drums, and blades mounted on or dragged by various types of power sources (Fig. 5-6). Mechanical site preparation frequently demands substantial effort and has limitations in that some terrain is too steep or certain soils are too subject to compaction or erosion for mechanical treatment. To guard against potentially harmful effects, site preparation methods must take into account ecological and physical site characteristics.

Sources of Reproduction

Establishing a stand involves making certain that there is an appropriate source of seeds or seedlings for a new generation of the desired tree species. This source, which must be determined well in advance of need, may already exist on the area itself or on a nearby area, or it may be necessary to transport seeds or seedlings from some distance. Tree reproduction can originate from seeds in nearly all species. Reproduction also originates vegetatively in many hardwoods, but in very few conifers. This vegetative reproduction will be discussed first.

Vegetative Reproduction   Although the cutting of mature trees frequently results in vegetative sprouting from *dormant buds* in the stumps (Fig. 5-7), sprouts from young or medium-aged trees are usually more vigorous. Stump sprouts are frequently poorly spaced (because of the wide spacing of the old stumps) and may be subject to early rot if they originate too high on the stump. However, in oaks and poplars this reproduction method may be useful. In some species, most importantly aspen and poplar, vegetative reproduction after cutting arises from root suckers that originate in *adventitious buds* on roots near the soil surface. Reproducing a stand by depending on vegetative reproduction of sprouting species is called the *coppice system* of reproduction and is sometimes used to regenerate even-aged stands of hardwoods.

Vegetative reproduction can also be used as a means of artificial regeneration in some tree species by means of *cuttings*. In this technique, a branch or shoot of the parent tree is cut off and planted, some-

FIGURE 5-7
Vegetative sprouting of dormant buds from hardwood stumps in a Michigan forest.
(Courtesy of Morbark Industries, Inc.)

times after a treatment with an artificial hormone to encourage root development, or after soaking in water for a few weeks. These cuttings range in size from that of a common lead pencil upward. Willow, sycamore, or poplar "whips" cut from branches or sprouts simply can be inserted in the soil with a good chance that they will develop roots and grow.

An important problem with vegetative reproduction is its apparent desirability as browse for herbivores. Rabbits and deer, for example, frequently eat sprout vegetation in preference to adjacent seed-originated reproduction of the same tree species. This may be due to the food reserves available to the sprout reproduction.

Seed Reproduction   Seed is the origin of most natural stands. After passing through a juvenile period, trees produce seeds. However, age of seed bearing, periodicity of seed crops, amount of seed, and exact characteristics of the seed in terms of longevity, resistance to environmental extremes, dispersal mechanisms including distribution by birds and mammals, and germination requirements, vary tremendously with species and environment. Some tree species produce frequent and abundant crops of highly germinable and thoroughly dispersed seeds at an early age; some produce abundant and frequent seeds of good mobility only after 50 to 70 years. Others produce infrequent, light crops of poorly germinable seeds that are not well dispersed. Still others are intermediate between these extremes. These characteristics greatly affect natural regeneration. For example, in the northern Rocky Mountains, ponderosa pine can be relied upon to have good seed-producing years fairly often, whereas western larch does not produce seed and regenerate so easily.

Seed already may have fallen to the forest floor from trees that were harvested or still may be attached to branches that remain in the area after cutting. More frequently, however, seed comes from trees that are left standing in the area for this purpose, as in the seed tree or shelterwood reproduction methods, or from adjacent uncut stands in the clearcutting method. If seed of the desired genetic constitution is not available as required, it may be collected elsewhere and the site artificially seeded (usually aerially). It may also be drilled in, in rows or some other configuration.

Because of the substantial loss of seed as the result of predation, infection, or inhospitable environments, and additional losses during germination and juvenile seedling stages, large amounts of seed are frequently required to establish a relatively small population of young trees. Because of uncertainties in securing reproduction by natural or artificial seedings, forest managers frequently rely on planting young trees or seedlings that have been grown either in outdoor nurseries or greenhouses. This technique, while perhaps requiring more effort and expense initially, has advantages in control of the timing and population characteristics of the new generation of trees. Also, planting well-developed seedlings gives the new forest a head start over competing vegetation.

Seed can be collected during years of abundant production and stored in special facilities until needed for use in later years. This can be done in a variety of ways, but usually there is an attempt to identify the source as accurately as possible. Parent trees, in stands of high quality reserved for seed production, or in seed orchards, are usually harvested of their cones by skilled employees using special techniques (Fig. 5-8). Seed is also bought on the open market. Frequently the pickers utilize tops from trees that have been harvested at seed-maturing time, or they may try to locate squirrel caches of fruit or cones.

After gathering the fruits or cones, the seed is extracted, cleaned, sorted, and stored in a refrigerated, oxygen-free environment. The seed of some species will maintain much of its germinability for many years if properly stored, but in other species, storage for even a few months leads to a substantial loss in viability.

When the seed is needed for use in a nursery, or for direct sowing in a forest, it is removed from storage, moistened, and kept at a temperature just a little above freezing for several weeks. This is called *stratification*. It simulates the moist, cool environment of the forest floor and readies the seed for immediate germination when exposed to warmer temperatures.

The seed is then sown in carefully prepared beds, at a spacing intended to produce a given density of seedlings. An intensive regimen of weed and disease control, together with equally rigid control of soil fertility and water, is applied to produce seedlings of uniform, desired quality (Fig. 5-9). Nursery sites are

FIGURE 5-8
The man in the bucket collects cones in a 20-year-old, genetically superior loblolly pine seed orchard in
Georgia. The seeds will be planted in a nursery to grow seedlings for reforestation. *(Courtesy of
Georgia-Pacific Corporation)*

FIGURE 5-9
A forest nursery in Oregon. Seedlings later will be lifted,
graded, root pruned, bundled, and shipped to the planting site.
(Courtesy of Weyerhaeuser Company)

carefully selected for correct soil and climate. Depending on species, location, and type of seedling, time in the nursery may vary from one to six years. Roots are often pruned to promote a compact root system. If a larger size is desired, the seedlings may be transplanted to a wider spacing after a year or two, and then allowed to develop for a given period of time.

After the seedlings have reached the prescribed stage of development, they are lifted from the nursery beds in a dormant season, sorted and graded for quality, and packaged in manageable numbers. Such seedlings are termed *bare root seedlings,* compared to *container seedlings.* It is important that these steps be carried out carefully and quickly to prevent dehydration. If possible, seedlings are shipped immediately to the field and planted. Frequently there is some delay in planting crew availability, or planting site conditions may prevent ideal timing. In these circumstances the packaged seedlings are placed in re-

FIGURE 5-10
An indoor nursery where conditions of temperature and humidity are carefully controlled. Here, nursery employees are tending young Douglas-fir seedlings at the Georgia-Pacific Forest Research Center at Cottage Grove, Oregon, where some 10 million genetically superior seedlings are grown annually for reforestation programs. *(Courtesy of Georgia-Pacific Corporation)*

FIGURE 5-11
Six-week-old red pine seedlings grown in container tubes. Note the growth of the roots at right. Superior National Forest, Minnesota *(Courtesy of USDA Forest Service)*

frigerated storage rooms. They can be safely held for several weeks or months, depending on their physiological condition when lifted from the nursery.

Seedlings may also be started in individual containers with special rooting medium in a greenhouse with a closely controlled environment (Fig. 5-10). After a relatively short time, often no more than four to six months, these "container seedlings" are planted at the site with the root medium intact (Fig. 5-11). This technique was originally established by foresters in cold, northern climates, where the difficulties involved in successfully producing and planting conventional bare-root seedlings were substantial. It is now used much more widely. Under some conditions it offers a number of advantages: flexibility of production because of shorter seedling rotations; improved success in seedling survival on some sites; and extension of the planting season. For some species that are difficult to propagate in bare-root nurseries, this method is often more successful. While costs of containerized seedlings may be 10 to 20 percent more, the additional cost may be more than compensated by relative ease of planting, survival rates, and seedling quality.

Planting of seedlings in the field is done both by hand (Fig. 5-12) and by machine (Fig. 5-13), depending on the availability of labor, the size of the operation, the characteristics of the soil, terrain, site preparation, and the seedlings. Generally modern planting is done at lower densities (400 to 600 trees per acre or 1,000 to 1,500 trees per ha) than was common several decades ago (700 or more per acre or 1,730 trees per ha in some species). This is a result of improved seedling survival from better quality seedlings and improved overall reforestation practices including site preparation and planting methods. Cost is also a factor, as the lower densities may allow delay of the first thinning until trees reach merchantable sizes.

INTERMEDIATE OPERATIONS

Cultural Practices

When a forest plant population consisting of both tree and nontree species is first established, the individual plants are small. However, as they develop rapidly in size and occupy all the growing space, interactions such as competition and mutualism come into

FIGURE 5-12
A planter loading his backpack frame with several hundred containerized seedlings while planting this Oregon clearcut. He works out of the styrofoam container hooked to his belt. *(Courtesy of Georgia-Pacific Corporation)*

FIGURE 5-13
Machine planting in the Chippewa National Forest, Minnesota. The tractor is equipped with a hydraulic plow that prepares the site for planting. The rear wheels close the furrow and compact the soil around the seedlings. *(Courtesy of USDA Forest Service)*

play, and changes in the characteristics of both individual plants and plant populations take place. These features of natural community dynamics may or may not be desired by the landowners in either detail or timing, and the cultural practices of thinning and pruning will probably be required to obtain the desired characteristics. The following cultural practices, known as *intermediate operations,* are some of the methods directed toward controlling the plant population or changing the physical environment.

Release Cuttings These have several forms. As the young forest community develops in the seedling and sapling stage, the desirable individuals may be crowded and thus at a competitive disadvantage relative to other members of the population. This usually

is caused by lack of control of seed sources or sprouting by undesired elements of the previous plant community. In either case a *cleaning* operation is required. This operation usually seeks to remove undesired species but occasionally is directed at poorly formed individuals of a desirable species. The objective is to improve the species composition and quality of the new stand. When undesirable older individuals have been left unharvested from the preceding stand and must be removed because of their depressing effect on the reproduction, the operation is referred to as a *liberation cutting* (Fig. 5-14).

Stand Improvement Cutting If a similar type of operation is required when the stand is older, it is referred to as a *stand improvement cutting* (Fig. 5-15).

FIGURE 5-14
A forest worker in the Lake states girdling undesirable oaks with an ax. The hardwoods eventually die, liberating the red pine reproduction desired by the landowner. (*Courtesy of USDA Forest Service*)

Each of these operations may be implemented by cutting if the undesirable individuals are small, or by girdling them if they are large. The latter operation consists of cutting through bark and phloem around the tree thoroughly enough to interrupt the flow of food to the roots (Fig. 5-14). This eventually causes the tree to die. However, cutting and girdling may result in sprouting from dormant buds below the cut in many angiosperms (hardwoods) and occasionally in a few gymnosperms (softwoods), particularly if the cut individuals are young. Such sprouts may develop in height so rapidly that competition with the selected species is equal to or greater than that in the precutting condition.

The use of chemicals, whether with a hormone or a poison, reduces both the effort required in release operations and the probability of sprouting. Such chemicals may be applied with a hypo-hatchet or other stem-injection device that introduces the herbicide into the undesirable tree, or by individual spray-ing. The chemical may be sufficiently specific to the target species to be applied on an area basis, usually by spraying from aircraft, if environmental constraints (usually water-quality standards) allow. Increasingly, application of chemicals is subject to environmental restrictions and public resistance, especially on public lands.

Pruning   Another intermediate operation is *pruning*. This is somewhat different from other practices in that it is aimed at altering the form or quality of individual trees rather than controlling their growth. Cutting off the lower branches close to the stem improves stem wood quality because it will eliminate knots, which are caused by branches as new wood is added to the stem. Clear lumber or veneer may then be processed from the clean boles.

Pruning or shearing the ends of branches gives trees a desired shape and can cause new shoot growth to be initiated by buds at lower positions on the branches on the tree. This type of pruning and shearing is commonly practiced on trees that are to be used for ornamental purposes as landscape or Christmas trees, giving them the desired conical and bushy form.

Sanitation Cutting   Still another intermediate operation is the *sanitation cut*. In any plant community, certain individuals, either because of genetic variation or competitive position in the forest stand, are more susceptible to disease or predator attack than are the remainder. Hence they may serve as the initiators of an epidemic outbreak that will severely affect the entire stand. If these susceptible individuals are eliminated, the chances of an epidemic will be greatly reduced and the overall health of the entire forest is improved and safeguarded. This cut also removes dead or already damaged trees. In some instances, removed stems must be burned or chipped to control the threat (Fig. 5-15). Sanitation cuts entail the same techniques used in release operations.

Salvage Cutting   Salvage operations remove dead or dying trees from the stand to recover their value. Thus, a salvage operation is a type of sanitation cut, but here the trees are commercially valuable. Removal can reduce the risk of infections by insects attracted to the dead and dying trees, but this must be weighed against the value of snags as potential cavity nesting sites for birds that will eat the damaging

FIGURE 5-15
Upper: A natural stand of mixed hardwoods in northern Michigan, prior to an improvement cut. *Lower:* The same stand after removal of unwanted, diseased, and deformed trees. A chipping machine disposed of the unwanted trees. *(Courtesy of Morbark Industries, Inc.)*

insects and animals that increase desired biodiversity in the forest. Snags play a positive role in the forest ecosystem, and some should always be left.

Thinning Thinning is the silvicultural adjustment of numbers and arrangements of trees so that the stand, and individual trees in the stand, will grow more rapidly and productively than otherwise would

be the case. After a forest stand has grown to a particular structure in terms of species composition, size of trees, and other features, there is still a great deal of variation both in the manner in which the individual trees may develop and in the way the entire forest community may grow. In particular, this is brought about by difference in size of trees, and therefore in the amount of growing space they need. Particularly in even-aged stands, an area of land that will support literally thousands of seedlings or saplings may be fully utilized by only a few dozen of the same trees 30 or 40 years later. This means there has been a tremendous loss of numbers due to competition for space, sunlight, nutrients, and water.

After a group of young trees develops in size sufficient to occupy all the growing space, continued growth and competition change individual tree morphology (shape) and physiological (growth) processes, compared with these same features in noncompetitive circumstances. For example, at any given age, forest-grown trees have shorter crowns than open-grown ones because lower branches die from lack of light, reducing overall crown size. They usually exhibit straighter and less tapered stems than open-grown trees. In addition, when trees are at too great a competitive disadvantage, their physiological processes may be so affected that they do not grow efficiently in proportion to the growing space occupied. As a population of trees fills the available growing space and continues to develop, it differentiates into crown classes. Some trees fall behind others in height and size, lose growing efficiency, and eventually may die, leaving their spaces vacant until refilled by the continued development of the adjacent individuals.

Thus, at a particular stage in the life of a forest stand, both efficient and inefficient individuals exist, and there are gaps in space occupancy. The proportions in each of these categories depend on the characteristics of the tree population (numbers, genetic makeup, spacing, age) and the physical environment. When there is a very large number of very similar trees and therefore little or no competitive advantage, almost the entire stand may grow very slowly and inefficiently, or growth increment may cease completely. This is called *stagnation.* Thinning thus seeks to increase vigor and promote growth efficiency. At the same time, thinning may provide an intermediate harvest of some of the trees that otherwise might be lost, and thus provide economic return sooner than if no thinning were undertaken until the final crop was mature and ready for removal.

There are several methods of thinning, based on the kinds of trees cut and the kinds of trees left. A thinning that removes the smaller individuals and leaves the larger is classified as *low thinning* or thinning from below, and most nearly simulates natural mortality (Fig. 5-16). This technique may result in

greatest total growth, since inefficient trees are generally removed, leaving the space they occupied available for utilization by efficiently growing trees. However, the cut stems are small and may not have any value.

In *selection thinning* the biggest trees are cut, and the smaller trees are left. This may be done because the smaller trees have some desired qualities, such as stem straightness or lack of taper, or because the big trees have undesired qualities, such as being too branchy. On the other hand, the larger trees have more value on the current market, and the trees left may well be genetic runts. Selection thinning is more likely to reduce total growth or lengthen the time required for the final crop to reach a certain size.

Crown thinning removes some of the large trees, leaving other large trees as well as the smaller trees. It provides space for crowns of the remaining trees to fill as they grow. This type of operation can be seen as a compromise between the advantages and disadvantages of low thinning and selection thinning.

In some circumstances an *arbitrary* or *mechanical thinning* is used. This method thins the stand according to some arbitrary pattern, with limited regard for any crown differentiation. For example, a tree shearer, bulldozer, or fellers with saws may simply cut through a stand at certain intervals to reduce competition so the remaining trees will have more access to light, moisture, and nutrients. Such mechanical thinning is most frequently used in stands where crown class is relatively unimportant—usually because the trees are young or lack genetic variation (for example, in clonal populations), or because of operational feasibility or simplicity.

When the trees removed in thinning young stands have no marketable value, the operation is termed a *precommercial thinning.* Thinnings of this nature are then investments to provide future values, as are most other intermediate operations. In older stands the trees removed usually have value as wood and are therefore called *commercial thinnings.* However, wood values may not necessarily be the management objective. For example, opening up the stand to promote development of understory species desirable for wildlife may be the main objective. Or the goal may be to reduce stand density in order to improve forest health by making moisture and nutrients more available to the remaining trees, which will safeguard the stand's vigor.

FIGURE 5-16
Upper: A 50-year-old red pine plantation in Michigan, before thinning. *Lower:* The same stand after a heavy, low thinning. *(Courtesy of USDA Forest Service)*

Fertilizing   As with genetic improvements, fertilizing on a commercial scale is a more recent and limited silvicultural practice in the United States. In spite of the fact that fertilizer was applied to forest communities in India and western Europe as early as the middle of the nineteenth century, the practice did not at once become widespread. Unfortunately, early attempts were made with little knowledge of the required nutrient elements. The measurement techniques used to assess response were also imperfect. As a result, it was commonly believed that forests were relatively insensitive to fertility levels in the soil.

In the 1930s, symptoms of nutrient deficiencies developed in the otherwise extraordinarily successful exotic pine plantations in Australia. This, together with a renewed interest in increasing wood yields of forests in Europe and North America, brought about an expanded research and development program dealing

FIGURE 5-17
The effect of nutrient deficiencies on seedling growth. Each of the three Douglas-fir seedlings at left was deprived of one nutrient: number 1 was deprived of nitrogen; number 2, of phosphorus; and number 3, of potassium. Seedling number 4 received all nutrients. *(Courtesy of University of Washington, College of Forest Resources)*

FIGURE 5-18
A modern method of applying fertilizer. Here the pilot triggers a streamer of forest nutrient from the hanging spreader over the 12-year-old Douglas-fir trees below. Such application increases tree growth. *(Courtesy of Georgia-Pacific Corporation)*

with augmenting forest nutrition through fertilization and other practices to conserve natural nutrient cycles.

Forest fertilization is undertaken to treat nutrient deficiencies. When these deficiencies retard tree growth and development, fertilizing will increase wood yield, improve tree or forest appearance and health, stimulate seed production, and increase production of subordinate plant species used as habitat and forage for domestic animals and wildlife. Deficiencies are frequently associated with scarcity of the so-called micronutrients, meaning those elements required in very small amounts, such as copper, zinc, and boron. Increases in productivity, on the other hand, often require additional quantities of macronu-

trients, or those elements needed in relatively larger amounts, such as nitrogen, calcium, phosphorus, and potassium (Fig. 5-17).

Substantial gains in productivity, perhaps 25 percent or more, have resulted from macronutrient additions, and relieving deficiencies on impoverished sites can mean even greater gains. As a result, fertilizing has become an economically attractive and increasingly utilized silvicultural practice, particularly in Australia, Scandinavia, and the southeastern and northwestern United States. When dealing with large areas, fertilizer is usually applied by planes or helicopters (Fig. 5-18). In some accessible areas, nutrients in the form of sewage sludge have been applied using tanks and hoses (Fig. 5-19).

Long-term sustained production of forest ecosystems is inseparably linked to nutrient-cycling processes. Each ecological situation has to be judged on its own merits. The cycle of each nutrient element should be assessed, and the importance of that portion removed in a timber harvest should be considered, as well as the role that the vegetation plays in nutrient cycling. Frequently it appears that nutrient problems in forest ecosystems arise, not so much from inadequate amounts of a particular element, as from problems with its chemical or physical availability for plant use.

FIGURE 5-19
Treated municipal sludge can dramatically improve tree growth. *Upper:* Sludge from Seattle's sewage treatment plant sprays on the forest floor through the nozzle mounted on the top of this tank truck. *Lower:* The effect of sludge on tree growth. After 17 years of slow growth, the tree more than doubled its diameter during the seven years of sludge treatment. *(Photos courtesy of the University of Washington)*

GENETIC CONSIDERATIONS

Of the economically important plant species, forest trees are among the least genetically altered by humans. In many locations in the world, their gene pools remain essentially intact. However, silvicultural practices frequently include introduction of seedlings genetically altered or selected to meet certain objectives, such as better individual tree characteristics, higher yields by populations, or resistance to some parasite or predator.

In silvicultural systems involving natural reproduction, efforts are made to select trees with desirable qualities as seed sources. It is also important that a large enough population of such trees remains and that it be properly spaced; otherwise negative inbreeding may result. In fact, an adequate population of seed producers may be more important than rigid selection for superior qualities, particularly in pioneer species; that is, those characteristic of early successional stages, such as lodgepole pine in the west or jack pine in the Lake states. This type of seed-producer selection is aimed at ensuring prompt regeneration of the site, and it does little more than maintain the existing gene pool in the succeeding generation.

Looking toward the long-term productivity of the forest, one advantage of artificial regeneration by seeding or planting is the opportunity to improve the gene pool of a forest stand. For that reason, enrichment plantings of genetically improved seedlings, or interventions by direct seeding, sometimes are used to supplement natural regeneration.

Research suggests that trees are genetically adapted to specific sites within a few hundred feet in elevation or a few miles from where they naturally grow. Thus, seedlings and seed are typed as to their origin or *provenance,* and, if purchased on the open market, obtaining seed or seedlings from near the area and elevation of the site to be regenerated is an important consideration. There is some evidence that soil variations within climatic zones may also be important. Some companies collect seed from trees as they are harvested and then grow seedlings from these seeds to regenerate that harvest site.

Another technique involves the use of selected forest stands to serve as seed-production areas. These are areas of unusual quality for a particular environment and are further cultivated by removal of the less-desirable individuals by proper spacing and possibly by fertilizing. Seed production may be increased by this cultivation, and the process of selection may provide certain genetic gains.

Serious attempts to improve or broaden genetic pools require selection of parent trees based on rigorous criteria. The superior individuals thus selected are called *plus trees.* These individuals are then transferred to seed orchards by grafting a cutting to a root stock or by rooting the cutting. When a number of such individuals are allowed to interbreed freely, they produce seedlings that may show some genetic gain. Controlled breeding between individuals may also be carried out and the ensuing progeny tested for desirable qualities. Matings that produce populations with abnormal concentrations of the desirable properties are then repeated on a production basis. It is important to continue the selection process in the progeny to realize further genetic gain. There is danger, of course, in narrowing the genetic base, particularly in relation to disease susceptibility. For example, in the production of genetically resistant white pine seedlings it has been found necessary to retain about one-third of the population in a natural state, without the resistance. Otherwise the blister rust fungus would mutate to prey on the resistant trees.

Vegetative propagation of superior individuals is also possible in some species. Exact genetic construction of stands can be repeated in this manner. The goal of micropropagation is the mass production of genetically identical populations of desirable seedlings, with accompanying concern for the susceptability of the resulting monoculture to some yet-to-be-experienced pest or disease.

ECOSYSTEM MANAGEMENT

Traditionally, silviculture has been focused on production of wood fiber for commodities such as lumber or pulp, and clearcutting was too often the silvicultural choice. Public opposition to clearcutting runs deep, not just for aesthetic reasons but because of many negative associations such as deforestation, environmental degradation, and exploitation (Bliss, 2000). Despite many warnings, resource managers were slow to respond to the explosion of environmental awareness that occurred in the 1970s, 1980s, and 1990s. The imminent loss of certain forest bird species such as the northern spotted owl, marbled murrelet, and red cockaded woodpecker played a key role in awakening managers and the public to the effects of reduced diversity across the landscape (Meslow, 1993; Paulson, 1992). All natural resource management, and perhaps especially forestry, has been deeply affected.

To meet these challenges, a greater emphasis on the management of the entire forest ecosystem and its diverse values has evolved. For example, in 1992, the chief of the U.S. Forest Service announced that harvests on the national forests would henceforth be guided by an ecosystem management approach in which clearcutting might be used to pursue ecologi-

cal objectives such as wildlife habitat improvement, or to improve forest health, but not for purely economic reasons.

This shift in forest management emphasis is partially in response to public concerns about clearcutting, but also addresses water quality and availability, wildlife and biodiversity, and the health and sustainability of forests. These concerns involve both public and private lands. Field techniques to apply this new philosophy, now commonly called *ecosystem management,* but referred to in earlier stages of its evolution as *new forestry* or *adaptive forestry,* have been tried in every region (Adams, 1992; Baker, 1989; Franklin, 1989; Lanasa, 1989; Marquis, 1989). The goal is forest management and timber harvest methods that minimize impacts and maintain healthy forests and biodiversity across the landscape as well as within individual stands. In addition to the public concerns, research results on how forests function are also driving changes in forest management. We now know more about relationships between the production of wood and the maintenance of other yields and amenities, such as wildlife habitat, visual aesthetics, and forest health, and the underlying science of ecosystem management (Kohm & Franklin, 1997).

A principal concept of these new ecosystem management approaches is that management will be directed toward maintaining healthy and holistic forest ecosystems and that it will be beneficial to the long-term production of wood fiber as well as other commodity and noncommodity values. Opportunities for silvicultural treatments over the entire life of the stand, such as thinning and selective cuttings, will be pursued. A dominant characteristic of this management is the fostering of *biodiversity,* from both the variety and distribution of different plant communities across the landscape, and the structural diversity of canopy layers and species variety in individual forest units.

Previously, the process of harvesting mature trees and regenerating a new stand commonly reduced diversity, especially in some clearcutting situations. The resulting new forest usually had fewer tree species, and efforts were often made to reduce other vegetation that might compete with the economically valuable trees for water, nutrients, light, and space. This simplified forest was easier to manage, and may have been more efficient for growing wood fiber in the short term. But as forest managers learn more about how forests function, they have developed a greater appreciation of the importance of retaining all elements of the system. The importance of fire, understory vegetation, and woody debris in nutrient cycling and the maintenance of long-term site productivity are now recognized, as is the value of species and age class diversity for reasons of forest health.

Among the new approaches that preceded ecosystem management, one called *new forestry* (for a few years) was developed in the West Coast Douglas-fir region. The predominant silvicultural system in the Douglas-fir region has been to clearcut, prepare the site (usually by burning), and replant with genetically improved trees. This rather simple system was apparently successful for managing wood production in mature, coastal Douglas-fir. However, as a result, the landscape is now defaced by huge, old clearcuts and dotted with artificial-looking patches of uniform single-species plantations. The forest no longer "looks like a forest," and much of the public rejects this, not only because of the visual impact, but also because they understand this forest has limited ecological diversity. *New forestry* (and now ecosystem management) prescriptions call for retaining a considerable number of snags and substantial amounts of large, woody debris on the ground, as well as a dozen or more mature trees per acre, which are planned for retention through all or a good portion of the next rotation (Franklin, 1989; Fig. 5-20). Research is showing the value of the downed woody material to the long-term nutrient budget, the value of snags for wildlife, and that leaving some standing green trees will ensure a future source of high-quality wood, snags, and coarse woody debris, as well as interconnected patterns of habitat. A variety of stand structures across the landscape can be produced in coastal Douglas-fir clearcut situations by careful layout of such harvest areas (Franklin, 1989). Diversity is a primary goal, with a focus on what is left after harvest, not just what is taken.

A similar new approach has been referred to as *adaptive forestry,* a term that also described new thinking in forest management in the 1990s. The term *adaptive* means forest management capable of adapting to social changes and demands on the forest, of adapting to characteristics of the ecosystems and sites where it is applied, of adapting to new scientific knowledge and techniques, and of adapting to conditions as yet not fully defined, such as global climate

FIGURE 5-20

A research area in Oregon, in which new forestry prescriptions are studied. Here elements of old-growth forest structure, such as large green trees, snags, and down logs, are retained to provide habitat for many organisms associated with late successional forests, including northern spotted owls. H. J. Andrews Experimental Forest. *(Photo by Jerry F. Franklin)*

change and industrial pollution (Adams, 1992). By maintaining diverse and fully functional ecosystems, both management and the forest will be better positioned to adapt and respond to future events and needs.

These new approaches to forest management, now generically called *ecosystem management,* will differ in the various forest environments, but some general implications are clear. There will be less clearcutting, and the clearcuts will be smaller and shaped to conform to the landscape. The alternatives to clearcutting require greater care in the planning and preparation process and may result in higher management costs, such as for individual tree marking, thinning, and pruning; greater care in logging to protect resid-

FIGURE 5-21
This innovative harvest pattern provides for wildlife and other ecological values. In this example, 15 percent of the green trees were retained. Maintaining small patches of intact forest incurs minimal interference with logging operations. *(Photo by Jerry F. Franklin)*

ual trees; and piling and burning excess logging slash (Fig. 5-21).

Ecosystem management will not replace established silvicultural practices—it will utilize them. However, clearcutting will surely be applied more judiciously, and silvicultural systems designed to produce multi-aged forest stands will be increasingly applied. Intermediate silvicultural operations, such as thinning, pruning, and salvaging dead and dying trees, will be more common and may provide employment in timber-dependent communities. Yields of wood products may fall in the short run, but the tradeoff will be sustainable harvests, improved stewardship of the land and its wildlife, and improved public relations. It can be applicable even to owners of small woodlands (Rickenbach et al., 1998).

RESTORATION AND MAINTENANCE OF FOREST HEALTH

Forest health concerns include symptoms reflecting weakened forests vulnerable to insect and disease infestations and to intense wildfires. Often, decades-long exclusion of fire has lead to fuel buildups on the ground and in understory vegetation that provide "fuel ladders," increasing the risk that any fire that ignites will become an uncontrollable crown fire. The increase in understory vegetation also weakens the stand by consuming moisture and nutrients that would otherwise be available to the dominant and codominant trees under more natural fire regimes. Past management practices may have also contributed to forest health concerns by reducing diversity at landscape or forest stand levels, thereby increasing vulnerability to insects and disease.

Attempts are underway to solve the forest health dilemma that has risen from past management practices, fire exclusion, and drought. For example, taking their cue from the area's natural history as well as from new ecosystem management thinking, many forest managers in the West have developed forest health strategies based on reducing forest density to alleviate drought stress. Stands are being precommercially thinned where necessary and commercially thinned where possible. Harvests are partial cuts with only a few small clearcuts to remove pockets of dead trees. Harvested areas are broadcast burned to eliminate the slash where possible, and most snags and dying trees are removed. The goal of this management strategy is to restore the resilience of the forest and improve its resistance to fire, insects, and disease. By thinning the forest and removing fuel, the moisture balance may be corrected, the cycle of uncontrolled wildfire broken, and forest health greatly improved. The current situation evolved over a period of six decades, and it may take that long for healthy conditions to be restored.

Thus, nationwide, especially in the West and in the national forests, a key focus of ecosystem management will be *ecosystem restoration.* A national fire plan is being implemented, the heart of which is restoring forests to a more natural condition. The plan will also focus on helping private landowners and communities prevent property damage from fire, and rehabilitating areas burned in the disastrous fire season of 2000 and several bad fire years in the late 1990s. Thinning overcrowded forests, planting grass and trees and controlling erosion on burned areas, and prescribed burning are all part of this large plan (Matthews, 2001). Thus, *ecosystem restoration*—management that returns the landscape to its historical stand-type and restores the related ecosystem—seems destined to be a major focus of forest ecosystem management in the early decades of the twenty-first century.

SUMMARY

Silviculture is the application of the principles of forest ecology to the task of creating and maintaining desired forest stands. Harvest or reproduction methods cover that part of silviculture dealing with replacing one forest plant community or age class with a new one, and include clearcutting, seed tree, shelterwood, and selection harvest methods. Intermediate forest management operations involve manipulating existing forest stands through such practices as release cuttings, pruning, sanitation and salvage cuts, thinning, and fertilization.

Genetic manipulation of forest tree species, carried out in tree improvement programs aimed at enhancing wood yields, quality, and resistance to insects and disease, has been in progress for several decades.

New approaches to forest management, whose concepts may appear under several names, have merged into the holistic name of *ecosystem management.* This new approach represents a response by agencies, resource managers, and landowners to public concerns over environmental and aesthetic impacts of timber harvesting and other forest health and wildlife issues. Ecosystem management may utilize all the classical silviculture techniques, but in combinations that seek to retain biodiversity in forests at landscape levels, and to an even greater degree in individual stands.

LITERATURE CITED

Adams, Dave. 1992. New and Adaptive Forestry in the Inland Northwest. Idaho Forest, Wildlife and Range Experiment Station. Misc. Pub. No. 16, Moscow, Idaho.

Baker, James. 1989. Alternate Silvicultural Systems South. P. 51–60. Proc: National Silvicultural Workshop Silviculture Challenge and Opportunities for the 1990s. USDA Forest Service Division of Timber Management. Washington, D.C.

Bliss, John C. 2000. "Public Perceptions of Clearcutting." *J. Forestry* **98**(12):4–9.

Franklin, Jerry. 1989. "Toward a New Forestry," *Amer. Forests.* **95**(11 & 12):37–44.

Kohm, Kathryn, and Jerry F. Franklin (editors). 1997. *The Science of Ecosystem Management.* Island Press, Covelo, Calif.

Lanasa, Mike. 1989. Alternative Silvicultural Systems in the Eastern Region. P. 29–35. Proc: National Silvicultural Workshop: Silviculture Challenge and Opportunities for the 1990s. USDA Forest Service, Division of Timber Management. Washington, D.C.

Marquis, David A. 1989. Alternate Silvicultural Systems— East. P. 36–45. Proc: National Silvicultural Workshop Silviculture Challenge and Opportunities for the 1990s. USDA Forest Service, Division of Timber Management. Washington, D.C.

Matthews, Mark. 2001, May. "The West Goes to Work Cleaning Up its Forest." *High Country News* **33**(9):1, 8, 9, 11. Paonia, Co.

Meslow, Charles E. 1993. Spotted Owl Protection: Unintentional Evolution Toward Ecosystem Management. *Endangered Species Update* (10)**3** and **4:**34–38.

Paulson, Dennis R. 1992. Northwest Bird Diversity: From Extravagent Past to Precarious Future. *NW Env. Jour.* (8)71–118.

Rickenbach, Mark G., David B. Kittredge, Don Dennis, and Tom Stevens. 1998, April. "Ecosystem Management: Capturing the Concept for Woodland Owners." *J. For.* **96**(4):18–24.

ADDITIONAL READINGS

Aplet, Gregory H., Nels Johnson, Jeffry T. Olson, and Sample Alaric. 1993. *Defining Sustainable Forestry.* Island Press, Covelo, CA.

Brunson, Mark, and Bo Shelby. 1992. "Assessing Recreational and Scenic Quality: How Does New Forestry Rate?" *J. Forestry* **90**(7):37–41.

Callicott, J. Baird. 2000. "Aldo Leopold and the Foundations of Ecosystem Management." *J. Forestry* **98**(5): 5–13.

Hansen, A. J., A. Spies, F. J. Swanson, and J. L. Ohmann. 1991. "Conserving Biodiversity in Managed Forests," *Bioscience* **41**:382–384.

Journal of Forestry. 1998, July. Theme issue on Multiaged Silviculture 96(7).

Journal of Forestry. 2000, August. Theme issue on Ecological Restoration 98(8).

Mutch, Robert W., Steven F. Arno, James K. Brown, Clinton E. Carlson, Roger D. Ottman, and Janice L. Peterson. 1993. Forest Health in the Blue Mountains: A Management Strategy for Fire Adapted Ecosystems. USDA Forest Service. General Technical Report PNW-GTR-310. February 1993.

Oliver, Chadewick D., and Bruce Larson. 1990. *Forest Stand Dynamics.* McGraw-Hill, New York.

Salwasser, H. 1990. Gaining Perspective: Forestry for the Future. *J. Forestry* **88**(11):32–38.

Swanson, F. J., and J. F. Franklin. 1992. New Forestry Principles. *Ecosystem Analysis of Pacific Northwest Forests Ecological Applications* **2**(3):262–274.

STUDY QUESTIONS

1. Distinguish between *forest ecology* and *silviculture.*
2. Describe the following methods for reproducing a forest stand: clearcutting, shelterwood cutting, selection system.
3. What three things are likely to be done in site preparation?
4. Explain the reasons for the following intermediate cultural operations.

   liberation cutting   sanitation cut
   pruning   thinning

5. Explain the reasons for the following kinds of thinning:

   selection thinning   precommercial thinning
   mechanical thinning

6. Give several reasons for fertilizing forests. What nutrients are involved?
7. Why should seed be collected from areas near where the seedlings will be planted? What term refers to this concept?
8. How do ecosystem management approaches differ from traditional forestry practices?
9. Define *forest health.*
10. How has efficient fire control affected forest health?

CHAPTER 6

Insects and Mammals

Chapter Outline

Introduction

Insects

Insect Populations

Destructive Forest Insects

Sucking Insects

Defoliators

Bark Beetles

Wood Borers

Terminal Feeders

Root Feeders

Gall Makers

Seed Insects

Methods of Managing Forest Insects

Silvicultural Controls

Biological Controls

Chemical Controls

Integrated Pest Management (IPM)

Mammals

Damage by Larger Mammals

Deer, Elk, and Moose

Bear

Rabbits, Porcupine, and Beaver

Damage by Smaller Mammals

Grazing Damage by Domestic Animals

Beneficial Effects of Animals

Summary

Literature Cited

Additional Readings

Study Questions

INTRODUCTION

Insects and mammals constitute an important area of consideration in forest and renewable resources management for several reasons. They form an integral part of the forest biota and, along with fungi and bacteriological components, comprise a vital part of forest ecosystems. Through their food and shelter needs, insects and mammals influence the decomposition of forest litter and its dispersal as humus in the soil. They help to fertilize the soil and to aerate it by their burrowing. Some act as pollinators, and others as seed-dispersing agents. Some are beneficial predators, controlling other damaging insects and mammals. Even the tree-destroying activities of insects and certain other species play a positive role in forest ecosystem processes, when forests are considered as a whole and not just for the production of wood products or other uses beneficial to humans.

However, for forestry aimed at increasing yields of wood and fiber, which must be viewed from an economic standpoint, certain insects and mammals in some situations can be termed *forest pests.* Likewise, forests managed for aesthetics or recreation purposes can have such purposes impaired by certain insects and mammals. By their numbers and activities, insects and mammals can reduce yields of commercial wood products, damage a forest's appearance or planned function, and otherwise disrupt management plans and block progress toward planned objectives.

The topic of forest ecosystem health has become a major concern in forest management and generally refers to an apparent loss of forest growth, vigor, and reproduction, and a noticeable increase in the numbers of dead and dying trees (see Chapter 5 on silviculture). The dead trees are often victims of insect attack and disease. But underlying causes are more complicated. Insects and disease frequently finish off trees weakened by drought, or by other factors such as atmospheric pollution and climate change. Nevertheless, insects are often the visible culprit and increasingly must come under consideration in forest management.

INSECTS

A major and successful component of the animal kingdom, insects are thought to have been evolving their present attributes and adaptations for over 390 million years. They constitute 90 percent of the species found in the animal world. More important for the purposes of this chapter, they are our main competitors for food and fiber.

The coverage of insects and disease might better have been combined in this book, since they are commonly referred to together as "forest pests," and mortality figures for them are usually combined. In future editions we will address forest insects and diseases together.

In 1998, a total of 54 million acres of forested land in the United States (about 8 percent) were affected by the seven leading insect and disease pests. For the most part, outbreaks of native insects and diseases have been episodic, with eventual collapse of populations/infestations by natural agents or fires. But approximately 19 of the 70 major pests in the United States are exotic species—the European gypsy moth and Dutch elm disease are prominent examples (USDA Forest Service, 2001).

Although fire is perhaps the most publicized enemy of our forests, insects destroy more trees each year in the United States than any other single factor (Fig. 6-1). They also disfigure trees, stunt growth, and by killing large numbers of trees can create a severe fire hazard.

Insects have developed several highly useful adaptations. They are tiny; thus they can conceal themselves readily. Their size also allows huge populations to be supported on a limited food source. They have the advantage of flight during the reproductive period, and their tremendous reproductive potential allows an efficient exploitation of the food source. It also gives them the ability to "solve problems" genetically, with their huge gene pool. Forest entomologists can almost watch genetic selection in action as populations, after building up and dying back in response to certain controls, come back again, this time resistant to the control agent. There seem inevitably to be a few immune individuals who survive to build up a new population.

Insect Populations

The forest has always been the natural home of thousands of insect species. Fortunately, not all of them are destructive, and those that damage or kill trees are usually held in check by their natural enemies or an unfavorable environment. The population of a

FIGURE 6-1
A stand of loblolly pine in Texas killed by the southern pine beetle, *Dendroctonus frontalis.* Insects destroy more timber in the United States than does any other factor. *(Texas Forest Service photo by Ronald F. Billings)*

destructive insect is said to be in an *endemic* stage if it is present in normal numbers and is doing little or no damage to trees. When an insect population reaches the point where annual losses due to insects in the forest exceed annual growth, the insect has reached *epidemic* proportions. The ability of an insect to multiply in the absence of any control is termed the *reproductive potential.* Take, for example, an insect species that has a very high ratio of females to its total population. If each female lays 100 female eggs, then the progeny of just two females hypothetically could reach 2 trillion individuals in only six generations. The force that works against the insects reaching such proportions is called the *environmental resistance.* Food, competition, parasites, predators, unfavorable temperatures, light, moisture, tree vigor, climate, and weather are all factors of environmental

resistance. If natural controls failed to maintain insects in some sort of balance through these factors, forests would never reach maturity. Should the resistance factors for a particular insect become appreciably lowered, this balance is temporarily upset. Relieved of whatever was holding its numbers down, the population literally explodes and an *outbreak* is said to occur. If the outbreak increases in destructiveness, then the forest manager is faced with an epidemic.

Studies must be made of endemic populations in order to understand and predict the occurrence of outbreaks. Forest entomologists employ qualitative and quantitative survey methods to monitor both endemic and epidemic situations. Insect attack is not a one-time battle ending in victory or defeat, but rather an ongoing problem, because of the cyclic nature of population crescendos. Characteristics of an outbreak, such as its strength, phase, preferred host species, preferred stand attributes, and the history of past incidence, must be identified in order to make decisions on appropriate management. Costs and benefits must be carefully weighed, not only in economic terms, but in relation to the health of the whole ecosystem.

Destructive Forest Insects

Forest managers do not have to know all the insects, but they must be familiar with the general groups and should be able to recognize characteristic damage. The seven categories in Table 6-1 contain most of the species that are potentially destructive to forests.

Adult insects can be distinguished from other small animals by their jointed bodies, which consist of three parts—the head, the thorax, and the abdomen. Adults generally are winged and have a single pair of antennae and three pairs of legs. Forest managers have no trouble recognizing the adult, but unfortunately they most frequently encounter the insect in some other, less easily identified phase of its life cycle. The beetles, butterflies and moths, wasps, and flies pass through four stages of development, and thus undergo *complete metamorphosis.* These four stages are egg, larva (the growing, feeding stage), pupa (resting stage), and adult (reproductive stage). The larvae of some beetles are called *grubs;* those of butterflies and moths are *caterpillars;* those of flies are called *maggots;* and the larvae of the wasps are either grublike or caterpillar-like. Scale insects, aphids, bugs, and termites undergo only

TABLE 6-1

SEVEN CATEGORIES OF DESTRUCTIVE INSECTS

Name	Order	Wing characteristics
Beetles	Coleoptera	Hard first pair of wings
Butterflies/moths	Lepidoptera	Scales on the wings
Wasps	Hymenoptera	Four membranous wings
Flies	Diptera	Only one pair of wings
Scale insects/aphids	Homoptera	Wings held "tentlike" over back
Bugs	Hemiptera	Half of front wings hard
Termites	Isoptera	All wings the same length

FIGURE 6-2
The aphid is one of several sucking insects found in North America. Control of aphids is difficult because of their enormous reproductive potential, which in turn is related to their ability to reproduce asexually. *(Photo by Ed Holsten)*

three stages—the egg, the nymph, and the adult—and have a life cycle that is called *gradual metamorphosis*. Here growth takes place in the nymphal stage, during which the juvenile is similar to but smaller than the adult.

The forest is the home of many potentially destructive insects that differ widely in the kind and amount of damage they inflict and in the period of their life cycle in which they do the damage. Some insects attack trees only when in their larval stage; others do their destructive work as adults. Some insects attack only flowers; others prefer cones, buds, or leaves; and still others damage only the trunk of the tree. Insects may be classified in several ways, but the most practical grouping, at least from the forest manager's standpoint, is based on the manner in which insects do their damage.

Sucking Insects These attack both foliage and stems. They are equipped with sucking mouth parts that pierce tissues and suck fluid from the tree. The sap-sucking insects of importance include certain aphids (Fig. 6-2), scale insects, tree bugs, and cicadas. Control with sprays of stomach poisons is difficult, because their sucking parts extend into the plant tissues, beyond the reach of the poison. Thus, contact or systemic poisons are the most effective. Damage to forests from sucking insects generally is not serious, and controls are seldom needed; the balsam wooly aphid is an exception in that it is deadly to true firs. Moreover, because of the immense biotic potential of this insect, no direct control method is applicable. Control of the less serious

sucking insects may be needed on ornamentals and shade trees.

Defoliators Thousands of forest insect species feed on the needles and leaves of trees. The most injurious ones are the caterpillars of moths and butterflies, some sawfly larvae, and a few beetles. Equipped with chewing mouth parts, they can chew down the center of the needles, skeletonize the leaves by eating the chlorophyll-containing tissue and leaving the veins, or consume the entire leaf. They can defoliate, not only one or two trees, but an entire forest. Defoliators and the next classification, the bark beetles, are the two most damaging insect groups. Deciduous trees can usually stand as many as three successive years of complete defoliation without serious harm; evergreens, on the other hand, will often die after one complete defoliation. The more vigorous trees, of course, have a better chance for recovery.

The most harmful members of the group are listed here. They are capable of defoliating hundreds of thousands of forested acres each year.

1. The spruce budworm and western spruce budworm attack all true firs of North America as well as Douglas-fir, but will defoliate any conifer in their path (Fig. 6-3).
2. The Douglas-fir tussock moth attacks Douglas-fir and all true firs.

FIGURE 6-3
The spruce budworm is a serious insect pest. *Left:* The nearly full-grown larva. Aerial spraying is directed at this stage of the insect's life cycle. *Right:* The newly emerged adult. *(Courtesy of USDA Forest Service)*

3. The pine butterfly attacks ponderosa, western white, and lodgepole pine.
4. The hemlock looper is destructive to hemlock, spruce, and balsam fir in the Northeast. In the West, it attacks western hemlock first, then any coniferous associates.
5. The gypsy moth attacks practically all hardwoods (Fig. 6-4). The larch sawfly attacks eastern larch (tamarack). Both insects were accidentally introduced from Europe.
6. The tent caterpillar defoliates many different species of deciduous trees across North America (Fig. 6-5).
7. The sawflies represent perhaps a dozen species, native and European, that attack many coniferous species throughout North America.

Direct control of these insects today is usually a large-scale operation involving the aerial application of insecticides. Timing is important if the spraying is to be effective. It should come just after the eggs are hatched, when the army of caterpillars is on the move.

Biological control also has been effective in reducing defoliator populations. For example, a small chalcid wasp was introduced to control the larch sawfly. This effective parasite is now distributed through the eastern United States and Canada and is reducing outbreaks of the sawfly wherever they occur.

Bark Beetles Considered to be more destructive than the preceding group, the bark beetles are potentially more damaging because they are not controllable in the mass by aerial spraying. The greater part of their life is spent in the bark of dying or dead trees, depending on the beetle species. There are too many kinds of bark beetles to cover their various life histories, but most of them follow a similar pattern. The emerging adults are in the open only long enough to find their particular host material, either standing trees or downed material such as logs or branches (Fig. 6-6). The first-attacking insects (e.g., females in the genus *Dendroctonus,* males in the genus *Ips*) enter the tree and sever the resin ducts. If the tree cannot repel the attack, the successful "pioneering" insect sends out a powerful chemical message—an attractant (pheromone)—which informs the dispersed population that new breeding grounds have been located. Thousands of males and females within range of the message are guided to the source

FIGURE 6-4
The gypsy moth was accidentally introduced from Europe. *Left:* The male gypsy moth is inconspicuous against a mottled, dark background. *Middle:* The larger female gypsy moth is nearly white and cannot fly. Adult moths do not feed, but the caterpillars can defoliate large areas of hardwood forest. *(Courtesy of Tennessee Valley Authority) Right:* Head-on view of the adult male showing the conspicuous feathered antennae. *(Courtesy of USDA Forest Service)*

FIGURE 6-5
The western tent caterpillar. *Left:* The egg mass on a twig. *Right:* The larvae just emerging from their protective tent. *(Photo by Ed Holsten)*

and congregate on the host material. The tree may try to "pitch out" the intruders by releasing resin, but the mass attack usually overwhelms the host and it succumbs. As the tree dies, the adults mate and the females bore egg galleries into the fresh phloem, or inner bark tissues.

Eggs are deposited along the margins of the egg galleries singly or in groups, or at the ends of the gallery, each species following its typical pattern. Soon the eggs hatch and minute larvae bore larval mines at right angles to the main egg galleries. As hundreds of larvae eat their way into the phloem

FIGURE 6-6
The adult Douglas-fir beetle. *Upper:* Just after emergence, with the thick forewings raised, exposing the hindwings, ready for flight. *Lower:* The adult on a tree trunk ready to bore through the bark to the phloem. *(Upper photo by Ernest Manewal; lower photo by Ed Holsten)*

layer of a single tree, the tree is girdled and killed from within (Fig. 6-7). Generally insects winter within the tree and emerge the following spring as adults to seek a new host and start the process over again. Both larvae and adults burrow into the phloem of trunks and larger branches of trees. The resulting patterns are called "engraving." The trees are killed by the girdling effects of engraving and by introduced blue stain fungi. Some of the most destructive bark beetles are the following.

1. The southern pine beetle, which inflicts periodic damage on most of the southern pines.
2. The turpentine beetle, which attacks all species of pine and spruce in North America.
3. The western pine beetle, an important insect enemy of mature ponderosa pine.
4. The mountain pine beetle, which attacks all species of western pine. It has wiped out thousands of acres of lodgepole and western white pine, and is also very destructive to ponderosa pine in the Black Hills and Rocky Mountain regions.
5. The Douglas-fir beetle, the most important insect enemy of Douglas-fir (Figs. 6-6 and 6-7).
6. The spruce beetle, which causes periodic damage to all North American spruces.

The six are species of the genus *Dendroctonus,* which means "tree killer." Next most important in the bark beetle group are the pine engraver beetles,

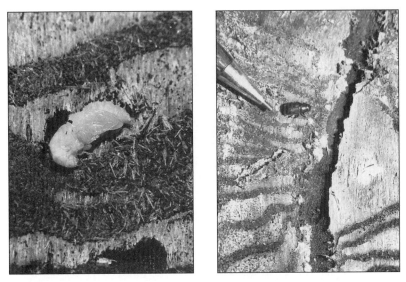

FIGURE 6-7
Insect larvae girdle a tree from within. *Left:* The Douglas-fir beetle pupa just before changing into an adult. (The larval mines in the phloem were made by a wood borer.) *Right:* The vertical egg gallery is shown at center, with larval mines at right angles on either side, all made by larvae of the Douglas-fir beetle. An adult Douglas-fir beetle (about actual size) appears at upper left. *(Photos by Ed Holsten)*

which belong mostly to the genus *Ips,* and the fir engraver (*Scolytus ventralis*) (Fig. 6-8). These and other Coleoptera are largely secondary enemies that attack severely weakened trees, or healthy trees when the insect population is high.

Traditional control of bark beetles includes most of the direct-attack methods, such as felling, burning, sun curing, or ground spraying by hand or power equipment.

Sanitation logging, salvage operations, and thinning regimes can be successful in preventing the buildup of large beetle populations. Blowdowns must be salvaged quickly before epidemic beetle populations occur.

New methods of control include behavioral chemicals, both attractants and repellants, that disrupt mating and host tree selection. These methods, used alone and in combination with silvicultural treatments, are already reducing losses to bark beetles (Goyer et al., 1998). In an over-simplified example, pheromone attractants are used to lure populations away from prime trees, to confuse their host tree selection process before damage occurs or to attract concentrations of insects to areas where they can be treated. Repellants are used to disrupt insect concentrations, to limit population expansion, and to minimize damage to host trees.

Wood Borers   The damage from wood borers usually comes after the trees have been felled but before they are utilized. Some wood borers attack standing but dying trees. A tree killed by insects, fire, or other causes becomes host to ambrosia beetles, which burrow into the sapwood. Rather than eat the wood themselves, they introduce a fungus into the tunnels upon which the adults and larvae feed. Another wood-boring group consists of the larvae of wood wasps, whose adult females insert an ovipositor as much as an inch into the newly felled tree and deposit their eggs (Fig. 6-9). The wasp larvae bore for one or two seasons in the solid, unseasoned wood and do considerable damage. The flat-headed and round-headed wood borers are the larvae of two families of beetles that mine into the inner bark or wood of forest trees. The eggs are laid in chewed slits or in bark crevices. The larvae then burrow into the tree, creating a series of rounded or flattened winding "wormholes," which make the wood worthless for lumber. The best control for such insects is to remove the felled tree from the forest as soon as possible.

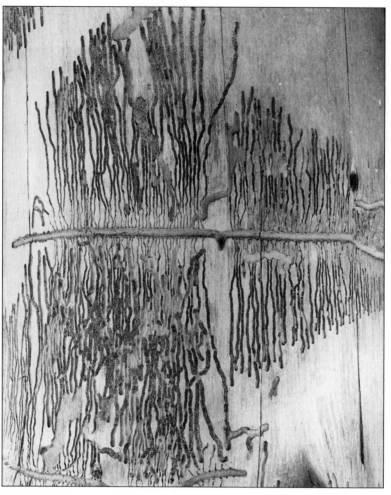

FIGURE 6-8
The egg gallery (horizontal line) and larval mines (vertical lines) after an attack by the fir engraver *(Scolytus ventralis)* in California. Such activity in the phloem girdles and kills the tree. *(Courtesy of USDA Forest Service)*

Once lumber becomes dried, it is fairly safe from attack by this group. If logs are to be used in the round form, for cabins, poles, or piling, for example, they should be peeled and treated with chemicals to kill the residual insects.

Even seasoned sapwood in the form of finished products is not safe from damage by powder-post beetles. Treatment with paint or varnish gives the wood a protective covering. Carpenter ants (in the West) and termites (in the South) tunnel into stumps, logs, snags, and heartwood of living trees, but their main damage, where humans are concerned, is in wooden poles, fences, and the framework of build-

ings. Prevention is effected by chemical pressure treatment of the wood with preservative. In the matter of protection from termites, keeping the wood dry and elevated above the ground is the best method of control.

Terminal Feeders These are serious pests, especially for seedlings and saplings. In nurseries and plantations, damage has been disastrous at times. Damage inflicted on older trees is serious only when repeated. However, a tree deformity such as forking of the top affects the value of the tree for timber. Injury to buds and other tender growing tips is caused by a wide

FIGURE 6-9
The wood wasp *(Urocerus spp.)* on the bark of a white spruce in Alaska. The ovipositor (indicated by the arrow) is extended into the bark and the eggs are deposited. The wasp larvae bore into the wood. *(Photo by Ed Holsten)*

FIGURE 6-10
The tip weevil *(Pissoides strobi)* is a devasting pest of white pine in the East, of Englemann spruce in the Rocky Mountain area, and Sitka spruce in the Pacific Northwest. *(Photo by Ed Holsten)*

variety of insects, such as weevils (Fig. 6-10), twig beetles, twig-boring caterpillars, aphids, midges, and scale insects. One of the most destructive members of the group is the Nantucket pine tip moth, which causes malformation of growing tips in pine nurseries and plantations in the southern United States. Control is accomplished by ensuring that the best pine species for a given site is planted. For example, in east Texas, damage by the moth is reduced when shortleaf pine is planted instead of loblolly pine on shadier, drier, upland sites. Similarly, damage by the tip weevil to Sitka spruce is reduced by planting in areas where environmental conditions favor spruce resistance.

Root Feeders   The group of insects that feeds on roots of small trees or their bark is composed of white grubs, certain weevils, termites, and wireworms. Damage is most serious in forest nurseries. To control these insects, the soil is fumigated by one of many available soil insecticides before planting or seeding. Test plots must be made to study the effect of the poison on the seedlings, since different soils and seedlings react differently to the various poisons. Also, fumigant incorrectly used could render the soil useless for several years.

Gall Makers   An insect gall is an abnormal growth formed on leaves or twigs by mechanical irritation or chemical stimulation from insects. The gall provides protection for the insect larvae while they are feeding on the tree's tissues or sucking its sap. Though these galls are commonplace, most people do not realize they are caused by insects. Examples are (1) the oak-apple gall, of several oaks, caused by a wasp; (2) the aphid gall on the stem of cottonwood leaves; (3) the Cooley spruce gall, a conelike structure formed on the western spruces by an aphid; and (4) gall midges of some pines and junipers of the West, balsam fir of the North, and a dozen or more different hardwoods of the East.

The damage to the forest from the gall makers is negligible except for the balsam wooly aphid, which also fits into this group. In spruce nurseries, plantations, or ornamentals, the presence of the Cooley spruce gall is of some concern, because it does tend to stunt and disfigure the trees. Cutting off and burning the young gall before the adult aphids emerge is one means of control. Sprays may be used in early spring if the infestation becomes serious.

Seed Insects   Cone and seed insects destroy trees even before they can germinate from seed. These pests are most destructive to seed orchards where crops represent years of accumulated care, and often represent significant investment in genetic selection for superior traits. In general, cone and seed insects lay their eggs on the outside of the cone or fruit. The developing larvae then bore through the seed coat and into the seed itself, where they feed on the endosperm. Seed-destroying insects often are controlled by systemic insecticides. These highly toxic materials are applied to the soil at the base of the tree and migrate down into the root system area. They eventually are "taken up" by the tree and translocated into the branches, and tend to concentrate in the seeds themselves. Invading insects are then killed when they bore through the seed coating into the seed.

Methods of Managing Forest Insects

There are several methods of reducing the amount of damage that insects may inflict on the forest. These include management with silvicultural methods to keep the forest healthy and less inviting to attack, encouraging the insects' enemies, interfering with the growth or reproduction of the insects, and applying specific pesticides to reduce pest populations during an outbreak. Certainly any control measures must be timed to avoid damaging natural enemies of the target population.

Silvicultural Controls   These forest management practices and treatments attempt to control insect damage by creating an unfavorable environment for damaging insects and encouraging favorable conditions for their predators. Data gained from surveys are applied to ascertain which insect "key pests" have been most troublesome in the past. What is their favorite host? When was the last outbreak? Why did the last outbreak occur? The answers to these and similar questions guide silvicultural controls. Sanitation and other types of cuttings can remove susceptible trees. These might include overmature, diseased, windthrown, and lightning- or fire-scarred trees. These cutting and thinning practices may serve other management purposes, such as releasing suppressed trees or generating funds to enable protection and improvement of other stands, but they also remove the insects' feeding and breeding grounds. Weakened or damaged trees may have little or no defense against insects. Some trees have defense mechanisms; in the pines, for example, vigorous specimens exude a quality and quantity of oleoresin that deters insect attack.

A forest with a mosaic of age classes and species is more resistant to insect attack, especially against disastrous outbreaks, than are even-aged forests. In managed forests, one way to provide for a healthy vigorous stand is by selecting, not only the correct species for replanting according to soil, aspect, and other factors, but also reproductive stock from the same area. In other words, care must be taken in selecting seeds that have similar provenances—that is, seeds from trees in the same area that have evolved under the same influences. When seed or root stock from out of the area is used, the resulting trees may be weaker, slow growing, and thus more vulnerable to attack by insects and pathogens.

Typical practices that discourage insect problems include piling and burning slash after logging, submerging logs in mill ponds or lakes, debarking the logs to expose the larvae, providing sprinkler systems on top of log decks, and piling and burning logs and slash from infested trees.

Biological Controls   Controls of this type are directed toward inhibiting insects through encouraging their natural enemies or through disrupting their reproductive and growth processes. Fungi, bacteria, and viruses are used, as well as parasitic and predatory organisms. These have proven to be very effective in the control of introduced pests.

Several exotic insects have been accidentally introduced into North America and have become the source of our most serious outbreaks. With no natural enemies to control them, they have spread rapidly—although natural enemies have been imported for control purposes, with some success. European parasitic wasps, for example, have been released to reduce gypsy moth populations. The idea is sound, but there are difficulties in rearing suitable numbers, providing sufficient food in the laboratory, and keeping the wasps in supply. The release of large numbers of sterile males at the peak of the damaging insects' mating periods has also been used in an attempt at population control.

There are many harmful insects that fortunately have not yet gained access to the forests of North America, but the chances of their eventual entry are depressingly good. For example, the proposed importing of logs from various parts of the former Soviet Union poses a significant risk from potential release of exotic insects or disease, which may be benign there, but for which no natural controls exist in North America. The result could be disastrous, as history has demonstrated with other imported pests such as Dutch elm disease, gypsy moth, white pine blister rust, Phytopthera root rot, and chestnut blight.

Birds can be an effective control on forest insects. For example, research has shown that birds can reduce spruce budworm populations up to 72 percent in a single summer (Torgerson & Campbell, 1982) and individual birds can eat as many as 25,000 larvae and pupae in one season (Takekawa & Garton, 1984). Thus, forest management practices that create favorable habitat for birds that eat damaging insects are an important biological control that can reduce the need for chemicals (Langelier & Garton, 1986). One novel application of this principle is in the Nanjing, China, city park, where birds have been trained to eat damaging insects. The trained birds are released in the park, where they are effective predators and also teach their young to prey on the insects.

A natural control on population explosion is intraspecific competition. At some point in the population buildup, certain species become more vulnerable to disease, or the food supply may run short. However, waiting for natural controls to function is not typical of intensive management strategies, and when outbreaks occur in valuable timber stands, some form of chemical control may be considered.

Chemical Controls   This is the most direct form of control, the most expensive, and the most controversial. Insecticides are emergency treatments designed to reduce damage to a stand rather than to eradicate the insect (eradication would probably be impossible).

Research continues for safer, less expensive, and more effective insecticides. Traditionally insecticides are categorized by the way in which the poison enters the insect's body. These include *stomach poisons,* taken in by mouth parts; *contact poisons,* absorbed through the body walls; and *fumigants,* which are taken in through respiration. Most of the modern organic pesticides act as both contact and stomach poisons.

Pesticides also are classified according to their chemical makeup. Five of the major groups are (1) inorganic poisons, (2) chlorinated hydrocarbons, (3) carbamates, (4) organophosphates, and (5) natural plant products.

Toxic chemicals may be applied to the soil whereby they are translocated into the tree (systemics), sprayed from the ground (a method not practical on large areas), or sprayed from above by low-flying aircraft (Fig. 6-11).

FIGURE 6-11
Aerial spraying by helicopter to control the Douglas-fir tussock moth in northeastern Oregon. *(Courtesy of USDA Forest Service)*

Another means of control involving chemical processes is the use of synthetic pheromones to lure populations away from highly valued stands of timber, to attract the males into traps, or to confuse males by spraying the appropriate sex hormone throughout the forest (Wood, 1982).

The use of chemicals, either as insecticides or herbicides, is important to intensive forestry, but chemicals are perceived by the public as threatening to water quality, fish and wildlife, and human health. In response to public concerns, more research, evaluation, and monitoring programs are needed to investigate the use of chemicals and develop information on their side effects.

Integrated Pest Management (IPM)

Attitudes about dealing with destructive insects in the forest have gone through a metamorphosis similar to that concerning fire. From the practice of identifying all potentially destructive insects (or mammals, for that matter) as "pests" and expressing the need to prevent an outbreak, or failing that, to counterattack immediately in force, the emphasis has changed to one of recognizing these organisms as part of the natural scene. In a forest dedicated to the primacy of natural processes, such as a national park or a designated wilderness, few, if any, insects would be considered pests. They are simply part of the forest ecosystem. Even in a wood-growing commercial forest, a greenbelt or windbreak forest, a pest only can be identified as a pest on the basis of its interference with established management goals.

Attempts are made to keep populations in balance. The system as a whole is now the focus of concern, and any action taken for one purpose must be considered in light of all effects on the entire ecosystem. A management plan must be developed that takes into consideration the numerous variables in each forest, such as policy considerations, public relations, history of past outbreaks and treatments, condition of the stand, and cost-benefit analysis, as well as possible effects on other organisms and the ecosystem as a whole. The various disciplines concerned with forest management in all its aspects must operate in concert.

This concept has been summed up as *integrated pest management* (IPM). It is a systematic approach to managing pest populations to keep levels below those causing economic injury, using all suitable techniques and methods in a compatible manner. For example, an integrated pest management strategy might combine silvicultural prescriptions to maintain forest biodiversity, good slash disposal, limited spraying of severe problem areas only, and salvage logging to remove dead and dying trees promptly. Integrated pest management is becoming more important as an integral part of ecosystem management, the more holistic approach to forest stewardship now being adopted.

MAMMALS

The forest is the natural home of many animals, and we have learned to respect this fact. It has also served for centuries as a pasture for domestic stock. Most animal species are of neutral or unrecognized benefit to humans, but some are viewed positively, such as game and fur-bearing animals. Others can be considered nuisances if their food is something that humans also use. The food demands of these animals may become so great at times that real damage to forests may occur. Heavy use by either livestock or large game animals can impact attempts at forest regeneration. Animal damage, whether by large mammals such as deer, elk, or bear, or by small mammals such as rabbits, mice, and gophers, can be a major problem that must be addressed. As with other forest pests, an integrated approach to forest management that combines silvicultural prescriptions with other methods targeting the pest is the favored approach (Black, 1992).

Damage by Larger Mammals

Deer, Elk, and Moose Several members of the deer family are known to cause some damage to the forest if present in numbers that exceed the carrying capacity of their habitat. Moose of Isle Royal National Park in Lake Superior have formed a "browse line" about 11 ft (3.3 m) from the ground. Elk have seriously injured reproduction in some western states by overbrowsing. Most of the damage from members of this family, however, comes from deer (Fig. 6-12). Excessive browsing actually has eliminated all palatable reproduction in some forest areas; in others that have been repeatedly overbrowsed, no reproduction under 30 years of age will be found. The entire acorn crop of some oak forests has been consumed by deer where these animals are too abundant. Deer also do further harm, such as deforming growth by nipping off the leaders of young trees, or damaging tree

FIGURE 6-12
A white-tail deer contributing to the formation of the browse line in a West Virginia forest. *(Courtesy of USDA Forest Service)*

FIGURE 6-13
Bear damage to young Douglas-fir on a tree farm in Washington State. *(Courtesy of Weyerhaeuser Company)*

trunks by debarking them when rubbing the velvet from their antlers.

Deer damage from browsing is greatest in the northern states and the Rocky Mountains, where deep winter snows confine the herds to small areas of winter range. Their damage is locally severe in West Coast Douglas-fir plantations, especially on sites where it is warm and dry in the summer (causing a moisture stress) and where deer congregate in winter. Controlling deer populations through hunting and intensive wildlife management is the best safeguard against deer injury. Intensive management may involve use of repellents, but this is usually not effective where deer are overly abundant.

Other measures include planting food crops to bait animals away from seedlings, temporarily fencing plantations, protecting terminal leaders with plastic tubes or paper caps, logging in the winter to provide tops for food, and reducing herds by special hunts. Any means of speeding up the rate of regeneration until the new trees reach a height of about 5 ft (1.5 m) is useful in controlling damage by deer. Much will depend on the relative importance of the recreation and hunting values of the deer, compared with the value of the timber crops. Hunting pressure can reduce deer populations below damaging thresholds.

In northern Wisconsin there is concern that white-tailed deer may be greatly reducing populations of native forest plants. White-tailed deer, which typically use forest edges, have gained access to forest interiors as forests have become increasingly fragmented. Preservation of plant species diversity in this instance may involve limiting the deer population by hunting, and planning for larger forest blocks (Alversen et al., 1988).

Bear   A surprising amount of second-growth timber is damaged or killed by black bears in West Coast forests. During the spring and early summer, bears expose the juicy cambium near the base of certain trees by scratching and biting off the tree's outer bark (Fig. 6-13). Analysis of bears' stomachs has proved conclusively that the cambium layer is a major food item. Studies on the Olympic Peninsula of Washington State, where damage is severe, indicate that bears have killed as many as 16 trees per acre (40 trees per ha) by girdling them. Douglas-fir is the bears' favorite tree in coastal forests, although they extend their sap-licking damage to western hemlock, Sitka spruce, and western redcedar. There has been a long history of bear damage to trees in the region extending from northern California to British Columbia, but the damage was never extensive until after World War II. The cutover lands supported a greater food supply, which resulted in a larger population of game animals, including black bear. Biologists speculate that, as the second-growth forests grew older, ground plants and berries became scarce and the increased bear population had to turn to eating the inner bark of trees to survive.

Trees in the 15- to 30-year-old range with an average diameter of 8 to 10 in (20 to 25 cm) have been hit the hardest. There is weak evidence that shows this behavior is learned, though most biologists still feel it is an instinctive trait. Damage became so extensive on the Olympic Peninsula that, for a few years, bears were declared predators. One method of controlling bear damage is to reduce the number of bears through special-license hunts or permitting extended sport-hunting seasons in areas with extensive damage. Relocation of problem animals has also been tried. A regular program of feeding has also been part of the management program for Weyerhaeuser and other owners in problem areas of coastal Douglas-fir.

Black bear can also be a problem in young to intermediate stands of western larch, lodgepole pine, and Engelmann spruce in the northern Rocky Mountains (Adams, 1991). Thinned or otherwise low-density stands with vigorous growth sustain the greatest damage. Since these are often trees selected for the final crop, the economic damage is compounded. Potential losses from bear damage must be considered in management plans for affected areas.

Rabbits, Porcupine, and Beaver Rabbit and hare damage to tree reproduction is a serious threat in New England, the Lake states, and the Pacific Northwest. Nipping and girdling the young shoots results in a deformed, bushy tree, if it survives at all. The control of rabbits is seldom practical because of the difficulty and expense involved. Fencing of small areas, such as forest nurseries, has been tried. Repellents have been used with some success.

The porcupine is another troublesome rodent that kills and damages trees. Damage is not confined to any one area, but in the Pacific Northwest porcupines do extensive damage each year. In the West, their favorite food is the bark of ponderosa and lodgepole pine, whereas in the Lake states and New England it is northern hardwoods. With no need to hibernate, they continue to girdle trees even during the winter months. If the porcupine does not kill the tree, it leaves much of the wood exposed to further attack by insects and disease. Instead of choosing to damage the poorest trees in a stand, porcupines pick the healthiest and most vigorous. As early as 1908, a bounty was offered for porcupines in New York State. The animal is becoming an increasingly greater menace in the Pacific Northwest (Fig. 6-14). Thousands are trapped each year in eastern Oregon trouble spots, and hunters are asked to shoot porcupines on sight. One reason for the porcupine's steady increase in numbers is that its old enemy, wildfire, has been reduced, as have its natural predators. One of these, the fisher, has been reintroduced to control the porcupine. In the Northeast, the recovery of the fisher has been matched by the decline of the porcupine.

By felling trees for food, lodge, and dam construction, beavers do some damage, but this is limited to small areas. Trapping is the best means of control. Since the animal is protected in most states, this can be done only by state wildlife department officials or trappers with special beaver licenses, if the decision is made to control its numbers.

Damage by Smaller Mammals

The tree squirrels, chipmunks, and white-footed mice do widespread damage to seed crops. But although seed crop loss from squirrels may be serious, control measures are seldom undertaken. Mice present a greater problem. Tree seed is one of the preferred foods of white-footed mice, and during years of sparse seed they can destroy an entire seed crop. Artificial seeding has often been a waste of time because mice eat most of the seed before it germinates. Seed treated with poison has been used to kill mice, but this control is controversial because of its impact on predators. The repellents can also train mice to change their eating habits. Finding the treated seed distasteful, the mice look for something else to eat.

Pocket gophers can be a major problem, especially in new plantations being established in open

FIGURE 6-14
An adult porcupine feeding on the phloem (inner bark) of a young ponderosa pine in a plantation in eastern Oregon. Girdling by porcupines kills thousands of trees each year. *(Courtesy of Weyerhaeuser Company)*

fields and in clearcut harvest areas. While they can do extensive damage to seedling plantations, alfalfa, and other crops, pocket gophers are also beneficial by loosening and mixing soil and thus making more air, water, and minerals available to the roots of plants. It is estimated that, during one year, an adult pocket gopher moves more than 2 tons of soil to the surface (Wentz, n.d.). Gopher control is accomplished by placing poison bait or traps in tunnels located with a probe, near the mounds of dirt that indicate the presence of gopher activity. Hand probes that allow poison bait to be released in gopher tunnels have been developed. In areas of severe problems, a "burrow builder" towed behind a tractor builds artificial tunnels and drops poison bait that is eaten by gophers as they explore the new tunnels.

Grazing Damage by Domestic Animals

Sheep grazing on the national forests of the Southwest destroy ponderosa pine reproduction. This is especially true in some areas during the early summer dry period, when sheep satisfy their thirst by browsing on the succulent shoots of the young pine.

Suppression of regeneration by competition from fast-growing grasses and brush in clearcut and replanted areas presents a problem in western coastal and inland forests. Carefully controlled sheep grazing, and even light grazing by cattle, may be compatible with seedling survival and reduce competing vegetation, thus eliminating the need for herbicides in some areas. Having just the right amount of grazing is key. Research continues on this potentially compatible and beneficial forest grazing.

Aspen reproduction is a favorite food of all livestock as well as of deer and elk on many western grazing lands. By careful regulation of the number and timing of forest grazing of domestic animals, reproduction is spared and damage can be controlled within acceptable limits.

In the small individual forests and farm woodlands of the South, cattle and hogs at one time were turned loose to roam unfenced forestlands, destroying reproduction and trampling the soil. Riding down saplings to obtain the tender foliage was a common form of injury by cattle. Hogs, in particular, did extensive damage by grazing on tree roots. The trend in the South today is to raise either livestock or trees, but not both on the same piece of land. If woodland owners want forests to reproduce, cattle and hogs are excluded.

Grazing injury is a distinct threat to forests and must be considered in management plans. However, forest grazing is a kind of multiple use attractive to livestock owners who also own forestland or who hold grazing leases on adjacent public lands. Research is showing that, with careful management, at controlled levels and under certain conditions, forest grazing is compatible and profitable. It can also reduce the amount of flash fuels provided by dry grasses, thereby helping reduce fire risk. Trampling has been found to be a bigger problem than browsing by cattle in a study in ponderosa pine, and to be less of a problem than other mortality factors (Kingery & Graham, 1991).

Beneficial Effects of Animals

Humans would be retreating, no doubt, before insect armies if it were not for the millions of birds that make their headquarters in the forest. Woodpeckers, in particular, are extremely important, and have been known to kill practically the entire brood of spruce beetles in some areas. Small mammals render a similar service. Skunks, opossums, raccoons, mice, chipmunks, squirrels, shrews, and voles eat gypsy moths and other forest insects, particularly in the insect's caterpillar phase. Mammals tend to be opportunistic in their feeding habits, however. Whether or not an insect will be consumed, and in what quantity, depends on the abundance and availability of other foods. Thus, mammals can hold a sparse population in a harmless range indefinitely. In an outbreak they do not eat enough to make a difference.

Plant succession after fire or other catastrophe is known to be closely connected with the habits of birds and rodents, and the latter are credited by some researchers with considerable service in cultivating and fertilizing poor sites. As most of the damage to the forest comes from local overpopulation, and since the interrelationships of plants and animals in the forest are complex and obscure, it is unwise to assume that widespread destruction of any animal is necessary. As in integrated pest management, decisions on control of destructive animals must be part of an overall management plan that reflects the objectives of the forest or park in question. Silvicultural approaches can be effective in maintaining losses within acceptable limits.

SUMMARY

Insects and mammals are a natural part of all forest ecosystems. As in all ecological situations, no insect or mammal interaction with a forested area presents a clear case of either beneficial or damaging results. Insects and mammals have their cycles of population growth and decline, and their interactions with a forest ecosystem during a particular part of these cycles will have mixed results. However, when forestry is being practiced in a forest ecosystem, and the production of wood and fiber, or aesthetic qualities, is the management goal, human plans are often frustrated or partially defeated by damage from insects or large and small mammals.

Contemporary forest management addresses such threats and attempts to control populations, or mitigate the depredations of the most damaging insects and mammals. Such measures must be socially, economically, and ecologically acceptable, especially under today's more holistic "forest ecosystem management" approach.

LITERATURE CITED

Adams, Dave. 1991. Personal correspondence. February 15, 1991.

Alversen, William S., Donald M. Walker, and Stephen L. Solheim. 1988. "Forests Too Deer: Edge Effects in Northern Wisconsin," *Conservation Biology* **2:**348–358.

Black, Hugh C. (tech. ed.). 1992. *Silvicultural Approaches to Animal Damage Management in Pacific Northwest Forests.* U.S. Department of Agriculture Forest Service, General Technical Report PNW-GTR-287.

Goyer, Richard A., Michael R. Wagner, and Timothy D. Showalter. 1998. Current and Proposed Technologies for Bark Beetle Management. *J. For.* **96**(12):29–33.

Kingery, J. L., and R. T. Graham. 1991, April. The Effect of Cattle Grazing on Ponderosa Pine Regeneration. *Forestry Chronicle.*

Langelier, Lisa A., and Edward O. Garton. 1986. *Management Guidelines for Increasing Populations of Birds That Feed on Western Spruce Budworm.* U.S. Department of Agriculture Ag. Handbook No. 653.

Takekawa, John Y., and Edward O. Garton. 1984. "How Much is an Evening Grosbeak Worth?" *J. Forestry* **827:**426–428.

Torgerson, Torolf R., and Robert W. Campbell. 1982. Some Effects of Avian Predators on the Western Spruce Budworm in North Central Washington. *Environmental Entomology* **11**(2):429–431.

U.S. Department of Agriculture Forest Service. 2001. 2000 RPA Assessment of Forests and Rangelands. Summary Report, Washington, D.C. Available online at http:www.fs.fed.us.pl/rpa/publications.

Wentz, W. Alan. n.d. *Pocket Gophers.* Cooperative Extension Service, South Dakota State University, U.S. Dept. of Agriculture FS 725.

Wood, D. L. 1982. The Role of Pheromones, Kairomones, and Allomones in the Host Selection and Colonization Behavior of Bark Beetles. *Ann. Rev. Entomol.* **27**:11–446.

ADDITIONAL READINGS

Edmonds, Robert L., James K. Agee, and Robert I. Gara. 2000. *Forest Health and Protection.* McGraw-Hill, New York.

Fredericksen, Todd S., Brad D. Ross, Wayne Hoffman, Eric Ross, Michael L. Morrison, Jan Beyea, Michael B. Lester, and Bradley N. Johnson. 2000. "The Impact of Logging on Wildlife. A Study in Northeastern Pennsylvania." *J. Forestry* **98**(4):4–10.

Gara, R. I., W. R. Littke, J. K. Agee, D. R. Geiszier, J. D. Stuart, and C. H. Driver. 1984. "Influence of Fires, Fungi, and Mountain Pine Beetles on Development of a Lodgepole Pine Forest in South Central Oregon." In Ed Baumgartner et al. *Lodgepole Pine: The Species and Its Management.* Coop Extension, Washington State University, Pullman.

Journal of Forestry. 1997, August. Theme issue on Wildlife in the Forest, 6 articles, pp. 16–41.

Mattson, William J., Jean Levieux, and C. Bernard-Dagan (eds.). 1988. *Mechanisms of Woody Plant Defenses Against Insects.* Springer-Verlag, New York.

Ross, D. W., and G. E. Daterman. 1997. Using Pheromone-bated Traps to Control the Amount and Distribution of Tree Mortality During Outbreaks of the Douglas-fir Beetle. *Forest Science* **43**:65–70.

Takekawa, John Y., Edward O. Garton, and Lisa A. Langelier. 1982. Biological Control of Forest Insect Outbreaks: The Use of Avian Predators. Wildlife Management Institute, Washington, D.C.

U.S. Department of Agriculture Forest Service. n.d. National Forest Health Monitoring Program.

STUDY QUESTIONS

1. How can insects damage forests? Why are they so difficult for forest mangers to keep in balance?
2. With respect to insects, define what is meant by:
   a. endemic stage
   b. epidemic stage
   c. reproductive potential
   d. environmental resistance
   e. outbreak
3. Tell how these eight categories of insects inflict damage on trees.
   a. sucking insects
   b. defoliators
   c. bark beetles
   d. wood borers
   e. terminal feeders
   f. root feeders
   g. gall makers
   h. seed insects
4. How do the following chemical controls work?
   a. stomach poisons
   b. contact poisons
   c. fumigants
   d. synthetic pheromones
5. Define integrated pest management (IPM).
6. How can deer damage forests? When is this likely to occur?
7. How do bear damage forests?
8. How can carefully managed grazing be beneficial to a forest?
9. Discuss two different methods of biological control. Can you give any examples?

CHAPTER 7

Disease and the Elements

Chapter Outline

INTRODUCTION

Chapter 6 pointed out that insects destroy more standing timber in a year in the United States than does any other factor. Disease, on the other hand, ranks number one in *slowing down the growth* of forests. Therefore, especially in young age classes, disease is more of a factor in growth loss than in tree mortality. This, coupled with damage by disease to finished wood products, makes it a serious problem for both forests and forest products. When forests were perceived as an unlimited resource, losses from disease were tolerated or else went unnoticed. Today, when the need for conservation of resources is obvious and forest health is a major concern, we are less willing to accept the undesirable effects of tree diseases. Something is known about most tree diseases, and although some cannot be controlled and new ones keep appearing, research continues and efforts are made to control and prevent many of them.

TYPES OF TREE DISEASES

Diseases can be classified as (1) weakening diseases, the kinds that cause growth loss, or (2) killing diseases. The latter types are mostly diseases introduced from other countries and are largely free of natural controls. They find our domestic trees, which have little natural resistance to them, an easy prey. The best example of a killer disease is the chestnut blight induced by *Cryphonectria parasitica,* brought accidentally into the United States in 1903 on nursery stock from Asia. During its first 20 years, the blight spread throughout the natural range of American chestnut (eastern United States), destroying millions of board feet of valuable timber. The blight effectively eliminated the American chestnut, which once was the dominant tree of many eastern forests. Though over a century has passed since its introduction in the United States, no effective means of control has yet been discovered. Many deep taproots of chestnut still live, sending up sprouts every year, but these too become infected and die after a few years. Perhaps the chestnut would come back if an effective treatment for the disease could be found. Blight-resistant Asiatic chestnut is now grown in this country, but it does not attain the size characteristic of American chestnut.

Another example of an importation, the Dutch elm disease, may eventually bring to native elms a fate similar to that which befell the American chestnut. Introduced in 1930, Dutch elm disease has killed millions of American elms, effectively eliminating them in some towns and woodlots. A third introduced disease, causing widespread destruction to the five-needled pines in North America, is the white pine blister rust.

Today jet airliners land in dozens of North American ports of entry from many parts of the world. This accelerated means of transportation makes it possible for fungal spores (and insects, for that matter) to arrive in a healthy condition. Each week prohibited plant materials are discovered in one port of entry or another, and frequently they contain a disease or insect not previously recorded in the United States or Canada. The gravity of the situation cannot be overestimated, and the enforcement and improvement of plant quarantine laws is essential. This type of legislation was first initiated in this country in 1912, with the Federal Plant Quarantine Act.

The Forest Pest Control Act of 1947, which was updated by the Cooperative Forestry Assistance Act of 1978, provided for better control of pests once their presence was known. By this act, the federal government accepted the responsibility of leadership in controlling all pests on federal lands and for making emergency funds available to states for detection and control of diseases and insects on state and private land.

The causes of plant diseases were poorly understood before the 1850s. Prior to that time, any loss not attributed to fire or insects was blamed on poor soil, bad weather, or even interference from supernatural forces. Today disease in plants is generally understood to mean any disturbance or interruption in the process of nutrition, or other growth processes, resulting in partial or complete stoppage of development, or causing death.

Occasionally, a disease appears with no known cause of mortality. For example, a disease called sudden oak death surfaced in 1995 in central coastal California, where tens of thousands of tanoak, California live oak, and California black oak trees have since died. Spreading at an alarming rate, it leaves forests littered with dead wood, resulting in extreme fire danger. Now, several years later, sudden oak death has been traced to phytopthora disease, but is still a major threat to these oak species in wildland and urban areas.

Another way of looking at tree diseases examines how they function physiologically. Manion (1991), who defines *tree disease* as any deviation in the normal functioning of a plant caused by some type of persistent agent, suggests three types: abiotic, biotic, and decline diseases. *Abiotic diseases* are caused by environmental factors or other conditions that reduce the vigor of a tree; *biotic diseases* are caused by certain insects, plant parasites, viruses, bacteria, and fungi; *decline diseases* arise from a combination of abiotic and biotic factors.

The study of tree diseases is called *forest pathology,* and it is *forest pathologists* who study them. In contrast, *forest entomologists* are concerned with forest insects. As we will see, often both insects and disease are involved in the infection of a tree, and the two forestry professionals must work together to determine the nature of the problem.

Abiotic Diseases

This term identifies growth problems induced by poor soil conditions such as mineral deficiency, drought, or excessively high or low temperatures. Other environmental factors involve the stress induced when a tree is growing at the limits of its normal range, as well as stress related to air pollution, pesticides, and mechanical damage. Though these agents are usually long range and beyond the control of the pathologist, he or she must be able to recognize them, since they often produce symptoms similar to diseases caused by fungi, parasitic plants, and other biotic agents.

Biotic Diseases

Biotic disease ranks as most important of the three categories. Within this grouping fall the rust diseases, canker diseases, wilt diseases, foliage diseases, root rot diseases, diseases causing wood decay, and the stains and molds that cause discoloration in wood. These biotic diseases are usually host-specific, and can be identified by the presence of specific fruiting structures.

Because many of these serious forest biotic diseases are caused by living organisms called *fungi,* located in the kingdom *myceteae,* a closer look at fungi follows.

The Plus Aspects of Fungi   Not all fungi act destructively. In fact, many species are essential, playing positive roles through fermentation such as in the making of bread, wines, beer, and the production of certain cheeses. Fungi, through their digestive enzymes, assist in the breakdown of tons of cellulose and lignin, converting this, along with the undecomposed piles of forest litter, into essential humus. Fungi create wildlife habitat for cavity nesters by stem decay organisms. They also serve as a source of food for numerous animals, and, in the form of edible mushrooms, as a gourmet delicacy.

Other fungi form *mycorrhiza,* or fungus–root associations, and serve most higher plants by assisting the plants to better absorb soil nutrients and water. This results from the hyphae of certain fungi invading the cells of the root cells, or in other instances, surrounding the roots, permitting the plant greater absorption capacity and facilitating growth.

Finally, there is the very important mold fungi *Penicillium,* used in the production of penicillin. This antibiotic agent destroys many of the bacteria that infect humans.

Lichens are dual organisms, resulting from a symbiotic relationship between a fungus and an alga. The fungus provides protection for the weaker alga, as well as moisture and some nutrients that the alga needs; the alga provides food for the fungus. Lichens cannot survive in areas where air pollution is strong, therefore they serve as indicators of pollution intensity (Alexopoulos & Mims, 1976).

Fungi as Agents of Disease   Fungi also have their negative aspects, particularly in forestry. For example, fusiform rust is a fungal disease that affects over 13 million acres of slash and loblolly pine in the South (USDA Forest Service, 2001). Every stand of timber has some kind of destructive fungus at work, although its presence may not always be detected (Fig. 7-1). As already noted, fungi frequently injure or destroy trees at various stages of growth, and certain ones attack logs or manufactured forest products. Fungi may reach epidemic proportions under certain favorable environmental conditions and then become a serious threat to forests and good forest management. For example, forest health problems in North America are characterized by abundant dead and dying trees, reflecting the results of attack by insects and disease on trees weakened by several years of drought. The continuous prevention of natural wildfires has also added to the stock of vulnerable trees.

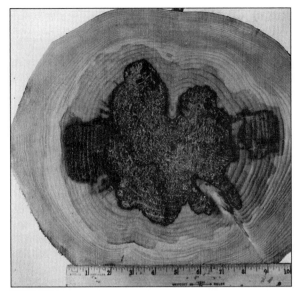

FIGURE 7-1
A section of a 50-year-old western hemlock, showing heart rot. No external symptoms were observable on the tree before cutting. The large dark area shows *advanced decay* (the fungus has decomposed both the lignin and cellulose). The lighter area surrounding the darker center is *incipient decay,* in which there is presently only a color change. Both were spreading outward to the point where the entire heartwood would have been consumed, leaving the tree standing but hollow. *(Photo by Terry Jordon)*

Starting with storage and germination of seed, there is a fungus threatening every stage of tree growth, from the *damping-off* diseases of tiny seedlings to destructive interior decays in the giants of the old-growth forests of the Northwest. Specialized diseases attack roots, stems, or foliage. Stains, decays, and molds attack as logs are taken from the forest and manufactured. Similar diseases show up in finished products such as pallets and boxes, lumber, posts, railroad ties, or buildings.

All fungi lack chlorophyll. They are composed of many threadlike structures called *hyphae,* which collectively are called *mycelium.* Not being able to manufacture their own food, as do plants with chlorophyll, the fungi must utilize other sources. Fungi grow *parasitically* on living material or *saprophytically* on dead material. Some adaptable fungi may live for a time as a parasite and later, if the host dies, as a saprophyte. Other fungi, such as the rusts, are *obligate parasites,* which means they grow only on a living host.

FIGURE 7-2
Fourteen or more conks *(Phellinus pini)* show on this western hemlock standing in the Chugach National Forest, Alaska. *(Photo by Dow V. Baxter)*

Without getting involved in the various kinds of reproduction and stages of development characteristic of fungi, let us look at the *spore,* the minute reproductive body common to many lower plants and central to our concerns here. The spore may be windborne, carried by insects or other animals, or transported in water to a suitable environment, where it germinates. The hyphae invade host tissue, growing between or into the cells, secreting enzymes that help digest the wood. After feeding for a sufficient time, the mycelium near the surface develops a fruiting body characteristic of the species, which emits millions of minute spores, in turn infecting other suitable hosts when the required conditions of transportation, moisture, temperature, and host tissue are met. A single large shelf fungus, or *conk,* is capable of releasing more than a million spores an hour for several months (Fig. 7-2). These billions of spores may infect trees in the vicinity, or travel long distances.

FIGURE 7-3
The large wound at the base of this tree now serves as a point where decay-fungus infection can enter. *(Photo by Lynn Mills)*

The fruiting bodies, and the destruction caused by the mycelia, serve as a means by which foresters can recognize the disease. The symptoms may be in the form of wilting, leaf spots, leaf curl, needle cast or discoloration, blight, scabs, blisters, lesions, cankers, discolored and cracked bark, resin discharges, galls, burls, conks, and drooping, dead, or broken tops. A certain species of conk on the outside of a tree may mean the disease extends 16 ft (5 m) above and below the conk; the presence of other types may indicate the entire heartwood is decayed.

Fungi gain entrance to the heartwood of living trees through fire scars, dead branch stubs, broken tops, ax blazes, insect holes, wounds from logging operations or from the toppling of an adjacent tree by wind, lightning strikes, storm breakage, and anything else that creates an entry point through the tree's protective tissues (Fig. 7-3). Still other fungi infect the leaves, buds, roots, cambium, flowers, and fruit of trees.

Typical Biotic Diseases of North American Forests

With perhaps as many as 50,000 individual diseases capable of affecting trees in one manner or another (Sinclair et al., 1987), it seems practical here to discuss only the most damaging ones.

Heart Rots   This disease occurs in the heartwood of a tree (the nonliving section of the trunk). One fungus is so common to many conifers that it may be known to persons other than forest pathologists by its scientific name (*Phellinus pini*). Some common names by which this disease's effects are known include red heart, red rot, ring scale, red ring rot, and white pocket rot. Its effects not only are widespread in North America, but have plagued European forests for many years and are common in other parts of the world. This rot offers an excellent example of a disease that reduces the interior of conifers to an almost pulpy mass. The tree may then blow over when exposed to strong winds. Loss from heart rot is particularly high in the old-growth forests of the Northwest, but is less prevalent in trees grown on rotation less than 80 years.

Heart rots cause more commercial wood loss than any other group of diseases in the Northwest. Of the numerous heart rot fungi found there, *Phellinus pini* (mostly in Douglas-fir) and *Echinodontium tinctorium* (mostly in western hemlock and true firs) are the most important. *Oligoporus sericeomollis* (formerly *Poria asiatica*) is the cause of hollowed-out trunks in living western redcedar.

Heart rot fungi may enter the tree at any wound, frequently where a branch has been broken off. Early accounts of heart rot in Germany indicate that it was most prevalent in forests near villages, where pilfering of branches was common, and in woods exposed to high winds where there were many broken branches. The cross-sections of logs, if taken at frequent intervals, indicate that the irregular cylinders or cones of decayed wood may extend for short distances in either direction from a wound entrance, up into the butt log from root infection and wounds near the ground, or for the complete length of the trunk. In the latter instance, several separate infected sections may join or overlap. The cross-sectional appearance, depending upon age and degree of the infection, is one of incomplete and irregular circles of reddish brown to very dark brown, sometimes merging into complete circular areas (Fig. 7-1). The conks, also called *sporophores,* are evident on the outside of the trunk in advanced stages of the disease. Obviously, the removal of the conk is not a method of control within the tree.

Forest management, including short rotation and intermediate salvage cutting, offers a practical con-

FIGURE 7-4
Heart rot on the butt end of a western hemlock. Though the rot extends throughout most of the tree, some solid wood was salvaged at the saw mill and some became pulp chips.
(Photo courtesy of Weyerhaeuser Company)

trol for this disease. Care to prevent damage to tree trunks from wildfire, prescribed burning, logging, or road building is worth the effort, as trunk wounds afford entrance for various diseases. Salvage of individual trees thus injured is advised.

Heart rot, caused by several species of fungi, is credited with being the greatest cause of growth loss in the older forests of the United States. Were it not for rot, a large number of conifers and hardwoods could be left to grow for many more years (Fig. 7-4).

Root Diseases Root diseases attack trees of all ages but are most important in young trees. They appear on conifers, and to a lesser extent on hardwoods, throughout North America. However, the most serious infestations are in the Pacific Northwest (Hadfield et al., 1986). *Phellinus weirii* is a real threat to the management of second-growth Douglas-fir. In a similar manner, *Heterobasidion annosum* threatens management of plantations of southern pine, and of western hemlock in the Pacific Northwest, and of true firs and ponderosa pine elsewhere in the West. On a worldwide scale, the *Armillaria* root diseases are extremely important.

Phellinus weirii enters the tree through wounded roots, and roots close to infected roots, causing a decay in the heartwood of the butt. An infected tree shows reduced height growth, loss of foliage, and increased cone crops. Some trees die standing up. The tree's anchorage is weakened when the roots decay, leaving it at the mercy of the winds. Old-growth interior western redcedar may be seriously affected at the base, often rendering the tree commercially useless for 15 ft (5 m) or more up the trunk. Removal of infected stumps and root systems is commonly used to control the disease.

An important root disease on pines in the Southeast is *littleleaf.* Shortleaf pine is hit the hardest and loblolly pine to a lesser extent. Littleleaf disease has been the subject of considerable study since it was discovered in Alabama during the mid-thirties. Researchers found the fungal pathogen *Phytophthora cinnamomi* already present in most Southeastern pine soils, but its ravages differed with the drainage capability of the soil. A root system in a soil with poor aeration develops poorly and becomes a favorable host for root fungi. As the fungus attacks the newer roots, it interferes with their growth and they lose their ability to absorb. The result is a yellowing of the foliage and, finally, death of the tree. The water content of the soil can be changed through drainage adjustments, but only over long periods of time. Planting species other than shortleaf or loblolly pine seems to be the best way to avoid the disease.

White Pine Blister Rust This destructive fungus (*Cronartium ribicola*) has probably been in the country since the late 1890s. The organism was first found on cultivated black currants at Geneva, New York, but it was not until 1909 that it was found on white pine seedlings imported from Germany. Unfortunately, shipments of these trees had been sent to many applicants for planting stock throughout the northeastern and Lake states and to eastern Canada; by 1915, in spite of control measures taken according to the best knowledge of the time, white pine blister rust was well-established in the eastern pine belt. Today it is found throughout the range of eastern white pine, from Newfoundland to northern Georgia, and west to Manitoba and Minnesota. In western North America the blister rust got started in 1910 on nursery stock sent to Vancouver, British Columbia, from France. All of the West Coast's five-needle pines are susceptible. The disease extends from British Columbia well into California and east to

FIGURE 7-5
The white pine blister rust. *Left:* The bark canker of an infected western white pine in Idaho. The lightly discolored area produces spores that are carried by the wind to the alternate host, any member of the genus *Ribes*. *Right:* Infected western white pine, showing the characteristic spike top. *(Courtesy of the University of Idaho)*

Montana. Death comes relatively soon to young pines; older trees are also killed after a number of years of severe attack, and may be deformed, spike-topped, and broken off at the top in a relatively short time (Fig. 7-5).

White pine blister rust control presently consists of managing white pine on those sites whose climatic conditions limit the ability of basidiospores to germinate and infect young pine needles, and to use seedlings specially bred from seeds of trees exhibiting a high level of natural resistance. Planting stock produced from such seeds is now restoring the five-needle pines to their former commercial importance. Interestingly, the goal of such breeding programs is not to produce totally resistant populations of blister rust-resistant seedlings, because then the rust would likely mutate. With a target of two-thirds of the specially bred seedlings rust-resistant, the results are satisfactory for the rust as well as producing rust-resistant trees for commercial use.

Dutch Elm Disease   This disease, induced by the fungus *Ophiostoma ulmi,* causes serious concern,

since elms are important native shade and ornamental trees. Spread from one city to another through the forest, this disease was first observed in the United States in 1930 on elms in Cleveland and Cincinnati, Ohio. Today the disease is found throughout most of eastern United States and Canada, with spot infections in some western cities where elms have been planted.

The Dutch elm disease pathogen is dependent on one of two insects, the native elm bark beetle and the European elm bark beetle, for its transportation from one tree to another. The latter beetle, imported into the United States in the early 1900s, was unimportant until the disease arrived. Once inoculated by contaminated beetles, the American elm is partly or completely killed by the disease and provides a suitable breeding ground on which increasing populations of beetles develop.

Wilting, often accompanied by a yellowing of the leaves, is the indication of an infected area on a tree. Although it can spread through roots, the disease is also spread by insects, and this aspect can be controlled. The method is expensive, since infected parts, or even entire trees, are burned to destroy the insects' breeding ground. Spraying the entire bark surface with an effective insecticide is another means of control. Most eastern cities have invested in portable power spray equipment, capable of reaching the tops of elms, to try to prevent further loss of their beautiful shade trees.

Elm Yellows A disease producing yellowlike symptoms on elms was first reported early in the twentieth century in the Midwest but has now spread into New York, New Jersey, Pennsylvania, Virginia, West Virginia, Maryland, and Delaware. Thousands of elms along the streets and in the parks of Washington, D.C. are under threat. The disease, caused by an obligate parasite transmitted by the white-banded elm leaf hopper, has little effect on nonnative elms, suggesting it was introduced from Europe or Asia.

Once infected, the elms exhibit bright yellow, drooping leaves in mid- to late summer, and trees will die by the next year (Fig. 7-6). Removal of the affected trees is the only management approach available (Sherald, 1999).

Canker Diseases The term *canker* usually refers to a disease that causes the death of relatively localized areas of a trunk or branch. Mechanical injury, such as a heavy blow, or frost and sunscald, can also result in a canker; however, most cankers are of fungal origin. Generally, infection that results in canker formation occurred when the stem was less than 10 years old.

The most serious result of canker diseases (which seldom kill their host) comes from heart rot fungi and wood-boring insects that enter the tree through the canker. The canker itself also reduces the quality of the wood through malformation of the stem and discoloration of the wood, and also makes the tree more susceptible to windthrow. Aspen is infected by both *Nectria* (Fig. 7-7) and *Hypoxylon* cankers, probably the most important cankers on hardwoods at the present time. Chestnut blight, mentioned at the beginning of this chapter under Types of Diseases, is also a canker disease. Conifers most frequently become infected by one of several species of *Dasyscypha;* the larch canker is currently the most serious.

Wilts and Diebacks Both types of disease are usually attributed to fungi. The fungi responsible for true wilt interfere with the upward conduction of water

FIGURE 7-6
An American elm showing bright yellow, drooping leaves. The inner bark becomes butterscotch brown. Elm yellows is a serious disease affecting elms in the midwestern and eastern United States. *(Photo by James L. Sherald)*

through the tree, resulting in the tree's death. The Dutch elm disease is an example of a true wilt. Oak wilt is presently one of the nation's most serious diseases both because of its widespread occurrence and the economic importance of oak species. Studies have shown that all species of oaks and closely related trees (chestnut, tanoak, and the chinkapins) are susceptible. Unfortunately, oak wilt has now spread throughout large areas of the eastern United States and Texas (Wilson, 2001).

Oak wilt can spread from an infected to a healthy tree through natural root grafts. The spores may be transmitted greater distances by birds and insects but

FIGURE 7-7
Nectria canker on quaking aspen in Minnesota. *(Courtesy of USDA Forest Service)*

can actually infect a tree only through a break in the bark. Control is possible by felling infected trees and spraying or burning the slash. Careful pruning of infected limbs may lengthen a tree's life.

The bacterium, bacterial leaf scorch *Xylella fastidiosa,* causes symptoms similar to those of wilt disease. It colonizes rapidly and thus physically clogs water-conducting tissues (xylem) of roots, branches, and leaves. Leaf hoppers, spittlebugs, and other xylem-feeding insects transmit this bacterium to healthy trees. Most problems occur in northern red oaks, although pin and scarlet oaks, sycamores, and elms also suffer. The disease has been observed from New York to Florida, killing young oak in four years, older trees in 10. Pruning out infected branches slows the spread, but will not prevent the tree from dying (Sherald, 1999).

Dieback is a condition in which a gradual deadening, starting with the tip, advances slowly down the stem. Fungi are responsible for most dieback, although there are other causes. The most serious manifestation of this condition is birch dieback, presently killing paper and yellow birches in the northeastern United States and southeastern Canada. The cause of birch dieback seems to be something other than fungi; stress factors related to prolonged drought and abnormally high temperatures apparently are involved, making it a classic example of a decline disease, discussed on page 146.

Blight Diseases Often any sudden dying of twigs, foliage, or flowers is called a *blight.* Its name frequently includes the affected part (needle blight, twig blight) or the name of the tree on which it grows (cedar blight, maple blight, sweetgum blight). Even the size or shape of the tree (pole blight) may be included in the name. Blight is accompanied by symptoms of wilting, dieback, discoloration, and spotting. The cause of the disorder may be difficult to determine. Blight, it seems, is a very general term. The fungi associated with blights usually appear on previously weakened or declining trees.

Pole blight, characterized by dieback of the entire crown, causes considerable damage to second-growth, pole-size western white pine in the Inland Empire region of Washington, Idaho, and Montana. Pole blight becomes more severe in areas where soil moisture is less abundant. Sweetgum blight, another of these nebulous blight diseases, is characterized first by dieback of the crown and then the death of the entire tree. Sweetgum blight in the South is also associated with a shortage of water in certain years. The origin of maple decline in the Northeast is also thought to be drought related. The disease appears to be prevalent along roadsides, lending credence to the assumption that there are several predisposing factors, including poor soil aeration, salt spray, and air pollution (see Decline Diseases, page 146).

Needle Diseases There are many diseases that infect the foliage of both hardwoods and conifers. A serious needle disease of ponderosa pine in much of the Northwest is *Elytroderma* needle cast (sometimes referred to as a *needle blight*), which can convert the entire crown into a series of witches'-brooms and stunted, swollen branches. Growth of the tree comes to a standstill. If it does not die from the disease, insects usually finish the job.

A second needle infection worthy of comment here is the brown spot needle disease of longleaf pine, present in various degrees over much of the Southern Forest region. It prolongs the "grass stage" of longleaf pine, often to the point where a loss of health and vigor may cause 100 percent mortality of the seedlings. Brown spot disease can be controlled by prescribed burning. The terminal bud of longleaf pine has a remarkable resistance to fire, and by burning off the young pine's needles, the disease is destroyed. The acreage burned must be large enough, 200 acres (80 ha) or more, to prevent a serious reinvasion of the disease from the surrounding unburned area.

Stains and Rots Blue stain and other sap stains in lumber are caused by various fungi. The organisms may start their staining in the forest or attack lumber. The colors associated with various decay fungi, for example, red heart or yellow ring rot, are caused by stains characteristic of that fungus.

Some staining fungi, although present in the wood, cause little change in the wood's original durability or strength properties and leave it merchantable. The stained wood is nonetheless depreciated in value because of its less desirable appearance. Also, conditions that are favorable to blue stain growth may favor the growth of decay-causing fungi.

The spores of stain fungi are carried by the wind, water, and bark beetles. Warm weather and ample moisture seem to promote development of most stains. Green lumber piled solidly in the hold of a ship or in a boxcar can arrive at its destination in a stained or moldy condition because of inadequate ventilation.

In warm, humid areas, stain can be avoided by removing the tree from the woods soon after felling. If this is not practicable or if logs must wait for several weeks at the mill before being sawn, it may be necessary to treat the logs with antistain solution, or store them in ponds or in log decks continuously sprinkled with water, which creates an environment too moist for the stain fungi. Reducing the moisture content of lumber, in kilns or air-drying yards, and dipping lumber in chemical baths before shipping, are other means employed to reduce the possibility of stain.

Damping-off in Forest Nurseries This trouble, which kills small seedlings just after they emerge

from the soil, may be the work of any one of 30 or more species of fungi. Some of the more common genera are *Pythium, Rhizoctonia,* and *Fusarium.* These fungi invade the tender root or stem soon after the seed germinates. The seedling may never appear above ground or it may cause the seedling to topple over just above the ground because of the destruction of tissue. Hartig, a distinguished German forester, spoke of the damping-off of beech seedlings as a disease observed but not explained by foresters in Europe as long ago as 1795. This disease results in a considerable loss and affects many species of American conifers planted in untreated nursery soil. Losses are likely to be heavy in broadcast or drill seeding, where there are larger numbers of seedlings per unit area of nursery bed. Risks such as these can hardly be taken, in view of the high cost of improved seed quality and the urgent need for maximum nursery output. The attack usually occurs under moist, warm conditions in nursery beds and may be recognized by examination of individual stems of seedlings for lesions or moist areas exhibiting brownish discolorations. Diagnosis and treatment at this point cannot save the crop. Its value lies in prevention of recurrence the following season, accomplished by treating the soil with applications of fungicidal chemicals, usually as soil fumigants.

Dwarf Mistletoes A close relative of the harmless mistletoe used in Christmas decorations, the dwarf mistletoe (*Arceuthobium spp.*), is a serious pest in western coniferous forests. So extensive is the reduction in growth due to this disease that it rates as one of ponderosa pine's worst enemies. It also infects western hemlock west of the Cascade Mountains, lodgepole pine, and Douglas-fir east of the Cascades, and black spruce in the Lake states.

All dwarf mistletoes are parasitic, leafless, seed plants, whose root systems, called *sinkers,* can invade the bark and become embedded in the sapwood of their hosts, extracting both dissolved nutrients, carbohydrates, and water (Fig. 7-8). The result is a slow-growing, weakened, poor seed-producing tree, which, after years of infection, may develop bark cankers or the characteristic tangled mass of twigs and needles called a *witches'-broom* that slowly starves and weakens the host. Dwarf mistletoe, without killing a tree, may greatly reduce its growth and so weaken it that it will fall prey to attacks by bark

FIGURE 7-8
Dwarf mistletoe plant parasitizing western hemlock in the Siuslaw National Forest, Oregon. *(Courtesy of USDA Forest Service)*

beetles and heart- and root-rotting fungi that a healthy tree can usually resist.

This parasite has a rather unusual method of spreading. The dwarf mistletoe produces small sticky seeds, which are discharged forcibly at maturity, and can travel up to 80 ft (25 m) from their source. By this method it is commonly spread to understory plants of the same host species. Therefore it is a major problem for young trees. On striking and sticking to a nearby young limb, the seed germinates, its roots penetrate the bark, and the plant becomes a parasite on the tree. The seeds are carried even greater distances by birds, squirrels, and porcupines.

Control of the disease is attempted by pruning out the infected limbs, and by thinning out heavily infected understory trees and cutting all infected overstory trees. Should dwarf mistletoe become established on the main trunk, the tree must be removed in the next sanitation cut.

Decline Diseases

As mentioned earlier, there are three types of diseases: abiotic, biotic, and decline diseases. *Declines* represent a combination of the other two causative agents. According to Manion (1991), a decline disease may involve three groups of factors. The first are static, such as climate, soil type, genetic potential of the tree, or the age of the tree. These are called *predisposing factors,* factors that cause weakness in the tree simply because it is growing in the wrong place, and predispose it to the actions of other factors. The second group, *inciting factors,* either physical or biological, cause drastic injury from which recovery is difficult. Examples would include insect defoliation, mechanical injury, a spring frost, drought, excessive salt, or air pollution. The third group is called *contributing factors,* usually fungi that may have little effect on a healthy tree, but can speed up the decline of a weakened tree.

Examples of declines would be birch dieback (of eastern Canada and northeastern United States), ash dieback (largely in New York), maple decline (northeast United States and eastern Canada), pole blight (found on western white pine in the Inland Empire), and littleleaf disease (southeastern United States). Other declines include cytospora canker on Colorado blue spruce, beech scale nectria canker (caused by the combination of a scale insect and a canker fungus), sweetgum blight, and oak decline.

As evidenced by these formulations, tree diseases, especially declines, refuse to fall into precise categories, and their classification tends to be somewhat difficult.

By now it should be apparent that knowledge of tree diseases is important as well as complicated. The forest protection specialist must have considerable knowledge of pathology, entomology, and silviculture practices.

As with insect and mammal damage, evidence of disease in the forest needs to be closely monitored, and its treatment considered in overall management objectives and strategies. Treatment will depend on management objectives. In some forested areas in parks or wilderness, the strategy may be to do nothing, letting nature, however modified by humans, take its course.

Silvicultural practices that produce healthy forest stands are presently viewed as the first line of defense

FIGURE 7-9
This area, once a showpiece forest, became a tangled mass of trees in a northwest windstorm. Note the two people. *(Photo by Andrew M. Prouty)*

against disease. Development of resistant strains of vulnerable tree species and research on silvicultural controls aimed at specific forest disease problems are continuing. But, as with insects, no one approach will solve disease problems. An integrated approach that takes into account the forest health implications of every management practice, as well as the objectives of the overall strategy, will be required.

OTHER DAMAGE TO THE FOREST

The Elements

Wind Like fire, wind has various effects, depending on many factors. From a forest manager's point of view, these effects may be favorable or unfavorable. Winds increase transpiration and thus are a major influence on the water regime of the plant. Photosynthesis is assisted by air movement, but salt spray and atmospheric pollutants are also conveyed to trees by wind. Distressed trees may be seen in many areas of strong prevailing winds and where sand- or salt-blasting occurs regularly. Wind disseminates seed and pollen for tree reproduction but also spreads fungal disease spores. Wind can topple seed

trees and cause the loss of valuable specimens, but it also opens up the forest to light and growth by felling overmature or diseased trees. Wind also contributes to disease entry into healthy trees through broken limbs and tops, acting as an inciting factor for decline diseases, as do the ravages of other elements mentioned here.

When, as a result of severe storms, wind velocity suddenly increases, breakage or windthrow is likely to occur, especially where trees may have been made vulnerable by a combination of harvest practices, topography, and disease. Observation of previous windthrow, care in the configuration of borders of clearcuts and shelterwood cuts, and prudence in the degree of thinning, can lessen local breakage and windthrow, with their attendant disease and insect entry problems, along with increased fire danger.

A major highly destructive storm such as occurs only a few times in a century can scarcely be guarded against. Hurricanes and tornadoes have historically laid waste large areas of forest in North America (Fig. 7-9). For example, Hurricane Hugo left 4.5 million acres (1.86 million ha) of timberland significantly damaged in the early 1990s, with softwood tree growing stock reduced by 21 percent and nearly one-third of the remaining volume damaged to some degree (Sheffield & Thompson, 1992). Some experts have predicted, that extreme weather events including hurricanes and tornadoes may be a by-product of global warming and may thus increase in frequency in the years ahead. Time will tell if this is true. .

It is vital that "hazard" trees be observed and possibly removed from campgrounds and other visitor use areas, since falling trees can be dangerous and lawsuits may result (Sharpe et al., 1994; Wallis et al., 1992).

Snow and Ice   Breakage and deformation occur occasionally in the forest from the weight of heavy snowfall. This usually happens in young stands or among trees that are crowded and spindly, or where the trees, in trying to recover an upright position, have assumed a bent and crooked form. Damage of this last type takes place at higher elevations, where accumulations of snow and ice may be most severe.

While snow and ice storm damage is a normally recurring forest disturbance force, sometimes a particularly severe ice storm will inflict noteworthy damage over a widespread area. For example, January 8 and 9, 1998, a severe ice storm damaged forests across 25 million acres (10 million ha) of Canada and the northeastern United States, even leading to a $48-million special appropriation from Congress to assist U.S. private landowners in their recovery efforts (Irland, 1998). The ultimate losses in downed and damaged trees, and subsequently from insects and disease, are not yet fully known.

Breakage is combated in commercial forests by silvicultural methods that encourage strong individual trees, and by favoring resistant species. Ice damage comes from tops and branches breaking under the weight of ice formed during glaze or ice storms. On damaged shade trees, pruning may help prevent entrance of disease. But such practices are not feasible in a forest, and damaged trees are generally removed during salvage operations.

Frost   Trees planted in low spots with insufficient air drainage may be killed or damaged by frost, and these pockets should be avoided in planting. Damage from frost is usually local and results from unseasonably low temperatures that injure buds and tender wood when the tree is not prepared for them. Reproduction can be somewhat protected from early and late frosts by use of silvicultural systems that retain some overstory, such as shelterwood and selection harvest methods.

An off-season cold snap, when the temperature stays below freezing for a sustained period, can cause widespread damage, especially to *exotic species* (those not native to an area). Such a freeze serves as a rather drastic means of determining the cold resistance of ornamentals.

High Temperatures and Drought   Droughts are natural phenomena, and the cost of watering forest plantations is usually prohibitive, but drought-resistant species and genetic varieties may be used to alleviate persistent problems. Watering of small plantations, shelterbelt plantings, or urban trees may be justified because of their high value. Sunscald, or injury to trees with thin bark by the heat of the sun, can be guarded against by taking precautions in thinning so that remaining trees will not be left open to the full rays of the sun for long periods. If from a nursery, tender coniferous seedlings, as well as those of certain hardwoods, require partial shade for a time. Prolonged drought is a concern of forest managers because of reduced growth and the fact that individual drought-stressed trees become vulnerable to insects and disease.

Lightning   Lightning serves as a catalyst for change. It damages many trees each year, and if it does not kill them, it wounds and weakens them so that they become prey to insects or disease. In some instances wildfires are ignited, and some of them cause extensive stand damage and mortality. Even if only a few trees are affected by lightning or its aftermath, there is increased susceptibility to windthrow. Lightning can be seen as a major agent of forest succession, often triggering fires that shape the residual forest or lead to replacement of the stand.

Air Pollution   Air pollution is one of several abiotic factors that contribute to forest disease. The major sources of air pollution are from ore smelters, electric power plants that burn coal or oil, and the toxic gases released from motor vehicles. Fluoride, sulfur dioxide, and nitrogen oxide are three of several plant toxicants that cause injury to hardwood leaves and conifer needles.

Ozone, a product of nitrogen oxide and hydrocarbons with ultraviolet light, is an airborne gas that can cause irritation of the eyes and nose. It also causes flecking and tip burn on leaves and needles. Ozone injury is particularly prevalent in the mountains of California.

Acid rain contains a heavy concentration of sulfuric and nitric acids. These are formed when toxic gases reach the clouds, mix with water vapor, and fall as rain, snow, sleet, or hail. This highly acidic precipitation damages fish and other aquatic life, and agricultural crops and trees. With pH values as low as 2 to 4, acid rain can leach nutrients from the foliage and the soil, leaving the trees susceptible to root rots. The problem is particularly prevalent in eastern North America, such as in the Appalachian and Adirondack mountain chains, and in Europe and Russia.

Catastrophic Geologic Events

Avalanches and Landslides   To suggest a remedy for an avalanche seems arrogant and futile, but avalanches and landslides starting at high points take large swaths from valuable forests lower down. Steep slopes with thin soils subject to heavy rains are conditions indicating possible avalanches and landslides. Little can be done except to protect grass, brush, and tree cover from fire, destruction by grazing, or improper timber harvest. The afforestation of high slopes or seeding with grass or herbaceous plants may be justified to reduce avalanche danger. Sheep particularly should be excluded from grazing on critical slopes. The Swiss have managed their forests with selection harvest systems to assure that tree cover will hold deep snows in place.

Volcanic Eruptions   Mexico, Hawaii, Alaska, California, and Washington State have recently witnessed volcanic eruptions that destroyed standing trees. On May 18, 1980, a Washington volcano, dormant for 123 years, caused the destruction of standing timber in a large, heavily forested area. Signaled by earthquakes, ash, and smoke for several weeks, Mount St. Helens erupted in a gigantic explosion in which an estimated 12 percent of the mountain was blown away. The force of the blast flattened trees in a fan-shaped area of 200 square miles (600 km^2) to the north of the mountain on federal, state, and industrial

FIGURE 7-10
Looking southwest from Independence Pass toward Mount St. Helens at the results of the blast of May 18, 1980. Searing hot ashes, traveling 300 mi/hr (480 km/hr), devasted more than 220 sq mi (60 km^2) of rivers, lakes, and forests. Approximately 3.7 billion board feet of timber were blown down by the lateral blast. *(Photo by Scott Shane)*

FIGURE 7-11
A truck bringing salvaged logs from the Clearwater Valley near Mount St. Helens.
Note the eruption-blasted timber on the slope above the truck. Gifford Pinchot
National Forest, Washington. *(Photo by Grant W. Sharpe)*

forest land (Fig. 7-10). Forest fires were started on the mountain's flanks from superheated steam and gases. Mud, from ash, melted glacial ice, and snow flowed down the mountain's sides to kill even more trees in the river valleys. Much of the damaged timber was salvaged within three years of the eruption, however (Fig. 7-11).

The eruption of Mount St. Helens occurred in an area of mature trees of high commercial value. The volcanic damage in Alaska, Hawaii, and Mexico has not been in stands of similar size or value. Although nothing can be done to prevent this manifestation of elemental fury, much has been learned about predicting volcanic activity, and protection of workers and equipment can be planned for in susceptible areas.

SUMMARY

As with insects, mammals, and disease, destruction of timber by the elements is part of the ecosystem's functioning, and human plans may be dramatically set aside by these events. In the battle against the ele-ments, humans and their enterprises almost invariably come off second best. Prudence and foresight must guide the forest manager in these matters.

LITERATURE CITED

Alexopoulos, C. J., and C. W. Mims. 1976. *Introductory Mycology* (3rd ed.). John Wiley and Sons, New York.

Hadfield, James S., Donald J. Goheen, Gregory M. Filip, Craig L. Schmitt, and Robert D. Harvey. 1986. *Root Diseases in Oregon and Washington Conifers.* USDA Forest Service, Pacific Northwest Region Forest Pest Management, Portland, Ore.

Irland, Lloyd C. (coord.). 1998. "Ice Storm 1998 and the Forests of the Northeast: A Preliminary Assessment." *J. Forestry* **96**(9):32–40.

Manion, Paul D. 1991. *Tree Disease Concepts.* (2nd ed.). Prentice-Hall, Inc., Englewood Cliffs, N.J.

Sharpe, Grant W., Charles Odegaard, and Wenonah F. Sharpe. 1994. *A Comprehensive Introduction to Park Management.* Sagamore Publishing, Champaign, Ill.

Sheffield, R. M., and M. T. Thompson. 1992. *Hurricane Hugo: Effects on South Carolina's Forest Resources.*

Res. Paper SE-284 Asheville, N.C. USDA Forest Service, SE Forest Experiment Sta. 51 pp.

Sherald, James L. 1999. "New Threat to Elms: The Yellows." *J. Forestry* **98**(10):6.

Sinclair, Wayne A., Howard H. Lyon, and Warren T. Johnson. 1987. *Diseases of Trees and Shrubs.* Comstock Publishing Associates, a division of Cornell University Press, Ithaca, N.Y.

U.S. Department of Agriculture Forest Service. 2001. *2000 RPA Assessment of Forest and Rangelands Summary* Report. Washington, D.C. Available on-line at www.fs.fed.us/pl/rpa/publ.

Wallis, G. W., D. J. Morrison, and D. W. Ross. 1992. *Tree Hazards in Recreation Sites in British Columbia.* B. C. Ministry of the Environment and Parks and Canadian Forestry Service. Joint Report No. 13. Victoria, B.C.

Wilson, Dan A. 2001. "Oak Wilt: A Potential Threat to Southern and Western Oak Forests." *J. Forestry* **99**(5): 4–11.

ADDITIONAL READINGS

Bega, Robert V. 1979. *Diseases of Pacific Coast Conifers.* USDA Forest Service Agr. Handbook No. 521. Washington, D.C.

Griffin, Gary J. 2000. "Blight Control and Restoration of the American Chestnut." *J. Forestry* **98**(2):22–27.

Smith, David M. 2000. "American Chestnut: Ill-fated Monarch of the Eastern Hardwood Forest." *J. Forestry* **98**(2):12–15.

Scharof, Robert F. 1993. Diseases of Pacific Coast Conifers. USDA Forest Service Agriculture Handbook 521. Washington, D.C. GPO

Tattar, Terry A. 1989. *Diseases of Shade Trees.* Academic Press, Inc. San Diego, Calif.

Youngs, Robert L. 2000. "A Right Smart Little Jolt: Loss of the Chestnut and a Way of Life." *J. Forestry* **98**(2):17–21.

STUDY QUESTIONS

1. State two different classifications of forest diseases.
2. What are some examples of introduced forest diseases?
3. Name at least one forest disease common in your area. Describe the signs.
4. Match the following with their associated terms and definitions with a line:

   A. sporophores — biotic and abiotic combination

   B. many hyphae — lack chlorophyll

   C. decline diseases — currant bushes the alternate host

   D. fungi — conks

   E. littleleaf disease — most damage of any disease

   F. heart rots — mycelium

   G. white pine blister rust — most severe in poorly drained southern soils

   H. canker diseases — death of localized areas of a trunk or branch

5. Why are introduced diseases (and insects) likely to be more damaging than native ones?
6. Suggest protective measures regarding avalanches, windthrow, and frost pockets.
7. Name some recent climate abberations or extreme weather events (anywhere in the world) that have caused significant destruction of trees.

CHAPTER 8

Fire Management

This chapter was prepared for the 6th edition by William C. Fischer (ret.), Fire Effects Team Leader, USDA Forest Service, Intermountain Fire Sciences Laboratory, Missoula, Montana. It was updated for the 7th edition by the authors.

INDRODUCTION

THE IMPORTANCE OF FIRE MANAGEMENT

Although natural resource professionals have been dealing with wildland fires for well over 100 years, today fire management presents greater and more complex challenges than ever before. Several decades of successful fire prevention and suppression have led to large and unnatural fuel buildups. Now, instead of frequent, light groundfires, unplanned ignitions are likely to erupt into intense conflagrations that burn so hot and are so widespread that they do irreparable harm to the soil and the affected ecosystems.

Because the typical, small groundfires that historically kept forests open and clear of debris have been suppressed, a thick understory of shade-tolerant trees has developed in many stands over large areas. Besides contributing to fuel loads, this dense understory depletes moisture and nutrients, weakening the stand and making trees more vulnerable to insects and disease. These factors in turn may kill many trees in a stand, further contributing to fuel loads.

The desire of citizens to build expensive homes in wooded areas has also compounded the difficulties of fire management, as has the recruitment of many new firefighters whose safety deeply affects management decisions.

The hallmark of fire management at the turn of the twenty-first century is new visibility and formal recognition of this important field, and commitment to a new federal, interagency, national fire plan of action to address all elements of the fire problem (www.fireplan.gov). Millions of acres of dense, overstocked stands are vulnerable to fire, and thousands of burned-over acres needing rehabilitation await assessment and treatment. Under the national fire plan, communities will be assisted in their fire prevention, suppression, and recovery efforts; millions of acres of overstocked forests will be thinned; burned-over areas will be restored; fire suppression forces will be expanded and their training strengthened; and new and better equipment will be made available (www.fireplan.gov).

The frequency of wildfire and acreage burned has changed dramatically over time, as shown in Table 8-1. Fire suppression resulted in relatively stable areas burned during the decades of the 1950s through the 1970s. But beginning in the 1980s, the

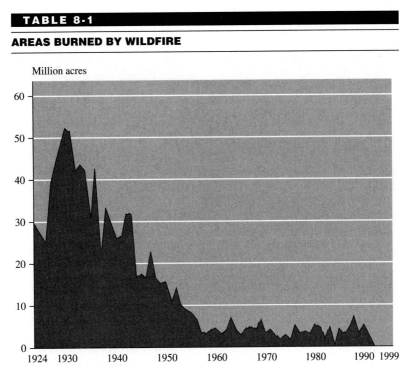

TABLE 8-1

AREAS BURNED BY WILDFIRE

Source: USDA Forest Service, 2001

area burned by wildfire began to increase again, due in part to the unprecedented success of fire prevention and suppression and its effects on forest conditions (USDA Forest Service, 2001).

The current 10-year average is 52,735 fires and 1.9 million acres (769,500 ha) burned (Hundrup, 2001). But drought or unfavorable weather can lead to "bad" fire years in which even more acreage may burn (see Table 8-1), and with fuel building and the weather warming, the risk, hazard, and acreage burned by wildfires are increasing. For example, during 2000, nearly 3 million acres (1,215,000 ha) burned in six northern Rocky Mountain states alone (USDA Forest Service, 2001).

All of these factors make fire management vitally important to renewable resource management in the twenty-first century, and a field ripe for employment and natural resource careers. Chapter 8 provides an historical overview of fire policy and delineates the elements of current fire management practices.

THE EVOLUTION OF FIRE POLICY

During presettlement and early settlement times, fire in North American forests and grasslands was a common occurrence. Scientific and historical literature, for example, includes up to 145 accounts of wildfire by 44 different observers between 1776 and 1900 for the Interior West—Montana, Wyoming, Idaho, Utah, Nevada, and eastern Oregon. Set by lightning, Native Americans, settlers, and lumberjacks, such fires were of little concern unless they threatened human life, structures, or livestock. Most fires did not, and fire and smoke were common features on the North American landscape.

A series of forest fires between 1871 and 1903, however, did more than just threaten life and property. Five great fire episodes occurring at roughly 10-year intervals burned across millions of acres of recently cut and uncut forest in Wisconsin, Minnesota, Michigan, and New York, causing extensive destruction of property and loss of several thousand lives (Table 8-2). It is little wonder that Gifford Pinchot labeled fire as the most terrible of all the foes that attack North American woodlands. Later as chief forester of the newly created U.S. Forest Service, he instructed his rangers to put out all fires occurring on their national forest districts.

Systematic fire control planning for the national forests rose from the ashes of the disastrous fires of 1910 in northern Idaho and adjacent western Montana (Table 8-1). Fire control became a major emphasis of the forest service not only in national forest management and research, but also in state and private forestry functions. Fire policy for the national forests was geared to remaining below the "maximum allowable burn" calculated for a given forest type, and attempted to remain consistent with economic objectives of balancing costs of prevention and suppression against loss of resource values and property.

National forest fire policy was changed in 1935 to reflect a different economic consideration—that it costs less to fight small fires than big fires. The "10 A.M. policy" was instituted, which required immediate and aggressive attack on all fires. If a fire was not controlled by initial attack forces, efforts on each succeeding day were expanded as required to obtain control before the start of the next day's burning period (10 A.M.). Each fire was to be attacked and controlled regardless of burning conditions or land and resource values.

The 10 A.M. policy was rigidly applied on the national forests until the mid-1970s, when the policy was modified to allow fire to play a more natural role in certain areas and under specified conditions. Fire policy for the national parks followed a similar evolution, except that experimental prescribed burning for vegetation management was conducted as early as 1951 in Everglades National Park. By the 1980s, the long-standing policy of *fire control* had gradually evolved into a policy of *fire management,* in which fire was allowed to play its natural role where possible and was prescribed and used as a vegetation management tool under certain conditions.

Among the driving forces behind this change was a growing awareness of the ecological effects of excluding fire from forests and rangelands. A new interest in the role of fire and its effects was generated by E.V. Komarek for the first of many Tall Timbers Fire Ecology conferences in 1962. Serving as a catalyst for the change was the Leopold Committee Report of 1963, "Wildlife Management in the National Parks," which advocated a reconsideration of the role of fire in national parks, with the goal of recognizing fire as a natural component. The Wilderness Act of

TABLE 8-2

SOME LARGE WILDFIRES OF NORTH AMERICA SINCE 1825

Name of fire	Location	Year	Area burned		Lives lost
			Acres	Hectares	
Miramichi	Maine &	1825	800,000	323,800	? est.
	New Brunswick		2,000,000	810,000	at 200
Pontiac	Quebec	1853	1,600,000	650,000	?
Silverton	Oregon	1865	990,000	400,650	?
Peshtigo	Wisconsin &	1871	1,280,000	520,000	1,500+
	Michigan		2,500,000	1,000,000	? 10+
Michigan	Michigan	1881	1,000,000	400,000	138–282
Hinckley	Minnesota	1894	160,000	64,000	418
Yacolt	Washington	1902	1,000,000+	400,000	38
Adirondack	New York	1903	637,000	257,800	0
The Big Burn of 1910	Idaho & Montana	1910	3,000,000	1,214,100	85
Cloquet	Minnesota	1918	250,000	101,000	432
Tillamook	Oregon	1933	300,000	121,000	1
Tillamook #2	Oregon	1939	220,000	89,000	0
Tillamook #3	Oregon	1945	230,000	93,000	0
Maine	Maine	1947	240,000	97,000	16
Black Saturday	New Jersey	1963	183,000	74,100	7
Wenatchee	Washington	1970	131,000	53,000	0
Laguna	California	1970	185,000	75,000	10
Marble Cone	California	1977	174,000	69,600	0
Canyon Creek	Montana	1988	240,000	97,130	0
North Fork/Wolf Lake	Idaho, Wyoming, & Montana	1988	504,025	203,980	0
Clover Mist	Wyoming & Montana	1988	319,575	129,330	1
Huck/Mink	Wyoming	1988	226,520	91,675	0
Snake Complex	Wyoming	1988	172,025	69,620	0
Foothills	Idaho	1992	257,000	104,000	0
Southern California Complex	California	1993	215,000	87,000	2
Rodeo-Chediski	Arizona	2002	468,638	189,800	0
Hayman	Colorado	2002	137,760	55,800	2
Biscuit	Oregon	2002	499,570	202,326	0

1964 was also seminal in this change of policy. Managers immediately recognized that they could not preserve natural conditions in wilderness without letting fire play its natural role, even though fire control was specifically authorized in the Wilderness Act.

ESSENTIALS OF FIRE MANAGEMENT

The concepts of fire management suggest that the ecological role of fire should be considered when establishing land and resource management objectives. Fire management oversees the inclusion of fire considerations in land and resource management plans, works to protect forest and rangelands from unwanted fire, and uses fire to accomplish desired vegetation objectives. The long-term objective of fire management is to minimize the destructive elements of fire while maximizing its benefits. The emergence of ecosystem management as the more holistic goal for public land management heightens the importance of fire management.

Fire is a natural event in forest and rangeland ecosystems and is an agent for change that can result in increased biological diversity. Many North American ecosystems evolved with fire; some are dependent on fire. Many lightning-caused fires, especially in large roadless and often rugged park and wilderness-type areas, tend to be self-limiting because of terrain or weather and fuel factors. This is not necessarily the case in those forests where fire has been systematically excluded for decades. In certain regions, protection from natural fires has permitted the forest floor to become a tangle of understory vegetation and accumulated woody debris. A fire of any origin in such conditions could be catastrophic both for the affected ecosystem and for management objectives, not to mention the human lives and property that might be involved.

Fire, even a big wildfire, does not destroy all wildlife in its path, since many species are able to flee or evade it. Fire alters habitats, and not all animals benefit from these changes, but the overall mosaic of vegetation stages resulting from fire favors wildlife diversity and generally supports a good representation of species.

Wildfire or fire out of control is still recognized as an enemy, both of humans and of their works, and of forest and rangeland (Fig. 8-1). If a fire reaches catastrophic dimensions, it can nullify the objectives of management, such as altering soil and stultifying regeneration of a new crop. Consequently, current fire management policies on most federal wildlands require that a decision be made as to the appropriate suppression action to be taken on

each wildfire. Fast and aggressive initial attack is required for all fires that threaten life or property, or that have the potential to become catastrophic. If a wildfire escapes initial attack, subsequent action is carefully considered. The fire's potential for resource damage is weighed against its potential benefits and the costs required for appropriate suppression methods. If, for example, the analysis indicates the escaped fire has a high potential for causing serious resource damage, an all-out suppression effort might be launched, using every tool and technique available. If, however, the potential for damage is low, the fire manager may elect to limit the suppression effort to the use of ground crews with hand tools, thereby trading off acres burned for the savings realized by not using more expensive fire control techniques.

"Light handed" suppression tactics and strategies are an integral part of current fire management policy so that fire control actions do not create more impact and disturbance than necessary. This does not preclude the use of mechanized equipment or any other tactic deemed necessary, but points in the direction of minimum impact where possible.

FIGURE 8-1

Wildfires are either acceptable or unacceptable, depending on risk to human life and property, as well as to land management objectives. Angeles National Forest, California. *(Courtesy of USDA Forest Service)*

The use of traditional prescribed fire is retained and, in fact, encouraged in current fire management policy. It also permits lightning-caused fires in wilderness areas to be managed as prescribed fires so long as they meet certain (prescribed) fire conditions.

Fire Weather

Climate, drought, and weather affect fire behavior, fuel loading, and vegetation. Extended drought, especially during the summer, increases the risk of wildfire, particularly of large conflagrations. The vegetation characteristics of different regions evolved under periodic drought conditions and regional drought cycles may last for several years. Large fires often occur during drought years both in areas that have high fire frequency and in areas where fires are infrequent. Thus, prevailing climatic cycles of drought and wetter periods are to be expected and have important implications for fire management planning and logistical preparations.

Wildland fires occur in, and are affected by, conditions in the lower atmosphere. Thus, atmospheric changes from one moment to the next may affect fire behavior. At times, fires may be affected only by changes in a small area at or near the surface; at other times, the region of influence may involve many square miles horizontally and several miles vertically in the atmosphere. All of these conditions and changes result from the physical nature of the atmosphere and its reactions to the energy it receives directly or indirectly from the sun.

Changes in weather elements (i.e., temperature, relative humidity, wind, precipitation, and atmospheric pressure and stability) influence the ignition, spread, and intensity of wildland fires. Fire management requires an appreciation of the volatile changes in fire behavior that can occur in response to even small and local changes in atmospheric conditions, and an understanding of the principles involved. Without such appreciation both safety and effectiveness are compromised.

Fire Response

Essentially, fire management requires a deliberate response to a fire through the execution of technically sound plans under specific prescriptions to achieve management objectives. Thus, a deliberate response is made to every fire with careful and thorough consideration of consequences. It is a planned response.

There are three general ways to respond to a fire: ignore it, attack it, or allow it to burn according to a predetermined plan. Ignoring a fire, or just "letting it burn" without careful consideration, is not management; hence, it is an unacceptable fire management response.

Fire Attack

Fire attack can be *aggressive, modified,* or *delayed. Aggressive attack* immediately follows discovery. Force must be sufficient to affect control at the earliest possible time with minimum acreage burned. *Modified attack* is less intense and suppression forces, techniques, strategy, or some combination of these factors are reduced somewhat from those designated for aggressive attack. For example, the use of power tools and motorized equipment may be prohibited when attacking wilderness fires. This would constitute a modified attack. *Delayed attack* means that attack does not immediately follow discovery. A fire that is discovered at night, for example, might not be attacked until daylight. Once such attack is made, it may be aggressive or modified. Once initiated, all modes of attack are fast, energetic, and thorough, although conducted with regard for the safety of fire fighting personnel.

Overall suppression strategy is another choice that fire management teams must make. Depending on the fire situation, three fire suppression strategies are identified by the Forest Service: *confine, contain,* and *control.*

To *confine a fire* means to limit fire spread within a predetermined area principally by the use of natural and preconstructed barriers or environmental conditions. Here, under appropriate conditions, suppression action may be minimal and limited to surveillance. The confine strategy must be reviewed each day to verify the self-confinement predication. If conditions change, another strategy, such as contain or control, must be employed.

To *contain a fire* means to surround it, and any spot fires, with a control line, as needed, which can be reasonably expected to check the fire's spread within a predetermined area under prevailing and predicted conditions. The contain strategy must be reviewed each day to ensure it remains applicable for the upcoming burning period. When the fire cannot be contained as predicted because uncontrollable conditions persist, the last suppression strategy must be employed (i.e., control).

To *control a fire* means to put it out. This is done by completing a control line around the fire, around any spot fires, and around any interior islands to be saved. Then any unburned areas adjacent to the fire side of the control line are burned out. Next all hot spots that are immediate threats to the control are cooled down, until the line can reasonably be expected to hold under foreseeable conditions.

The other strategy in responding to fire is to allow the fire to burn according to a predetermined plan. Even though the source and time of ignition were uncontrolled, the decision to let the fire burn rests on a plan setting forth acceptable conditions for such a response, so fires allowed to burn are also considered prescribed fires. Thus there are two kinds of fires: *wildfires* and *prescribed fires.*

Prescribed Fire

A *prescribed fire* is a fire ignited or allowed to burn in a certain area under prescribed environmental conditions to accomplish specified management objectives. *Prescribed burning,* the act of creating a prescribed fire, is the controlled application of fire to wildland fuels, in either their natural or modified state, under favorable conditions of weather, fuel moisture, or soil moisture, that allows the fire to be confined to a predetermined area. Heat intensity and rate-of-spread must be considered in order to meet the planned objectives.

Simply stated, a prescribed fire is any fire burning according to prescription to meet an objective. A *fire prescription* is a written direction for the use of fire on a specific piece of land to accomplish certain land management objectives.

There are two kinds of prescribed fire: manager-ignited prescribed fire and prescribed natural fire.

Manager-ignited Prescribed Fire   A *manager-ignited prescribed fire* is started at a predetermined time when current and forecasted fuel moisture and weather conditions indicate a high probability of achieving land management objectives and controlling the fire (Fig. 8-2). Other terms used to describe this type of fire are *traditional prescribed fire, artificially ignited prescribed fire, planned ignition,* and *scheduled prescribed fire.* As deliberate ignition is a sensitive policy issue, a careful distinction is made between manager-ignited and natural prescribed fire.

FIGURE 8-2
A prescribed fire specialist igniting an underburn with a drip torch. Bitterroot National Forest, Montana. *(Courtesy of USDA Forest Service)*

Uses of Manager-ignited Prescribed Fire   Manager-ignited prescribed fires are used to accomplish many land management objectives:

1. To dispose of logging residues and other woody debris
2. To remove physical barriers to tree planting and to travel by big game animals
3. To reduce fire hazard by removing or reducing unnatural fuel buildups and by maintaining firebreaks
4. To control wildfires by burning out and backfiring to expand and strengthen firelines during a suppression operation
5. To prepare seedbeds for natural or artificial regeneration by removing litter and duff, thereby exposing mineral soil
6. To reduce competition for tree seedlings by eliminating shrubs

7. To recycle nutrients in order to maintain site productivity
8. To sanitize sites against insects and disease
9. To eliminate less-desirable plant species in a type conversion operation
10. To maintain seral (early successional stage) species on a site that would otherwise be taken over by less-valuable (late successional stage) tolerant species
11. To mimic a natural fire regime
12. To kill invading trees and shrubs in order to maintain grasslands
13. To thin dense stands of pole-sized trees so moisture and nutrients can be used by fewer stems
14. To increase amount of palatable grass and forbs for animals (domestic and wild)
15. To rejuvenate decadent shrubs by top killing, in order to induce sprouting and thus improve deer and elk forage
16. To maintain or restore "natural" conditions, such as in a national park or wilderness area

The use of *aerial ignition* techniques has expanded prescribed burning capabilities for hazard reduction, site preparation, silvicultural treatments, range improvements, and for burning-out and backfiring in the control of wildfires. The most popular methods of aerial ignition are the helitorch and various capsule systems. The *helitorch* is essentially a large drop torch with a gelled fuel system suspended beneath a helicopter. *Capsule systems* are used in fixed-wing aircraft as well as helicopters. They consist of a dispenser that drops ignition devices from the aircraft onto the area to be burned. Ignition devices vary from different types of fusees to ping pong balls filled with flammable chemicals that are activated just before dropping.

A variation of the helitorch developed particularly for range burning is the *terra torch.* It is a less costly application of the helitorch technology for application from a ground vehicle.

Types of Manager-ignited Prescribed Fires Manager-ignited prescribed fire takes many forms, including *slash burning, piling and burning, underburning, stand replacement burning,* and *range burning.*

Slash burning is a method of disposing of logging residues and other woody debris and vegetation that would otherwise impede reforestation and create a serious fire hazard. It is probably the oldest use of fire in American forestry. Many of the early catastrophic fires (Table 8-1) began in logging slash or similar debris resulting from land clearing. Consequently, safe and effective slash disposal was an early fire control method on state and federal lands. Two widely used methods of slash burning are *broadcast burning* and *piling and burning* (Fig. 8-3).

Broadcast burning involves burning large concentrations of slash usually on clearcuts, in a single fire. *Piling and burning* involves piling the slash, by hand or by tractor, and igniting each pile.

Underburning, also called *understory burning* or *light burning,* is burning beneath a forest or woodland canopy. This technique was pioneered in the South to reduce fire hazard and to control competition from understory hardwoods in pine stands. Underburning is increasingly used in many western forest types, for example, ponderosa pine, larch–Douglas-fir, and giant sequoia, to accomplish fire hazard reduction as well as a variety of silvicultural and wildlife objectives.

Stand replacement burning refers to manager-ignited prescribed fires that destroy an existing stand of trees in order to replace it with a new stand. In Michigan, older jack pine stands are burned to create young stands needed by the endangered Kirtland's warbler for nesting habitat (Fig. 8-4). In Colorado, fire has been used to renew lodgepole pine stands heavily infested with mistletoe and to restore traditional migration routes for bighorn sheep, and other related objectives. Overmature aspen stands have also been burned to stimulate reproduction favored by deer and elk.

Range burning applies prescribed fire to grasslands or shrublands, usually to stimulate production of palatable forage and browse for livestock and wildlife and to control unwanted trees or shrubs. Other uses include maintaining natural vegetation on restored prairies in the Midwest, improving eastern and southern wetlands for wildlife, and reducing flammability, such as in the burning of chaparral in southern California.

Prescribed Natural Fire This starts from a natural force such as lightning or a lava flow from a volcanic eruption; the timing of such ignitions is not known in

FIGURE 8-3
Slash burning in the Pacific Northwest. *Upper:* Burning tractor-piled slash. Plastic
sheeting is used to keep the pile dry until burning conditions are declared
appropriate. Drip torches are used to ignite the material. *Lower:* A dozen or more
slash piles may be ignited simultaneously. Gifford Pinchot National Forest. *(Photos
by Loretta Sharpe)*

advance. Other terms used to describe this type of fire
are *managed natural fire, natural fire, wilderness fire,
management fire, unplanned ignition, unscheduled prescribed fire,* and *prescribed natural fire.* A naturally ignited fire is designated as a prescribed fire only after it
has been determined that the situation, existing weather
conditions, and forecasts agree with conditions suitable

for a prescribed fire. Natural fires are most commonly
prescribed in national parks and wilderness but are not
necessarily limited to such locations.

Prescribed Fire Planning

From a fire management perspective, a successful
prescribed fire is one that is executed safely, burns

FIGURE 8-4
A prescribed burn in a stand of jack pine in central Michigan. The objective is to provide habitat for Kirtland's warbler. This specialized habitat, once maintained by natural fires, was disappearing with effective fire suppression. *(Photo by Robert E. Martin)*

under control, accomplishes the prescribed treatment, and attains the land and resource management objectives for the area involved. Successful prescribed fires result from careful planning and knowledge of the land management objectives, expected fire behavior, and probable fire effects.

Prescribed Fire vs. Wildfire

Any fire that is not a prescribed fire is a wildfire. It is an unwanted fire. A prescribed fire that deviates irreversibly from prescribed conditions; that is, escapes prescription and cannot be quickly brought back, becomes a wildfire. If a fire is receiving delayed or modified attack or any type of low-intensity suppression action, it is a wildfire, not a prescribed fire.

Wildfires that are not successfully suppressed by initial attack, and prescribed fires that escape prescription and burn as wildfires, are called *escaped fires*. Subsequent action on such fires is selected on the basis of total cost effectiveness, public safety, probability of success, protection of property, and the effects of fire and fire suppression on the resources.

In the case of prescribed fire, some suppression action may be taken to keep the fire in prescription, such as keeping it away from developments or out of new, unprescribed terrain.

PROTECTION FROM WILDFIRE

In the emerging era of ecosystem management, the concept of fire protection is changing because fire has been recognized as an essential component in most ecosystems. Protection of wildlands and their associated resources from damage by wildfire is the goal of a fire suppression program. Effective wildfire protection, in turn, is characterized by well-planned programs of fire prevention, presuppression, and suppression.

Fire Prevention

Fire prevention activities are intended to decrease the number, size, and behavior of wildfires and the damage they cause. Consequently, such activities are directed at reducing both wildfire risk and wildfire hazard. *Risk* refers to the direct causes of fire (ignition source); *hazard* refers to the fuels ready to ignite and burn.

Wildfire Risk Wildfires occurring in the United States are categorized according to nine risk factors.

1. **Lightning**—wildfires started by lightning (Fig. 8-5).
2. **Arson** (incendiary)—wildfires willfully set by anyone to burn, or to spread to, vegetation or

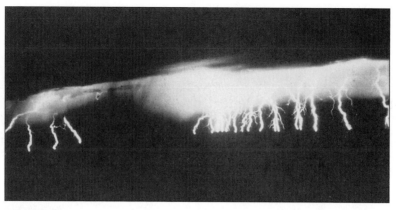

FIGURE 8-5
A dazzling multiple-strike system photographed over the mountains, from Missoula, Montana. *(Courtesy of USDA Forest Service)*

property not owned or controlled by that person, and without consent of the owner or his or her agent.

3. **Campfire**—a wildfire resulting from a fire started for cooking, heating, or for providing light or warmth.
4. **Children**—wildfires started by children less than 12 years old playing with matches, lighters, or other sources of ignition.
5. **Debris burning or fire use**—a wildfire spreading from burning trash, clearing lands, range, stubble, meadow, on right-of-ways; from logging slash burning; or other prescribed burning.
6. **Equipment use**—wildfires caused by any and all mechanical equipment other than railroad operations.
7. **Railroad**—wildfires caused by all railroad operations, including burning right-of-ways and ties.
8. **Smoking**—wildfires caused by smokers, thus ignited from matches, lighters, cigarettes, or other smoking materials.
9. **Miscellaneous**—wildfires that cannot be properly classified under other statistical causes.

Wildfire statistics are compiled according to calendar year and for respective landownerships by the U.S. Forest Service, the state forester (or comparable state official) in each of the 50 states (for state and private lands); by the officials of Puerto Rico and Guam; by the U.S. Department of the Interior for National Parks, Fish and Wildlife Refuges, Indian Reservations, and BLM lands; and by the Tennessee

TABLE 8-3		
CAUSES OF WILDFIRES AND ACRES BURNED NATIONALLY BY PERCENT FROM 1984 TO 1990		
Statistical cause	**Number of fires**	**Acres burned**
Arson	29%	21%
Debris burning	25%	9%
Smoking	5%–6%	3%
Equipment use	5%	4%–5%
Children	5%	1%
Campfires	3%	1%–2%
Railroads	2%	2%
Miscellaneous	14%–15%	11%
Lightning	11%	48%

Valley Authority. Table 8-3 shows the percentage of fires and the acres burned in each category (USDA Forest Service, 1992).

According to forest fire statistics reported by the U.S. Forest Service, the major causes of fires in the United States for the 50-year period from 1917 to 1966 were: incendiary (arson) fires, 26 percent; smoking, 19 percent; debris burning, 18 percent; machine use, 8 percent; campfires, 6 percent; lightning, 9 percent; and miscellaneous, 14 percent. Comparing these figures with those reported in Table 8-3 for the 1984–1990 period, we see a large reduction in the number of fires attributed to smoking and a 50 percent reduction in those caused by campfires. Arson fires show a slight increase, and debris burning shows the largest increase from the long-term averages.

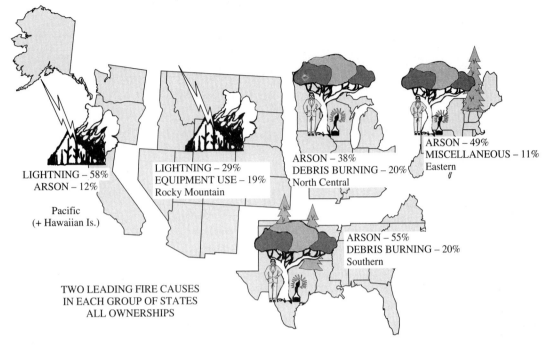

LIGHTNING – 58%
ARSON – 12%

Pacific
(+ Hawaiian Is.)

LIGHTNING – 29%
EQUIPMENT USE – 19%
Rocky Mountain

ARSON – 38%
DEBRIS BURNING – 20%
North Central

ARSON – 49%
MISCELLANEOUS – 11%
Eastern

ARSON – 55%
DEBRIS BURNING – 20%
Southern

TWO LEADING FIRE CAUSES
IN EACH GROUP OF STATES
ALL OWNERSHIPS

FIGURE 8-6
Wildfire differences by regions. Arson and debris burning are leading causes of fires in the eastern half of the United States; lightning causes most fires in the western part.

Differences between regions are illustrated in Fig. 8-6. Arson is the leading cause of fires, by a large margin, in the entire eastern half of the country, while lightning causes most of the fires in the West.

In the north central and the southern group of states, debris burning was the second-leading cause of fires; in the eastern states causes classified as miscellaneous were second. In the 12 Rocky Mountain states, the second-leading cause of fires was equipment use, and in the Pacific group of states, which includes California, arson was the second greatest cause of fires.

All of the causes but one, lightning, deal with fires caused directly or indirectly by people. Historically, such people-caused fires account for 90 percent of the total number of wildfires. The objective of risk reduction, then, is to reduce the number of wildfires that result from accidental, careless, and willful *human acts.*

Wildfire Hazard As previously noted, *wildfire hazard* refers to the combustible materials or fuels available for burning. The wildfire hazard of a fuel complex is largely a function of its flammability, and the physical impediments it presents to suppression actions or its resistance to control. The hazard of any given fuel complex will vary over time, depending on fuel moisture and weather. We are concerned here, however, with the variation in wildfire hazard that results from differences in fuel size, amount, arrangement, and location.

Small fuels burn faster than large fuels (Fig. 8-7). Fast-drying dead fuels less than ¼ inch in diameter, and other fine fuels with a comparatively high surface-area-to-volume ratio ignite readily and are consumed rapidly by fire when dry. Examples of such fine fuels are grass, ferns, leaves, dead pine needles on the ground and needle drape (dead but still hanging in the tree), tree moss, and slash consisting of branches with small twigs and needles. These fuels are also called *flash fuels* because of their extreme combustibility.

Very large or heavy fuels, such as large-diameter downed logs and old, compacted slash without needles, are difficult to ignite and burn slowly but can burn intensely if they are dry. Ordinarily, the greater

FIGURE 8-7
Small fuels burn faster than large fuels. *Upper:* This fire in the understory of a ponderosa pine stand illustrates surface fuels consisting of pine needles and low shrubs. *Lower:* This fire in logging slash shows the extreme flammability of fine, dead fuel, and the slow-burning character of large logs. *(Courtesy of USDA Forest Service)*

the amount of fuel, the more intense the burning compared to lesser fuel loads of similar type. Heavy fuels can also impede suppression actions, especially the building of fire lines.

The vertical and horizontal distribution or arrangement of fuel also affects wildfire hazard. There are three levels of fuels that must be considered when evaluating vertical arrangement: *ground fuels* at the lowest level, *surface fuels* above them, and *aerial fuels* above the surface fuels.

Ground fuels consist of combustible materials lying beneath the loose surface litter including deep duff, roots, rotten buried logs, peat, mulch, and other subsurface woody fuels. Fire in ground fuels usually spreads slowly but can be difficult to extinguish. Ground fuels, especially duff, may aid fire spread by carrying the fire between patches of more flammable surface fuels.

Surface fuels are all materials lying on or located immediately above the ground, including needles or

FIGURE 8-8
A fire spreading through branches, often with explosive force, is called a crown fire.
It is extremely dangerous and difficult to control. Boise National Forest, Idaho.
(Courtesy of USDA Forest Service)

leaves, duff, grass, forbs, small dead wood, downed logs, stumps, large limbs, low shrubs, and small trees. Most fires start and spread in surface fuels. Understory vegetation may either aid or hamper fire spread. Deciduous tree reproduction seldom will carry fire, but coniferous seedlings will, especially where there is grass or litter around to ignite them.

Aerial fuels are all green and dead materials located in the forest canopy, including tree branches and foliage, snags, mosses, lichens, epiphytic plants, and high shrubs. Concentrations of dead tree crowns, especially recently killed conifers, are an important aerial fuel and can carry fire from tree to tree. While green leaves of hardwood trees ordinarily will not carry fire, foliage of some deciduous and nondeciduous western oaks and tall shrubs can be quite flammable, especially when their crowns contain a lot of dead branches and leaves. *Crown fires* burning and spreading in the crowns of trees are extremely dangerous and difficult to control, and they can move very fast (Fig. 8-8).

Fuel continuity affects how fast and how far a fire will spread, and the difficulty of suppression. Horizontally continuous fuels are those in which the fuel loading remains nearly the same over large areas. Noncontinuous fuels are often too scattered to support fire spread. Vertical continuity occurs when understory vegetation extends more or less uniformly into the lower parts of the overstory. These materials, such as understory vegetation, heavy accumulations of moss and needle drape, and leaning trees, are called *ladder fuels* because they allow a surface fire to climb into the overstory and become a crown fire.

Reduction of Risk and Hazard

The objective of fire prevention is to decrease the number of fires and the damage they cause by reducing wildfire risk and hazard. Successful fire prevention depends on education, enforcement, engineering, and fuel treatment in logical, well-planned combinations that are designed to counteract those fires that can do the most damage.

Education Programs   The objective of fire prevention education is to influence people to be careful with fire so as to reduce or eliminate both risk and hazard. The intent is to regulate the use of potential ignition sources in such a way that wildfires do not occur. Special efforts should be made to reach the people who could start wildfires, especially the most damaging wildfires. The message should emphasize specific ways and means of preventing wildfires, and why it is important to do so. The central approach is a positive one of educating people how to do things safely.

Education is disseminated through posters, newspapers and magazines, leaflets and brochures, radio and television, billboards, and exhibits, as well as through personal contact. Other opportunities to spread the message arise at public meetings, fairs, parades, sporting events, and similar gatherings. Changing the attitudes and behavior of children by conducting comprehensive fire prevention programs in the schools receives special emphasis. Many opportunities for fire prevention education are available through working with organizations such as 4-H, scouts, and campfire girls, and by presenting special programs at youth camps and gatherings that occur every summer.

The Smokey Bear Program   Without a doubt, this has been the most successful attempt to influence public opinion about forest fires in the United States. Smokey Bear (not Smokey the Bear—Smokey has no middle name!) was created in 1945 as a poster character in an effort to reduce the number of forest fires when most experienced firefighters and other able-bodied men and women were in the military. This was the first time a nationwide forest fire prevention campaign was launched, and it utilized the talents of a professional advertising firm, which volunteered its services. During the subsequent 60 years, Smokey and his fire prevention message, "only you can prevent forest fires," have become part of modern American consciousness, and this trademark is one of the best known symbols in the United States (Fig. 8-9).

A live Smokey was added to the campaign in 1950. He was a badly burned bear cub that survived a wildfire in the Lincoln National Forest in New Mexico. The bear was then named Smokey and housed at the National Zoo in Washington, D.C., until he died in 1976. A new Smokey carried on the tradition until

FIGURE 8-9
"Smokey the fire prevention bear" played a major role in reducing wildfire losses from 30 million acres per year in the 1930s to 3 million acres by 2000. *(Courtesy of USDA Forest Service)*

its death in 1990. The idea of a "living" symbol of Smokey was then discontinued.

During the changeover in fire policy from a fire control emphasis to one of fire management, some questioned the relevance of Smokey Bear, while others suggested a change in his message. Some even want to replace his ever-present shovel with a drip torch in order to extol the many advantages of prescribed fire management options. However, Smokey's image and message remain intact. Trying to eradicate this appealing and popular image would be futile, and successful fire management programs will always seek the elimination of unauthorized and unexpected fires. Based on the fire statistics presented earlier, there is still plenty of room for Smokey Bear messages.

The Keep Green Program   This fire prevention campaign began in Washington State in 1940 under the slogan *Keep Washington Green!* It was adopted by Oregon a few months later and then spread throughout the country. Some Keep Green support comes from state forestry departments, but most is from private forest owners and the timber industry. The Keep

FIGURE 8-10
"Come to the forest, come without fire." The award winning entry for the 52nd Wildfire Prevention Poster Contest, sponsored by Keep Washington Green. Through this competition, students become aware of what uncontrollable wildfire can do to forests and animals. *(Contribution by Julie Weber; courtesy of Keep Washington Green)*

Green program reaches the public through newspapers, radio and television, stamps and decals, leaflets, posters, and novelties. The annual Keep Green poster contests for school children are a very popular part of the program (Fig. 8-10). As they depend heavily on local support, Keep Green organizations thrive best in states with a strong forest industry. In recent years they have become more closely aligned with the Smokey Bear program and serve effectively in distributing the many and varied materials generated by both programs. This has served to solidify that element of the prevention program directed at wildfires caused by human actions.

Violations and Enforcement   The objective of enforcement is to reduce violations of fire laws, fire safety requirements in forest and range use permits, fire clauses in timber sale and other contracts, fire ordinances that apply to construction and land development, and regulations governing public use during periods of high fire danger. Such enforcement deals with two types of violations: violation of fire regulations and illegal actions. The first results from a person's failure to perform a legally required action; for example, failure to have a burning permit; failure to have an approved spark arrester on a tractor, chain saw, or trail bike; failure to clear a railroad right-of-way; failure to dispose of slash; failure to have a shovel, ax, and bucket; failure to extinguish a campfire; or failure to observe smoking restrictions or fire closures. The second type of violation deals with illegal fire starts. Criminal or civil court action for the collection of suppression costs and damages may be necessary, depending on the type and circumstances of the violation.

Another aspect of enforcement involves the investigation of wildfires to determine the cause and to attempt to identify the responsible party, if it was person-caused. It is important to determine the precise cause of all fires so as to properly direct prevention programs. Successful identification and, if appropriate, prosecution of responsible parties, can serve to emphasize the often serious consequences of carelessness when dealing with fire.

Enforcement of fire laws and regulations educates the general public, as well as those who work or live in wildlands, about specific fire safety measures. Deterring negligent or malicious persons from destructive behavior also constitutes an important objective of fire prevention.

Engineering Solutions   Fire prevention engineering seeks to reduce or eliminate fire risks and hazards. Many fires can be prevented by the application of mechanical, electrical, chemical, or industrial engineering techniques that modify or eliminate ignition sources or ignitable fuels. There are many examples of engineering technology that have been successfully developed and applied to the prevention of wildland fires:

- Exhaust spark arresters for locomotives, trucks, tractors, logging and roadbuilding equipment, powersaws, and other portable gasoline-powered equipment

- Exhaust system heatshields for vehicles
- Improved brakeshoes for railroad cars
- Portable fire extinguishers
- Improved fire grates and fire circles for camp and picnic grounds
- Lightweight, portable propane campstoves for backpacking
- Herbicides and soil sterilants for fireproofing railroad and electric transmission line right-of-ways
- Improved receptacles for outdoor burning of trash
- Chimney caps and screens for wood-burning stove stacks
- Cutters, chippers, choppers, crushers and compactors, and other mechanical equipment to reduce the ignitability and flammability of logging slash and land clearing residues

Many fire prevention engineering solutions are products of teams of engineers, wildfire experts, and resource specialists working together at the U.S. Forest Service Technology and Development Centers located in San Dimas, California, and Missoula, Montana. The research and development at these centers have led to many "engineered" improvements in equipment for fire prevention, suppression, and personal fire fighter safety.

Fuel Treatment *Fuel treatment* refers to the manipulation or removal of fuels to reduce the likelihood of ignition and to lessen potential damage and resistance to control. Fuel treatment as used here is part of, but not the same as, fuel management. Modern fuel management has an ecological objective in addition to a fire-hazard reduction objective.

Fuels can be modified in many ways to reduce their hazard. The techniques used most often are isolation, removal, and rearrangement.

Fuel Isolation The objective of fuel isolation is to minimize the spread of a wildfire by containing it within an isolated area, or to keep it from entering an isolated area. The most frequently used methods of fuel isolation are *firebreaks, fuelbreaks, greenbelts,* and *greenstripping.* A *firebreak* is a strip or track of mineral soil from which all fuels and vegetation have been cleared (Fig. 8-18). A firebreak must be recleared every year, sometimes twice a year. A *fuelbreak* differs from a firebreak in that the cleared area

is permanently converted to a cover of low fuel volume and low flammability, such as perennial grass or low shrubs. While a firebreak is designed to stop a fire, the fuelbreak is meant to provide a line that can be safely occupied under any burning conditions, and where a fireline can be built and backfired. A *greenbelt,* a strip that has been cleared and converted to a nonflammable cover type, is maintained in that state by irrigation and mechanical treatment. *Greenstripping,* gaining use on rangelands, involves planting grass that maintains its greenness later than other vegetation. Greenstrips break up massive areas of fuel continuity and tie together natural firebreaks, such as rock outcrops.

Fuel Removal Fuel removal for hazard reduction purposes most often involves logging or thinning residues; that is, slash disposal. One way of removing logging slash is through intensive utilization. Larger residues can often be converted to pulp chips or fuelwood. Unmerchantable tree boles may make poles, posts, or house logs. Opportunities for intensive utilization depend on demand and proximity to markets. Utilization by itself may not be sufficient to reduce fire hazard to acceptable levels. The most common method of slash disposal, with or without intensive utilization, is prescribed burning. As indicated earlier, fuels may be piled and burned or broadcast burned. Broadcast burning may be applied to slash on clearcut areas, as an underburn in partially cut areas, or in a natural forest. Prescribed burning of large tracts of hazardous fuel may be accomplished with the aid of aerial ignition. During recent decades, air quality considerations have become an important part of decisions on all prescribed burning.

Fuel Rearrangement In some cases, acceptable hazard reduction can be accomplished by modifying the arrangement of the fuel, especially its compactness and continuity. For example, fire hazard may be abated by lopping and scattering slash resulting from partial cuttings, light thinnings, and pruning operations. Lopping involves cutting the limbs from the tree stems to make the slash more compact and closer to the soil, where it will decompose more quickly. Scattering the detached limbs about the area helps reduce fuel continuity. In heavy slash, a variety of large machines may be used to crush and compact or otherwise rearrange fuels by crushing, chopping, chewing, chipping, and mixing the fuel into the surface soil.

PRESUPPRESSION OR PREPAREDNESS

Presuppression or preparedness activity includes all the numerous and varied actions undertaken before a fire occurs, to help ensure safe and effective fire suppression. These activities relate to fire preparedness or readiness, fire danger rating, fire detection, and plans for cooperation between fire suppression organizations when needed.

Fire Preparedness

Fire preparedness means being prepared or ready to take appropriate action on fires when they occur. Preparedness activities can be relatively simple or quite complex, depending on the nature of the land to be protected. Its size, terrain, vegetation and fuels, value, fire potential, and the land management objectives all enter into preparedness. For a large, multiple-use area like a national forest these activities might include the following:

Integrating fire considerations in land management plans

Developing fire management plans such as for slash burning, and prescriptions for the use of prescribed natural fire

Planning appropriate fire organizations and their budgets such as for fire suppression or slash burning crews

Designing, installing, and maintaining an adequate communications system

Procuring an adequate cache of equipment and supplies

Recruitment and training of personnel

Developing and implementing preattack plans

Developing and implementing fire control plans containing work force guidelines, dispatching procedures, and other routine fire readiness activities

Fire Danger Rating

Anyone who travels through wildlands during fire season has undoubtably seen large roadside signs featuring Smokey Bear and "today's fire danger" (Fig. 8-11). As the sign changes from low to moder-

FIGURE 8-11
The Smokey Bear "Prevent Forest Fires" sign, reminding locals and forest and park visitors of today's fire danger. Dixie National Forest, Utah. *(Photo by Grant W. Sharpe)*

ate to high and on to extreme fire danger, fire prevention activities are also increasing incrementally, as is the level of fire preparedness. The general public is informed so people can use extra caution in all areas, but particularly during travel in remote areas. The level of fire comes from a fire danger rating system that numerically integrates weather, fuel, terrain, and fire risk factors into a series of indexes reflecting current protection needs. In the United States, this National Fire Danger Rating System (NFDRS) is used by federal, state, and private wildfire protection agencies (Fig. 8-12).

Computation of fire danger rating indices requires daily weather data collected from the protection area. On large ownerships, this requires the installation and maintenance of a system of fire weather stations. Increasingly, manually operated stations at fire lookouts and guard stations are being replaced by "remote automatic weather stations" (RAWS) that bounce their data off satellites to a receiving computer on the ground (Fig. 8-13).

The NFDRS provides fire potential information on a local basis. Another fire presuppression information system, METAFIRE, forecasts the probability of 500-acre (200-ha) or larger fires in 49 states up to two days in advance (Paananen, 1991). METAFIRE uses the National Weather Service

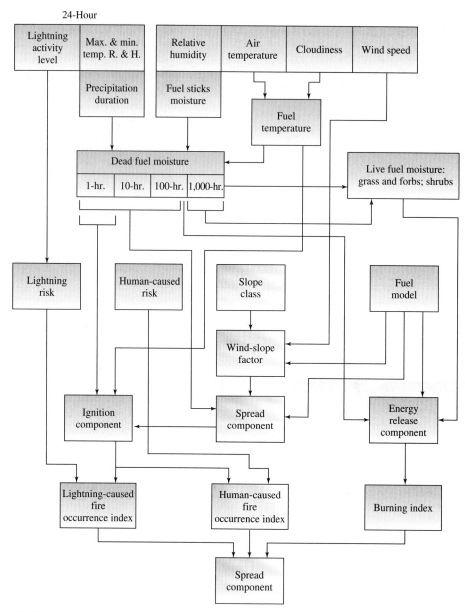

FIGURE 8-12
The structure of the National Fire Danger Rating System. Fuel, weather, and fire conditions are translated into indexes that reflect the probable occurrence and potential severity of fires.

network to produce reliable and consistent reports of the severity index for all 390 climatic divisions in the country. With onsite software, fire managers can produce regional and national maps of the current day's fire severity at 1:00 P.M. local time and predict the fire severity probability for the next day and the day after (Fig. 8-14). Managers can use the report and maps to monitor large-fire potential, perhaps to implement a burning ban, to inform the public about the fire danger, and to preposition suppression teams for strategic attack before a major fire occurs.

FIGURE 8-13
To calculate fire danger, daily weather observations must be collected from the protection area. Solar-powered, remote automatic weather stations are rapidly replacing traditional, manually operated stations for this purpose. *(Courtesy of USDA Forest Service)*

Fire Detection

Early detection and fast attack are traditional means of reducing damage from wildfires. This requires adequate surveillance of the protection area. In the "old days" this was done by placing people in lookouts on mountain tops, or in towers, or at guard stations from which they would patrol by horseback or vehicle. Although lookouts and fire patrols are still utilized for fire detection, especially during periods of high or extreme fire danger, use of aircraft is the most common method of fire detection today. Aerial detection systems may involve both routine patrol flights based on the level of fire danger and special patrols following lightning storms or during periods of high fire activity.

The utility and effectiveness of aerial detection has been greatly enhanced by the installation of airborne infrared (IR) line scanning systems and "forward-looking infrared" (FLIR) systems. The U.S. Forest Service began using infrared radiation systems in 1966 for fire detection and fire mapping. Infrared systems can detect fires in the dark and through smoke, but not too well through clouds. Nonetheless, they are a powerful fire detection tool, especially when teamed with information from lightning detection systems. Lightning detectors can detect ground strikes and transmit their location, via satellite, to a central dispatch center. Maps showing major concentrations of lightning strikes can then be used to plan IR detection flights as soon as the storm passes, day or night (Fig. 8-15).

Fire Cooperation

Cooperation has always been an important element of wildfire protection. In the West, private forest owners have traditionally formed Fire Protection Associations or cooperatives for the common purposes of slash disposal following harvest operations, and for fire suppression. State fire protection agencies often participate in such associations. Similarly, federal land management agencies often enter into agreements with the associations, cooperatives, and state agencies to authorize initial attack by the closest firefighters, regardless of agency and landownership. In the Lake states and Northeastern states, forest fire protection compacts with adjoining Canadian provinces have been in place for many years for the purpose of mutual preparation and response to wildfires even across the U.S. and Canadian border (Fig. 8-16).

The National Wildfire Coordinating Group An unprecedented era of cooperation among wildfire protection agencies in the United States began in 1976 when the National Wildfire Coordinating Group (NWCG) was formed by interagency agreement between the secretaries of the interior and agriculture. The NWCG is made up of representatives of the Forest Service, Bureau of Land Management, Fish and Wildlife Service, Bureau of Indian Affairs, National Park Service, and two representatives from the National Association of State Foresters. Liaison is maintained with the U.S. Fire Administration and the National Fire Protection Association (Fig. 8-17).

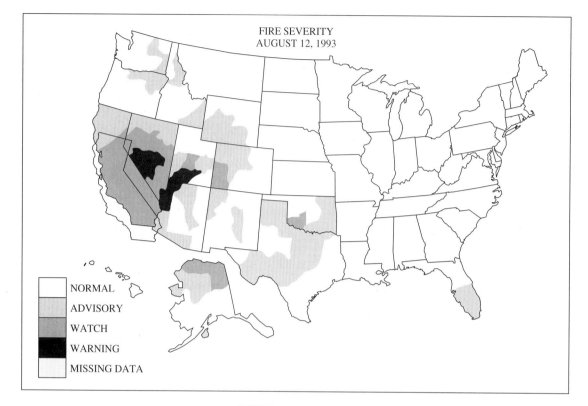

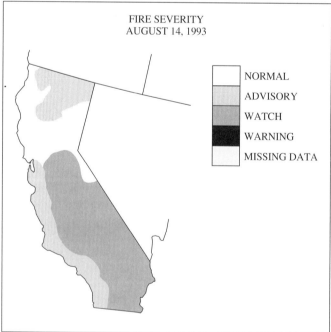

FIGURE 8-14

The METAFIRE system provides maps of large-fire potential. *Upper:* Nationwide fire severity for August 12, 1993, was mostly normal, with advisories for parts of the Rockies and the Southwest. Large-fire potential was more serious in California, the Great Basin, and interior Alaska. *Lower:* Fire severity map for California on August 14, 1993, showing the differences in fire severity in the state. *(Courtesy of USDA Forest Service)*

FIGURE 8-15
Computer monitor showing lightning strikes in the western United States for a
specified date and time. Data is generated by the Automatic Lightning Detection
System. *(Courtesy of National Interagency Fire Center, Bureau of Land
Management)*

FIGURE 8-16
An example of international cooperation in controlling wildlife. Here, a Canadian
DC-6 makes a retardant drop on a wildfire in the Wenatchee National Forest,
Washington. *(USDA Forest Service, photo by Don Seabrook, Wenatchee World)*

FIGURE 8-17
Upper: The entrance sign to the National Interagency Fire Center at Boise, Idaho. The Administration Building can be seen in the background. Symbols on the sign represent participating agencies: Bureau of Land Management, USDA Forest Service, National Park Service, National Oceanic and Atmospheric Administration, Bureau of Indian Affairs, and U.S. Fish and Wildlife Service. *Lower:* Aerial view of the National Interagency Fire Center. *(Photos courtesy of NIFC, Bureau of Land Management)*

WILDFIRE SUPPRESSION

Wildfire suppression includes all the tasks and activities involved in attacking, fighting, and controlling a wildfire. Firefighting is hazardous and requires well-trained, physically fit men and women. In addition, firefighters must be outfitted with proper safety clothing and modern, well-maintained firefighting equipment. Of course, firefighters must be fully qualified through experience and training for their assigned jobs.

The essential fire suppression task is to successfully break the *fire triangle.* The fire triangle repre-

FIGURE 8-18
Building a fire line with hand tools to stop the advance of an approaching wildfire.
Angeles National Forest, California. *(Courtesy of USDA Forest Service)*

sents the three things—heat, oxygen, and fuel—that are required in proper combination for ignition and combustion. To suppress a fire, the firefighter must, therefore, remove the fuel, reduce the heat, and/or exclude oxygen.

Suppression Tactics

Sizing Up the Fire   The first step in the suppression process is to size up the fire so as to determine the best way to break the fire triangle and achieve control. The person in charge, or fire boss, must choose the method or combination of methods that best satisfy the fire suppression objective established for the area. This decision requires an estimate of probable fire behavior.

Predicting Fire Behavior   Predicting potential fire behavior constitutes an important task in sizing up a fire to determine the appropriate attack strategy and tactics. Computerized fire behavior prediction and fuel modeling systems are used to accomplish this task. For example, a popular program named BEHAVE is a set of interactive computer programs that draws together fire behavior prediction technology into a single package (Andrews & Chase, 1990).

Information on the fuel type and weather conditions is integrated, and initial estimates are made for potential fire behavior, including spread rate, intensity, and size. This information is made available to the fire boss leading the initial attack.

Attacking the Fire   Next, the type of attack must be decided. *Direct attack* requires building a control line at or very close to the leading edge of the fire. *Indirect attack* means building the fire line some distance from the flaming edge of the fire.

Direct attack is selected when the fire is small, spread is slow, heat intensity is low, and access to and line building at the fire's edge are easy with the tools available. Direct attack may involve *hotspotting, line building, cold trailing,* and *burning out.*

Hotspotting involves moving quickly around the fire's perimeter, attacking only the worst flaming parts, or *hotspots.* It is a delaying tactic to slow the fire's spread until the fireline can be built. In difficult terrain, hotspotting is often accomplished using air tankers that drop fire retardant on the hotspots (Fig. 8-16).

Line building is often a two-step process. First a scratch line is built to stop the fire's advance. This temporary line is then improved to become a solid fire line that is wide and clear enough of fuel so that the spread of the fire is stopped (Fig. 8-18). When aerial attack is utilized, the retardant liquid is applied directly to the flaming edge so as to knock down flames and stop fire spread.

Cold trailing is done when the fire has ceased to burn actively and the perimeter is cold to the touch. In this situation a minimum fire line or trail is constructed to assure that no residual fire will flare up at a later date and cause the fire to spread further.

Burning out is the deliberate burning of any unburned fuel concentrations within or near the fire line.

Indirect attack is selected when a fire is large, spread is rapid, heat intensity is high, and access to and line building at the fire's edge are difficult with the tools available. Indirect attack involves line location, backfiring, and burning out.

Line location for indirect attack takes advantage of existing barriers to fire spread such as roads, trails, rocky outcrops, streams, firebreaks, fuelbreaks, and other natural or constructed barriers, as well as changes in vegetation and fuel type. Aerial attack involves applying liquid retardant to unburned fuels that are in the path of the fire.

Backfiring is used to remove the fuel between the fire front and the indirect line. This technique is often the only practical means of stopping a fire's spread in indirect attack. Backfiring is dangerous and should be performed only by qualified firefighters with advanced knowledge of fire behavior.

Burning out the fuel between an indirect line and a flaming flank of a fire is also risky, but not as dangerous as backfiring.

Mopping-up and Patrolling the Fire Area   The final tactic for fire suppression is mop-up and patrol. Before a fire is abandoned, all hot and smoldering material must be put out, or *mopped up*. Mop-up involves traversing the entire fire area for smoke and embers, inspecting the entire fire line for breaks, checking snags on either side of the fire line for smoldering embers, and checking stumps just inside the fire line to negate the possibility of roots burning beneath the fire line (Fig. 8-19). Mop-up, done correctly, is a tedious, unglamorous, dirty job, but an extremely important one. Following mop-up, a fire area is usually patrolled for awhile to make sure it is truly "dead out."

FIGURE 8-19
Mopping-up is the tedious, unglamorous, dirty job of suppressing all smoldering embers in and near the fire line. *(Courtesy of USDA Forest Service)*

Suppression Forces

In earlier days, pick-up labor from among the unemployed was used to fight fire, but now fire protection agencies increasingly rely on highly trained, organized crews of experienced firefighters to handle the brunt of their fire suppression responsibilities. These "hotshot" crews are managed under a concept of total mobility. They are sent anywhere they are needed, when they are needed. If their home unit gets in fire trouble while they are away, an available unit from elsewhere is dispatched to handle the problem. Consequently, each agency and unit can maintain fewer crews than would be needed to cover their worst-case scenario. During major fire emergencies and certain other situations, less well-trained crews may have to be utilized where appropriate.

Smokejumpers are the elite of professional firefighters. Only experienced firefighters are trained for

FIGURE 8-20
A smokejumper descending on a fire near Grand Junction, Colorado. Air escaping through the steering slats and three tail lobes gives the jumper maneuverability. *(Courtesy of USDA Forest Service)*

this dangerous duty. They are used for initial attack, often in conjunction with air tankers, in situations where they can get to a dangerous fire quicker than ground forces (Fig. 8-20). Jumpers may be dispatched to large fires by ground transport, as they are also trained in several specialty areas such as fire line explosives, aircraft operations, and first aid, or as sawyers. Thus, they serve in other places as well as in the initial attack.

A more recent, specialized suppression force, helitack crews, is a group of highly trained firefighters who rappel to locations near fires from 200–300 ft above the ground. Like smokejumpers, helitack crews can be deployed quickly to remote areas where they can attack fires while small.

Aircraft The use of aircraft in fire suppression is widespread and varied. Both fixed-wing aircraft and helicopters transport firefighters and their supplies to and from fires; they drop fire retardant chemicals, water, and foams on fires (Fig. 8-16). They are used also to reconnoiter fires, and some are equipped with infrared equipment for making precise maps of the fire and its perimeter through smoke and darkness.

Firefighting aircraft and airborne equipment are constantly being improved to take advantage of new technology (Fig. 8-21). Old airtankers are being replaced by newer, state-of-the-art aircraft equipped with variable flow retardant systems and capable of delivering 3,000 gallons of retardant per trip at superior speeds with improved performance and safety. Similarly, new airborne infrared scanner systems utilizing advanced technologies are constantly being improved to accomplish their mission more efficiently and completely.

Safety Concerns

Personal safety is a top priority in all fire suppression operations. To this end, fire equipment and development specialists have provided today's firefighters with a variety of personal protective gear that includes fire shelters, hardhats, goggles, flame-resistant gloves, leather boots, wool socks, shirts, jeans, coveralls or jumpsuits, and chain-saw chaps (Fig. 8-22). The fire shelter (a lightweight, compact, fire-resistant tent) has proven to be a vital safety item for line fighters and undergoes continuous evaluation to ensure effectiveness (Fig. 8-23). Currently, the use of respirators by wildland firefighters is under evaluation in response to health concerns regarding fire line smoke. Crews of healthy men and women in good physical condition, well trained in firefighting techniques, represent the most important factors related to safety. Regardless, firefighting is difficult, dangerous work and injuries, and even deaths, regularly occur. The danger from fire is dramatically portrayed in Norman Maclean's *Young Men and Fire* (Maclean, 1992), and more recently in accounts of the 2001 Thirty-mile Fire that claimed four lives in northern Washington State, where lack of experience and training were factors.

FUELS MANAGEMENT

The most important fire management challenge today is to develop and implement programs that successfully address fuels management issues. There are actually three separate but closely interrelated fuels management problems: the continued buildup of fuels, the decline in ecosystem health, and wildland-urban interface situations. These fuels management problems have a common fire ecology background.

FIGURE 8-21
New aircraft are constantly being converted for fire fighting purposes. *Upper:* An Orion P-3 with 3,000-gallon tank. *Middle:* Erickson Skycrane with a 2,000-gallon tank. Intake hose allows taking on water while hovering over a body of water or portable tank. *Lower:* Bell UH-IF helicopter with 340-gallon Bambi bucket, shown here cooling down flank of a prescribed fire in California chaparral. *(Courtesy of California Division of Forestry)*

FIGURE 8-22
A firefighter wearing the latest in personal protective equipment (PPE). This includes flame-resistant clothing, a fire shelter, hard hat and safety goggles, leather boots, and wool socks. *(USDA Forest Service, photo by George Johnson)*

FIGURE 8-23
A cut-away view of the fire shelter, showing a firefighter properly positioned inside. Developed by Forest Service equipment and development specialists, this shelter has saved the lives of several hundred firefighters who were overrun by fast-spreading fires. *(USDA Forest Service, photo by Jim Boyd)*

Fire Ecology Background

Fire prevention and suppression, and other humanly induced changes that have limited the natural role of fire in wildland ecosystems, have produced successional changes in wildland vegetation and fuels, increased the risk of severe fires, and reduced resource values. Over most of North America, fire was a principal reason that shade-intolerant (or early successional) tree and shrub species remained abundant. Fire also kept shrubs and forests from encroaching into grasslands, and favored herbaceous species and sprouting shrubs in the semi-arid steppes. Historically, most grasslands, southern pinelands, and western ponderosa pine forests burned every 25 years or less. That was their natural fire cycle (Fig. 8-24).

Fire suppression, as practiced until recently, was relatively easy in these vegetation types because of their generally low elevation and accessibility. The resultant exclusion of fire, along with the effects of poor logging and grazing practices, have brought about undesirable changes in plant communities. Interruption of the natural fire cycle in ponderosa pine, for example, has resulted in the development of dense understories of Douglas-fir, and several species of shade-tolerant true firs (Gruell, 1983; Fig. 8-25). The new forest is more crowded, and moisture is insufficient under drought stress. Trees are unable to retain vigor and are vulnerable to insect and disease attack. This results in an unhealthy forest, low in vigor, with many dead and dying trees. These forests are now experiencing large, stand-destroying fires, where the presettlement norm would have been frequent, low-intensity surface fires that periodically cleansed the forest floor, killing back dense reproduction and maintaining an open, often parklike stand of large pine trees. A similar situation is occurring in the southern pines, as well as in other forest types in various areas of the country.

The Fuels Buildup

After fire management succeeded fire suppression as the prevailing fire policy, its initial efforts were directed primarily at establishing programs and developing plans for the prescribed use of natural fire. These programs have attained some success in restoring a mosaic of vegetation and fuels to near "natural" conditions in some large wilderness-type areas. However, even in large wilderness or roadless areas, many

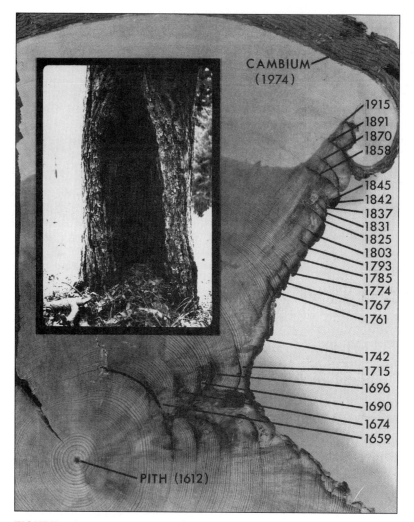

FIGURE 8-24

A cross-section of a ponderosa pine. Shown is the chronology of 21 successive low-intensity, short-interval fires between 1659 and 1915, an average of one every 12 years. Bitterroot National Forest, Montana. *(Courtesy of USDA Forest Service)*

lightning fires, when occurring under severe fire hazard conditions, must be suppressed because of escape risk and potential damage to property and resources outside their boundaries. Also, lightning-caused fires that could spread into a wilderness-type area from outside multiple-use or private lands will continue to be suppressed. Consequently, even in wilderness or natural-type areas, fuels continue to accumulate and natural fire regimes have not been restored (Parsons & Landres, 1998).

Similarly, fuels on multiple-use wild lands continue to increase faster than they are being recycled through harvesting, fire, and decomposition. This is especially evident in low-elevation forests. In many such areas, tree harvesting, and, consequently, associated fuel treatment, have been curtailed because of environmental regulations, endangered species restrictions, old-growth concerns, and other issues. These include excessive smoke, the possibility of fire escapes, massive fuel accumulations, steep terrain, and difficult access.

Vegetation and fuels management utilizing prescribed fire or other methods also has been hampered by lack of public awareness of the fuels buildup problem and by shortages of funding and personnel. In contrast, vast sums of money are spent attempting to control severe wildfires in untreated fuels. In 1988, for

FIGURE 8-25

These photographs show the dramatic changes in a western Montana ponderosa pine stand as a result of disturbance from timber harvest and virtual exclusion of wildfire. All three photos are from an identical photo point. *Upper:* 1909. This stand was selectively cut in 1907 or 1908. Luxuriant grass/forb cover reflects prelogging conditions. A low-growing shrub can be seen between the forester and stump. *Middle:* 1927, 18 years later. Douglas-fir regeneration has resulted in marked change in understory. Grass/forb ground cover persists, but now shrubs are more evident in foreground. Pine stand is somewhat less dense because of cutting or windfall. *Lower:* 1948, 39 years later. Original view is now screened out by growth of young Douglas-fir. Ground cover in foreground now has considerable numbers of low shrubs. *(USDA Forest Service, photo by Gruell et al., 1982)* example, $145 million was spent to suppress the Yellowstone-area fires. These fires burned mostly in old-growth forests that had a natural fire return interval of 100 years, and in vegetation types where stand replacement fires are the norm. Thus, while fire protection for decades contributed to heavy accumulations of fuel, and prescribed fire might have reduced fuel loading, the Yellowstone fires were a natural event in the ecosystems where they occurred. Another $4 million was spent the same year battling the intense Red Bench Fire in and adjacent to Glacier National Park, which burned through heavy beetle-killed forest fuels. Again, during 1992, $18 million was spent fighting the Cleveland Fire in the timberlands of California's Sierra Mountains, and $16 million was spent on suppressing the Foothills Fire, which burned 257,000 acres (104,085 ha) of southern Idaho forest and range. Such evidence, continuing through the mid- and late 1990s, led to the current emphasis on restoring ecosystem health and avoiding undue fuels buildup to prevent the monumental efforts required to stop large conflagrations.

The Decline in Forest Health

The continuing fuels buildup is related to and compounded by a serious decline in forest and other ecosystem health. The problem is especially prevalent in the fire-dependant ponderosa pine forests of eastern Oregon and Washington, where insects and disease have killed trees across millions of acres, but this decline extends through the entire western pine regions.

The composition of the forest at lower elevations has shifted from historically open stands of widely spaced ponderosa pine and western larch to stands with dense understories of Douglas-fir and grand fir. The preponderance of dead and stagnant trees has a profound effect on forest health by fostering epidemic levels of disease and insect infestations. They also provide fuel accumulations of dead and dying trees, and large wildfires are now causing widespread mortality, adversely affecting visual quality and wildlife habitat, as well as stream sedimentation, after the big burns.

The suggested management strategy to restore forest health in lower-elevation forests involves a combination of partial cutting and prescribed fire on a large scale to produce healthy, open, or more open, parklike forests in the future such as were present before European settlement when fire played its natural

FIGURE 8-26

Many forest health problems have been caused by years of excluding fire from fire-adapted ecosystems. Scientists and managers hope that reducing fuels through silvicultural treatments and reintroducing low-intensity fire through prescribed burning will restore the health of long-needled pine forests. *(Courtesy of USDA Forest Service)*

role (Mutch et al., 1993; Fig. 8-26). See Chapter 5, Silviculture, for additional material on forest health.

The Wildland–Urban Interface

The fuels buildup and forest health problems are compounded by the proliferation of residential and recreational dwellings on private lands adjoining and intermingled with large tracts of state and federal wildlands. On the one hand, people residing on the forest edge may oppose prescribed forest and fuel management activities because such activities might change the landscape they view, but when fire occurs, protection of their homes often diverts the work force of firefighters needed to protect that very landscape.

Perhaps more than anything else, the almost yearly loss of life and property in this wildland–

FIGURE 8-27
The wildland–urban interface is a dangerous place to build a home. *Upper:* The slope, nearby flammable vegetation, and the wood shingle roof make fire survival very improbable. *Lower:* One of several homes destroyed in a canyon fire near Missoula, Montana. All combustible materials in and around this home were burned, illustrating the vulnerability of forest homes to fire. Note the closeness of the now blackened forest. *(Photos courtesy of USDA Forest Service)*

urban interface testifies to the folly of ignoring the buildup of wildland fuels and allowing unregulated residential use of such hazardous areas.

The increasing number of major incidents in such areas is making fire in the urban–wildland interface a major national issue. For example, in 1990, the Paint Fire in southern California burned 648 structures, caused $240 million in damages, and cost over $10 million to suppress. In 1991 the worst fire in California history, the Oakland/Berkeley Hills Fire, started when high winds carried burning embers from a duff fire to overgrown vegetation, which in turn ignited

tree crowns and then spread to adjacent homes (National Fire Protection Association, 1992). The result was 2,449 homes and 437 apartment and condominium units destroyed, 1,600 acres (650 ha) burned, $1.5 billion in damage, and 25 lives lost. Again in 1993, in less than two weeks, over 215,000 acres (87,000 ha) were burned and 1,000 homes lost in a southern California wildfire before the winds ceased and control was achieved.

Suppression efforts in the interface are often hampered by poor roads, inadequate water supplies, flammable vegetation close to homes, and flammable

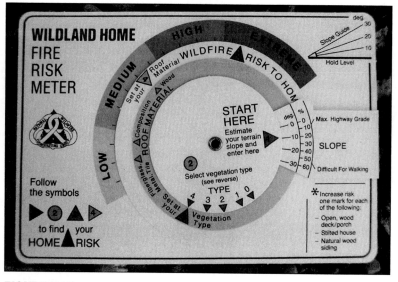

FIGURE 8-28

The major fire risk factors for wildland homes are incorporated into this easy-to-use meter developed by fire research foresters. The meter is often used as a handout when making fire prevention contacts with wildland home owners. *(Courtesy of USDA Forest Service)*

construction materials such as wood-shingle roofs (Cohen, 2000; Fig. 8-27). Coordination with local fire companies is essential, but often complex.

One strategy, *defensible space,* has proved effective in wildland–urban interface fire presuppression (Bailey, 1991). This involves clearing areas between structures and the surrounding flammable vegetation that are sufficient to allow firefighters to battle an oncoming wildfire before it reaches structures, or to stop a structural fire before it ignites the adjacent vegetation. With defensible space and the prudent use of less-flammable building materials, the structure has a chance to survive on its own when firefighting forces are unavailable. A woodland home fire risk meter has been developed by fire research foresters (Fig. 8-28).

SUMMARY

Paradoxically, successful long-term fire prevention and suppression have changed forests and increased risk of severe and costly wildfires. As a result, fire policy in the United States has gradually evolved

from one of fire control to one of fire management. Fire management may include a system of vegetation and fuels management based on wildland fire ecology. As the century turns, a national fire plan is being implemented (www.fireplan.gov) to restore burned areas and reduce fuel loads on millions of acres, primarily in the West.

Generally, more manager-ignited prescribed fires are needed in park and wilderness-type areas. In multiple-use areas where timber harvest, grazing, and other resource management activities are allowed, fire is being recognized as an important ecological process to maintain diversity and productivity. In residential wildlands, managing vegetation and fuels to avoid hazardous accumulations and provide defensible space is the accepted goal.

This change of emphasis from fire control to fire management has provided managers with more latitude to respond to wildfires and to use prescribed fire in such a manner as to minimize the destructive elements of fire while maximizing its benefits. As in the past, protection from most humanly caused wildfires through traditional and updated prevention, presuppression, and suppression techniques remain a major and coordinated effort of resource managers. Fire management capabilities are greatly enhanced by the seemingly continuous arrival of new and improved technological devices for fire detection, presuppression, suppression, and worker safety.

LITERATURE CITED

Andrews, Patricia L., and Carolyn H. Chase. 1990. "The BEHAVE: Fire Behavior Prediction System," *Compiler* **8**(4):4–9

Bailey, Dan W. 1991. "The Wildland–Urban Interface: Social and Political Implications in the 1990s. *Fire Management Notes* **52**(1):11–18.

Cohen, Jack D. 2000. "Preventing Disaster: Home Ignitability in the Wildland–Urban Interface." *J. Forestry* **98**(3):15–21.

Gruell, George E. 1983. Fire & Vegetative Trends in the Northern Rockies: Interpretations from 1871–1982 Photographs. General Technical Report INT-158. Ogden, Utah: U.S. Department of Agriculture, Forest Service, Intermountain Research Station.

Hundrup, Wyatt. 2001, September. "Fire Plan Hearing Heats Up." *The Forestry Source*, p. 7. www.safnet.org.

Maclean, Norman. 1992. *Young Men and Fire: A True Story of the Mann Gulch Fire.* University of Chicago Press, Chicago.

Mutch, Robert W., Stephen F. Arno, James K. Brown, Clinton E. Carlson, Roger D. Ottmar, and Janice L. Peterson. 1993. Forest Health in the Blue Mountains: A Management Strategy for Fire-adapted Ecosystems. Gen. Tech. Rep. PNW-GTR-310. Portland, Oreg.: U.S. Department of Agriculture, Forest Service, Pacific Northwest Research Station. (Quigley, Thomas M. ed; Forest health in the Blue Mountains: science perspectives).

National Fire Protection Association. 1992. The Oakland/ Berkeley Hills Fire. National wildlands–urban interface fire protection initiative. Quincy, Mass.: National Fire Protection Association.

Paananen, Donna M. 1991. "What METAFIRE Can Do for Managers." *Wildfire News & Notes* **5**(2):5–6.

Parsons, David, and Peter Landres. 1998. Restoring Natural Fire to Wilderness. How Are We Doing? In: Teresa Pruden & Leonard Brennan, eds. Fire in Ecosystem Management: Shifting the Paradigm from Suppression to Prescription. Tall Timbers Fire ecology, pp. 366–373. Conf. Proc. No. 20 Tallahassee, Fl. Tall Timbers Res. Station.

U.S. Department of Agriculture, Forest Service. 1992. 1984–1990 Wildfire Statistics. Washington, D.C.

U.S. Department of Agriculture Forest Service. 2001, January. Toward Restoration & Recovery: An Assessment of the 2000 Fire Season in the Northern & Intermountain Regions.

U.S. Department of Agriculture Forest Service. 2001. 2000 RPA Assessment of Forest and Rangelands. P.O. Box 96090, Washington, D.C. 20090-6090. FS-687 on-line at www.fs.us/pl/rpa/list.htm.

WEBSITE

www.fireplan.gov. USDA and USDI Interagency National Fire Plan.

ADDITIONAL READINGS

Arno, Stephen F., and James K. Brown. 1989. "Managing Fire in our Forests—Time for a New Initiative," *J. Forestry,* **87**(12):44–46.

Chandler, Craig, Phillip Cheney, Philip Thomas, Louis Trabaud, and Dave Williams. 1983. *Fire in Forestry,* Vol. II: *Forest Fire Management and Organization,* A Wiley-Interscience Publication. John Wiley & Sons, New York.

Cottrell, William H. 1989. *The Book of Fire.* Mountain Press Publishing Company, Missoula, Mont. in cooperation with the National Park Foundation.

Edmonds, Robert L., James K. Agee, and Robert Gara. 2000. "Forest Health and Protection." McGraw-Hill, New York.

Fuller, Margaret. 1991. *Forest Fires: An Introduction to Wildland Fire Behavior, Management, Firefighting, and Prevention,* Wiley Nature Editions. John Wiley & Sons, New York.

McIver, James D., and Lynn Starr. 2000. *Environmental Effects of Postfire Logging: Literature Review and Annotated Bibliography.* General Technical Report PNWGTR-486. Pacific Northwest Research Station, Portland, Ore.

Miller, Don, and Stan Cohen. 1993. *The Big Burn: The Northwest's Great Forest Fire of 1910.* First Printing, revised. Pictorial Histories Publishing Co., Missoula, Mont.

Pyne, Stephen J. 1982. *Fire in America: A Cultural History of Wildland and Rural Fire.* Princeton University Press, Princeton, N.J.

Society of American Foresters. 1990. *Glossary of Wildland Fire Management Terms Used in the United States.* Guy R. McPherson, Dale D. Wade, and Clinton B. Phillops (compilers). The University of Arizona Press.

Walstad, John D., Steven R. Radosevich, and David V. Sandberg (eds.). 1990. *Natural and Prescribed Fire in Pacific Northwest Forests.* Oregon State University Press, Corvallis.

Wright, Henry A., and Arthur W. Bailey. 1982. *Fire Ecology: United States and Southern Canada.* A Wiley-Interscience Publication. John Wiley & Sons, New York.

STUDY QUESTIONS

1. Describe the evolution of fire policy. How does current policy fit into fire ecology and ecosystem management concepts?
2. Describe a situation in which manager-ignited prescribed fire might be used.
3. How does a naturally occurring fire become a prescribed natural fire?
4. What is the difference between fire risk and fire hazard? Give two examples of each.
5. What do you perceive as the major cause of wildfires in your state or province?
6. Describe the components of the fire triangle. Which side of the triangle does line-building seek to remove? Backfiring? Shoveling soil on coals? Water or slurry drops?
7. What role has fire historically played in forests and grasslands?
8. Why are there now large-scale problems with fuels accumulation and management?
9. How does this fuels management problem become a forest health issue? How does it affect wildlife?
10. What measures would you suggest to reduce deaths and destruction of property at the wildland–urban interface?

CHAPTER 9

Wildlife Conservation and Management

This chapter was authored by Dr. R. Gerald Wright, Professor of Wildlife and Principal Wildlife Scientist, USGS Idaho Cooperative Research Unit, College of Natural Resources, University of Idaho.

INTRODUCTION

Humans have always revered animal life even while fearing it. Some of the earliest drawings, found in caves in Europe, are characterizations of animals. Animals such as the eagle and bear, admired for their strength, fearlessness, and beauty, remain important symbols in the religious ceremonies of Native Americans.

However, for most of recorded civilization, a reverence for animals was tempered by a realization that animals must be utilized in order to sustain and protect human life. All men were hunters in early human cultures, as they relied on animals for many essentials including meat for nourishment, skins for clothing, sinews for rope and thread, fat for fuel, and antlers and bones for tools, weapons, and needles. Comparatively, humans have only recently begun to appreciate animals for aesthetic reasons as reflected in nonconsumptive activities such as viewing, photographing, or listening to animals, and through painting, identifying, studying, writing, and reading about animals.

FIGURE 9-1
A hunter surveys the landscape in search of prey. Some sport hunters associate hunting with mastery of skills; others simply enjoy the chance to spend time outdoors. *(Courtesy of USDA Forest Service)*

UTILIZATION AND APPRECIATION OF WILDLIFE

Consumptive Activities

Consumptive wildlife activities involve the harvest of animals through hunting and trapping, and are a direct outgrowth of our primitive past when life depended on such harvest.

Hunting Beyond obtaining meat for the table, which still powers rural hunting to some degree, hunting for recreation and sport remains a tradition among millions of Americans (Fig. 9-1). In 1996, 14 million people, or about 7 percent of the adult population, hunted. Proceeds from the sale of hunting licenses, as well as taxes on arms and ammunition, support state wildlife management agencies. These proceeds are used to acquire and protect habitat as well as to regulate harvest. U.S. hunters spend more than $12 billion per year on hunting activities and equipment, and, since 1923, hunters have paid more than $6 billion in license fees, all of which has gone to finance wildlife conservation through state wildlife agencies.

People hunt for meat, sport, and contact with nature (Kellert, 1996). Several reasons have been advanced to justify the sport of hunting. Klein (1973) summarized the traditional justifications for hunting in this way.

1. It is more humane to kill animals directly rather than allow death by starvation or by predation.
2. Man has evolved as a hunter, and thus sport hunting reenacts this traditional drama and brings people closer to nature.
3. Hunting keeps wildlife populations healthy, generally replacing other sources of mortality.
4. Hunting does not endanger populations; in fact, hunted species are usually the most abundant.

Some of these points are debatable; certainly, the role of hunting and trapping is now challenged on ethical, moral, and philosophical grounds. Peek (1986) questioned whether modern-day hunting serves to reenact the original hunter-gatherer lifestyles, when one considers that the extensively equipped hunters of today rarely need the game to survive. Likewise the "humane" aspects of hunting are an imposition of human values and judgments on a natural process. This debate will undoubtedly widen and will have considerable impact on the way wildlife will be managed in the future.

For the past two decades, the proportion of hunters in the U.S. population has been declining. The most recent survey data available from the U.S. Fish and Wildlife Service (www.fws.gov) indicate that the actual number of hunters has also been declining slightly, dropping from 14.1 million in 1991 to 14 million in 1996. These diminishing numbers lessen the political clout of sport hunters. One of the primary reasons for the decline in the sport of hunting is that access to private lands is increasingly denied to hunters throughout the country, either through closures or leases of land to membership-only clubs, or through urbanization. There is concern that average hunters may thus find themselves excluded for economic reasons.

This has important economic and political implications for wildlife management. License sales and excise taxes on sporting goods for both hunting and fishing are primary sources of funds for most state wildlife management programs as well as for some federal wildlife programs. As the opportunity for hunting declines, so do hunter numbers and, proportionately, revenues to support wildlife management and the political support wildlife agencies receive from their constituencies.

Programs to encourage public recreational access to private lands have met with limited success despite a variety of approaches, including various forms of economic compensation to the landowner. Too many instances of misconduct by hunters, as well as trespass, property damage, personal liability lawsuits, and other problems dissuade landowners from permitting public access.

Trapping   Trapping is pursued both as a sport and a vocation because, unlike hunters, successful trappers can sell their products. Trapping primarily involves furbearing animals such as muskrat, raccoon, red fox, and beaver, whose pelts are used mainly for clothing and decoration. Three types of traps are commonly used for capturing furbearers; steel or leghold traps, snares, and killing mechanical traps. Leghold traps are by far the most popular among trappers because they are portable, relatively inexpensive, and can be used in a variety of situations. Unless a leghold trap is set in or near water in such a position that the trapped animal drowns, the victim is held alive until the trapper arrives. With both leghold traps and foot snares, animals unintentionally caught can be released, whereas neck snares usually strangle their victims.

Aside from its claim to sport status, trapping serves several management functions including reducing or controlling nuisance populations of certain furbearers such as striped skunk, raccoon, beaver, and martin, all of which are generally hard to harvest by any other means. The history of trapping in the United States shows sporadic attempts by humane groups to ban trapping because they perceive it as cruel. More moderate groups, and the fur industry itself, have long sought and even offered a monetary prize for the development of a humane trap (Gentile, 1987). Regulatory actions have also been strengthened over the years to include established trapping seasons and restrictions on the size and type of trap. Other regulations have required daily visits to the trap lines, restrictions on trap location, name tags on traps, and reports on harvest.

Fishing   Available to young, old, rich, poor, able-bodied, and infirm alike, this form of consumptive use does not seem to arouse the disapproval that hunting incurs, and remains a popular activity within various social groupings. In its 1996 survey, the U.S. Fish and Wildlife Service estimated that there were 35.2 million anglers in the United States, a slight decline from 35.6 million anglers estimated in the 1991 survey.

Fisheries are a broad, renewable natural resource, and include both freshwater and marine species. With some exceptions, freshwater fisheries concentrate on recreational fishing, and marine fisheries on both recreational and commercial fishing. States manage the freshwater fisheries resource. They must work with the Native American tribes and the federal agencies where jurisdictions and interests overlap.

Marine fisheries include the renewable natural resources of oceans and estuaries—those plants and animals that live in everything from diluted saline waters at the mouths of rivers and streams, to deep mid-ocean areas.

Saltwater fishing stocks have declined precipitously in recent years, following decades of spawning-habitat degradation (see Chapter 10, Watersheds and Streams). Increased numbers of commercial boats, the efficiency of harvesting techniques, and the complex social and political relationships are also factors in the saltwater fisheries management crisis. While habitat degradation has affected freshwater systems also, supply and demand seem to remain in better balance in many areas, bolstered by state and federal stocking and habitat enhancement funds, and controlled by state regulations.

Nonconsumptive Activities

Wildlife recreation based on nonconsumptive activities is one of the most popular pastimes in the United States (Fig. 9-2). In 1996, 62.9 million adults participated in activities such as feeding, observing, and photographing wildlife, and spent over $29 billion doing so. These activities frequently coincide with other pursuits such as hiking or traveling, and surveys reveal that an increasing number of people travel to specific areas such as wildlife refuges and national parks with the express purpose of viewing animals.

The traffic jams that result when even the most common large mammal is spotted along a park road leave little doubt about the appeal of animals, and there seems to be a direct correlation between visitor satisfaction and wildlife observed. Hastings (1986) surveyed over 4,000 visitors to Cades Cove in Great Smoky Mountains National Park, an area noted for its history and scenery. Even there he found that the expectation of seeing wildlife was a major reason people visited the area, and 92 percent of all visitors reported that actually seeing wildlife increased the enjoyment of their trip. Similarly, Miller and Wright (1999) found that of the visitors surveyed, 43 percent visited Denali National Park to see wildlife—more than for any other purpose, and 86 percent of the visitors said that seeing wildlife contributed to their trip satisfaction. There is strong evidence that spiritual and emotional interest in wildlife will continue to grow. One reason for this appears to be the fact that in their daily lives, Americans are increasingly isolated from nature, since more than three-fourths now live in urban areas.

Not all animals are equally appreciated. Most people prefer mammals and birds over reptiles, amphibians, and invertebrates, and large mammals such as deer, bear, and elk seem more popular than smaller mammals (Kellert, 1996). Predatory animals have a special and growing appeal, as evidenced by the intense interest in wolf reintroduction to areas where they are now absent, and efforts in California to protect cougar.

FIGURE 9-2
Viewing and photographing wildlife is a popular nonconsumptive pursuit. The photographer probably enjoyed capturing these sea otters on film. *(Photo by Ed Klinkart)*

THE RELATIONSHIP BETWEEN WILDLIFE AND HABITAT

The Concept of Habitat

An animal's habitat is, in the most general sense, the place where it lives. In a formal sense, *habitat* can be defined as an area with the combination of resources such as food, cover, water, substrate, topography, temperature, precipitation, and security that promotes occupancy by individuals of a given species and allows them to survive and reproduce (Morrison et al., 1992). The quality and extent of an animal's habitat in large measure governs its ability to survive, and loss of habitat is the greatest single threat to wildlife in the United States. Understanding a species' habitat requirements and preferences allows management to increase desirable components, such as preferred forage plants.

Important Components of Habitat

Food   Above all else, habitats sustain animals by providing food. The numbers of animals an area will support is ultimately limited by the amount of food available, particularly during seasons of scarcity, which is usually the winter. This concept is referred

FIGURE 9-3
Cover provided by vegetation offers protection from predators through concealment, and protection from extreme cold. Can you find the deer? *(Photo by Grant W. Sharpe)*

FIGURE 9-4
A snag, purposely left behind after timber harvesting, provides nesting, foraging, and perching sites for birds and small mammals. *(Photo by Grant W. Sharpe)*

to as the *carrying capacity* of the given habitat. Birds and mammals have evolved a wide range of adaptations for utilizing potential foods found in their environment. They have developed ways of eating and digesting such diverse plant foods as lichens, leaves, roots, stems, nectar, nuts, seeds, and fruits. In addition, all animals, either living or dead, serve as food for animals that are predators and scavengers. Some animals are highly specialized and feed on a limited range of foods. Others such as crows, bears, and raccoons have adapted to an omnivorous diet that includes a wide variety of plant and animal foods.

Cover *Cover* is an inclusive term that includes shelter from the extremes of wind and temperature (*thermal cover*), and protection from predators either through concealment or escape routes (*hiding cover* or escape terrain) (Fig. 9-3). Cover is provided primarily by vegetation, water, and topography.

Agricultural practices that leave cover in the form of hedge rows and crop residues are important to the over-winter survival of upland game birds. A major concern in the management of upland game birds such as the ring-necked pheasant is *clean-farming,* a practice that encourages the cultivation of crops right up to fence lines and roadsides, and the burning of drainage ditches, with few fields left vacant. Some states provide economic incentives to farmers to leave strips of cover along roadsides.

Forests and woodlands also provide important thermal cover, particularly in winter. In northern

parts of their range, white-tailed deer traditionally seek areas of dense timber, such as conifer swamps, where they spend the harshest part of the winter. In hard winters, given a choice between available cover and available food, deer have been found to choose cover. The best cover in these areas, called *deer yards,* consists of a closed canopy forest of mature spruces and firs that reduce windchill and intercept falling snow (thermal cover).

Habitat Necessary for Reproduction The quality of the cover used by waterfowl and upland game birds when nesting and incubating eggs can be critical to a species' success. Eggs and nestlings are eagerly sought by most predators. Concealment can take many forms. Some shorebirds such as terns place their nest on bare ground, but in sites where the color of the sand and rocks makes the speckled eggs virtually invisible, but most birds must use surrounding vegetation to conceal the nest.

Snags as Habitat Standing dead and dying trees provide nesting, foraging, and perching sites for a variety of birds and small mammals (Fig. 9-4). Until a few years ago, forest management called for removal of snags, as they were seen to be fire hazards and breeding sites for insects and fungi. Now managers recognize the value of snags for avian cavity-nesters. Further, they see these insectivorous cavity-nesting species as important to the health of the forest by helping regulate damaging insect populations such as the spruce budworm. Now forest management routinely requires that a certain number of snags be left standing after harvest operations.

Habitat Requirements of Selected Species

All species have unique habitat requirements but can be broadly categorized based on their habitat requirements as *generalists,* meaning they are adapted to a wide range of habitats, or *specialists,* meaning they are adapted to a narrow range of habitats.

Generalists These species are common and widely distributed and share a number of attributes. They usually can tolerate a wide variety of climates, have broader dietary, nesting, or breeding habitat requirements, and they often adapt more easily to the presence of humans. Generalist species include white-tailed deer, coyotes, raccoons, pheasants, fox squirrels, cotton-tailed rabbits, and cowbirds.

Exotics, a special category of generalist species, are organisms introduced into places they have not previously occupied, usually through human assistance. Many North American garden plants, as well as livestock species, are exotic in origin. Wildlife managers are primarily concerned with two categories of exotic species: those that have been introduced with the hope of establishing wildlife populations for recreational purposes, and those that directly or indirectly affect native wildlife or its habitat. Examples of species in the first category include the ring-necked pheasant, chukar partridge, nutria, barbary sheep, and axis deer. Exotic species that have negatively affected native species either through predation, competition, hybridization, or habitat destruction are starlings and English sparrows, the mongoose in Hawaii, sika deer in Maryland, feral burros in the southwestern United States, and mountain goats in Olympic National Park (see Fig. 17-10, page 398).

Specialist Species These have many characteristics that are the exact opposite of generalist species.

Their distribution is usually limited by narrow habitat tolerances or because preferred habitat has been diminished, and, as a result, they are usually less common. Narrow habitat requirements may be the result of an inability to tolerate a wide variety of climates, the need for specialized nest or breeding habitats or structures, because of requirements for specific kinds of food, or because of an inability to tolerate disturbance. In most instances, specialist species do not adapt as easily to the presence of humans or their associated disturbances. As a result, the management of specialist species is more difficult, and many have become rare, threatened, or endangered. Examples of such species include grizzly bear, northern spotted owl, condor, ferret, and Kirtland's warbler. This is discussed in greater detail under "Endangered Species Legislation."

Migratory Species These are animals with periodic or regular movement from one spatial area to another for access to food supplies, to find suitable breeding habitat, or to avoid predators or climatic extremes. Migration patterns are extremely diverse. For example, the migration route of the arctic tern covers 10,000 miles (16,000 km) between the polar regions of the northern and southern hemispheres. Conversely, species like elk in the western United States migrate relatively short distances, perhaps only 30 miles (50 km) between winter and summer ranges (Fig. 9-5).

The primary management challenge in dealing with species migrating between seasonally different habitats is the crossing of state, provincial, or even international boundaries. These political units often have different laws and different habitat management strategies that make integrated management difficult. Some of the first international wildlife laws, such as the Migratory Bird Treaty Act of 1918, were enacted to deal with the difficulties of protecting and managing migratory species.

Ecosystem Management

An important trend here recognizes that effective resource management must look beyond the needs of individual species towards an *ecosystem approach* in order to preserve a region's plant and animal communities. The ecosystem approach recognizes all elements of an area—not only the animate components

FIGURE 9-5
Elk have a relatively short migrating distance compared to other large animals. This herd, in western Wyoming, grazes on its summer range. *(Courtesy of Bureau of Land Management)*

but the physical features of the landscape such as the air, soil, and water. Ecosystems can range in size from an area as small as a pond or woodlot to systems as large as a rain forest or an entire ocean. The essential feature of an ecosystem is that all components are related to, and dependent on, one another via intricate networks that allow the flow of energy and the cycling of nutrients. These interrelationships mean that changes or manipulations in one component may have significant and perhaps unexpected effects on other components.

Emphasis on the ecosystem approach to management, particularly for large land areas, has developed for two reasons. One is a growing understanding of the deleterious effects of habitat fragmentation on wildlife—effects that are often obscured when only a small area is considered. The second is a recognition that the traditional approach to conservation that has centered on rescuing individual species from the brink of extinction has proven biased, inefficient, too difficult, and too expensive. Conservation funds are limited, and biologists tell us recovery will be exceedingly difficult for the growing number of species classified as threatened or endangered. In the United States alone, the list of candidate species is now in the thousands.

Biodiversity

This refers to preservation of all facets of ecosystems, that is, their biological diversity or *biodiversity,* including the variety and variability among living organisms and the environments in which they occur. Land managers recognize that, under the best of scenarios, on only a relatively small portion of the total land area in the United States will biodiversity flourish in all its aspects. These lands presently include national parks and designated wilderness areas. Yet, less than 6 percent of the conterminous United States is protected in nature reserves, and these are most frequently at higher elevations and on less productive soils. Conversely, the majority of plant and animal species occur at lower elevations and on more productive lands (Scott et al., 2001). Thus, an ecosystem approach that includes both private and public lands is needed to ensure the long-term protection of biodiversity.

Habitat Corridors

Given that extensive fragmentation of habitat now exists, there has been considerable interest in providing corridors for wildlife to travel between habitat patches, thus minimizing isolation. Such corridors can provide less mobile species with cover, protecting them from predators and harsh environmental conditions, thus allowing them to move between areas for feeding, breeding, and resting. On the grand scale, corridors have been proposed to link national forests, national parks, and wilderness areas within and between physiographic provinces. One of the most ambitious plans has been termed "Y2Y" and is envisioned as a protected corridor connecting the Yukon Territory with the Greater Yellowstone ecosystem (Posewitz, 1998). It seeks to facilitate the dispersal of grizzlies and other large carnivores in the region.

Despite the obvious appeal of corridors, few have been developed or analyzed. The effectiveness of corridors depends on the types of organisms involved, the type of movement, and the type of corridor. Narrow strips of trees are a regular feature of many landscapes, and these shelterbelts, hedgerows, and road rights-of-way can be effective in expediting the movements of some species. But most are too narrow to provide the environment needed for several competing species, especially those species that need to live, grow, and reproduce along the way.

MAJOR WILDLIFE HABITAT TYPES

Forests

The numbers and kinds of wildlife on forested lands depend on the quantity and quality of food and cover available, and these conditions in turn depend largely on the way forests are managed. Each type of forest community under a certain management regime provides a different mixture of food, cover, water, and other conditions needed by wildlife. Some forest types support a rich abundance of wildlife, others support only a few species. These differences are primarily due to variations in the structure and composition of forest communities. Years of field studies have provided evidence that forest communities with a high structural diversity contain a greater assortment of microhabitats and thus have the richest variety of animal life.

Forest management influences wildlife habitat in a variety of ways, but harvest methods, particularly clearcutting, often have the most dramatic influence. Clearcuts have both detrimental and beneficial effects, depending on the species of wildlife and the size, shape, and position of the cut, as well as its proximity to other clearcuts. Wildlife species that need openings may be helped, whereas those species that need dense tree cover will be harmed. Species that require shrubby growth for browse may be helped for many years after the cut, but only if adequate shelter is available nearby. For example, clearcutting can provide new growth of nutritious forage that may persist for up to 20 years after the cut and provide food for animals like deer, elk, and moose, while at the same time it would eliminate thermal cover to protect them from cold and deep snow.

Forest management practices that speed the establishment of a new stand, accelerate tree growth, and shorten the rotation period can have profound effects on wildlife. When natural regeneration occurs, grasses, forbs, and shrubs dominate the site for several years, greatly increasing the production of forage for herbivores like deer, rabbits, and some types of mice, and they provide a variety of habitats for birds. When a site is artificially planted with fast-growing conifers, the young trees soon shade out the grasses and shrubs. The use of herbicides to kill shrubs and hardwood sprouts in order to minimize competition also diminishes the quality of the habitat for many species.

Shorter rotation of forest harvests decreases the proportion of old trees. Southern pines grown for pulpwood are sometimes harvested in as little as 25 to 30 years. Douglas-fir in the Northwest is managed with rotations as short as 50 years. Wildlife species that depend upon older trees during part of their life will therefore decline under such management strategies. Other species requiring younger trees may benefit. In the Pacific Northwest, the short rotation period for Douglas-fir forests has doubled the number of black-tailed deer in some areas.

A particularly controversial aspect of forest management in recent years has been the retention of old-growth stands as wildlife habitat. The characteristics of old-growth forests are a moderate to high canopy closure, dominant trees in the stand with relatively

FIGURE 9-6

Left: The red-cockaded woodpecker is endangered because of the gradual loss of old-growth longleaf pine forests. *(USDA Forest Service photo by Jim Hanula) Right:* The northern spotted owl, now endangered because of its dependency on old-growth forests in the Northwest. *(USDA Forest Service photo by Eric Forsman)*

large diameters, dead or decaying snags or fallen trees, and heavy accumulation of logs and other woody debris on the forest floor (Thomas et al., 1988).

Numerous species of wildlife favor old-growth (Meslow et al., 1981). Such species include marten, fisher, red tree vole, northern flying squirrel, brown bear, bald eagle, northern spotted owl, marbled murrelet, red-cockaded woodpecker, and several species of amphibians (Fig. 9-6). However, the true importance of old-growth to most species has been hard to measure. Lacking this understanding, old-growth advocates have argued that the wisest course of action is to preserve all remaining old growth (Thomas et al., 1988).

This issue is most contentious in the Pacific Northwest, where estimates are that only about 17 percent of old growth that existed in the early 1800s remains today (Spies & Franklin, 1988). Conflicts over preserving old-growth forests spurred a several-year effort that resulted in the Northwest Forest Plan, approved by President Clinton and passed in 1993. Its aim was to cut logging levels significantly on millions of acres of national forests. The forests would be managed less for timber than for wildlife, fish,

and recreation. Many conservationists, however, remain concerned about the plan's goal and the reality of continuing harvests in the Northwest.

These concerns are also raised in the southern United States, where the endangered status of the red-cockaded woodpecker has been associated with a loss of old-growth longleaf pine forests (Conner & Rudolph, 1991). Across northern New England and the Lake states, white-tailed deer seek old-growth conifer stands for protection during winter. Even relatively small clumps of older trees can be important to wildlife. In the Southwest, suitable roosting sites for wild turkeys may consist of a group of as few as 12 large ponderosa pines.

The concept of a *featured species* plan allows a land manager to address a species that has special significance in an area while still taking into account the presence of other wildlife. For example, habitat-specific species such as the red-cockaded woodpecker can be provided for in management for turkeys or quail simply by leaving some decaying trees to serve as nest sites. Management actions that retain bald eagle nesting trees often serve a dual role for other species requiring mature conifer cover.

FIGURE 9-7
The barren ground caribou, an animal of the arctic tundra. Unlike other deer, both sexes have antlers; males shed in early winter, does in spring. Adak Island, Alaska. *(Courtesy of U.S. Navy)*

Rangelands

Rangelands occupy about one-third of the area of the United States, mostly west of the Mississippi (see Fig. 11-4, page 242 and Chapter 11 on rangelands). Their common attribute from the wildlife manager's viewpoint is the utilization of forage by grazing animals. Rangelands encompass a wide variety of habitats, from arctic tundra grazed by caribou (Fig. 9-7) to the grasslands of the Great Plains formerly grazed by great herds of bison, to the cold and hot desert shrublands of the Great Basin and the Sonoran and Chihuahuan deserts grazed by bighorn sheep and mule deer. Open areas within forests, as well as coastal marshes, are also important grazing resources.

Growth in grasses comes from the base of the plant, with new leaves replacing upper leaves consumed by grazing. Because of this, in most grasses, light grazing actually stimulates regrowth and can, over the grazing season, produce more forage than will areas that are ungrazed. In comparison, shrubs and trees grow new shoots from the tips of the older stems. When these tips are eaten or browsed, regrowth of the shoot is prevented or impaired. However, not all grass species are equal in their ability to withstand grazing. Excessive grazing, or grazing at times when plants are stressed, can injure or kill grasses. Overgrazing for extended time periods can deplete the forage resource beyond its ability to replace itself.

In most instances, excessive use by domestic livestock is detrimental to wildlife, as the best forage is removed and wildlife and domestic stock compete directly for what is left. The diets of sheep, goats, cattle, white-tailed and mule deer, bighorn sheep, and elk may overlap substantially, depending on the area and season. Excessive grazing can also reduce cover necessary for nesting and predator avoidance.

A balance between proper use of the forage resources by domestic livestock and the production of wild animals can be achieved with proper management. However, it is a balance often achieved with great difficulty and controversy continues over the extent of grazing to be allowed on publicly owned rangelands.

Riparian Areas

The interface between an aquatic environment (usually a stream) and an adjacent drier terrestrial zone is termed a *riparian zone*. These areas are often no wider than the floodplain of the associated river. Because of their proximity to water and their rich alluvial soils, riparian zones have long been attractive sites for agriculture, timber harvest, and grazing, and have provided ready access routes resulting in trails, railroads, and highways concentrating here also. Riparian zones are exceptionally important in the West, where the interface between the riparian and xeric (dry) upland vegetation can be abrupt. In such regions, the riparian zones are often the most productive timber and forage sites. These factors, plus the influence that riparian vegetation has on water quality and quantity and the associated aquatic environment, mean riparian zones are the focus of both human and animal communities.

Their use has been contested in recent years as the importance of riparian zones, particularly in the West, has become better understood, and the degradation that some have undergone becomes more widely recognized. The grazing of cattle in riparian zones, on either publicly or privately owned land, has long been a common practice. The cattle graze heavily there because the vegetation is lush, but they trample stream banks, increasing erosion and degrading water quality. Overuse of streamside vegetation also increases runoff and results in further erosion, deeper channel cutting, and therefore, higher stream flows. Over time the water table is lowered, depriving adjacent streamside vegetation of water, which further degrades the riparian community. A cycle can develop that often results in the total loss of the riparian community, leaving a deeply incised narrow stream channel that may flow only seasonally, and then with water of marginal quality. At this stage, as noted in Chapter 10, wildlife associated with the riparian zone would have long since gone, as would the fish associated with the stream.

Wetlands

A *wetland* is defined as any area of land covered by water, usually shallow, for at least part of the year, such as estuaries, swamps, marshes, and bottomlands.

Depending on the kind of soil, water quality and availability, and temperature patterns, the dominant plants in wetland areas may be grasses, sedges, bullrushes, cattails, shrubs, or trees, in any combination.

Wetlands are extremely important habitat for numerous species of wildlife. Muskrat, mink, beaver, otter, and raccoon are all associated with wetland ecosystems. About 44 species of waterfowl occur in the wetlands of North America (Bellrose, 1980). Many other species of birds are also dependent on wetlands and are termed *waterbirds*. These include loons, grebes, coots, rails, and white pelicans. In addition to their role as wildlife habitat, wetlands serve many other important functions. They help control floods by moderating the effects of runoff, operating as giant sponges to absorb water. Their high plant productivity also allows them to absorb large amounts of nitrogen, phosphorus, and heavy metals that would otherwise pollute waterways. In this respect, wetlands have been called the kidneys of the land (see Chapter 10, "Watersheds and Streams").

MANAGEMENT OF WILDLIFE

State and Federal Government Responsibility

Most wild animals in the United States are considered to be publicly owned and are primarily managed by state agencies for the public good. The idea that wildlife is the collective property of all citizens, rather than the property of individual landowners, has its roots in English common law and was brought to America by the colonists. Early in the history of the United States, individual states took over responsibility for the ownership and management of wildlife, focusing primarily on seasons and bag limits of hunted species. However, over the years, federal involvement in wildlife management has increased. As a general rule, except in dedicated federal areas such as national parks and U.S. Fish and Wildlife refuges on federal lands, the states manage wildlife populations and the federal agency manages the habitat.

The Federal Aid in Wildlife Restoration Act

Popularly known as the Pittman-Robertson Act, this Act was passed in 1937 and levies an excise tax (now 11 percent) on the sales of sporting arms, ammunition, and archery equipment. These funds are pooled and redistributed to the states for use in approved projects related to wildlife research, land acquisition, associated construction, and hunter safety.

Endangered Species Legislation

The Endangered Species Preservation Act of 1966 authorized the secretary of the interior to determine which wildlife species were facing extinction. In 1969, this legislation was expanded through the Endangered Species Conservation Act to include all species of vertebrates and some invertebrate groups as well, to prohibit the importation of endangered species or their products, and to provide for the addition of foreign species to the United States list. In 1973, the Endangered Species Act (ESA) was again modified to include plants as well as animals. This Act made distinctions between "endangered species," those faced with extinction in all or much of their range, and "threatened species," those that seem likely to become endangered. This Act also recognized separate populations and subspecies. Recently, the agencies administering the ESA have established a policy that would narrowly define distinct population segments as "evolutionarily significant units" based on morphological and genetic distinctiveness between populations. This offers protection for the Florida panther, a subspecies of cougar that is not threatened or endangered, and for salmon populations that spawn in specific streams along the West Coast.

To "list" a species as endangered or threatened, the process may be initiated by any individual or group by submitting a petition to the secretary of the interior. Evidence for the submission is judged by the U.S. Fish and Wildlife Service, except for all marine species, which are considered by the National Marine Fisheries Service. The listing process is lengthy, requiring public hearings, and in the instance of controversial situations, environmental impact statements.

The ESA has been judged to be one of the most powerful wildlife laws in existence. Its major provisions are that it prevents the "taking" of any endangered species and requires protection of its critical habitat. The term *taking* means to harass, harm, pursue, hunt, shoot, wound, kill, trap, capture, or collect such a species. The biggest impact on society comes from the fact that, when a species is listed, its critical habitat is also designated for protection, even if on private land. This has been defined as the area occupied by the species at the time of its listing and essential to its conservation (Coggins, 1991). The law prohibits actions that will adversely affect the critical habitat of the species in question. These prohibitions have made the law very controversial and have been a major reason why the addition of some species to the endangered list has been so controversial. For example, the critical habitat of the northern spotted owl overlays large areas of valuable old-growth timber in the Pacific Northwest and the red cockaded woodpecker needs mature pine stands in the southern United States. To resolve these conflicts, Congress added a provision to the ESA in 1982 allowing landowners to receive permits to "take" imperiled species on their lands provided that such takings were associated with approved activities such as construction or timber harvest. To receive a permit, the landowner had to prepare a "Habitat Conservation Plan" describing how much "take" would occur as a result of the proposed activity and the steps to follow to minimize and mitigate the impacts. However, such plans, which have become more common in recent years, have evoked considerable controversy with respect to their effectiveness and efficacy (Watchman et al., 2001).

As of September 2001, there are 367 animal species, 593 plant species, and 519 foreign species on the U.S. endangered species list. In addition, there are 128 animal species and 142 plants on the threatened species list. The total number of species listed has grown from 281 in 1980 to 1,230 in 2001. The distribution of species by state is shown in Fig. 9-8.

Federal Land Management Acts

Two relatively recent federal laws pertaining to federal lands have had a major influence on wildlife conservation and management. The National Forest Management Act (NFMA) of 1976 provided specific guidelines for the preparation and implementation of forest management plans. These plans must provide for maintenance of diversity in plant and animal communities, ensure that fish and wildlife interests are included in all resource management decisions, and ensure critical habitat is protected during timber harvest plans. The requirements of NFMA have substantially affected how national forests can be man-

FIGURE 9-8

Endangered and threatened plants and wildlife in the United States, in 2001, totaled 1,230 species. The state-by-state breakdown is noted here. Several species are now endangered or threatened in more than one state. *(Courtesy of U.S. Fish and Wildlife Service)*

aged and have increased the protection of most flora and fauna (Cubbage et al., 1993).

The Federal Land Policy Management Act (FLPMA), also passed in 1976, requires that the Bureau of Land Management take into account all resource uses in its planning process. It also directs that half of the funds received by the United States as fees for grazing livestock on public lands be spent to improve rangeland conditions, including fish and wildlife habitat.

A thorough description of all federal wildlife laws is found in Bean (1983).

Management by Private Landowners

The majority of land in the United States is privately owned, including 99 percent of cropland, 60 percent of pasture and rangeland, and 57 percent of forest land (see Chapter 3, Distribution of North American Forests). The value of private lands as wildlife habitat is important and will undoubtedly increase in the future. As a result, most states offer some type of technical and financial support to encourage habitat improvement on private lands (see Chapter 18, Forest Management by the States). They also help in evaluating habitat and wildlife populations and assist in the development of management plans. Technical assistance provides landowners with fencing and nest structures, and planting materials such as seeds or seedlings. Other programs may involve cost-sharing in the development of ponds and wetlands. Some states offer landowners tax incentives for managing habitat or permitting public access.

FIGURE 9-9
The sale of wildlife products is increasing in the United States. This dealer carries his message on the sides of his pickup truck. *(Photo by John C. Hendee)*

Larger private landholdings in many areas of the United States are becoming increasingly involved in the commercialization of wildlife products and hunting rights. These initiatives generally fall into three broad categories: 1) the sale or lease of rights to hunt native species, 2) the sale or lease of rights to hunt nonnative species, and 3) the sale of wildlife products. Activities are often grouped under the heading *game ranching,* which is defined as the intentional raising of wild animals, primarily ungulates, for commercial purposes. The rising popularity of hunting native big game on privately managed game ranches or hunting club lands is primarily connected to the decreasing success rates and deteriorating quality of the sport-hunting experience, and the reduced size of trophies on public lands. Some game ranches actually sell the rights to hunt native species.

Another type of game ranching involves the sale of rights to hunt nonnative species maintained on privately owned lands. This activity is particularly common in Texas, where many ranches now stock and breed several species of African ungulates. Hunting exotic big game species differs from hunting native game in that exotic species are generally not subject

to state regulations and can be harvested year-round. However, hunting seasons are beginning to be required in some states.

Hunting for exotics in the United States is growing in popularity mainly because it is less expensive than hunting in foreign countries. Other advantages cited are the absence of import/export fees, reduced shipping charges for trophies, and the opportunity to keep the meat. Language, passport, and immunization delays and difficulties are bypassed as well.

The sale of wildlife products such as meat, velvet antlers, teeth, ivory tusks, claws, and pelts is also growing in the United States, following a practice common in other countries (Fig. 9-9). In China, Korea, and Siberia, deer ranching for antlers is traditional (Goss, 1983). In New Zealand, ungulates are raised on small, fenced, intensively managed parcels of high-quality agricultural land. Meat is sold locally in stores and restaurants, and an export market exists for dried velvet, antlers, hides, and glands. Game ranching for meat products is also increasing in the Canadian prairie provinces.

In the United States, most state game management agencies are opposed to game ranching of species

FIGURE 9-10
Canada geese in a wildlife refuge. In some urban areas of North America these geese have ceased to migrate and have become resident pests. *(Photo by R. G. Wright)*

such as Rocky Mountain elk because of fears that they may carry diseases that could spread to the wild populations should the captive animals escape. However, most state agencies have little legal authority to control such activities.

Wildlife Management in Urban Situations

As North America has grown more urbanized, the management and protection of wildlife in towns and cities has become a recognized responsibility. Most urbanized wildlife lives within parks, greenbelts, woodlots, gardens, and cemeteries where vegetation provides habitat. Although more species exist within the limits of large cities than people realize, urban and suburban environments generally support fewer species than do adjacent rural habitats (Shaw, 1985).

Highly competitive species with broad habitat tolerances tend to increase, while those with specialized habitat requirements decrease. Successful urban species, such as starlings, English sparrows, and Norway rats, often reach such high levels of population density that they become pests. White-tailed deer, as a result of protection from hunting and natural predators, a favorable climate over the last several decades, and abundant food from suburban gardens and plantings, have reached such high densities that they have often become pests. They also cause problems be-

cause they can carry Lyme disease and are involved in numerous vehicle collisions. In Pennsylvania, for example, over 55,000 deer are killed on the state's highways annually.

Nonmigratory populations of Canada geese have also established themselves in many urban and suburban areas of North America during the last 50 years, often to the point of becoming pests (Fig. 9-10). At first, most landowners like seeing the geese, but eventually the extent of goose grazing and the accumulation of goose feces on lawns and gardens cause many landowners to view geese as a nuisance. Unfortunately, these problems are difficult to alleviate. Hunting in suburban areas is usually restricted, and techniques to scare geese, such as harassment, decoys, and noise, have not met with much success.

In the final analysis, for most species that become habituated or that may thrive in suburban or urban environments, management options are often restricted, and controlling undesirably high populations may be difficult, if not impossible.

Management Strategies

Hunting Regulations Virtually all wildlife species are subject to some form of artificial mortality, hunting and trapping being the most common. Other factors such as road kill, pesticides, and poisons also decimate populations in some situations. Artificial mortality should be distinguished from natural mortality, which is the result of factors such as predation, starvation, and disease. Wildlife populations respond to artificial mortality in various ways, depending on their habitat conditions, age structure, and reproductive rate. These factors greatly complicate management plans.

Sport hunting and, to a lesser degree, trapping, are the primary management and population control tools for most species of game animals. A major goal of wildlife management is the production of acceptably high yields of game and fur-bearer species while ensuring that enough animals are left to replenish the population. The term *harvestable surplus* describes the yield available to hunters and trappers that still maintains the base population. This surplus is generally considered to be that proportion of the population that would invariably die of other causes if not taken by hunters and trappers. Implicit in this definition is the idea that hunting and trapping mortality compensates for, or replaces, some natural mortality and therefore is not an additional source of mortality (Shaw, 1985).

State wildlife agencies are faced with the task of setting regulations that avoid overharvests while at the same time satisfying the demands of hunters for both harvestable game and the long-term maintenance of game populations. As a result, out of practical necessity, most state wildlife agencies in the United States set harvest regulations (season lengths and bag limits) that are typically conservative, so that any error favors underharvest rather than overharvest.

To more effectively monitor hunting pressure and set more realistic regulations, wildlife agencies typically collect information on habitat conditions, the status of game populations, and the number, distribution, and success rate of hunters. Surveys of habitat conditions can alert managers to the fact that populations may be too large for the habitat to support, thus requiring a reduction. Check stations are often set up along highways to provide an estimate of hunter success as well as a measure of the condition of individual animals that can be related to the quality of the habitat. Many states survey hunters through questionnaires attached to hunting licenses, or through permits that provide data on success rates, effort, the number of animals seen, and the distribution of hunters.

Predator Control   Throughout history, predatory animals have fascinated human societies, arousing emotions ranging from fear and loathing to more recent feelings of appreciation of their beauty, charismatic nature, and their role in natural ecosystems (Mattson et al., 1996). As a result, predator control has had a long and controversial history. Bounties were placed on some predatory species soon after European settlement. Control methods include shooting, trapping, and poisoning. For example, during the late 1800s, wolves were killed throughout the West when they scavenged carcasses that had been treated with strychnine. Unfortunately, this also killed all other scavenging animals. These and other practices were so efficient that all species of wolves in the continental United States were virtually exterminated, and the grizzly bear was eradicated from all but the most protected and isolated areas.

Historically, a primary reason for predator control was the protection of domestic livestock, particularly sheep, which are especially vulnerable. Early in this century, predator control was also used to protect big-game populations (Fig. 9-11). These activities even took place in national parks. For example,

FIGURE 9-11
Control of predators to protect domestic livestock and big game species was a common practice in the United States in the first one-third of the twentieth century. Here, a hunter, with his two dogs, packs out an 11-foot mountain lion, *circa* 1936. Olympic Mountains, Washington. *(Photographer unknown)*

wolves were eliminated from Yellowstone by the 1920s in an effort to protect resident elk and mule deer. However, as these populations increased, justification for predator control diminished.

Today, predator control as a game management tool is rarely used, and the practice has come under increasing attack in livestock programs. There is ample evidence that individual predatory animals can become selective consumers of livestock. To deal with this, livestock ranchers have had to become more selective in their controls, including the use of guard dogs to protect sheep, and insurance policies or compensation programs for losses, which are paid for by wildlife protection organizations.

The role of predators, as opposed to habitat factors, in regulating native wildlife populations continues to be debated. This is particularly true in areas where there have been large increases in ungulate populations such as Rocky Mountain elk and white-tailed deer. However, despite the large number of studies that have looked at the role of predation on such species, there is still no agreement on whether predation does or does not play a significant role in limiting ungulate population growth.

One of the best opportunities scientists have had to examine the relationship between a predator and prey has been at Isle Royale National Park, as island in Lake Superior (Fig. 9-12). Because of its isolation, the park has provided a unique laboratory where the effects of wolf predation on moose have been continuously studied since the mid-1950s, showing that wolves are a major source of moose mortality and have played an important role in the long-term equilibrium of the moose population. However, wolves have not completely controlled

FIGURE 9-12
The predator-prey relationship of the wolf–moose has been studied for over 50 years in this island park. *Upper:* Moose came on to the island early in the century, and with no predators to control them, their population increased to over 700 before a food shortage occurred. *Middle:* The gray wolf, driven from most of the United States, arrived after the moose by crossing the winter ice from Ontario or Minnesota. Over the years, the wolves have kept the moose population in check. *Lower:* Two wolves feed on a moose. Predation on the weakened animals keeps the moose herd strong.
(Courtesy of Isle Royale National Park)

moose, and wolf predation has varied over the years depending on winter severity. Plant succession and changes in food availability have also played an important role in the health of the moose population and therefore its vulnerability to predation (Wright, 1992).

Stocking and Translocation   Moving individual animals from one area to another (*translocation*) has long been a common wildlife management practice. Animals are typically moved to restock an area in which the original population has become depleted or to expand the range of a given species. Stocking has been most commonly employed for fish populations (primarily trout), but it has also involved several other species of animals.

Elk were widely translocated from Yellowstone National Park in the early part of the twentieth century in order to restock areas where the native elk had been eradicated, and to help decrease the elk population in the park. Such translocations are often controversial because of potential genetic differences between the native animal and translocated species, which in turn might cause displacement or inbreeding and loss of the original genetic stock. This was the case in Mount Rainier National Park, where introduced Rocky Mountain elk displaced the native Roosevelt elk in some areas (Schullery, 1984). However, other species such as wild turkeys have been widely relocated throughout the United States without harming ecological relationships in the new sites.

One of the most widely transplanted species is the bighorn sheep, particularly the desert subspecies. Bighorn sheep populations have been decimated over much of their range due to past overharvests and by diseases, sometimes contracted through contact with domestic sheep. Successful transplanting of bighorn sheep has proven to be difficult, as the transplants are often subjected to the same diseases that killed the native species.

THREATS TO WILDLIFE

Habitat Fragmentation and Loss

The single most important factor in wildlife population decline in most areas of the world is the outright loss of habitat and/or the fragmentation of existing habitat into parcels too small or too isolated to support viable wildlife populations (Morrison et al., 1992). Fragmentation, or the excessive subdivision of contiguous areas of homogeneous habitat, affects the habitat quality of wildlife in subtle ways, such as by changing the types and quality of the food base. It also changes microclimates by altering temperature and moisture regimes, and changes availability of cover and brings species together that normally have little contact, and thus may increase rates of parasitism, competition, disease, and predation. Fragmentation of habitat can also increase contact with and exploitation by humans.

Until recently, wildlife managers viewed the edges, or *ecotones,* resulting from fragmentation as desirable wildlife habitat because of the high floral and faunal richness typically associated with edge environments. The value of woodland openings and clearings to wildlife has long been recognized. In recent decades, however, as many once contiguous forests have been cut into ever-smaller pieces by timber harvests, the adverse effects of fragmentation on fauna are being noted. These include increased predation on the nests of birds and a general decline in the numbers of passerine birds associated with forest interiors. As a result, the traditionally advocated advantages for creating edges and forest openings are under increasing scrutiny by wildlife professionals (Reese & Ratti, 1988).

Fragmentation can also result from the housing development in rural areas. Such housing developments may block traditional migration routes of animals like Rocky Mountain elk, as has been the case in Jackson Hole, Wyoming. The vegetation changes as the result of plantings on residential parcels can enhance the habitat for some species while making it less suitable for others. For example, changes in cover on residential lands in the Gallatin Valley of Montana are resulting in an increase in white-tailed deer and a decline in the numbers of mule deer (Vogel, 1989).

Habitat fragmentation commonly creates small islands of contiguous habitat surrounded by areas of unsuitable habitat that isolates populations, limits genetic interchange, and can ultimately cause the localized extinction of some populations. For example, bighorn sheep require habitat that has good visibility, areas of escape terrain, and abundant continuous forage. This pattern gives sheep the option to respond to variations in weather, forage conditions, and human disturbance. The maintenance of traditional migration corridors providing sufficient visibility and escape terrain is critical to the maintenance of large, mobile sheep populations. In contrast to this pattern, many Rocky Mountain bighorn sheep herds are small, isolated, and sedentary. Traditional migration corridors have been cut, making the sheep less mobile and, thus, more susceptible to disease, predators, and disturbance (Risenhoover et al., 1988).

Decline in the health and vigor of forest stands in the South, Northeast, and northern Rocky Mountains, plus a marked increase in acres burned by wildfire in affected areas has prompted great concern over forest health. Insects and disease kill these vulnerable trees, already weakened by drought stress brought on by overcrowding. The resulting deadfalls, excess standing snags, and general stagnation can result in a forest difficult for some wildlife to travel through.

Conflicts Over Habitat Management

The quality of a particular habitat is dictated by the activities of its owner, and the government often has no means of control when privately owned habitat is adversely modified. (The only exception to this is in the instance of an officially listed endangered species, where adverse modification of habitat can be declared illegal even on private land.) Thus, the rights of landowners to do as they wish with their lands frequently conflict with the desires of those who seek to conserve public wildlife resources. This conflict can also exist on public lands dedicated to certain uses, although the public may have the opportunity to appeal adverse use of public land.

Human Disturbance of Wildlife

Wildlife species vary widely in their tolerance to disturbance by humans or human activities. Some species are very sensitive to any human activities, and to some, human disturbance can be fatal. Many of the so-called wilderness-dependent species require solitude and are intolerant of disturbance (Mattson, 1997). In other species, disturbance may trigger an aggressive or hostile response, while some animals tolerate or even benefit from association with humans. In general, species' tolerance of humans can

FIGURE 9-13
Herring gull chicks and egg, both vulnerable to predators when the adults are absent. Little Duck Island, Maine. *(Photo by Grant W. Sharpe)*

be viewed as a continuum from highly tolerant to highly intolerant. In many cases, the reaction of individual mammals to human activities most likely reflects their previous experiences with humans (Yost & Wright, 2001).

Colonial nesting birds are particularly vulnerable to disturbance, because breeding populations concentrate in small areas and eggs and young are defenseless when adults are absent (Fig. 9-13). Human disturbance of waterbird colonies has been shown to cause nest losses through predation, trampling, and nest abandonment, often forcing the birds into less-preferred habitats. Because of this, access to areas used by colonial species is often closed during the nesting season.

Most species appear to tolerate nonthreatening human activities with minimal stress, commonly moving away from the source of disturbance. However, even a benign reaction may be accompanied by subtle responses like elevated metabolism, and when persistent may cause lowered body weight, reduced fetus survival, and a withdrawal from suitable habitat.

The timing or seasonality of a disturbance factor is often a major determinant of its effect on a species. Disturbance of ungulate species in winter, when the animals can be under physiological stress from cold and marginal food supplies, is often far more traumatic than at other times of the year. Feeding of animals, either by visitors to natural areas or at residential locations, can cause otherwise shy animals to become habituated to human activities and so dependent on artificial food sources that, if they are removed, the animals may not be able to survive.

The variety of recreational impacts are many and varied. Rock climbers may disturb nesting raptors, mountain sheep, and other cliff-dwelling species at particular times of the year. Recreational cave exploration has been implicated in the decline of some bat populations. Off-road vehicle use and snow machines have caused major wildlife problems through harassment, noise, and habitat destruction. All of these issues must be considered in planning for recreational use of wildlands in order to minimize impacts on wildlife.

Poaching

Poaching, or the illegal killing of wildlife, has been common as long as laws protecting wildlife and regulating harvest have existed. Poaching includes killing animals in areas where they are protected; killing game animals out of season; killing animals that are protected because of their sex, species, or age class; and killing animals by illegal methods. Typically, poachers are poor, relatively uneducated persons from rural areas, who consider poaching to be an integral part of their way of life, passed down from one generation to the next. These poachers often do not consider their actions to be illegal, but rather an assertion of the "right" of an individual to use wildlife for utilitarian purposes. It has been very difficult for game wardens to apprehend and prosecute this type of violator (Nelson & Verbyla, 1984).

In recent years, however, the extent of poaching activity has broadened considerably. It now includes individuals who are out for an easy take; these may include wildlife "vandals" or hunters who are too lazy to pursue game by normal or legal means. To monitor the level of illegal take, experiments have been conducted in parks and other areas in which decoys such as stuffed deer are placed along roadsides and monitored. The number of individuals who have stopped and taken shots at these decoys, usually from vehicle windows (often several each night at some locations), has yielded startling evidence on the extent of this type of poaching.

Even more insidious has been the organized poaching of selected species, often by criminal groups, for selected body parts. Of particular importance are the gall bladders and other glands of bears and the antlers of elk and deer, which are valued as aphrodisiacs in many cultures and worth considerable sums of money.

The Animal Rights Movement

The concept is not new, but it was not a focus of public attention until the late 1970s. Since then, the issue has come to dominate many groups who originally merely advocated the humane treatment of animals. Some rights advocates go beyond this, believing that domestic and wild animals have certain rights, including most importantly the right to exist apart from humans and their activities (Gentile, 1987).

The rights movement takes many forms, with different segments supporting different venues such as the abolition of the use of animals in scientific studies, the elimination of hunting and trapping, and even an end to the raising of animals for food. All of this has been particularly disconcerting to many wildlife professionals, because it seems inimical to the underlying assumptions and precepts upon which the profession has been based.

The fundamental problem with wildlife management, as animal rights enthusiasts see it, is that it does not respect the right of individual animals in a population. Wildlife management is also criticized for its emphasis on promoting high populations of game animals with little attention given to other species' needs (Decker & Brown, 1987). Advocates hope the abolition of hunting and trapping will be an evolutionary process through increasing restrictions. They also work for political changes such as shifting the financing of state wildlife management programs away from hunters and trappers to the general public.

To date, advocates have achieved some moderate successes, mostly against trapping. Legislation to ban trapping has been proposed in several states, and leghold traps are banned in several. Efforts to discourage the wearing of furs have weakened the market, making trapping and fur ranching less economically attractive. Legal challenges to sport hunting have been limited, to date.

SUMMARY

Attitudes toward wildlife have changed dramatically in recent decades. As people become more urbanized and removed from the natural environment, conflicts over the habitat needs of wildlife and the economic and recreational needs of humans seem likely to increase.

We have learned that all silvicultural methods affect wildlife habitat in some way. Most practices will benefit some species and be detrimental to others, and no single management activity benefits all forms of wildlife. Forest managers must not only decide which forest management practices are appropriate for the preferred tree species on the site, but how these will affect the habitats required for indigenous wildlife. As forests become more intensively managed, and their resources become more valuable, there is an ever-greater need for foresters and wildlife managers to work together in planning and applying management actions.

LITERATURE CITED

Bean, M. J. 1983. *The Evolution of National Wildlife Law.* Praeger, New York.

Bellrose, F. C. 1980. *Ducks, Geese, and Swans of North America* (3rd ed.). Wildlife Management Institute, Washington, D.C.

Coggins, G. C. 1991. Snail darters and pork barrels revisited: Reflections on endangered species and land use in America. In K. A. Kohn (ed.), *Balancing on the Brink of Extinction: The Endangered Species Act and Lessons for the Future.* Island Press, Washington, D.C., pp. 62–74.

Conner, R. N., and D. C. Rudolph. 1991. Effects of midstory reduction and thinning in red-cockaded woodpecker cavity tree clusters. *Wildl. Soc. Bull.* **19:**63–66.

Cubbage, F. W., J. O'Laughlin, and C. S. Bullock. 1993. *Forest Resource Policy.* John Wiley and Sons, Inc., New York.

Decker, D. J., and T. J. Brown. 1987. How Animal Rightists View the "Wildlife Management-Hunting System." *Wildl. Soc. Bull.* **15:**599–602.

Gentile, J. R. 1987. "The Evolution of Antitrapping Sentiment in the United States: A Review and Commentary." *Wildl. Soc. Bull.* **15:**490–503.

Goss, R. J. 1983. *Deer Antlers.* Academic Press, New York.

Hastings, B. C. 1986. Wildlife-related perceptions of visitors in Cades Cove, Great Smoky Mountains National Park. Ph.D. Thesis. University of Tennessee, Knoxville.

Kellert, S. R. 1996. *The Value of Life.* Island Press, Washington, D.C.

Klein, D. R. 1973. The Ethics of Hunting and the Anti-Hunting Movement. *Trans. N. Am. Wildl. and Nat. Resour. Conf.* **38:**256–266.

Mattson, D. 1997. "Wildness Dependent Wildlife: The Large and the Carnivorous." International Jour. of Wilderness 3(4):34–38.

Mattson, D. J., S. Herrero, R. G. Wright, and C. M. Pease. 1996. "Science and Management of Rocky Mountain Grizzly Bears." *Conservation Biology* **10:**1013–1025.

Meslow, E. C., C. Maser, and J. Verner. 1981. Old-growth forests as wildlife habitat. *Trans. N Am. Wildl. and Nat. Resour. Conf.* **46:**329–335.

Miller, C. A., and R. G. Wright. 1999. "An Assessment of Visitor Satisfaction with Public Transportation Services at Denali National Park and Preserve." *Park Science* 19(2):18–20.

Morrison, M. L., B. G. Marcot, and R. W. Mannon. 1992. *Wildlife-habitat Relationships.* University of Wisconsin Press, Madison.

Nelson, C., and D. Verbyla. 1984. "Characteristics and Effectiveness of State Anti-poaching Campaigns." *Wildlife Society Bulletin* **12:**117–122.

Peek, J. M. 1986. *A Review of Wildlife Management.* Prentice-Hall, Englewood Cliffs, N.J.

Posewitz, J. 1998. Yellowstone to the Yukon: Enhancing Prospect for a Conservation Initiative. *Int'l Jour. of Wilderness* 4(2):25–27.

Reese, K. P., and J. T. Ratti. 1988. Edge effect: A concept under scrutiny. *Trans. N Am. Wildl. Nat. Res. Conf.* **53:**127–136.

Risenhoover, K. L., J. A. Bailey, and L. A. Wakelyn. 1988. "Assessing the Rocky Mountain Bighorn Sheep Management Problem." *Wildl. Soc. Bull.* **16:** 346–352.

Schullery, P. 1984. A history of elk in Mount Rainier National Park. Report to Mt. Rainier.

Scott, J. M., F. W. Davis, R. G. McGhie, R. G. Wright, C. Groves, and J. Estes. 2001. "Nature Reserves: Do They Capture the Full Range of America's Biological Diversity?" *Ecological Applications* **11:**999–1007.

Shaw, J. 1985. *Introduction to Wildlife Management.* R. R. Donnelley & Sons Company, Chicago.

Spies, T. A., and J. F. Franklin. 1988. "Old Growth and Forest Dynamics in the Douglas-fir Region of Western Oregon and Washington." *Nat. Areas Jour.* **8:** 190–201.

Thomas, J. W., L. F. Ruggiero, R. W. Mannan, J. W. Schoen, and R. A. Lancia. 1988. "Management and Conservation of Old-Growth Forests in the United States." *Wildl. Soc. Bull.* **16:**252–262.

Vogel, W. O. 1989. "Response of Deer to Density and Distribution of Housing in Montana." *Wildl. Soc. Bull.* **17:**406–413.

Watchman, L., M. Groom, and J. Perrine. 2001. "Science and Uncertainty in Habitat Conservation Planning." *American Scientist* **89:**351–359.

Wright, R. G. 1992. *Wildlife Research and Management in the National Parks.* University of Illinois Press, Urbana.

Yost, A. C., and R. G. Wright. 2001. "Moose, Caribou, and Grizzly Bear Distribution in Relation to Road Traffic in Denali National Park, Alaska." *Arctic* **54:** 41–48.

ADDITIONAL READINGS

Boyle, S. A., and F. B. Samson. 1985. "Effects of Nonconsumptive Recreation on Wildlife: A Review. *Wildl. Soc. Bull.* **13**:110–116.

Hunter, M. L. 1990. *Wildlife, Forests and Forestry.* Regents/Prentice Hall, Englewood Cliffs, N.J.

J. Forestry. Special issue on wildlife in the forest. **95**(8), Aug. 1997.

Mighetto, L. 1991. Wild animals and American environmental ethics. University of Arizona Press, Tucson.

Noss, R. F. 1990. "Indicators for Monitoring Biodiversity: A Hierarchical Approach." *Conservation Biology* **4**: 355–364.

Regan, T. 1985. The case for animal rights. In P. Singer (ed.), *In Defense of Animals.* Basil Blackwell, New York.

Robinson, W. L., and E. G. Bolen. 1989. *Wildlife Ecology and Management* (2nd ed.). Macmillan Publishing Co., New York.

Rosenthal, M. 1990. *North America's Freshwater Fishing Book.* C. Scribners Sons Publishing, New York.

WEBSITES WITH WILDLIFE INFORMATION

Endangered species: (http://endangered.fws.gov.us)
Fish and wildlife recreation: (http:/fws.gov.us)

STUDY QUESTIONS

1. Describe the two types of wildlife-related activities: *consumptive* and *nonconsumptive.* Can you think of two better words to describe these approaches?

2. Choose a familiar wild mammal and discuss what you would assume to be its habitat needs. Do the same for a bird. Try to be specific in terms of vegetation and forage, but do it all from your present knowledge.

3. What is the concept behind the ecosystem approach to management of wildlife? What has brought it forward at this time?

4. How does the idea of biodiversity tie in with ecosystem management? Where would you go to find a richly biodiverse situation? Its opposite?

5. Locate a local or regional example of each of the major habitat types, with an animal (this includes birds) species you would expect to find there.

6. How do the states and the feds divide up authority over wildlife?

7. Why is the ESA such a powerful and controversial law? Describe the controversy over protection of critical habitat for a species near where you live. Be specific. Who are the various interests taking sides, and what are their arguments for or against the protective measures being proposed?

8. Explain the following terms: *taking, thermal cover, harvestable surplus, riparian zone, threatened species,* and *specialist species.*

9. What species of animals exist in the urban area you know best? Coyotes and cougars prey on domestic animals and pets in some urban and suburban areas. Do you think these predators should be controlled?

10. Distinguish between *artificial* and *natural mortality.*

Watersheds and Streams

This chapter was prepared by C. Michael Falter (Aquatic Ecology), College of Natural Resources, University of Idaho

Chapter Outline

INTRODUCTION

Chapter 10 introduces students to the terminology and controlling ecological processes of watersheds and their streams. Management requires an understanding of these ecological forces. A *watershed* is a geographic unit of land defined by geology and topography into a drainage pattern. Watersheds can include a wide variety of forest, rangeland, or alpine vegetation. Every location on the globe is a part of a watershed; all land uses and activities have some impact on the watershed. Thus, all renewable resource managers are involved with watersheds.

Watersheds have distinctive patterns of water yield, water quality or composition, and associated stream systems with unique physical, chemical, floral, faunal, and ecological characteristics. Each of these unique patterns is further shaped by land use, which in turn impacts options of stream water use for households and industrial and agricultural purposes as well as for in-channel uses like wastewater dilution, recreation, wildlife, and fisheries—all termed *beneficial uses.* Using human anatomy as a model, streams in any watershed are the veins draining the land. The *riparian (streamside) zones* and adjacent

streamside wetlands serve as the kidneys of the landscape, protecting the watershed from excess loss of soil and nutrients, as well as buffering the streams from overloading of materials eroded from the watershed. Thus, watershed ecosystems depend on these critical streamside buffer zones to absorb impacts and maintain beneficial uses. Healthy streamside vegetation and waterside edge zones help protect streams from negative effects of improper land use management practices in the watershed. Stream water quality and biota are clearly controlled by the watershed.

WATER YIELD

The Hydrologic Cycle

Water in the biosphere is distributed among reserves in the air (as humidity, clouds, and fog), beneath the surface of the ground (in aquifers and deep rocks), on the land surface (in stream channels, lakes, puddles, or near-surface soil), and in plant and animal biomass. The routing of water through and among these sectors is known as the *hydrologic cycle* (Fig. 10-1). Each sector of the hydrologic cycle is a reservoir of

FIGURE 10-1

The hydrologic cycle, illustrating the two-way flow between the atmosphere and the earth, with the excess water passing through the watershed as groundwater flow and surface runoff. E-T stands for evapotranspiration. *(Illustration by Michael Falter)*

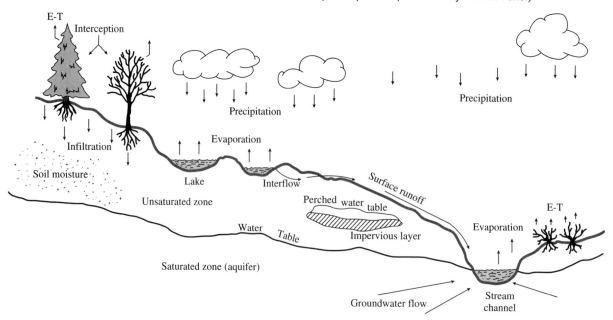

water (as liquid, vapor, or ice) with continual flow into and out of other sectors. Rather than the one-way circuit traditionally depicted, there is actually a two-way flow between each possible pair of sectors. For example, there is a two-way routing between atmospheric water and plants, between animals and the atmosphere, between soil and surface channels, and so forth.

Nonetheless, the source of most of a watershed's incoming water is precipitation, either as rain or snow. Precipitation may (1) enter groundwater to either the shallow, unsaturated *vadose zone* or to the deeper saturated groundwater zone known as the *water table;* (2) evaporate directly back to the atmosphere after reaching the ground, surface water, plants, or animals; (3) be taken up and stored by bacterial, plant, or animal biomass; (4) be transpired back to the atmosphere after uptake by plant or animal biomass; or (5) leave the watershed as runoff. *Hydrology* is the science that quantifies the dynamics of these water reservoirs and the vectors between them, thereby producing a budget of moisture intake and yield for a watershed. A simple water budget can be expressed as:

$$P = RO + ET + S$$

where P = precipitation, RO = runoff, ET = combined evapotranspiration, and S = storage. Over the long term, storage may be considered a constant in watershed budgets, so:

$$P - ET = RO$$

This equation correctly highlights runoff as the residual after evapotranspiration needs are met (Satterlund & Adams, 1992) and correctly depicts runoff as the available surface water leaving a watershed. It is apparent that runoff is a function not only of precipitation or of water use by plants and animals, but also of climatic factors that control evaporation. These factors include temperature, wind, exposure, shade, topography, soil and rock type, time of year, vegetation type, and elevation.

The Watershed

Watersheds are defined by their stream channels. A watershed is all the land area supplying water to a certain downhill point in a stream channel. Moving that point upstream encompasses a new, smaller watershed, while moving the point downstream defines a new, larger watershed. It is helpful, then, to supplement the traditional concept of a watershed as a catchment basin with the additional reality of a watershed as a *drainage* basin.

Stream Channels   Water flowing down a slope tends to follow defined channels, cutting and deepening these channels with time. Streams high in the watershed (drainage) are typically *interrupted,* meaning water flow is sporadic. Occasionally, during drier seasons, water may flow only beneath the streambed surface. The smallest stream drainages near the top of the watershed are called *first-order* streams with no tributaries. Two first-order streams combine to form a *second-order* stream; the combination of two second-order streams is required to form a *third-order stream,* and so on. This hierarchy is a convenient way to categorize and compare stream channels because there are similarities of *stream power,* channel shape, and structure between streams of the same order within a common *geomorphic region.* First-order streams make up most of a watershed's total stream channel length. Most streamflow thus originates as contributions to first-order streams. This fact serves to emphasize the importance of bank stability (generally contributed by soil and vegetation) as a controller of runoff on these small, first-order streams.

Watersheds that drain via surface channels to the ocean are called *exorheic,* in contrast to the *endorheic* watersheds, whose surface streams lack sufficient flow to reach the sea. Endorheic basins have average precipitation—that is, evaporation ratios less than 1, resulting in sporadic surface runoff. In some endorheic streams, years may pass between surface flows in the channel. Surface waters of these closed basins typically have high dissolved solids content because of the combined effects of minimal freshwater dilution and very high evaporation rates. Examples of endorheic basins in the United States are the Great Salt Lake basin of Utah, the Humboldt River basin of Nevada, and the Salton Sea basin in southern California. Notable worldwide examples are the Lake Eire basin in central Australia, the Caspian Sea of south-central Asia, the soda lakes of northern Kenya, and the Qattara Depression in western Egypt.

Groundwater   Precipitation, as either rain or snow, enters surface soils by infiltration. Infiltration is controlled by the amount of water falling on the soil surface and by soil conditions that determine how fast this water is absorbed. Soils above the water table lie in the *unsaturated zone* with respect to water content (also called the *vadose zone*). Water in the unsaturated zone may be retained in pore spaces or drain from surface soils down through the soil column via macropores.

Since capillary action retains water in the pore spaces of the unsaturated zone, soil water retention capacity is a function of particle surface area and, hence, is inversely related to soil particle size. Clay soils have a greater water-holding capacity than sandy soils because clay particles are very fine. Once clay soils are water-saturated, however, resultant swelling reduces pore size and restricts further water uptake. When coarser materials are mixed with clay soils (such as in loam soils), a greater water-holding capacity actually results. Water movement (*permeability*) through pure clay layers is essentially zero. The small pore size and conversely high particle surface area give a high surface tension to pore water so that clay soils hold their water content *more tightly* than do more porous, coarse soils. Thus, at a given water content, clay soils actually have less available water for plant uptake than do coarser soils. This is obvious in a dry period when plants in clay soils wilt earlier than plants in coarser soils.

Macropores in the *soil column* are relatively large, vertical channels that conduct water rapidly through otherwise fairly impermeable surface soils down to the water table. These can be small voids created by decomposed plant roots or fungal mycelia or even the larger channels created by earthworms, insect larvae, and mammals such as marmots, ground squirrels, badgers, and moles. The very deep clay soils of the Palouse region in the northwestern United States had long been thought to be largely impervious to downward movement of water, but the observed rapid response of the deep saturated zone to surface-applied fertilizer and agricultural chemicals has since been explained by soil macropores conducting water downward through the otherwise impermeable heavy clay soil.

The interface between the unsaturated zone and the saturated zone of groundwater is the *water table.* The saturated material below the water table is the *aquifer.* Material in the saturated zone ranges from porous, unconsolidated sands to solid rock, where free water is present only in fissures and channels. When saturated, a subsurface zone holds all the water it can at ambient pressure. The water table depth changes in response to surface infiltration, seasonal draw by plants, and evaporation from the soil surface. Shallower aquifers typically drain laterally downslope to surface streams. As noted in Fig. 10-1, most surface streams are, in fact, a surface exposure of the saturated zone. Streams can be recharged by groundwater when the water table level is higher than the stream. Conversely, streams can recharge the aquifer as the stream rises at flood or as the water table falls. These changes in water supply occur normally with seasonal cycling of precipitation, evapotranspiration, and runoff. An aquifer may *recharge* a stream at summer low flow, for instance, or *be recharged* by that stream at spring flood when the stream level can be many feet higher than in summer. A stream can alternate between reaches that lose water to groundwater and reaches that gain flow from groundwater. In eastern Washington state, the Spokane River flows across deep deposits of alluvial gravels that receive water from the river. Thirty miles to the west, the gravel beds are interbedded with basalt layers that channel groundwater flows to the surface as emerging springs on the river bank. These springs lower the river temperature by up to 10°C in August.

The Spokane River example illustrates that water pathways between water sources and surface streams are not uniform throughout the watershed. The land varies in its contributions of water to streams. The concept of *variable source area* recognizes these variegated patterns of water supply to surface streams. For reasons of geology, aquifer shape, slope, and soils, certain land areas are more consistent suppliers of water to streams. Source areas are typically either saturated zones or zones where soils are more porous near the surface. Riparian zones, a type of wetland with spongy, organic-rich soils, are often saturated and, thus, can be seen as good source areas to surface streams from ground storage.

Although riparian areas are generally critical source areas with rapid insoak and little surface runoff across them, important source areas may also exist in the watershed far removed from stream channels. Such an area might occur where an up-sloping, sandy layer intersects the land surface, intercepting overland flow and rapidly carrying the water underground to a distant stream channel. The existence of variable source areas in a watershed requires land managers to rank watershed sensitivity according to how tightly various land areas are linked to streams.

Building roads in a low-groundwater-source area might be acceptable whereas those same actions in a high-source area may be deleterious to long-term stream quality and, thus, totally unacceptable.

Deep geologic water, which lies too deep for much interaction with surface streams, may recharge streams via artesian (pressurized) flows to the surface. However, the usual interaction with streams occurs when *perched streams* (sealed stream channels perched on unsaturated zones) leak water through the streambeds downward to a deep saturated zone of groundwater. In most cases, in the absence of pressurized artesian systems, water from an overlying stream flows into, not out of, a deep groundwater aquifer.

Streams   At any instant, streams of the world are estimated to contain 300 cubic miles of water, or 0.029 percent of the total free fresh water (liquid or gas) in the biosphere (including groundwater). The water in those stream channels is estimated to be replaced approximately 20 times per year, for an average hydraulic retention time of 18 days (U.S. Geological Survey, 1969). Even though stream water is typically the second-largest sector of water in a watershed, it remains a small component of the total water in a watershed, ranking far behind the larger water volumes found either in the flowing or static groundwater storage.

Water in stream channels could be thought of as having a *base flow* component derived from groundwater that has seeped directly into the stream channel, either from the unsaturated or saturated groundwater zones. Base flow varies over time as water infiltration changes with precipitation, snowmelt, and vegetation demands. A *surface runoff* component is derived from overland flow. A stream with a year-round base flow component, ensuring 12-month flow, is called a *perennial stream* (Fig. 10-2). An

FIGURE 10-2
The stream channel with its downstream flow of water consists of substrate, riffles, pools, chemical components, stream biota and debris, and adjacent riparian zone. North Cascades National Park, Washington. *(Photo by Grant W. Sharpe)*

intermittent or *ephemeral* stream only has sufficient base flow to provide surface water in the channel during the wetter months; it dries up when overland flows cease in the dry season. As previously noted, in most instances, water remains in the "dry" stream channel, but flows out of sight inches or feet beneath streambed sediments. Many aquatic insect larvae rely on this deep *hyporheic* water for habitat during the months or even years that a stream channel may lack surface water.

Lakes   Sometimes called the "eyes of the earth," lakes are resting places for surface waters flowing out of a watershed toward the ocean. Lakes are an example of the *conservative nature* of watersheds. Compared to streambank storage of sediments, lakes are an even more effective mechanism for material and energy trapping and subsequent storage within the watershed. At any instant, lake basins of the world contain 2.9 percent (30,000 cubic miles of water) of the total free fresh water in the biosphere.

Lake waters are not quiescent or stagnant, but are in constant motion with their persistent currents, turbulence, and river-like internal flows. These in-lake water velocities are simply less than those found in rivers. Even lakes with no surface outlets have patterns of internal water movements and currents (Fig. 10-3).

There is no definitive boundary between a "lake" and a "stream." A stream flowing into a lake gradually loses its riverine characteristics of strong downstream flow, high turbulence, and high transport power. These streamflows assume *lacustrine* (lake-like) characteristics of multidirectional velocity (i.e., a predominance of random, chaotic, but low-velocity turbulence) and low transport power. Particle size of lake bottom materials declines from coarse cobbles, pebbles, and sands of streams to the uniform, clay-sized particles of bottom ooze in a lake's central basin. Only in the wave-washed shore zones do coarse bottom substrates characteristic of streams persist.

The temporary retention of water (hydraulic residence time), whether in a widened stream channel or lake basin, permits development of physical, chemical, and biological systems different from those in a turbulent stream system. Lakes may thermally stratify with warmer water on the surface. Chemical equilibria in water with a longer hydraulic residence

FIGURE 10-3
The lake, a resting place for surface water flowing out of a watershed to the ocean. Lakes have their own chemical and biological system, basically similar to, but different in some ways, from stream systems. Bear Lake, Oregon. *(Courtesy Oregon State Highway Department)*

time, for example, may be different. Evaporation from the lake surface increases the dissolved solids concentration of lake waters over that of streams. Plant communities transition from the attached *benthic* (living on the bottom) algae and mosses of streams to free-floating phytoplankton in the lake water. More of the delicate or weakly swimming biota (such as filmy-leaved aquatic plants or floating plankton) can develop in lakes as compared to streams. Animal communities shift from the diverse, strongly muscled forms of turbulent stream bottoms to passively floating, open-water zooplankton and lower diversity worm and clam communities of lake bottom ooze.

The degree of vertical thermal stratification is proportional to the hydraulic residence time of water in a channel or basin. Stratification inevitably develops in lakes with water retention times greater than

15 days. Lakes, therefore, have a far greater propensity for vertical thermal stratification and the resulting seasonal isolation of cold, deep waters from warmer surface conditions of high turbulence, light, oxygen, and chemical recharge. Most of a lake's heat content is found in surface layers where abundant light energy, high temperatures, and water interaction with the atmosphere predominate. Most biological activity occurs in these surface waters. As this stratification continues for a few months through the summer, deeper waters of a lake are isolated, quiet, and dark, and remain cool from the previous winter. A long list of chemical changes can occur in those deep, isolated layers, including deoxygenation and low pH, as well as increases of soluble plant nutrients, heavy metals, carbon dioxide, methane, hydrogen sulfide, and ammonia.

Until relatively recently, lakes were viewed as microcosms or isolated systems with their unique assemblages of plants and animals functioning quite independently of their surrounding watersheds. That perception resulted in a fatalistic approach to lake management in which lake function was assumed to follow a set pattern regardless of watershed activities. The controlling linkages (water, nutrients, and sediment runoff) between watersheds and their surface water bodies gradually became recognized when lake managers began to quantify the relationships between watershed actions and surface water function. A specified logging road density in a granitic watershed, for example, could be expected to deliver X kilograms of phosphorus per hectare of watershed surface to a lake or stream. The water manager could thus project algae response to that predicted input of phosphorus, often a critical growth-limiting nutrient in surface waters. Where phosphorus is the first limiting nutrient in a water body, the result is a projection of algae, zooplankton, and insect growth in the water body for assumed scenarios of human activities (watershed use and management). Much research emphasis is now placed on defining those controlling linkages between a watershed and its water bodies in order to expand our ability to predict water impacts from land management activity.

Riparian Zones and Wetlands   Riparian zones, with their associated wetlands and floodplains, are the *ecotones,* or edges, between terrestrial and aquatic habitats. These are the waterside lands directly influenced by water—lands transitional between terrestrial and aquatic systems. The water table in riparian zones is usually near the soil surface, or the land may even be covered by shallow water. Riparian zone soils are typically saturated for at least part of the year. One or more of the following attributes typify wetlands in general (Cowardin et al., 1979):

1. Most plants are *hydrophytes* (water-loving plants) at least part of the year.
2. Soils are predominantly undrained *hydric soils.*
3. Soils are saturated with water at least part of the growing season.

Wetland soils have a high organic content provided by the high rates of plant production in these water-rich and nutrient-rich environments. The high organic content, in turn, provides tremendous capacity for near-surface water retention. Much surface runoff will be retained by riparian soils before reaching the stream channel—another example of the *conservation of materials* in a watershed. Wetlands, floodplains, and riparian areas may store a large portion of the shallow groundwater in a watershed. Periodic flooding of these waterside lands allows water to be stored for later slow release to the stream channel during low streamflows. When high streamflows are permitted to move out onto adjacent floodplains, the stream's water, sediment, nutrient, and energy loads are dissipated throughout the riparian zone.

These riparian zones and floodplains serve as safety valves for energy release through *aggradation* (sediment deposition and accumulation) high in the drainage. Erosive channel cutting would otherwise be much greater. Management actions such as stream channelization and diking (which restrict stream access to floodplains) always heighten downstream peak flows by speeding time of travel through the altered reach and preventing dissipation of stream power in the riparian zone. Fully functioning riparian zones probably offer the most efficient, cost-effective flood-control measures possible. After a century and a half of being isolated from their streams by "flood-control levees," filled in, or paved over, riparian wetlands have at last been officially recognized for the natural flood-control systems that they are. Following the disastrous floods of the midwestern United States in 1993, federal agencies responsible for flood control and relief programs announced for the first

TABLE 10-1	
SUMMARY OF RIPARIAN ZONE EFFECTS ON STREAM HABITATS	

- Flow recovery through:
  - increased stream channel capacity
  - raised water table
  - slower runoff
- Temperature control:
  - summer cooling
  - winter warming
- Streamed aggradation
- Decreased turbidity, nutrients, and pesticides of stream water
- Increased terrestrial/aquatic "edge" provides enhanced food supply to stream
- Sediment, nutrients, and agricultural chemicals filtered before entering stream . . . in the riparian zone
- Increased bank storage of sediments and decreased bank erosion
- Increased channel length
- Physical structure of stream enhanced for fish habitat
- Increased plant biomass and structural diversity
- Movement and migration corridors for wildlife
- Breeding, cover, and feeding habitat for terrestrial wildlife
- Improved floodplain forage production

time, programs to buy up farms, developments, and even entire towns in riverine floodplains to let the natural water-storage function of these wetlands once again operate. Long-term study of floods has shown that these floodplains can be more effective and less expensive than structural alternatives (dams and levees). A summary of riparian zone effects on water quality and stream systems is shown in Table 10-1.

FOREST STREAMS

Yield from a Watershed

Measuring Yield from a Watershed The influences on water delivery to a stream channel vary with time of day and season, from year to year, and between stretches of streams. The water volume in a stream channel will change over time. Water in a stream channel is expressed as a volume (cubic feet or cubic meters) moving downstream per unit of time, usually cubic feet per second or cubic meters per second (cfs or cms). Water flow per unit time equals *dis-*

charge. Discharge should not be confused with *water velocity,* which is linear distance traveled per unit of time such as feet per second (fps). Discharge is measured by partitioning the stream cross section into segments of known width and depth. The average velocity of each section multiplied by the area of each section corrected for friction with the stream bottom (roughness coefficient, or R) equals the discharge for that section (Fig. 10-4).

$$\text{Width} \times \text{depth} \times \text{feet/second} \times \text{roughness coefficient} = \text{Cubic feet per second}$$

Summing up the sections' discharge across the channel gives total water discharge for the stream at that point.

Water velocity in a stream channel is greatest above the bottom and toward mid-channel. In the open-water areas, friction-induced turbulence with bottom substrate is minimal, so gravity can "pull" the water downhill more efficiently. At the stream bottom and sides, velocity decreases sharply (Fig. 10-5, *upper*). Velocity is also slowed or even reversed on the downstream side of submerged rocks and logs (Fig. 10-5, *lower*). Aquatic biota use these *microhabitats* in the calm-water zones that exist in otherwise uninhabitable, high-velocity waters.

The reduction in water turbulence with stream depth (the water mass is in relatively less contact with the friction-inducing surface and bottom) makes larger, deeper streams flow faster than small, shallow streams. The average water velocity of an upland whitewater brook may be about 1 foot per second (fps) while the placid-appearing, meandering, lowland stream may actually have an average velocity of 3–4 fps. Turbulence limits the speed of water outflow from a stream section, so water added to a stream reach beyond the section's outflow capacity makes the stream deeper. The water-surface elevation then rises with discharge. This water-surface elevation to discharge relationship is unique for each stream and for each point in a stream channel.

If water-surface height rises with discharge, it follows that there must be a downstream slope to the water surface in streams. Water-surface slope results as water-surface height drops with elevation of the streambed. It is this water slope that is measured to give *stream gradient* (vertical drop per unit of horizontal distance). Gradient of upland mountain streams generally ranges from 2 percent to 7 percent,

$$\Sigma \, (WDV_r) = Q$$

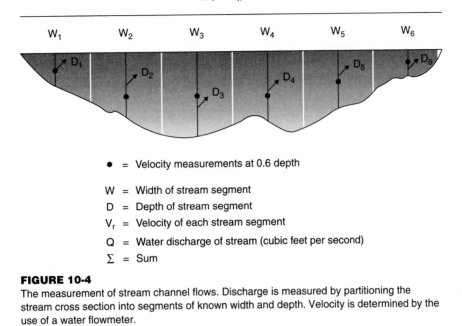

● = Velocity measurements at 0.6 depth

W = Width of stream segment

D = Depth of stream segment

V_r = Velocity of each stream segment

Q = Water discharge of stream (cubic feet per second)

Σ = Sum

FIGURE 10-4

The measurement of stream channel flows. Discharge is measured by partitioning the stream cross section into segments of known width and depth. Velocity is determined by the use of a water flowmeter.

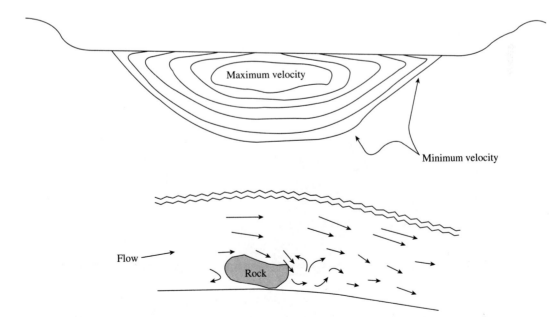

FIGURE 10-5

Water velocity in a stream channel cross section. *Upper:* Water velocity is greatest away from the channel sides and bottom. *Lower:* Submerged objects in streams create microhabitats by slowing or reversing downstream flow.

with 1 percent to 2 percent gradients in lower foot-hill streams. Flatland stream gradients may be in the range of 0.1 percent to 0.2 percent or less. The lower 100 miles of Australia's River Murray, for example, drops 0.4 inch per mile, for a gradient of 0.00001 percent.

Time of travel for water in a stream channel varies as a function of bottom substrate (because of turbulence with the rough bottom), stream gradient, suspended solids load, and temperature (which controls viscosity). Time of travel (measured in hours or days) is the average travel time required for a water particle to traverse the channel from point A downstream to point B. It is an important controller of in-stream processes because stream chemical cycling and biological production are a function of reaction time in the volume of water, not distance traveled. Travel time is surprisingly slow down a stream channel. A turbulent, upland stream of 1 foot/second average velocity (0.7 mph) would have an average time of travel of only 16 miles/day. Even at high water discharge, a flood pulse moves downstream relatively slowly. Average time of travel slows even more if

channel configuration permits rising streamflows to move out onto adjacent floodplains. These slowing effects are all desirable interactions of the stream with its floodplain.

The Hydrograph of a Stream A plot of water discharge against time (generally a calendar year on the horizontal scale) at a particular point along the stream is the stream's *hydrograph* at that point. Climatic, seasonal, and daily patterns of streamflow produce the stream hydrograph so that the annual hydrograph showing water export from a watershed is unique for each stream (Fig. 10-6).

Streamflow varies with time of year and with the distinctive climate of each watershed. A stream on the moist western slopes of the Coast Range of the northwestern United States, for example, has peak discharge in the winter. The moderate maritime climate ensures winter-long runoff because the watershed usually does not experience the extreme cold that locks in water and snow over the winter. Most floods in that coastal region occur in the winter, often after sporadic, mid-winter rain-on-snow events. In-

FIGURE 10-6

The hydrograph of a stream, illustrating the cfs discharge over a calendar year (the 365-day Julian calendar). The peak discharge of this stream (144 to 174) is in early June. Squaw Creek, Salmon River Basin, Idaho. October 1, 1990, to September 30, 1991. *(Courtesy U.S. Geological Survey, 1992)*

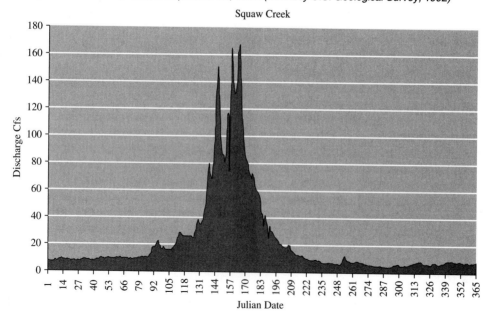

land, more continental climates produce a late spring pulse of snowmelt that dominates annual stream discharge. These hydrographs have an annual peak in early summer, as illustrated by Squaw Creek in the Salmon River drainage of central Idaho (Fig. 10-6). Streams in the midwestern United States, on the other hand, usually have a hydrograph dominated by mid-summer streamflow peaks as a result of warm-weather, monsoonal precipitation patterns. The most severe flooding in the Midwest typically occurs in mid-summer as a result of high runoff in the warmer summer months of high rainfall. Eastern U.S. streams seem to express all of these features (i.e., with snowmelt but not as dominant as that in the northern Rockies, some flashiness from thawing, and monsoonal runoff common in the summer).

A hydrograph shows peaks and valleys as streamflow rises and falls in response to water supply. First-order streams (streams high in the drainage and with no tributaries) are but a short water path from water source to stream, so these small streams show rapid flow responses to precipitation or snowmelt. They even vary with time of day—for example, more snow melts on a spring afternoon than at night, or as more water runs into the stream channel on a cool night when evaporation is low compared to a hot summer afternoon when evaporation is high and runoff is low.

Water-stage changes are rapid in a small stream. In a first-order stream, a storm event produces a very sharp rise in the hydrograph, perhaps up to 10 times. Flows then abate as runoff rapidly diminishes following the storm. An analogy is seen during a summer storm, when runoff from a paved parking lot rapidly peaks and then drops off. Further downstream, runoff contributions from throughout the progressively larger drainage merge so that hydrographs become less and less flashy as stream size increases. Thus, in-channel water storage (i.e., the in-transit water volume moving down the stream channel) increases downstream.

Stream channel banks and bottoms tell the history of water flows in streams. Substrate size, bank erosion, and vegetation are all indicators of water flow history. A stream channel with cut and steep eroding banks indicates recent deliverance of higher flows to a channel. The long-term historical relationship of the watershed and stream channel produced a certain-sized channel, but now the watershed and stream channel are "out of synch" with each other. The wa-

tershed sometimes delivers water to the stream channel faster than the channel can carry it without bank erosion. This could be the result of too much vegetation removal, such as from heavy logging, overgrazing, or urban development. Either the peaks of water delivery from the watershed to the channel must be evened out, or flows must be allowed to overflow the banks onto floodplains to dissipate the energy of the high flows. It is apparent that resource management practices and development in the watershed must be coordinated carefully with the unique characteristics of the watershed to produce a stable stream hydrograph of minimal flashiness. Only then will a stable stream channel persist.

Stream Power and Sediment Loads *Stream power,* the capability to erode and transport bedload sediments and suspended sediments, increases with channel slope and stream discharge. High water velocities at high stream flows erode material from stream banks and bottoms, carrying materials either in suspension or bouncing along the stream bottom as bedload. *Bedload* is the mass of sediments pushed along the stream bottom by the current. *Suspended load* is the mass of sediments carried in the moving water column. Water turbulence keeps suspended load from settling. Material shifts between these two forms of sediment load as velocity changes down the stream channel. Suspended materials in a riffle reach may become bedload in the lower velocities of a pool and even settle to the bottom along the stream banks as *sediment deposits* where water velocity drops (Fig. 10-7). Localized velocity increase, such as from flow deflection due to a fallen log or boulder in the stream, might erode those sediment deposits into a bedload mode or even to a suspended mode once again. Thus, sediments are moved downstream in an irregular fashion. The result is a "leap-frogging" or spiraling of sediments down a stream channel. A particle may move 100 feet or hundreds of miles in a year, depending on peak flows and stream power. Material remains in suspension until stream power drops below transport velocity for a given-size particle. Such phase changes of particles in streams illustrate the dynamic nature of stream channels wherein a change in one environmental factor causes changes in others as the stream shifts to a new equilibrium. An increase in velocity at a point in the stream may cause deposited sediment to become suspended sediment.

FIGURE 10-7
A riffle reach where suspended sediments have been deposited as a result of the velocity reduction in the stream channel. North Fork, Clearwater River, Idaho. *(Photo by J. M. Skille)*

An increase in sediment load, causes velocity reduction and an increased rate of sediment deposition, but increased stream power increases the stream's capacity to transport sediment.

Total sediment load of a stream increases geometrically with discharge, being small to nonexistent at low summer flow and very large during annual peak flow. In small upland streams, more than 90 percent of the stream's total annual sediment load may come down the channel in a single, several-hour storm event. Stream channel shape is clearly determined by the catastrophic and occasional high flow. Stream power is controlled by slope, so streams with greatest power tend to be found either in high mountain regions or very low in the drainage where high water volume creates a powerful stream. Eastern streams tend to be of moderate power while smaller, flatland streams in the Southeast, Midwest, or tundra streams of northern forests have low power.

Stream Sediment and Stream Power Depending on localized hydraulics, streambanks can be either a source or a repository of sediments. Streambanks are a major source of sediment if previously stored sediments are eroded from a channel-cutting reach and carried downstream. A stream reach can still be in dynamic equilibrium if sediment export from a reach equals sediment input to the reach. Streams are dynamic entities, however, and the equilibrium of a reach can shift to either deposition or erosion if stream power is altered, sediment input changes, or streambeds and streambanks change physically.

Since stream power can be changed in a stream by increasing stream discharge, washout of a debris dam can result in a dammed pool being flushed out and causing localized increase of stream power and channel-cutting in that reach. Effects of that "flood" might be dissipated over the next several miles downstream before the combination of floodplain and bank storage of the sediment torrent gradually absorbs the pulse of stream power. Much stream channel shaping occurs in this pulse-like manner. This was illustrated in a recent analysis of stream response to flooding and landslide events in a number of second- to fifth-order streams in the Clearwater National Forest of northern Idaho.

After flood/landslide events, upstream reaches of a stream tended to actually show improved stream habitat (narrower width, deeper channels, coarser sediments, and reduced cobble embeddedness by fine sediments) while larger downstream reaches were degraded after the same event (wider stream channels, shallower channels, and increased cobble embeddedness by fine sediments). Upstream, stream power was high because the flood pulse was high relative to channel size; downstream, the flood pulse was diluted in larger stream channels resulting in stream power insufficient to move delivered sediments out of those reaches. For that point in time at least, the lower reaches became repositories of sediment.

Increased sediment supply to a reach beyond the stream's equilibrium level will cause more deposition so that the streambed builds up or aggrades. Relatively low-gradient Appalachian streams may have less self-cleansing ability than steeper gradient, western streams, but the eastern watersheds export lower sediment loads to the stream. Silt-laden irrigation runoff is a common source of sediment to streams receiving irrigation return flows in the western United States. Intensive livestock trampling of streambanks may contribute enough sediment so that downstream streambeds aggrade, usually with fines (as in high meadows of the western United States) (USDI, 1991).

Finally, the character of the streambed and banks in a reach may be altered, usually by a flood event or a change in stored woody debris. The stair-stepped nature of mid-elevation stream profiles is largely provided by woody debris dams that trap pockets of sediment. Removal of a debris dam alters the streambed.

Sedimentation on the Streambed Declining flows, with their lower water velocities, permit deposition of silt-, sand-, and gravel bars. A rising limb on a hydrograph indicates cut stream channels; a falling limb shows deposits of material in stream channels. Streambeds actually deepen at high flows and fill back in during the subsequent deposition period on the declining limb of the hydrograph. Streambed material is therefore constantly changing from point to point in a stream over space and time. Rocks of a riffle are likely to change during each high flow peak even though the riffle feature may remain in the same location from year to year. Practically all substrate movement occurs during high water flow. Channels are relatively stable during dry periods when runoff is low.

An overhead view of a stream channel shows a variegated pattern of water velocity as rocks, logs, bumps, and dips in the stream bottom interact to pro-

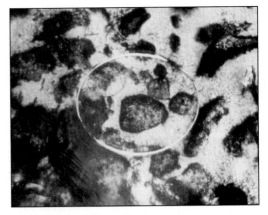

FIGURE 10-8
An armored streambed where stable deposits of large cobbles remain on the streambed surface but are embedded by fine sediments. Cobble embeddedness sampling ring shows in center. North Fork Clearwater River, Idaho. *(Photo by J. M. Skille)*

vide a constantly changing array of stream cross sections and depths and, therefore, a changing array of water velocity and sediment. In this three-dimensional scale, stream power changes from point to point across the stream channel. Think of stream substrate as the material that past high velocity could not remove and was left behind as bottom deposits. As successive high flows remove increasing amounts of *fines* (substrate material ≤ 0.25 in or 6.25 mm), deposits of larger cobbles remain, leaving a stable streambed surface. Such a streambed is said to be *armored*, or protected from further degradation. An armored streambed surface protects substrate beneath it so that fines can accumulate just below the protective armor layer and effectively seal deeper sediments to prevent intragravel water flow (Fig. 10-8).

Low stream power during low flows of summer and droughts permit deposition of fine particles and lumps of organic materials on streambeds. These fine sediments normally transported out of a stream can accumulate as streambed deposits even reaching several feet in depth after a long absence of flood flushing. Near the end of the seven-year drought (1985–1992) in the western United States, streams of all sizes were clogged with deposits of ooze following many low spring flood peaks. The high spring runoff of 1993 removed at least some of those long-term deposits, but several additional higher flow years will be required to completely remove the ooze deposits (Falter & Carlson, 1993). Even in higher flow years, one can find summer deposits of fine sediments in stream backwaters and pools.

Sediment Sources Sediments are eroded from their storage locations in a watershed and moved downslope by moving water or gravity, a natural process in watersheds. Land management activities can dramatically increase sediment export rates over those normal in geologic aging of a given watershed. Almost any activity in watersheds increases sediment production over background export rates. Resource management attempts to minimize these sediment losses.

Most dislodged sediment does not immediately leave the watershed; it is simply moved downhill a bit, deposited downslope on floodplains, in riparian zones, in streambanks, or channel backwaters. Trimble (1975) estimated that of all eroded sediments from 10 river basins in the southeastern United States over the last 300 years, less than 5 percent has been transported to the sea.

Overland transport is the oversurface movement of sediment, typically from rill or gully erosion on cultivated and grazed lands. It can be a factor in forested watersheds where the protective vegetative cover has been removed by fire, road cuts and fills, landslides, or timber harvest. Vegetation controls overland erosion, even eliminating it where cover is heavy, by several mechanisms:

1. Absorption of falling rain to break down the size of raindrops so that soil particles are not broken into smaller, more erodible particles;
2. Obstructing and slowing surface flows with organic debris;
3. Increasing soil permeability by improving *tillage* and organic content; and
4. Root masses contributing to soil cohesion and erosion resistance.

Vegetation management is thus critical for protecting watersheds from unnaturally high erosion as well as riparian zone destruction.

Mass wasting describes the wholesale downslope movement of sediments where cohesive attachment to the slope has been reduced. Such slumps, slides, and debris avalanches are natural features of mountain erosion and are always a possibility in steep terrain. Human activities can accelerate mass wasting through (1) cutting the toe from a soil slope, such as

in building roads or houses; (2) diverting high volumes of surface drainage to a groundwater source area (water-saturated soil masses are more prone to slumping) such as in irrigation or storm drainage systems; or (3) removing vegetation whose root masses provide soil stability (as in road building or construction).

A number of studies relating roots to soil cohesion and, therefore, resistance to mass wasting emphasize the long-term role of root masses in providing soil stability, even many years after the death of the vegetation (Satterlund & Adams, 1992).

Stream Water Chemistry and Quality

Solution in Water From the first contact of rainfall or snowmelt with rocks, soil, and vegetation of a watershed, the chemical composition of the incoming water is altered. Water is considered to be the "universal solvent" because of two unique characteristics:

1. Water ionizes into H^+ and OH^- ions, which ionize a myriad of chemical compounds in solution; and
2. The water molecule exhibits hydrogen bonding, which links with and holds nonionizing compounds, such as sugar molecules, in solution.

The result is that, given enough time, anything in the watershed will eventually dissolve, to some extent, in water. The chemical composition of the runoff water will therefore reflect the chemical makeup of watershed soils and vegetation.

Natural Patterns of Water Chemistry Natural patterns of the physical and chemical loads of water are complex. Relative amounts of various components at a point in a watershed are controlled by upslope conditions of geochemistry, soil structure, climate, exposure, vegetation, and the length of time water has been on the watershed (controlled by slope and relief). A granitic watershed tends to produce runoff water low in dissolved load, nutrients, and aquatic productivity. Runoff from sedimentary watersheds is comparatively "richer" with nutrients. Basalts and metamorphosed rocks are intermediate in their dissolved contributions to water. Waters in a low-elevation watershed could be expected to contain a greater dissolved load than a higher-elevation watershed simply because the water has had more time on the watershed for rocks and soil to dissolve. The organic complexity of stream water increases

TABLE 10-2

THE DISSOLVED AND SUSPENDED LOAD OF STREAM WATER

1. Inorganic salts . . . Na^+, K^+, $SO_4^=$, Mg^{++}, etc.

2. Organic solutes . . .
   - carbohydrates
   - fatty acids and oils
   - alcohols and aliphatics
   - ring compounds, phenols, benzenes, etc.
   - amino acids and proteins
   - acids, sugars, and vitamins
   - complex organic compounds
   - metals

3. Colloidal . . .
   - clays
   - S^0

4. Gases . . .
   - oxygen, nitrogen, argon
   - carbon dioxide
   - methane, ammonia

5. Particulates . . .
   - silts and clays
   - detritus (organic residues)
   - bacteria and algae
   - zooplankton and aquatic insects
   - eggs and larvae of worms, insects, and fish
   - fecal matter
   - dust (airfall)

with the biomass of vegetation in the watershed so that a heavily timbered watershed delivers a higher, more complex load of dissolved organic compounds to runoff waters than does a lightly vegetated watershed. These factors all come together in the stream to produce a unique combination—that stream's chemical makeup at a given point in space and time (Table 10-2). This explains the saying "You cannot put your foot in the same stream twice." Chemical variation between streams is generally greater than between various samplings of the same stream, however. Upstream-migrating salmon home-in on this unique chemical "fingerprint" of their natal stream to reach the waters they left as seaward-migrating *smolts* some two to five years previous.

Water Quality   Given this limitless combination of physical and chemical components, how can we reasonably compare and evaluate the water quality of different streams? We do it by targeting the relatively few parameters shown to be effective indicators of the more important aspects of aquatic ecology. The following list of water-quality parameters correlates well with those biotic aspects of streams and lakes.

Water temperature (°F or °C) determines rate of biological activity. Water temperature responds to timber harvest and streamside vegetation removal by increasing in summer and reaching colder wintertime lows. Most upland stream and riparian biota have relatively narrow preferred temperature ranges and, thus, are sensitive to temperature change.

Dissolved oxygen (mg/l concentration or percent saturation) is required by all aerobic aquatic organisms. Normally levels of upland streams approach 100 percent saturation or equilibrium with the atmosphere (~14 mg/l at 4°C), and demand is balanced by supply. Increased temperature, organic loading, or instream organic production from excess nutrient loading can combine to increase oxygen demand over availability from the atmosphere or photosynthesis. Low oxygen values result.

Suspended sediment (NTU turbidity units or mg/l total suspended solids) is normally low in upland streams except during a few days high flow each year. Extended sediment loading increases the content of fines in substrate as explained previously, reduces habitat diversity, blankets and smothers bottom algae and benthos communities, and physically abrades delicate organism tissues. Higher peak flows usually increase turbidity and suspended sediment concentrations.

Water transparency (feet or meters visibility) reflects the rate of light absorption by the water column and inversely relates to suspended sediments. An easily measured parameter, water transparency is also indicative of algae and aquatic plant production in summer months when abiotic turbidity drops.

pH is the measure of hydrogen ion concentration or acid-base balance of water. Excess decomposition may lower the pH (release of H^+ during organic matter mineralization) while high overland runoff often raises pH (solution of exchangeable bases from the soil). The prevailing acidity (low pH) of precipitation in many watersheds around the world has accelerated loss of exchangeable bases from surface soils. The H^+ in precipitation leaches divalent bases (calcium and magnesium) and buffering agents (HCO_3 [bicarbonate ion] and CO_3 [carbonate ion]) from the soil. When soil buffering capacity is depleted, the pH of runoff to streams and lakes drops precipitously. Low pH enhances solution and bioavailability of metals, alters biological membrane permeability, and increases toxicity response of aquatic biota to many toxicants and dissolved metals.

Alkalinity (mg/l as $CaCO_3$ [calcium carbonate]) is a measure of the buffering capacity of water (i.e., the resistance to pH change as H^+ are either added to or removed from water).

Nutrients (especially NO_3 [nitrate], NH_3 [ammonia], PO_4 [phosphate], HCO_3 [bicarbonate ion], Ca [calcium], and K [potassium] as mg/l) control production (new growth) of algae and rooted aquatic plants. Increased soil disturbance and sediment loading are usually accompanied by higher nutrient loading because of greater contact of runoff water with soil particles and greater opportunities for chemical dissolution.

Toxicants (heavy metals such as Cu [copper], Pb [lead], Zn [zinc], Co [cobalt], and As [arsenic], as well as excess NH_3 [ammonia] and CO_2 [carbon dioxide], all measured as mg/l or μ/l) are present in natural waters, but may be toxic to aquatic biota if present in high concentrations. Land management materials such as pesticides, herbicides, fuels, and fire retardants may not normally be present in pristine, upland streams, but can enter stream water via management activities such as timber harvest, road building, fire fighting, and range management. Low concentrations of these toxicants are tolerated by most biota but toxicity ensues at species-specific threshold levels as concentrations increase.

These parameters are all measures of the stream water quality, but users of such data must remember that point samplings are but snapshots of conditions in the stream at a moment in time. Such point sampling may miss effects of short-lived storm or pollution events. Time- and spatially integrated sampling are necessary to provide a complete picture of stream water composition. Continuous, low-volume pumped sampling or streamside samplers that automatically collect samples at hourly or daily intervals can address this deficiency with time-integrated coverage.

Stream ecologists are always seeking measures of the stream environment that *integrate* habitat conditions over a period of time. Measures of substrate are useful, since stream bottom substrate develops from a long sequence of hydraulic scenarios. For example, cobbles become gradually embedded with fines or a sedimented pool may require many high flows for the fines to wash out and cleanse the cobbles. Measures of cobble embeddedness or loss of gravel open spaces are easy and integrative measures of sedimentation (Skille, 1990). Biota also integrate effects of controlling factors over time. After scouring, stream algae take months to reach full colonization levels while benthic insects may have life cycles of two to three years. Compared to a "point" sample of water, these biotic communities act as continuous monitors of stream conditions, 24 hours a day, 365 days a year.

The Concept of Water Quality The question of water quality—"good" or "bad"—is always before the watershed or stream manager. Most people usually have a very narrow interpretation of water quality. The forest manager or river rafter, for example, might consider a stream is high quality if it runs clear and has low cobble embeddedness (e.g., an intragravel environment with low percent fines), regardless of the chemistry of the water. A fisheries biologist might target high dissolved oxygen as an indicator of suitable habitat for target fish species even if the stream carries a high nutrient load. Most fishery biologists, in fact, don't complain of higher nutrients as fish production usually increases, while the rafters complain about green water.

In the 1950s and 1960s, fishery biologists tended to consider upland streams of the northwestern United States healthy if they were physically open to the sea (i.e., free of salmon-blocking debris dams or impassable cascades). Streams with worsening conditions of sedimentation were overlooked until the precipitous declines of anadromous fish runs throughout the western United States in the 1970s through 1990s. Much of the West's upland stream habitat was in a serious state of deterioration because land use activities impacted the watersheds for many decades with little investment put back into sustaining their viability as salmon nursery streams. The streams at the bottom of their watersheds, with their increasing accumulations of fine sediments, poor salmonid egg survival, and deteriorating, low-diversity benthos communities, announced this bad news. So, too, did the warmer, more nutrient-rich stream waters with degraded riparian streamside zones. Resource managers of all disciplines were not listening to the streams or were not asking the relevant questions about stream condition. Had a broader concept of stream water quality been in use, we would have sooner recognized declining habitat quality and lessened the need for the emergency, very expensive salmon-saving measures now required in the twenty-first century. Water quality and the stream environment are reliable indicators of watershed health. Stream and lake water quality measures can alert managers to a deteriorating watershed long before other watershed measures such as soil fertility, vegetation, or wildlife populations reflect change. Managers must listen, however.

Stream Structure

Spatial Structure In cross section, a stream does not present a gradual slope, but an irregular, stair-stepped series of drops. Water from a low-gradient reach or *pool* drops over some channel structure resistant to erosion that retards streamflow, such as a debris dam, gravel bar, or rock outcropping. A plunge pool below the dam can quickly absorb the energy of the falling water. Flow then speeds up over the higher gradient *riffle* reach. If the high-gradient reach is deeper, a high-velocity *run* may comprise part of the channel. The resulting complex, stepped channel, dissipates erosive energy regularly and efficiently, maximizing sediment and debris deposition. A stream with regularly alternating pool-riffle-run

reaches maximizes variation of velocity and, hence, of substrate. This situation also provides a variety of niches for stream biota.

Organic Debris Organic debris (logs, branches, stems, and leaves) enters the stream from adjacent lands. Riparian zones are particularly good suppliers of debris. Prior to 1980, stream management attempted to minimize organic residues such as logs and slash in stream channels. Chief concerns were that debris would build stream blockages to fish migration and that the excess organic matter would cause oxygen depletion. Land managers were told to keep *all* debris out of the stream. Many timber harvest operations not only removed debris added by the immediate operation, but further scoured the streams of old debris deposits. Channels stripped of all their large organic debris (LOD), old and new, are likely to scour badly at the next high flow. Beschta et al. (1987) showed that removal of large organic debris from an 820-foot (250-m) reach of a coastal Oregon stream resulted in loss of more than 6,580 yd^3 (5,000 m^3) of sediments in the following high flow (Fig. 10-9).

Debris dams encourage retention of other organic debris in headwater areas where debris like leaves, twigs, and grasses can slowly decompose and serve

FIGURE 10-9
Organic debris in appropriate amounts is vital as a long-term food supply to in-stream biota. *Left:* Note abundant leaf litter in stream. *Right:* Organic debris in the form of logs, branches, stems, and needles. *(Photos by Michael Falter)*

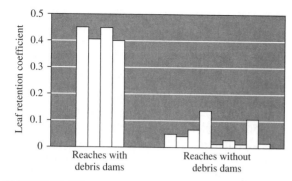

FIGURE 10-10

A comparison of leaf retention in two stream reaches, with and without large organic debris dams. *(From Speaker et al., 1984)*

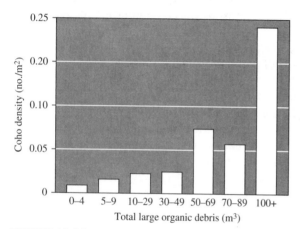

FIGURE 10-11

The benefits of large organic debris for fish habitat in Oregon streams. As the debris increased in volume, so did the juvenile coho salmon numbers. *(From Speaker et al., 1984)*

as a long-term supply of food to the stream biota. Speaker et al. (1984) demonstrated the dramatic increased leaf retention in stream reaches with debris dams (Fig. 10-10). The benefits of LOD carry over to fish habitat—coho salmon numbers showed a dramatic positive response to increasing LOD in Oregon streams (Fig. 10-11).

Unfortunately, some LOD may worsen channel scour if oversized, unstable debris dams divert flows into unprotected banks. When debris dams wash out, they can cause devastating debris torrents that roar down the channel, gaining debris, sediment, water, and, thus, scouring power. Moderate amounts of

LOD in the stream, however, serve a valuable role of enhancing structural diversity, channel stability, and food supply by providing more underwater substrate surface. Bilby (1984) developed a classification system for LOD that helps managers decide whether to leave debris or remove it, depending upon its size, placement in the stream channel, and potential longevity. Managers can learn much about debris suitability in a given forest type by observing existing debris deposition in terms of species composition, expected longevity of the dam, effects on stream channel structure, and other factors. They can then formulate a debris policy suited to the watershed characteristics and the given land management situation.

In-stream organic detritus tends to be greater in lower-gradient streams than in high-gradient streams as it is not flushed out as readily. LOD is not as essential to organic matter pooling in lower-gradient eastern streams as it is in steeper, cascading streams.

Stream Biota   Organisms living in upland stream environments are mostly associated with the substrate. The high water velocity and short hydraulic residence time within a stream work against development of a biotic community in the water column. Despite this restriction to using underwater surfaces, stream biota are fairly diverse, so long as there is sufficient structural diversity provided by a varied stream substrate and LOD.

Decomposers are those largely microscopic flora that break down previously fixed organic matter into simpler organic compounds and eventually to CO_2, inorganic nutrients, and energy. Aquatic bacteria, fungi, and protozoans dominate this group of *heterotrophs* (cannot produce their own plant food), which rely on the stream's organic supply. Decomposers are not spatially isolated from, but are intermingled with, the *autotrophs,* the green plant components of the attached periphyton film.

Forested watersheds, especially deciduous forests of eastern biomes, are especially rich suppliers of fine particulate organic particles (*detritus*) to the stream. From the moment it enters the stream, detritus begins to break down to simpler, smaller particles, to dissolved organic matter, and finally to CO_2 and inorganic end products of complete oxidation (e.g., NO_3, PO_4, SO_4, SiO_3, etc.). Fungi initiate breakdown when their mycelial filaments penetrate the detritus and exude enzymes, causing extracellular digestion. For this reason, fungi are particularly effective decomposers of leaves and woody debris. The stage is then set for a myriad of aquatic bacteria to attack the newly released organic molecules.

Animal components of the decomposer community are primarily protozoa, rotifers, nematodes, and crustaceans that graze on fungal and bacterial mats as well as on algae. These grazers ensure continual high rates of organic breakdown by preventing overpopulation of fungi and bacteria.

The autotrophic *periphyton* community is analogous to a multistory climax forest community in miniature, with its hundreds of species of varied life forms forming "canopy, understory, shrub, and forest floor" layers, all in the thickness of 1 or 2 mm above the substrate. Periphyton is the film of attached algae, bacterial, and fungal communities that cover submerged surfaces in the stream with the brown-green film (and that make wading over slippery stream rocks so difficult). Periphyton communities function at a very high production rate because of their high surface area-to-volume ratio and the high rate of food and nutrient supply provided by the always-present stream current. Even though the biota of flowing waters are largely limited to the substrate, the rich variety of particulate and dissolved organics flowing by in the water constantly nourish the periphyton. Some algae, such as certain diatoms and blue-green algae, may also function as heterotrophs in very low light situations. One can expect major differences in community diversity and production among the acidic, nutrient-poor waters draining coniferous forests and the more alkaline, nutrient-rich waters of deciduous forests and, finally, the most productive prairie streams of grassland biomes.

Attached benthic algae are the green-plant, photosynthetic components of the periphyton. Primarily comprised of diatoms, green algae, blue-green algae, and some dinoflagellates, these autotrophs fix new organic carbon from CO_2, nutrients, and sunlight via photosynthesis. Their abundance in streams is strongly related to light and nutrient availability. Where riparian cover is dense, autotrophic green-plant production is very low, with most of the stream's food supplied as organic debris from the forest canopy. This is the most common situation in forested first- to third-order streams. Given adequate sunlight, autotrophic production will increase up to

FIGURE 10-12
A shaded first-order stream showing natural debris dams. Blue Mountains, Oregon. *(Photo by Michael Falter)*

the threshold of limiting nutrients, usually imposed by dissolved phosphorus (PO_4) or perhaps nitrate (NO_3), the principal dissolved nitrogen form. Timber harvest, thinning, or removal of the riparian zone by either timber harvest or grazing often increases stream production sharply because of the superabundant light then available. That scenario would happen *if* there were no sediment increase accompanying the vegetation removal. Sedimentation is especially deleterious to attached algae communities because it can completely cover the algae, retarding light and nutrient resupply. Because of these relationships, vegetation management in riparian zones is very important.

Rooted aquatic plants are the second component of the stream's autotrophic community. Mosses (e.g., *Fontinalis* and *Drapanocladius*) are often the most common rooted aquatics in first-order streams where shade is heavy and the groundwater influence provides clear, cool, and CO_2–rich water (Fig. 10-12). Other common upland, stream-rooted aquatic plants in North America are *Ranunculus* (water buttercup), *Elodea, Potamogeton* (pondweed), and *Myriophyllum* (milfoil). These can be important in mid-order streams (third to fifth order) where stream channels are deeper with finer sediments so that rooted plants can develop, and wide enough so that the streamside canopy does not shade the stream. High water velocity often restricts rooted aquatic plants to the slower stream margins. Rooted aquatic plants also prefer deeper, finer sediments as a rooting medium, so they will often flourish after increased sedimentation, attaining very high densities in excess of 2,000 g/m^2 in rich, fine sediments (Falter & Carlson, 1993).

Benthic macroinvertebrates (BMIs) are the principal animal component of upland streams. Like attached algae and plants, most BMIs live on or in the stream bottom for protection against being swept downstream. Aquatic insects, mostly juvenile larval forms, live on top of and within the coarse stream bottom substrate. Water flowing through the interstitial environment supplies this complex community of life (*benthos*) with essential nutrients, oxygen, and flushing currents. These aquatic animals fall into four functional groupings based on their food supply. Each functional grouping includes many different taxa of benthic macroinvertebrates:

1. *Grazers* subsist on the film of attached bacteria, fungi, and algae on submerged surfaces. Mayflies and midge larvae, snails, and freshwater shrimp (amphipods) are typical stream grazers.
2. *Shredders* consume the organic debris such as twigs, grass stems, leaves, and bodies of any terrestrial or aquatic organism in a stream. Principal shredders are caddisfly larvae, stonefly larvae, and crayfish.
3. *Collectors* scavenge the fine particulate organic matter from the flowing water. Fine detritus particles released by the shredders, mechanical breakdown of plant and animal matter in the stream, and washings from the watershed contribute to the collectors' food base. Major forms are caddisflies and midges, which filter the streamflow with nets spun from body secretions; blackfly larvae, which filter the streamflow with specialized antennae; clams and mussels, which strain organic particles from the water with sticky respiratory surfaces; and the mud-dwelling freshwater earthworms, which ingest fine sediment.
4. *Predators* consume each other, organisms from the other functional groups, and small fish. Most of the four kinds of aquatic animals have predaceous members, except the clams and snails.

The total stream community is a beehive of activity as upwards of 150 kinds of BMIs might be living, eating, and reproducing in a stream reach. Numbers wax and wane with the season, typically peaking in late summer and reaching annual lows at high water flow when the community is scoured out. Distribution of the functional groupings of BMIs along a stream course is largely dependent on the nature of the riparian zone and stream size. A continuum of stream communities results, over the gradient of stream environments from headwaters to mouth (Fig. 10-13).

Upstream in forested areas with heavy stream canopy, the supply of organic debris from overhanging riparian vegetation dominates stream trophic (food) relationships. Shredders and predators dominate BMI structure in these dark, cool, high-gradient streams. As the riparian canopy opens downstream, light-requiring green plants do well, providing grazers with a food base. Collectors gradually increase downstream as the organic particles become smaller and smaller with continual processing. Small, upland streams in midwestern prairies or treeless sagebrush steppe of the western United States are not tree-covered and ligh-limited; their trophic base is in-stream photosynthesis from the smallest through mid-sized streams.

Nutrients and food in streams are repeatedly taken up by biota, stored as plant or animal tissue, released by decomposition, and taken up again, albeit farther downstream. The process of stop-and-go utilization by stream biota is likened to a coiled spring stretched parallel to the stream channel. The spring is stretched when these uptake-release cycles are long (stretched out along the stream channel) and compressed when the cycles are short. This concept of *nutrient spiraling* illustrates the conservative nature of streams (i.e., bank, channel, and organism storage of energy and material slows their seaward rush).

IMPACTS OF WATERSHED USE AND FOREST MANAGEMENT ON WATER YIELD AND QUALITY

Controllers of Streamflow

In small watersheds, streamflow responds rapidly to rainfall because of the short link between rainfall and the stream. Snowmelt usually delivers water to a stream channel more slowly than does rainfall by the buffering provided by cold weather storage. Prevailing

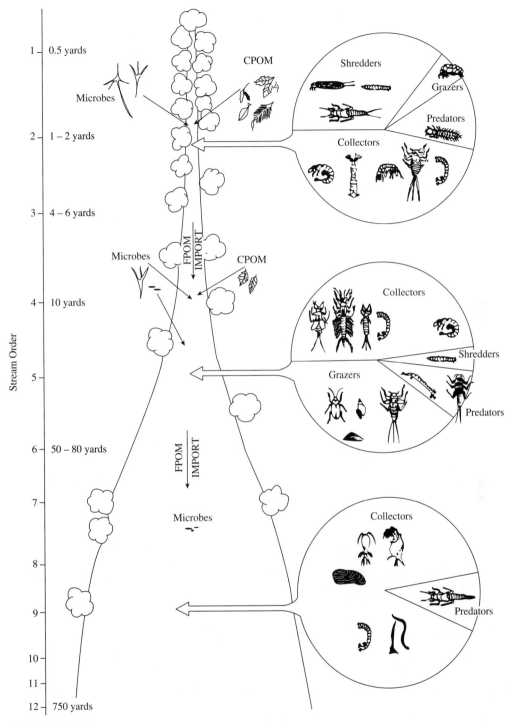

FIGURE 10-13
The stream continuum concept as a model depicting aquatic invertebrate distribution. CPOM = coarse particulate organic matter; F = fine. *(From Cummins, 1977)*

weather patterns also regulate runoff via control of solar input, rate of heat gain, and evaporation. We have little control of the delivery of rain or snowmelt to the land surface, so management attention centers on soil conditions, namely the surface vegetation as well as the characteristics of surface soils that control the amount of water and the rate of water entry into the soil. Tools such as cloud seeding and snowpack management (control of deposition and snowmelt) may offer some possibilities of runoff control, especially in higher-elevation zones.

There are several streamflow controllers subject to large modification by human management activities in wildland watersheds. *Vegetation removal* through harvesting will usually reduce evapotranspiration, leaving more water available on the watershed to enter the soil and flow into stream channels. *Soil compaction* results in greater overland runoff with less insoak to the groundwater. Equipment operation and overgrazing compact the soil so that most runoff might occur very rapidly as overland flows. *Burning* and *soil erosion* remove organic matter in topsoil, so that soils compact easier and have decreased water-holding capacity. This self-perpetuating cycle causes even less insoak, more erosion of the soil surface, and more topsoil loss. *Water storage* in such soils is greatly reduced.

Two things happen to the stream runoff pattern in a watershed: (1) Total annual water flow from the watershed increases (primarily as a result of less evapotranspiration); and (2) the runoff pattern is flashy (i.e., with very high peak flows relative to low flows because of little water storage on the watershed). Much of the total annual runoff passes down the stream channel in a short time. In watersheds where a high percentage of the timber cover is removed, perennial streams may flood in spring, only to dry up in the summer. The small amount of water left in the stream channel may flow beneath the sediment surface. In either case, the result is the same—a dry stream channel. The extreme high flows (relative to stream channel capacity) increase stream power and streambank erosion. Stream sediment load increases, as do fines deposited in the substrate. Fine sediments, especially, accumulate in the lower stream reaches where gradient levels off, accentuating the tendency of summer flows in a small stream to flow underground close to the stream mouth.

Water Quality and Biological Impacts of Land Uses on the Stream Environment

The most obvious impact of high sediment loading to the stream (i.e., loading a stream beyond its transport ability) is buildup of fine sediments in the substrate. Average particle size declines from the original gravel-cobble substrate to sand and silt-sized particles. Interstitial spaces become smaller, retarding intragravel water flow. The intragravel environment becomes more isolated from surface waters. The respiration of intragravel biota, primarily insects, developing fish eggs, and bacteria (decomposing the stream's organic load) depletes dissolved oxygen to lethal levels. In a low-oxygen environment, dissolved products of decomposition can accumulate to toxic levels. Ammonia and carbon dioxide are the most critical, but hydrogen sulfide and methane gases can also reach toxic levels. Cumulative mortality eventually changes a diverse biota of many species to a biota dominated by fungi and bacteria.

The intragravel community beneath the streambed is normally nourished by through-gravel water flows, but sedimentation fills the interstices and restricts flow, thereby isolating and changing the intragravel environment. With the resulting oxygen depletion and accumulation of toxic conditions, insect production declines and taxonomic composition changes to small burrowing forms, more acclimated to life in fine-sediment, low-oxygen conditions. Insect *diversity* declines and the normally heterogeneous community becomes dominated by only a few taxa.

In the sedimentation situation, *resilience,* or the capacity of aquatic communities to absorb environmental disturbances with little change, diminishes. Population numbers become more variable, resulting in an unstable benthic community. Such sedimentation in many forested streams of the northern Rocky Mountains has resulted in replacement of communities dominated by stoneflies, caddisflies, and large mayflies to communities dominated by small midge and mayfly larvae.

Developing fish eggs in sedimented gravels suffer from oxygen depletion and accumulation of respiratory products, namely carbon dioxide and ammonia. Even in an unproductive stream where organic production is so low that oxygen limitation does not occur, the filling-in of intragravel spaces by sedimentation reduces living space and traps emerging fish embryos in the sediments. Mortality of rainbow trout

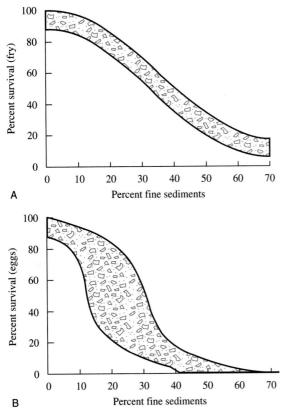

FIGURE 10-14
Salmonoid embryo survival versus percent fines in the gravel environment. A = fish fry; B = fish eggs. *(From Everest et al., 1987)*

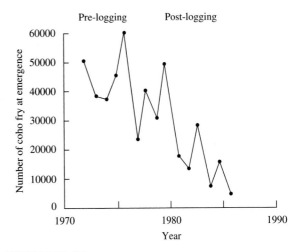

FIGURE 10-15
The comparison of coho salmon fry emerging from gravel before and after logging, over a nine-year period. Carnation Creek, British Columbia. *(From Hartman & Scrivner, 1990)*

and coho salmon embryos is a function of fines in incubation gravel (Fig. 10-14). Above 35 percent fines, embryo survival declines precipitously. Older, juvenile salmonids also depend on coarse sediments in the first two to three years of stream life for food production and over-winter habitat. As water temperatures drop below 3°C, juveniles move into the gravel and rest there in a state of semi-hibernation. Substrate embeddedness with fines reduces that essential winter habitat.

The long-term response of a salmonid population to logging was followed in the Carnation Creek watershed, British Columbia (Hartman & Scrivner, 1990). They found that coho salmon fry emerging from the gravel declined over a nine-year period following logging in the drainage (Fig. 10-15). One im-

pact of logging was that year-to-year variability of emergence increased, suggesting a more variable stream environment. A fish population already at dangerously low numbers would stand a greater chance of declining to extinction in such an environment of greater variability.

Nutrient export shows a response pattern similar to that of sediments. Stream waters carry ions exported from a watershed, whether at baseline rates or at rates enhanced by management practices. The Hubbard Brook Experimental Watershed in the northeastern United States showed markedly increased ion export rates in the second and third years following deforestation as indicated by stream water chemistry. Net losses in the second year after deforestation exceeded baseline conditions by 6.9 times for magnesium, 22.4 times for potassium, 9.8 times for calcium, and 29.7 times for nitrate (Likens & Bormann, 1975). Enhanced ion mobilization by increased decomposition was the main reason for the increased ion losses from the watershed; increased water yield from the watershed following harvest was a contributing cause.

Table 10-3 illustrates the generalized response of nutrient export from a watershed as a function of land use. Vegetation and soil surface disturbance

TABLE 10-3

REPRESENTATIVE MEAN VALUES FOR PHOSPHOROUS AND NITROGEN EXPORT RATES (KG/HA/YR) FROM VARIOUS WATERSHEDS

Land Use	Total Phosphorus	Total Nitrogen
Forest	0.2	11.0
Pasture	0.3	8.0
Row crops	2.4	17.2
Feedlot	315.0	850.0
Residential urban	0.85	6.2
Industrial urban	2.5	8.0

(*Source:* NALMS & U.S. EPA, 2001)

dramatically increase natural nutrient export from watersheds of all sizes.

The ultimate negative impact of management practices on aquatic biota is, of course, drying or dewatering of the stream. Low-flow drying of a stream reach is totally devastating to biota—some forms, such as most insects and fish, may require several or more years for total recovery. Replacement of coarse substrates with fines clearly leads to reduction of required salmonid habitat in streams. In the Pacific Northwest, these losses contributed to the situation in 1993 wherein 9 of the 10 major stocks of anadromous salmonids in the coastal, Columbia, and Snake River drainages were considered either extinct or at risk of extinction across most of their historical range (Snake River Salmon Recovery Team, 1993). The combination of land management practices cumulatively impacting spawning and rearing habitat over the past 120 years, mainstream dams interfering with both juvenile and adult fish migration, and harvest (sport and commercial) is universally accepted as the principal cause of these precipitous declines.

SUMMARY

The hydrologic cycle depicts various reservoirs of the biosphere's water and the pathways between them. Hydrology quantifies patterns of rainfall, snowfall, and surface storage and routing of water as well as groundwater storage and routing. Watersheds, defined by their stream channels, are the hydrologic units that store and distribute water on the earth's surface. Within a watershed, there is a constant infiltration of water from the surface through soils to deeper storage in geological layers and the water table. Most streams and lakes are localized surface exposures of the water table, although perched surface water bodies above the water table are not rare. Although lakes and streams were viewed historically as being isolated from their watersheds, a more functionally correct view now recognizes the many waters supply, energy, nutrient, and trophic linkages that force streams and lakes to operate in synchrony with, and at the mercy of, their watersheds.

Streamflow from a watershed is not a constant. It varies not only with precipitation, but also with weather, soil, and vegetation variables. Management actions control water yield by altering the latter two. As a result, there is a seasonal pattern of streamflow (the hydrograph) unique for a watershed as well as for the present condition of the watershed. Water moving through a watershed, both as surface flow and groundwater flow, mirrors the watershed, reflecting watershed geology in the sediment and chemical composition of runoff water. Water quality is a concept defined by the end user of water but may correctly refer to any one of many physical or chemical parameters. Different water quality parameters are targeted by different user groups. The wide array of physical and chemical controllers shape biotic communities of lakes and streams in a watershed so that each surface water body has a unique assemblage of aquatic communities.

Land uses such as timber harvest, road construction, livestock grazing, wildland habitation, and recreation use, control streamflow by altering evapotranspiration and water insoak rates. Impacted watersheds usually have more total runoff delivered to the streams with higher mean and peak flows. Perennial streams may become intermittent as surface water flows beneath new deposits of fine sediments. The stream biota are variously impacted by resulting stream drying and loading of fine sediments, but population numbers, community diversity, and stability of aquatic communities nearly always decline following land use impacts. Patterns of functional groups of stream biota provide effective indicators of watershed and stream health.

The cumulative adverse effects of many decades of land management practices on stream systems have contributed to catastrophic declines of anadromous salmon stocks in the Pacific Northwest, with some of the runs pushed to the verge of extinction.

LITERATURE CITED

Beschta, R. L., R. E. Bilby, G. W. Brown, L. B. Holty, and T. D. Hofstra. 1987. Stream Temperature and Aquatic Habitat: Fisheries and Forestry Interactions. In: Salo, E. O. and T. W. Cundy (eds.), pp. 191–232. Streamside Management. University of Washington, Institute of Forest Resources, Contribution No. 57. Seattle, Wash.

Bilby, R. E. 1984. "Post-logging Removal of Woody Debris Affects Stream Channel Stability." *Jour. of Forestry* 82:609–613.

Bisson, P. A., R. E. Bilby, M. D. Bryant, C. A. Dolloff, G. B. Grette, R. A. House, M. L. Murphy, K. V. Koski, and J. R. Sedell. 1987. Large Woody Debris in Forested Streams in the Pacific Northwest: Past, Present, and Future. In: Salo, E. O. and T. W. Cundy (eds.), pp. 143–190. Streamside Management. University of Washington, Institute of Forest Resources, Contribution No. 57. Seattle, Wash.

Cowardin, L. M., V. Carter, F. C. Golet, and E. T. LaRoe. 1979. Classification of Wetlands and Deepwater Habitats of the United States. Office of Biological Services Report FWS/OBS-79/31. U.S. Fish and Wildlife Service, Washington, D.C.

Cummins, K. W. 1977. "Form Headwater Streams to Rivers." *American Biology Teacher* 39:305–312.

Everest, F. H., R. L. Beschta, J. C. Scrivener, K. V. Koski, J. R. Sedell, and C. J. Cederholm. 1987. Fine Sediment and Salmonid Production: A Paradox. In: Salo, E. O. and T. W. Cundy (eds.), pp. 98–142. *Streamside Management.* University of Washington, Institute of Forest Resources, Contribution No. 57. Seattle, Wash.

Falter, C. M., and J. W. Carlson. 1993, July. Sediment, nutrient, and aquatic macrophyte interactions in the Middle Snake River, Idaho. Report submitted to Idaho Division of the Environment, Twin Falls.

Hartman, G. H., and J. C. Scrivner. 1990. "Impacts of Forestry Practices on a Coastal Stream Ecosystem, Carnation Creek, British Columbia." *Can. Bull. Fish. Aquat. Sci.* 223, 148.

Likens, G. E., and F. H. Bormann. 1975. An Experimental Approach to New England Landscapes. In: Hasler, A. D. and L. Tonolli (eds.), *Coupling of Land and Water Systems.* Springer-Verlag, New York.

McClelland, D. E., R. B. Foltz, C. M. Falter, W. D. Wilson, T. Cundy, R. L. Schuster, J. Saurbier, C. Rabe, and R. Heinemann. 1999. Relative effects on a low-volume road system of landslides resulting from episodic storms in northern Idaho. *Transportation Research Record* 1652, 235–256. (National Research Council).

North American Lake Management Society, The Terrene Group, and the U.S. Environmental Protection Agency. 2001. Managing Lakes and Reservoirs: The Lake and Reservoir Restoration Guidance Manual (3rd ed.). ISBN 1-880686-15-5. EPA 841-B-01-006. NALMS, Madison, Wisc.

Satterlund, D. R., and P. W. Adams. 1992. *Wildland Watershed Management* (2nd ed.). John Wiley and Sons, New York.

Skille, J. M. 1990. Stream and Lake Nutrient Loading from Burned Logging Slash. Water Quality Summary Report No. 26. Idaho Department of Health and Welfare, Bureau of Water Quality, Boise, Id.

Snake River Salmon Recovery Team. 1993, October. Draft Recommendations to the National Marine Fishery Service.

Speaker, R., K. Moore, and S. Gregory. 1984. "Analysis of the Process of Retention of Organic Matter in Stream Ecosystems. *Verh. Int. Ver. Limnol.* 22:1835–1841.

Trimble, S. W. 1975. Denudation Studies: Can We Assume a Steady State?" *Science* 188:1207–1208.

U.S. Department of the Interior. 1991. Riparian-wetland Initiative for the 1990s. Bureau of Land Management. BLM/WO/GI-91/001+4340. Washington, D.C.

U.S. Geological Survey. 1969. *Water of the World.* Pamplet O-348-604. U.S. Government Printing Office, Washington, D.C.

U.S. Geological Survey. 1992. *Water Resources Data— Idaho: Water Year 1991.* U.S. Geological Survey Water-Data Report ID-91-2, Boise, Id.

Vannote, R. L., G. W. Minshall, K. W. Cummins, J. R. Sedell, and C. E. Cushing. 1980. "The River Continuum Concept." *Canadian J. Of Fisheries Aquatic Science.* 37:130–137.

ADDITIONAL READINGS

Allan, J. D. 1995. Stream Ecology: *Structure and Function of Running Waters.* Chapman and Hall, New York.

Cushing, C. E., and J. D. Allan. 2001. *Streams: Their Ecology and Life.* Academic Press, New York.

Giller, P. S. S., and B. Malmqvist. 1999. *The Biology of Streams and Rivers.* Oxford University Press, Inc., London.

Gordon, N. D., T. A. McMahon, and B. L. Finlayson. 1992. *Stream Hydrology.* John Wiley & Sons, Inc., Somerset, N.J.

Horne, A. J., and C. R. Goldman. 1994. *Limnology.* McGraw-Hill, New York.

Hunter, C. J. 1991. *Better Trout Habitat.* Island Press, Washington, D.C.

Mitch, W. J., and J. G. Gosselink. 1993. *Wetlands* (2nd ed.). Van Nostrand Reinhold, New York.

Naiman, R. J., and R. E. Bilby (eds.). 1998. *River Ecology and Management: Lessons from the Pacific Coastal Ecoregion.* Springer, New York.

National Research Council. 1992. *Restoration of Aquatic Systems: Science, Technology, and Public Policy.* National Academy Press, Washington, D.C. USGPO. ISBN 0-309-04534-7.

Thorne, C. R. 1998. *Stream Reconnaissance Handbook: Geomorphological Investigation and Analysis of River Channels.* John Wiley & Sons, Inc., New York.

STUDY QUESTIONS

1. What watershed do you live in? Does the water you drink come from the same watershed? If not, where does it come from? Is it surface water or groundwater?

2. What does the equation P – ET = RO express? Give an example of this from real life.

3. Draw a simple diagram showing first-, second-, third- and fourth-order streams.

4. Define the following terms: *evapotranspiration, water table, aquifer, runoff, benthic, mass wasting, hydrographs, periphyton film, deoxygenation, stream reach, stream collectors, shredders and grazers, riparian zone, wetland,* and *sediment transport.*

5. How could we ever have thought that lakes functioned independently of their watersheds? What societal and economic factors fostered this view?

6. Why do first-order streams often exhibit "flashy" hydrographs? How does increasing watershed size affect the hydrograph?

7. Contrast stream insect communities likely found in headwater streams of eastern deciduous forests with those in headwater streams of western rangelands.

8. Define *fines* (with respect to sediments). In what way can they affect fish habitats? Contrast sedimentation impacts on bass, sunfish, or catfish communities of eastern and southeastern U.S. streams relative to salmonid streams of the western United States.

9. Build a case for maintaining ecologically diverse streamside riparian zones.

10. Discuss major potential impacts of timber harvest on streams. Now, add road construction to the equation. How can roads be constructed to minimize watershed and stream impacts?

11. In what ways do streams reflect their watersheds?

CHAPTER 11

Conservation and Management of Rangeland Resources

This chapter was prepared by Dr. Kendall Johnson, Head, Department of Range Resources, College of Forestry, Wildlife and Range Sciences, University of Idaho.

INTRODUCTION

The term *range* was developed in the United States to describe the extensive, unforested lands dominating the western half of the continent, probably from the idea of freely roaming or ranging for long distances. These lands are characterized by limited precipitation, either seasonally or annually, resulting in relatively sparse herbaceous or shrubby vegetation. Sharp climatic extremes, highly variable soils, frequent salinity, and diverse topography are also characteristic features. Western range is thus a kind of land, and is properly referred to as *rangeland.*

Rangeland is also described in more specific terms. For example, prairie, plains, grassland, shrubland, savanna, steppe, desert, semidesert, sward, tundra, and alpine are all types of rangeland. Some areas are both rangeland and forest, such as near the forest edge, or within a forest where the tree overstory is sparse. Although not strictly rangeland, natural openings within the forest (meadows), or areas where the trees have been removed by fire or harvest, also produce vegetation characteristic of rangeland. Even some forestlands, to varying degrees, support an understory of herbaceous and shrubby vegetation and are described as *forest range.*

Globally, the many types of rangeland together form the largest part of the earth's land surface—about 44 percent. For comparison, forestland comprises about 28 percent and tilled cropland 10 percent. Ice caps, fresh water, and permanent snow cover 15 percent; the small remainder is taken up by the world's urban, industrial, and transportation lands.

Considering the North American continent alone, the extensive rangelands of Canada, the United States, and Mexico form a large portion of the continent. Within the United States, rangelands make up about one-third of the nation's total land area. As rangelands contain a principal share of the earth's natural resources, their proper use and management become a matter of vital importance to people everywhere.

For thousands of years, rangeland vegetation has provided food for a wide variety of grazing animals known as herbivores. Although livestock grazing is clearly an important use presently as well as historically, rangelands produce a wide variety of goods and services including wildlife habitat, water, mineral resources, wood products, many forms of wildland recreation, open space, and natural beauty.

SOCIAL AND ECONOMIC STATUS OF U.S. RANGELANDS

Historical Development

Before the advent of European settlement in the United States, millions of acres of varied rangeland supported native wildlife. The native herbivores of rangelands included an incredible variety of insects, of which the grasshoppers and crickets were probably the most visible and dramatic; numerous burrowing animals, notably ground squirrels and prairie dogs; and the large, charismatic grazers of fact and legend—bison, elk, deer, pronghorn, and bighorn. The major grazing animals in the North were caribou, reindeer, and musk ox (Fig. 11-1); in the South, deer were the major species. With settlement came plowing, logging, mining, and grazing in the familiar pattern of frontier exploitation and development of resources. The large grazers were eliminated from the eastern Great Plains, due to conversion of the grasslands to agriculture, and were generally reduced elsewhere. Bison were largely eliminated, elk and bighorn were pushed into the Rocky Mountains, and pronghorn antelope were confined to the intermountain basins and the western Great Plains, which were too dry for agriculture. Deer (whitetail and mule) generally seem to have adapted to the new living conditions throughout their former range, while elk continue to expand in their new forest range habitats. In some areas, especially in the southeastern United States, populations of feral hogs have become important foraging animals.

After the first transcontinental railroad was completed in 1869 (thus opening eastern markets for livestock products), huge herds of cattle, and later sheep, were driven to the western rangelands. The rise of the livestock industry initiated a colorful phase in American history. Although trail herds were known before the Civil War, they increased dramatically when many soldiers returning to Texas in 1865 saw an opportunity to make money marketing cattle. Expanding eastern markets, extension of railroads to the west, settling of native claims (from the European viewpoint), money invested from abroad, and millions of acres of free forage on western public land

FIGURE 11-1
Two of the original grazers of western rangelands. *Upper:* The pronghorn antelope, once 40 million strong, today numbers about 165,000, largely confined to the intermountain basins. *(Photo by John Schecter) Lower:* The American bison, or plains buffalo, numbered in the 60 millions when the first Europeans arrived. It ranged from the Appalachians and Rockies south to the Gulf of Mexico. The bison faced extinction when, around 1900, action by cattlemen and conservationists led to its protection on government and private reserves. *(Photo by Grant W. Sharpe)*

allowed a phenomenal rise in the cattle industry over a 20-year period. However, a series of catastrophic winters coupled with droughts—1883–1884 in the southern Great Plains, 1889–1890 in the Great Basin and Pacific Northwest, and especially 1886–1887 in the central and northern Great Plains—all but wiped out the industry. During the next few years, the survivors completely reestablished the livestock busi-

ness on wholly different premises. External investment money was largely gone. So was the concept of free-roaming winter grazing; range operations were restructured on the basis of winter feeding when necessary. This entailed the raising, harvesting, and storage of hay. Land ownership in the Great Plains states became almost entirely private through the several laws regulating the transfer of the public domain to private ownership. Thus, livestock raising in the grasslands became much like any other agricultural enterprise, a situation that continues to this day.

In the forested South, land tenure also became almost entirely private through acquisition of large blocks of land by timber companies. These tracts were then cut over and most were essentially abandoned. Without direct competition from trees, herbaceous and shrubby vegetation quickly produced a significant forage resource that became the basis of a major livestock industry in the South. Private livestock operators built the industry largely through rental or lease of cut-over lands. Forage production was maintained for many years through a program of periodic burning, but intensifying use of Southern Forests for wood production, together with increased use of the forests for hunting and recreation, made livestock raising mostly a secondary use.

But in the Rocky Mountain, Great Basin, and southwestern states, great blocks of public land remained because substantial portions of the land were too arid to be put to effective agricultural use under the land disposal laws. From about 1870 to the 1930s, the public lands became a general commons, in which access to natural resources fell to whomever claimed them first. Cattle ranchers generally secured rights to well-watered lands in the valleys under the land disposal laws, and then exercised domain over vast areas of adjoining arid lands and high country. Battles were often fierce between ranchers competing for scarce forage, especially in drought years. Other ranchers, especially those running sheep, chose to become itinerant herders, moving their animals to wherever forage was available. Sometimes the sheep herds were shipped in by train, grazed over the surrounding countryside, and shipped out again. The cattle raisers, forced to remain with nonmigrating herds and permanent ranch facilities, resented sheep and sheepherders. The bitter range wars that sometimes resulted are familiar to all who read novels, attend movies, or watch television shows based on tales of the Old West.

By the turn of the twentieth century, however, unrestrained exploitation of rangeland resources, especially on public lands, had begun to affect the land itself, with the result that soils eroded, streams became silted and some dried up, and floods increased. Early stockmen understood very little about the relationships between grazing and rangeland condition, and there were no public means to regulate grazing. Most observers agree that as a result, rangelands everywhere were in a very damaged condition in the early years of the twentieth century.

These and other natural resource problems prompted the establishment of the U.S. Forest Service in the early 1900s. Only then did grazing of forested public lands begin to come under management. However, not until passage of the Taylor Grazing Act in 1934 did the vast areas of open western rangeland acquire similar management by the U.S. Grazing Service, which later became the Bureau of Land Management (BLM). Also in 1934, the Soil Conservation Service (now the Natural Resources Conservation Service (NRCS)) was established to assist private landowners in land conservation measures.

With management, the condition of public and private rangelands has steadily improved. The first national range survey (U.S. Senate, 1936) showed 16 percent of rangelands to be moderately depleted (good condition), 5 percent materially depleted (fair condition), and 58 percent severely or extremely depleted (poor condition). Due to scientific advances over the last decade, rangeland condition is no longer described in simple classes, but on the basis of ecosystem function or sustainability, region, and as federal or nonfederal rangelands. For example, the NRCS reported that range condition improved from 1982 to 1992 on nonfederal rangeland in regions with significant area (Pacific Coast, Rocky Mountain, Plains, and South). Condition of lands managed by the BLM in the Pacific Coast and Rocky Mountain regions, on an overall basis, held steady during the period from 1986 to 1996. Data from Forest Service lands also support the conclusion that rangeland condition has remained essentially static over the same period (Mitchell, 2000). Most rangeland observers agree, and all data show, that range condition has materially improved since the 1930s (Box, 1993). At the same time, a decline in utilization of grazing lands by livestock is anticipated in the western and northern regions of the country for the period 2000 to 2050, but is expected to remain stable in the southern region. Wildlife utilization is projected to increase in all regions (Van Tassell et al., 2001).

Social Context

There is no question that the health of rangelands has improved—and is still improving—yet the future use

FIGURE 11-2
Recreation is an important use of rangelands. Here, a fly fisherman casts for the elusive trout on a western national forest. *(Courtesy of USDA Forest Service)*

and management of range resources, especially of those on public lands, are now frequently a matter of contentious debate. Many people are no longer ready to accept the idea that the primary value of public rangelands lies in contributing to the production of food and fiber. These citizens focus on the nonconsumptive uses of rangeland—wildlife habitat, water, recreation, open space, and natural beauty (Fig. 11-2).

Thus, a substantial portion of the American population is deeply interested in, and concerned about, the use and management of rangeland, although few understand the ecological processes involved. Consequently, much concern directed toward livestock grazing centers not on the continued improvement of grazing methods and the use of livestock grazing as a vegetation management tool, but on the question of whether livestock should graze public rangelands *at all* (see Donahue, 1999).

Livestock producers, as a group, are very concerned about how changes in planning objectives, standards, and grazing fees will affect the stability and long-term viability of their operations. For some of them, the issue is nothing less than economic survival, but as with their critics, their ranks are divided. Despite legal opinions to the contrary, some ranchers feel livestock grazing on public land under long-

standing permit is a right, not a privilege, and loss of a grazing permit constitutes a "taking" of private property. While some demand freedom from government "interference" of any kind, many others wish to be in the forefront of developing ecologically sound and socially acceptable solutions for rangeland problems. Several examples are described by Dagget (1995).

Although many of the public rangeland concerns are not present—or are different—on private lands, changing social values also affect the use and management of private rangelands. Thus, current rules and regulations may require maintenance and improvement of water quality, preservation and improvement of wetlands, best management practices, soil conservation plans, and preservation of threatened and endangered species. To an increasing degree, private landowners face major new requirements and restrictions in their uses of the land and its water.

Economic Status

The numbers of livestock (cattle, sheep, horses, and some goats) grazing on public and private lands have decreased substantially from a post-World War I peak, but the use of rangelands by livestock remains a significant economic enterprise (Fig. 11-3). The common economic unit for measuring rangeland use by livestock is *animal unit months* (AUMs), meaning the amount of forage needed for one cow–calf pair for 1 month, or equivalently, for five sheep. Use on public lands is administered through a permit system under which fees for grazing a certain number of livestock are collected. Use on private lands is determined by market arrangements.

The BLM has about 167 million acres of land under grazing permits and leases. In 1991–1992, approximately 19,000 operators held grazing rights for about 13.5 million AUMs, including 3.4 million AUMs of authorized nonuse. Authorized use of 10.1 million AUMs generated about $18.5 million in receipts (USDI-BLM, 1993). About half of the national forest system lands plus the national grasslands are in grazing allotments, with permits issued to more than 9,000 livestock producers. In 2000, the system accommodated nearly 8 million AUMs of grazing for nearly 7,500 operators (USDA-FS, 2001). Both the BLM and Forest Service issue free-use permits for recreational and research livestock. Thousands of wild horses and burros, and hundreds of

FIGURE 11-3
The principal western rangeland species are domestic sheep and cattle. *Upper:* A sheepherder and his sheep in the sagebrush-grass country of Wyoming. *Lower:* A cowboy trailing a herd of cattle to spring pasture in the high plains of Wyoming.
(Courtesy of the University of Wyoming)

thousands of big game animals, also graze on the national forests and BLM's national resource lands.

Grazing Fees Livestock producers who use and are dependent on public lands (permittees), and members of Congress from the West, generally feel that the existing grazing fee formula is reasonable (currently $1.97 per AUM). Some environmental groups and members of Congress believe the current grazing fee constitutes a subsidy to permittees, since fees on private lands are much higher, as they are on most state lands.

The relatively low fees charged for grazing livestock on national forests and national grasslands administered by the Forest Service, and on the public lands administered by the BLM, are a long-standing issue. Livestock permittees are concerned about the uncertainty over future grazing fees and would like some assurance of stability and predictability for future fees. They support a continuation of public land grazing as a "way of life" important to rural area stability. Environmental groups, however, believe that there is a direct relationship between low fees and poor range conditions, and that fees should at least equal agency costs for the livestock grazing program. Some groups would remove all livestock grazing from public lands.

The issue of fees for private livestock grazing on public lands will continue to be a bitterly contested issue, but fees are sure to increase in the future.

RANGELAND RESOURCES IN THE UNITED STATES

Location and Extent

Although estimates vary, most observers agree that rangeland vegetation occupies roughly 800 million acres (324 million ha), or 36 percent of the total land base of the United States. A recent assessment by the USDA Economic Research Service (ERS) ascribed 589 million acres (238 million ha) to grassland pasture and rangeland in the 48 contiguous states. Adding cropland used for pasture (66 million acres) and grazed forestland (145 million acres) brings the total to slightly over 800 million acres (see Table 11-1). These figures do not include Hawaii, with very little rangeland, and Alaska, with a great deal of rangeland; the rangeland of these states is not expected to change.

There is less rangeland today than there was in pre-settlement times, due to the extensive conversion of native rangeland into agricultural uses (especially in the grassland types of the Great Plains), as well as through urban and transportation uses. As noted by Mitchell (2000), the rangeland/pasture area of the United States will probably continue to decline slowly due to the social and economic forces currently in place, but will remain nearly one-third or more of the U.S. land base in the forseeable future. Following is a description of the range resources in key geographic regions of the United States.

TABLE 11-1

DISTRIBUTION OF RANGELAND IN THE 48 CONTIGUOUS STATES*

Assessment region	Total rangeland/ pasture	Cropland pasture	Grazed forestland
	(million acres)		
Northeast	3.0	2.0	1.4
Lake states	5.3	2.6	3.1
Corn Belt	12.3	10.1	6.6
Northern Plains	69.7	10.6	1.6
Appalachian	6.0	9.1	5.2
Southeast	9.8	4.2	7.3
Delta states	6.4	4.3	15.9
Southern Plains	118.7	15.5	11.6
Mountain states	303.5	5.7	66.7
Pacific Coast	64.5	2.6	25.6
Total, 48 states	589.0	66.8	145.0

*Distribution may not add to totals due to rounding.
Source: USDA, ERS, 1997.

Atlantic Southeast Rangelands of the Atlantic Southeast, aside from the tallgrass outliers and other natural grasslands, are better described as forest range because forests are the dominant vegetation type of the southeastern United States. Longleaf and slash pine forests on the lower coastal plain from eastern Texas to North Carolina occupy about 18 million acres (7 million ha). Farther north on the upper coastal plain, loblolly and shortleaf pines with hardwood species form a broad zone of transition to the eastern deciduous forest of the northeastern states. The zone of about 70 million acres (28 million ha) borders the pine forest from eastern Texas and Arkansas to Virginia and Maryland. Aside from several national forests, nearly all of the southeastern forest range is under private ownership for timber purposes.

The region has characteristically warm temperatures and high precipitation amounts. Rainfall averages 50 in (125 cm) spread evenly through the seasons except for autumn. Soils of the region tend to be highly leached and somewhat acidic, but due to generally warm and moist conditions, the potential productivity of southern forest range is the highest of all rangelands. The open pine forests produce abundant herbaceous growth, making them the most important of all range types for livestock production (Fig. 11-4). Important understory species include little bluestem and other bluestem grasses, pineland threeawn, and

FIGURE 11-4

A trial approach to grazing on the Atlantic Southeast Range is the silvopasture system. Here pine are planted in rows separated by a 40 foot grazing alley. Left: Livestock grazing in a loblolly pine stand in Mississippi. (*Photo by Steve Grado, College of Forest Resources, Mississippi State University*). Right: Sylvopastural system of slash pine and bahiagrass in Florida. (*Photo by Michael Bannister, Center for Subtropical Agroforestry, School of Forest Resources and Conservation, Institute of Food and Agricultural Sciences, University of Florida.*)

several species of oak. However, the tall grasses do not provide nutritious forage year-round, making use of improved pasture and supplemental feed necessary.

During the 1890 to 1930 period, the southern pine forests in the Atlantic Southeast were extensively harvested for timber; the resulting cut-over lands became prairie-like open ranges, producing abundant forage. After World War II, pine regeneration efforts through planting and direct seeding increased rapidly, resulting in declining forage production as the overstory of the new forests developed. Most of the pine forests are now under intense management for timber production, with forage production dependent on timber management. Grazing by domestic livestock is a secondary land use, although both uses are improved by periodic prescribed burning to control hardwood regeneration and reduce mulch accumulation.

The high productivity of Southern Forest range also supports large numbers of wild herbivores, especially whitetail deer together with upland birds and some feral hogs. This has induced opposition to livestock grazing from sportsmen and hunting groups, citing reputed livestock competition with deer for winter browse. Opposition to grazing also arises from hunting clubs, which lease large tracts of land from timber companies, and therefore from the companies themselves. Consequently, management of the range resource today centers on maintenance of productivity and nutrition of the forage, and on conjoined use of wildlife and livestock grazing. Due to favorable growing conditions, recovery from disturbance can be fairly rapid.

Great Plains Occupying roughly the central third of the continent from the boreal forest in Canada to the woodlands of Texas, the Great Plains stretch westward from the ragged, deciduous forest margin to the Rocky Mountain front. The Great Plains form the largest continuous grassland on the continent, about 530 million acres (215 million ha), and are divided into two major zones: tall grass or true prairie, and mixed grass or mixed prairie. These zones reflect a precipitation gradient of about 10 in (25 cm) at the

FIGURE 11-5
The true prairie of the Great Plains, grassy uplands dissected by wooded draws. Most of the type has now been converted to cultivation. The Flint Hills of eastern Kansas. *(Photo by Walter H. Fick)*

mountain front to about 35–40 in (89–102 cm) at the forest margin (increasing southward), occurring mostly within the growing season. The dark brown, fertile grassland soils tend to increase in depth along the same gradient. Temperature extremes, high winds, and periodic droughts occur commonly. Nearly all the land is under private ownership, mostly for agricultural purposes.

Adjacent to, and integrating with, the southern extensions of true and mixed prairie is the semidesert grassland, characterized by sparse summer and winter distribution of 8–20 in (20–50 cm) of rainfall, and shallow soils highly variable in texture. Mixed public and private ownership is common in the semidesert grasslands.

True Prairie   This part of the Great Plains lies west of the eastern forest margin to about the 100th meridian and extends from southern Manitoba to central Texas. The true prairie originally comprised about 250 million acres (101 million ha). Units of tall grass outside the central distribution are the Nebraska sandhills, the Texas coastal prairie, the black prairie of Mississippi and Alabama, and the tall-grass prairie

north of the Everglades in Florida. These are clearly tall-grass communities, but these differ somewhat in vegetal composition.

Most of the true prairie is now converted to cultivation, but in the true prairie of 150 years ago, tall grasses dominated other vegetation, creating a uniform appearance over thousands of square miles (Fig. 11-5). Many combinations of species occurred based on topography and drainage, but by far the most important dominant grasses of the prairie were big bluestem on lowlands and little bluestem on the uplands; together they constituted as much as 75 percent of the vegetation by weight. Other major grasses were switchgrass, Indiangrass, and, in the northern areas, porcupine grass. (See Appendix E for scientific names of range vegetation.) Although flowering herbs were common in the prairie, woody species such as eastern redbud and American elm were few and mostly confined to bottomlands.

The true prairie was maintained by fire and grazing as the most productive grassland in North America. The deep, fertile soil and abundant rainfall of the true prairie made cultivation inevitable; the region is now called the "Breadbasket of the Nation." Today,

FIGURE 11-6
The mixed prairie, as the name suggests, is composed of mid- and short grasses.
Here it is short grass, at the western edge of the largest remaining grassland on the
continent, stretching from the Rocky Mountain front to about the 100th meridian.
Albany County, southeastern Wyoming. *(Photo by Kendall L. Johnson)*

only about 18 million acres (7.3 million ha) of true prairie vegetation remain, mostly in the rocky Flint and Osage Hills of Kansas and Oklahoma, together with numerous small pastures within the cultivated areas elsewhere.

In the past, the true prairie supported a wide variety of native herbivores, including several large grazers, especially bison. Early stockmen grazed cattle from Texas on bluestem ranges in summer. Where extensive livestock grazing remains today, year-round utilization predominates with management centered on controlled grazing and burning.

Mixed Prairie   West of the true prairie, from about the 100th meridian to the Rocky Mountain front and stretching from the boreal forest in Alberta and Saskatchewan to the woodlands of Texas and New Mexico, the mixed prairie originally comprised about 280 million acres (113 million ha). Cultivation has now reduced it to about 200 million acres (81 million ha), but it is still the largest remaining grassland on the continent. The mixed prairie is composed primarily of short- and mid-stature grasses in varying combinations (Fig. 11-6). The shortgrass blue grama and buffalo grass occur everywhere, with other important species varying along a climatic and soil gra-

dient from north to south. As in the true prairie, flowering forbs are common; woody species, especially plains cottonwood, are generally confined to bottomlands.

In the mixed prairie, vegetative production is controlled by wide annual variation in rainfall. Other factors include severe stress from fire, insect outbreaks (especially grasshoppers), temperature extremes, and protracted drought. Yet, the mixed prairie was very well adapted to supporting herbivores, including periodic overgrazing by the huge bison population. The region was the scene of the great cattle drives from Texas in the late 1800s; today it constitutes the center of the modern range livestock industry.

Strong attempts were made to cultivate mixed prairie, but the undependable rainfall resulted in dryland farming during wet cycles, followed by abandonment in dry years. During the severe drought of the mid-1930s, millions of plowed acres were abandoned. Consequently, much of the mixed prairie is in various stages of recovery following those failed attempts at farming. Planting windbreaks and seeding crested wheatgrass have helped stabilize the soil. Cultivated areas are now farmed with improved methods to conserve soil. However, the great bulk of the mixed prairie remains a grassland used by graz-

FIGURE 11-7
The intermountain basins, dominated by shrubby vegetation, lie between the
Cascades and Sierra Nevada on the west and the Rocky Mountains on the east.
Sweetwater County, southwestern Wyoming. *(Photo by Kendall L. Johnson)*

ing animals, wild and domestic, with management centering on controlled grazing.

Semidesert Grassland   Extending as a discontinuous belt from southwestern Texas to southeastern Arizona and south into Mexico, the semidesert grassland type involves 60–90 million acres (25–36 million ha), depending on boundary criteria. Originally dominated by black grama and tobosa grass, with scattered shrubs typical of surrounding deserts, much of the area has been invaded by mesquite. In southwestern Texas, a definite mesquite savanna or woodland has formed, representing a major ecological change. The proper name of this rangeland type should now be semidesert grass-shrub land, because half of the original grassland is now shrub dominated.

Reasons suggested to explain this change include (1) a warmer and drier climate favoring shrubs, (2) reduction of herbaceous competition by livestock grazing, (3) increased mesquite seed dispersal by livestock, and (4) direct reduction of fire frequency by suppression and reduced fire hazard through removal of fuel by grazing. Probably some combination of these reasons initiated the change, and once mesquite trees are established a positive feedback loop—more trees, less grass—develops.

Perennial vegetative production in the semidesert is highly variable because of the amount and distribution of rainfall; production of annuals can vary from virtually nothing to a dominant amount. In the past, wildlife use of semidesert grasslands was significant and included both small and large herbivores. Because it was the foundation of the Spanish livestock industry, this area has been grazed by livestock longer than any other rangeland on the continent. The semidesert provides valuable forage in both summer and winter, so grazing occurs year round. Cattle are now the predominant livestock, but sheep and goats are abundant locally. Forage in this semidesert grass-shrub type is in reduced condition due to weed and brush invasion, so management centers on brush and weed control, reseeding, grazing control, and water management.

Intermountain Basins
The general mountain uplift in the western third of the continent created a landform extending from Canada into Mexico. It runs through all or parts of the 11 western states and is best described as a series of mountain ranges separated by valleys or basins (Fig. 11-7). Between the Cascade Mountains and Sierra Nevada on the west and the Rocky Mountains

on the east, four major regions of this "basin and range" topography were formed:

- Columbia Plateau—central Washington and Oregon, northern Idaho, and parts of western Montana;
- Great Basin—Nevada, western Utah, southern Idaho, southeastern Oregon;
- Wyoming Basin—western Wyoming and northwestern Colorado and parts of northwestern Wyoming and southeastern Montana;
- Colorado Plateau—the Four Corners area of Utah, New Mexico, Colorado, and Arizona.

Characteristically, the majority of precipitation in these basins falls on the surrounding or internal mountains. Annual precipitation in the basins commonly ranges from 5 to 12 in (12–30 cm), increasing with elevation and latitude. A higher proportion of snow occurs at higher elevations and northerly latitudes. There are cold winters in the north and hot summers in the south. Soils vary in depth and texture from shallow, coarse soils at basin edges through moderately deep, well-drained loams at midslopes, to clayey, salty silts in basin bottoms.

Controlled by these gradients, vegetation in the intermountain basins is arranged into four major types: (1) salt-desert shrub, (2) sagebrush–grass, (3) pinyon–juniper, and (4) Pacific bunchgrass.

Salt-Desert Shrub One of the smaller rangeland vegetation types (38–40 million acres, 15–16 million ha), salt-desert shrub areas are distributed irregularly throughout the intermountain basins in warmer, drier, and more saline areas, frequently in drainage bottoms. Northern distributions occur primarily in valleys or on adjacent low, gentle slopes; southern types occur more on steep foothill slopes of low mountains. Much of the salt desert was formerly covered by lakes of glacial melt water such as Lake Bonneville in the Great Basin. Ownership is primarily federal, including a few large reserves managed by the Departments of Defense and Energy.

The vegetation communities consist of a mosaic pattern dominated by one or two shrub species, primarily shadscale and winterfat, with few herbaceous plants and sparse ground cover. The mosaic frequently adjoins or intergrades with sagebrush–grass stands. Together, the low shrubs comprise 65 percent to 90 percent of the vegetation, with the remainder made up mainly of perennial grasses such as squirrel-

tail and inland salt grass. To an increasing extent, annual forbs and grasses, especially cheatgrass and Russian thistle, have invaded many salt-desert shrub stands, not only reducing forage conditions, but putting the shrubs at risk through creation of a fire type.

In the past, wildlife use of the salt desert has consisted mostly of numerous small mammals, especially black-tailed jackrabbits. Today, the dominant use is winter range for sheep and other livestock. Although salt-desert communities are among the best winter sheep ranges in the world (Fig. 11-8), they produce only half or less of their potential, due to past grazing abuse. Grazing management since the 1930s has improved ecological conditions substantially, but unfortunately, the harsh environment means recovery from disturbance comes very slowly.

Sagebrush–Grass Perhaps the classic western range type, sagebrush–grass occupies plains, plateaus, and valleys over huge areas and occurs in small stands on foothills and mountains of the intermountain basins (Fig. 11-9). The type usually lies just above the salt-desert shrub and just below the pinyon–juniper communities, intergrading with both. Depending on boundary criteria, a reasonable assessment of size may be 140 million acres (57 million ha), with 65 percent to 90 percent in public ownership.

Big sagebrush dominates here, from nearly pure stands of stunted shrubs in the south and at lower elevations to increasingly more savannalike stands in the north and at higher elevations, with up to 75 percent herbs. The many species and subspecies of sagebrush serve as useful indicators of site conditions and proper management along the gradient. Within the big sagebrush complex of subspecies, basin big sagebrush indicates deep, well-drained loams, a characteristic the early settlers used as a rule-of-thumb indicator of arable soils. Mountain big sagebrush, occurring more often at higher elevations, is adapted to deep, well-watered soils, ordinarily in association with abundant herbaceous growth. Wyoming big sagebrush is found primarily on shallow, rocky, and very dry soils, with a corresponding lack of herbaceous growth.

Substantial acreage has been reseeded to crested wheatgrass, intermediate wheatgrass, and Russian wildrye as a means of increasing herbaceous production and of expanding ground cover. For the most part, the use of introduced grass species has been positive; on the other hand, the accidental introduction of

FIGURE 11-8
The salt-desert shrub type, represented here by a shadscale community, provides important winter and early spring forage for both sheep and pronghorn. Desert Experimental Range, southwestern Utah. *(Courtesy of USDA Forest Service)*

FIGURE 11-9
Sagebrush-grass is the hallmark of the western range, distributed over plains, valleys, and foothills, providing important forage for both wildlife and livestock. Cassia County, southern Idaho. *(Photo by Lee A. Sharp)*

several grass and forb annuals, especially cheatgrass, now poses serious weed problems throughout the sagebrush region. These have tended to dramatically increase fire frequency in sagebrush stands leading to establishment of annual grasses—a wholly new plant community in the intermountain basins.

Lower elevations in the sagebrush–grass type serve as important winter range for deer, elk, and antelope, and also provide a spring/fall range for livestock. Fire and grazing are major historical factors, site by site, in this range type. Although many areas with deeper soils or access to irrigation water have been converted to cultivation, the greater part of the sagebrush–grass type remains as rangeland. Guided by the amount of herbaceous growth present, management centers on controlled grazing, reduction of overly dense sagebrush, and reseeding of damaged areas.

Pinyon–Juniper Best developed in the Great Basin and Colorado Plateau, pinyon–juniper is distributed in an irregular belt on both sides of the Rocky Mountains from southwestern Texas to southern Idaho and along the east slope of the Sierra Nevada–Cascade Mountains. Only scattered stands occur north of the 42nd parallel. Pinyon–juniper occurs as a "pygmy forest" on the rougher terrain of foothills and lower mountains, or on lower plateaus and mesas in the Southwest. It forms a transition zone between sagebrush–grass or desert types at lower elevations and the coniferous forests above (Fig. 11-10). The commonly accepted size of this type is 76 million acres (31 million ha); about two-thirds is publicly owned.

As the name suggests, the pinyon–juniper type is dominated throughout its range by several species of small-statured juniper trees, with pinyon (pine) less frequent than the juniper. The most important species, Utah juniper, is widespread through the Great Basin, while Rocky Mountain juniper occurs at higher elevations, alligator juniper in the southwest, and western juniper in the northwest. True and single-leaf pinyon are usually most abundant at higher elevations or on better-watered sites. There is general consensus that pinyon–juniper stands have become denser woodlands than they were in their original distribution, and have expanded into lower elevation communities, especially sagebrush–grass. Probable causes include the reduced competitive ability of the herbaceous understory due to exploitive early grazing, decrease of fire frequency resulting from direct

FIGURE 11-10
Pinyon–juniper forms a "pygmy" forest at elevations below Ponderosa pine, and the juniper component often expands into the lower sagebrush–grass zone over large areas of the intermountain region. *(Courtesy of USDA Forest Service)*

suppression and lack of herbaceous fuel, and the greater physiological efficiency of coniferous trees. These combine to set up a definite feedback loop: the more trees, the less competing vegetation, resulting in denser stands and continued expansion. This process is still underway.

Historically, the pinyon–juniper type provided important winter range for wildlife, especially deer, and was an important habitation site for early human populations. Today, pinyon–juniper stands continue to provide winter habitat for wildlife, yield wood products, and provide livestock forage. These uses require sensitive management such as new grazing systems, direct rehabilitation of degraded stands, and reseeding of adapted herbaceous species.

Pacific Bunchgrass Best developed on the Columbia Plateau, the Pacific bunchgrass type, locally known as Palouse grasslands, stretches from eastern Washington into northern Oregon. Smaller stands are scattered across southern Idaho and into southwestern Montana as inclusions within the larger sagebrush–grass type (Fig. 11-11). The total area of the Pacific bunchgrass type is difficult to determine; the original Palouse occupied about 36 million acres (14.6 million

FIGURE 11-11
Bluebunch wheatgrass and Idaho fescue dominate a high-condition stand of Pacific bunchgrass. Most of the type, commonly known as the Palouse, has now been converted to cultivation. Colville Indian Reservation, eastern Washington. *(Photo by Stephen C. Bunting)*

ha) but is now almost entirely under cultivation. Nearly all of the land is privately owned.

The Pacific bunchgrass region was characterized by bluebunch wheatgrass, with important associate species varying by site and elevation, including Idaho fescue on better-watered sites. True Palouse grassland formed on deep, wind-deposited soils (loess) under a Mediterranean-like climate of 12–25 in (30–65 cm) rainfall, with about two-thirds falling from October to March. The largely winter rainfall in the Palouse made possible wheat cultivation, but also supports many weedy species today. Noncultivated portions of bunchgrass ranges often bear major populations of cheatgrass, medusahead, knapweed, or starthistle, especially where previous land use has lowered the competitive ability of native plant communities.

In historic times, the Palouse did not support large populations of herbivores, possibly due to lack of water in the normally dry summer. The deep, fertile soils of the region were soon put under cultivation for dryland wheat, pea, and lentil production. Soil erosion from these cultivated lands remains severe and effective control measures are difficult to attain. Where extensive livestock grazing remains, management centers on controlled grazing and exotic weed control.

Southern Deserts   One of the world's major desert regions (including (1) the Mojave, (2) the Sonoran, and (3) the Chihuahuan deserts) is located in the southwestern United States and northern Mexico. Low annual precipitation is a characteristic feature of all deserts, but major differences in rainfall distribution result in three divisions of the southwestern desert. Each of these deserts has topographically defined plant community types, usually dominated by characteristic woody plants, with some species, such as creosote bush and ocotillo, common to all three. Herbaceous species are mainly winter and summer annuals. Boundaries between the deserts are not distinct and transition zones between them generally contain vegetation characteristics of both, as controlled by elevation, soils, or aspect.

The Mojave Desert   This desert in southern California is about 35 million acres (14 million ha), occupying an interior drainage basin. Its rainfall (2–8 in) (5–20 cm) occurs predominately in winter. About 70 percent of the Mojave Desert is dominated by creosote bush, which often occurs in nearly pure stands on well-drained soils of slopes and valleys. Joshua

FIGURE 11-12
One of the three southern desert types of rangelands is the Sonoran Desert. The saguaro cactus stands prominently among the other species of drought-resistant plants. Note the desert bighorn sheep in the foreground. Pusch Ridge Wilderness, Coronado National Forest, Arizona. *(Photo by Jim Gionfriddo)*

tree characterizes the well-drained mesas and slopes at higher elevations.

The Sonoran Desert   This desert covers about 65 million acres (26 million ha), and is composed of flat plains and basins straddling the border with Mexico at California and Arizona (Fig. 11-12). A combined summer and winter (bimodal) rainfall totals 3–11 in (7.6–28 cm). Vegetative communities are arranged along a soils and elevational gradient, with coarse, well-drained soils on upper slopes supporting the trademark of the desert—saguaro—plus palo verde, ocotillo, and other cacti.

The Chihuahuan Desert   This desert occupies 270 million acres (109 million ha), and has a predomi-

nantly summer rainfall pattern (7–16 in) (18–40 cm) on high tableland located mostly in Mexico but edging into southern New Mexico and Texas. Vegetation and zonation of the Chihuahuan Desert are similar to those of the Sonoran, except for a greater abundance of perennial grasses associated with the shrubs, reflecting higher summer rainfall. The northern portions of the desert intergrade with semidesert grassland at the southern reaches of the Great Plains (Fig. 11-13).

All three deserts have mild winters and very hot summers with high sunlight intensities. Soils tend to be shallow and vary in texture with topography, from coarse, gravelly soils on mountain slopes to clayey, salty soils in the basins. In the United States, the Mojave and Sonoran deserts are primarily publicly owned, while the Chihuahuan Desert is both public and private.

Vegetative production, especially for the annuals, depends entirely on timing and amount of rainfall, varying from virtually nothing to several hundred pounds net annual production per acre. Indeed, the availability of water and competition for water are the transcendent factors for all living things in these deserts. Relatively few large grazers inhabit the deserts, but they have historically been winter range for livestock. Recreation is probably now the largest resource use, especially in the Mojave Desert, including such pursuits as dirt-bike racing, 4-wheel driving, dispersed camping, and recreational prospecting. All resource uses in the desert require careful management because recovery from disturbance is very slow.

Pacific Southwest   Rangelands of the Pacific Southwest are located in California, primarily in the central valley, south central to southern coastal lands, and southern mountains. Originally comprising about 42 million acres (17 million ha), California rangelands divide into two major types: (1) Pacific bunchgrass, now known as California annual grassland and (2) California chaparral. Both types are a function of the prevailing Mediterranean climate, characterized by cool, rainy winters and hot, dry summers. Most of the area receives 10–40 in (25–100 cm) of precipitation annually, increasing south to north and with elevation, but variable by location and season. Soils of the central valley are typical deep, fertile, grassland soils; those of the chaparral tend to be shallow, rocky, and relatively infertile. Nearly all of the central valley type and about half of the chaparral are under private ownership.

FIGURE 11-13
A semidesert grassland, dominated by black grama and agave. Higher summer rainfall supports the grassland within a desert environment. Guadalupe Mountains, western Texas. *(Photo by Stephen C. Bunting)*

California Annual Grassland Perennial bunch-grasses originally composed about 25 million acres (10 million ha) of the central valley and formed the herbaceous cover under about 7 million acres (2.8 million ha) of adjacent oak woodland. The perennial bunchgrass type has been transformed into an annual grass type through a combination of overgrazing, drought, cultivation/abandonment, and ready availability of introduced annual species ideally adapted to a Mediterranean climate. Although over 500 species of annual plants are now found in California, the grassland is dominated by a handful of species, principally the grasses soft chess and slender oats, and the forbs filaree and bur clover. Only scattered perennial grasses remain in what is now generally regarded as a permanent annual grassland (Fig. 11-14). A similar change in herbaceous cover from perennials to annuals occurred in the adjacent oak woodland. Reversion to perennials seems unlikely. Today, much of the central valley is intensively farmed for a wide variety of crops; only about 10 million acres (4 million ha) of grassland remain around the fringes of the valley.

In the past, California grasslands were moderately used by numerous herbivores, including herds of elk, deer, and pronghorn. After settlement, the bulk of the central valley was converted to intensive cultivation, with livestock ranching in the foothills. Year-round grazing is possible under moderate stocking, although both quantity and quality of forage are often deficient, requiring supplemental feeds. Management centers on maintaining end-of-year mulch levels for optimum production, lengthening the green forage period, providing supplemental feed, and some reseeding.

California Chaparral Although stands of chaparral are distributed irregularly along the coast and the western slope of the Sierra Nevada, the type is best developed in the southern California mountain ranges, generally above oak woodland and below coniferous forest. The dense and often impenetrable stands of low to tall shrubs dominate extensive areas of rugged terrain in the foothills and lower mountains (Fig. 11-15). Chaparral occupies 8–11 million acres (3.2–4.4 million ha), with about 60 percent under federal management within national forests.

Chaparral vegetation is composed of both sprouting and nonsprouting species of shrubs with small, hard, evergreen leaves. Differences in moisture conditions, elevation, and exposure produce variation in vegetation composition. Three main types are evident, each dominated by different shrubs. At lowest elevations and rainfall, the most common and most

FIGURE 11-14
A characteristic representation of California annual grassland, once known as Pacific bunchgrass, in the foothills west of the central valley. Note the firebreak in the foreground. *(Photo by James Clawson)*

FIGURE 11-15
California chaparral type is extensive in southern California. The major use of these lands is not for grazing, but for watershed protection. The dense shrub forming the type is adapted to survive periodic fires. *(Photo by James Clawson)*

FIGURE 11-16
The Alaskan north slope tundra, an area of low rainfall; long, hard winters; and
frozen subsoil, supports reindeer, caribou, and musk ox. Northern Alaska. See also
Fig. 9-7. *(Photo by Ronald Robberecht)*

important type is chamise, typically forming over
80 percent of stand cover. Above this is the ceanothus type, composed of several species. At still
higher elevations and rainfall, the mountain chaparral
composed of species of manzanita, ceanothus, and
chinquapin occurs. All three types tend to consist of
a single dominant shrub layer, with very limited, if
any, subshrub and herbaceous layers. Stands tend to
be uniform, and can become very dense. This tendency increases with aridity.

Chaparral is a fire-dependent vegetation type requiring periodic burning to maintain healthy stands
that will control erosion. The absence of fire has a
greater impact on the plant community than fire itself, because shrub recovery following fire allows a
brief opening for understory species. For several
decades, large wildfires have raged through the chaparral type in areas of southern California occupied by
suburban homes, causing damage calculated in the
millions.

Before settlement, chaparral was used by a variety
of small herbivores; mule deer were the only large
grazers. Today, watershed values are the primary
consideration, and require stabilization of steep
slopes following fire. This involves reseeding and occasional construction of erosion barriers and diver-
sions. Management efforts include fire control for
watershed and home site protection, and maintenance
of firebreaks through goat browsing.

Alaska Rangelands A majority of the Alaskan
rangelands lie on the northern and western coastal
plains known as *tundra,* a Russian word describing
the often featureless terrain characteristic of the arctic (Fig. 11-16). Most of the remaining rangelands
are distributed in the valleys and lowland basins of
interior Alaska within and between rugged mountain
ranges. Both regions, especially the tundra, have severe growing conditions, marked by extremely long,
hard winters and surprisingly low annual precipitation, often less than 10 in (25 cm). The black organic
soils normally have a permanently frozen subsoil
known as the *permafrost layer.*

Essentially a cold desert, Alaskan rangelands support a dense growth of mosses and lichens, known as
reindeer moss, with dwarf herbs and shrubs in boggy
areas known locally as *muskeg.* Over extensive areas
of the tundra, cottongrass and other sedges form a

continuous cover. Dense mats form on wet sites and well-developed tussocks on moist sites. On the Aleutian slope, bluejoint reedgrass and heath-like shrubs dominate the vegetation. On drier sites, low shrubs such as cranberry and dwarf willow predominate, also forming an extensive heath-like cover. Taller shrubs like alder, willow, and salmonberry often form thickets of varying density.

Alaskan rangelands support wide-ranging herds of reindeer, caribou, and the more local musk ox. Due to harsh conditions and lack of readily available markets, livestock husbandry is not extensively pursued in Alaska. Management centers on maintenance of vegetative cover because recovery from disturbance is very slow.

RANGELAND MANAGEMENT

Bases of Management

Rangeland management has developed over the past 50 years or so into a new profession and body of knowledge. Based on the concept of land stewardship— preventing or avoiding damage and restoring previously damaged lands—it can be defined as the management of rangeland for the optimum production of goods and services while maintaining the productive potential of the basic soil resource (Fig. 11-17).

Guided by ecological principles, rangeland management reflects the concept that the status of a plant community at any given time is a function of its past development and its response to recent stresses or disturbances. Rangeland ecosystems evolved under different sets of soil, topographic, and climatic growing conditions. They also responded to different sets of recurrent disturbances, such as fire, grazing, drought, wind, flood, disease, and insects. In some situations, lack of disturbances, through strict fire suppression or prohibition of grazing, created stresses.

When an ecosystem such as a rangeland plant community is disturbed, it normally responds by following paths of recovery in soil, vegetation, and animals. If growing conditions are similar to those preceding the disturbance, and if basic environmental factors such as soil have not been altered, the community may recover along the same path it followed in original development. On the other hand, if growing conditions or environmental factors have changed, the community may progress along another path of recovery, perhaps to a different end. Both processes are known as *succession,* and the different

FIGURE 11-17
Maintaining the productivity of the soil resource is a basic objective of rangeland management. Here a range manager carries out a prescribed burn in a sagebrush stand. Lincoln County, southwestern Wyoming. *(Photo by Kendall L. Johnson)*

plant and animal communities along any path of recovery represent various successional stages.

Recently, rangeland studies have indicated the presence of multiple steady states along successional pathways, especially in drier rangeland communities, produced by different combinations of growing conditions. These can produce ecological states wherein no change is evident within a state for extended periods of time. Normally, a steady state would continue in place until a sufficiently strong change in growing conditions—climate, fire, disease, animal action, or human activities—starts movement along a successional pathway again, perhaps to a new steady state. Thus, an ecological approach to range management involves an understanding of how individual ecosystems respond to stress.

Rangeland ecosystems before European settlement most likely formed a mosaic of communities in different successional stages or steady states. For example, sagebrush–grass communities probably passed through a cycle of stages, from mostly herbs to mostly shrubs, depending on the length of time since the last fire.

An understanding of these ecological principles becomes necessary to managing rangeland ecosystems for multiple uses, because specific outputs may be best provided by different successional stages of the same community. For instance, wildlife habitat is often best served by early successional stages following fire.

Thus, the first objective of rangeland management is to maintain the growing conditions and environmental factors that facilitate successional processes. Where necessary, a number of interventions designed to initiate, advance, or restrain succession can be employed. These include prescribed fire, mechanical or chemical brush control, seeding, weed control, land treatments for water conservation and development, and grazing management. Joint treatments are sometimes necessary, as in using prescribed fire to remove an overabundance of brush, followed by reseeding of adapted herbs in order to restart a stalled successional process.

In summary, the many objectives and functions of range management include the following (Joyce, 1989):

- determining suitability of vegetation for multiple resource uses;
- designing and implementing range vegetation improvement practices;
- understanding social and economic effects of management alternatives;
- controlling range insects;
- determining the combined wildlife, recreational, wild horse and burro, and livestock carrying capacities;
- protecting soil stability;
- reclaiming disturbed areas (Fig. 11-18);
- designing and controlling livestock management systems;
- managing and controlling undesirable range vegetation;
- coordinating management activities with other land and resource managers;
- maintaining the environmental quality proper to the soil, water, and air.

FIGURE 11-18
A helicopter lifts seed to sites burned by a range wildfire in Montana. Rehabilitation will help prevent erosion and benefit both domestic livestock and native wildlife. *(USDA Forest Service photo by Karen Westly)*

Application of range management to achieve these goals and functions has brought about a significant improvement in condition of both private and public rangelands. Because of wider knowledge of the characteristics of the range resource, combined with improved management practices, livestock use today generally takes place under a grazing system suited to the site, with shorter grazing seasons and periodic rest. These changes have often brought about startling improvement across western rangelands (Dagget, 1995). Although signs of past damage still persist in many areas, especially in the driest portions, and there are still too many instances of improper grazing, there is strong evidence of generally improved rangeland conditions. Indeed, most range managers agree that rangelands generally are in their best overall condition of the past century (Box, 1979). They are working to improve them still further.

Improved management results in more productive rangeland, not only for livestock use, but for wildlife habitat, water quality, recreational opportunities, and open space, as well as maintainance of soil stability. The huge increases in many native wildlife populations that have occurred all across the South and West within the last few decades are, in part, a reflection of substantial improvement in rangeland condition and productivity. Several examples are presented by Jensen (2001).

Given the necessity that rangeland, public and private, must provide multiple outputs desired by society, range management of the future must become more and more a cooperative and coordinated effort among the several interests involved. Coordinated resource management (CRM) programs are now a significant component of cooperative efforts to improve resource management. Under CRM, all federal, state, county, and local entities plus local interest groups are involved in long-term efforts toward environmental improvement, integrated and improved management of all ownerships, and long-term stability of the economy through improved management. Only a joint effort involving all parties interested in, or affected by, the results seems likely to be successful over the long term (Cleary, 1988).

Concerns and Trends

The management concerns associated with rangeland resources have traditionally focused on livestock grazing, but are now much broader. Providing the type of vegetation necessary for multiple use outputs on public and private lands is a particular challenge now and in the future. Several concerns and trends should be noted.

RIPARIAN VEGETATION

One of the most explosive issues for range managers today concerns the management of riparian or streamside zones—lands with vegetation directly influenced by surface water fed by runoff or groundwater (Fig. 11-19). Although such lands occupy only a small portion of total rangeland area, water makes them exceptionally productive and important in the overall ecological relationships of the landscape (see Chapter 10, "Watersheds and Streams").

There is no doubt that ill-managed livestock grazing has negatively affected many rangeland riparian areas, both historically and presently. Riparian areas have frequently failed to match the general improvement in upland vegetation achieved through management in recent decades. This is because livestock, especially cattle, tend to congregate in riparian areas for the same reasons other animals do: presence of water, shade and protective cover, and more diverse, higher-quality forage. In addition, riparian areas attract a wide variety of human activities.

Livestock grazing negatively affects riparian zones in four main ways (Platts, 1989): (1) changing streamside vegetation and soils through grazing and trampling; (2) degrading streambanks through caving and sloughing; (3) altering channel cross sections to either wider, more shallow streams or deeply entrenched streams; and (4) altering the characteristics of the water column itself.

Solutions to these problems through improved livestock management keyed to the characteristics of individual stream reaches are now part of the science of range management (Elmore, 1992; Winward, 2000). Techniques are generally available to reduce or prevent riparian damage; in some instances, livestock grazing can be used as a tool to improve riparian functions. But the larger consideration is the social, economic, and cultural restraints on range management created by public concern.

Undesirable Plants The spread of undesirable plants is a major concern. Undesirable plants or weeds are plants "out of place," that is, unacceptable

FIGURE 11-19
Proper management of the riparian zone's soil and vegetation is a major function of
the rangeland manager. Because of the presence of water, riparian areas tend to
attract all forms of animal life in an otherwise dry region. Medicine Bow Valley,
southern Wyoming. *(Photo by Kendall L. Johnson)*

for planned land uses or unwholesome for rangeland ecology. Many undesirable plants are species exotic to native communities, such as cheatgrass or leafy spurge. Many such species have been introduced accidentally or perhaps deliberately for forage, fiber, erosion control, or ornamental purposes. Others are native species, such as tall larkspur or broom snake-weed, whose dominance in the community becomes undesirable through natural succession, climatic fluctuations, or local denudations of land. Disturbances caused by human activities, such as agricultural cropping or transportation rights of way, virtually ensure plant invasions by both exotic and native weeds.

Because the spread of undesirable plants crosses all political and ownership boundaries, the problem becomes very complex. In the past, management programs have relied almost exclusively on chemical control through application of herbicides. But rising public awareness and environmental concerns have resulted in the withdrawal or restriction in use of many herbicides, creating a need for alternative control methods such as prescribed fire, biological control, or grazing management. In addition, there is general realization that simply controlling weeds by any effective method is insufficient because the

undesirable plants will return in the absence of further management. Effective weed management requires establishment of a replacement plant community of sufficient stability to inhibit post-control invasion by either the treated species or by other undesirable plants.

Efforts to develop effective and environmentally sound control methods, biologically or chemically, are hampered by a lack of knowledge of the physiology and ecology of undesirable plants. Most likely, effective range weed control programs of the future will be integrated pest management efforts employing fire, highly effective but short-lived herbicides that break down rapidly, host-specific biological agents, short-term intensive grazing, and reseeding.

Wilderness and Rangeland Use   Range landscapes, often rich in historical, archeological, scenic, and aesthetic values, can satisfy the human need for open space and natural beauty. Such areas may be legislatively set aside as parks, wilderness, or other special areas with associated restrictions on commodity-producing uses. Because most designated and proposed wilderness areas have been grazed by livestock for many years, continuation of grazing was provided for in the Wilderness Act of 1964. But for many reasons, ranging from general political opposition to livestock economics and difficulty in maintaining grazing facilities such as water points and fences, livestock grazing presents problems in many wilderness and other specially designated areas. This in turn tends to render grazing permits less valuable or untenable. Even when public policy is clear in its intent to allow grazing in such areas, the difficulties tend to discourage it.

Grazing on Public Lands   Much controversy has arisen within the last three decades over livestock grazing on public lands.

Federal lands provide about 10 percent of the feed requirements of livestock in the United States. Although this amount seems small, it is important because federal lands form a critical link in the annual forage cycle of western rangelands. Public lands, often located in high country and inaccessible in winter, are usually grazed in summer when forage on private lands has matured and lost quality or been harvested for winter feed. Thus, the 10 percent of the national feed requirements from public lands is essential to the western livestock industry.

An assertion frequently made is that livestock raised in part on public ranges contributes only about 2 percent of the nation's red meat production, and can therefore be easily discontinued. While this may be true from a national perspective, from a regional, state, or local perspective livestock may be the largest part of the agricultural economy, and its loss would be hugely disruptive.

The question of the desirability of livestock grazing on public lands intertwines with economic considerations, especially the question of public benefit vs. public cost. Although the economic valuation of rangelands for alternative uses such as wildlife or recreation is very complex and poorly understood (Joyce, 1989), public attention now centers on the question of grazing fees paid for livestock grazing on public lands. The popular view holds that public land grazing fees are too low. Ranchers who graze livestock on public lands are perceived as receiving a subsidy. The affected ranchers point to the public benefits they provide such as reduction of fire hazards, development of watering opportunities also used by wildlife, other rangeland improvements, and

the expense they incur through grazing in remote, inaccessible locations.

Much attention centers on the apparent disparity between public and private land grazing rates, both locally and regionally. Several studies have shown that the differences are not so large as popularly supposed (Torell & Fowler, 1992), due to the differences in services and facilities provided by public grazing permits compared to private leases. Typically, few services or facilities are provided with public land permits; the permittee must provide all inputs necessary to proper management, whereas private leases, while more costly, often include these. Nonetheless, the economic implications, whether greater or less than perceived, will continue to be a highly contentious issue, fueling intense debate over national forest and public land policies for rangeland practices and grazing fees. As with forest and wildland managers, rangeland managers have not done a good job of communicating their goals and methods to the public.

Livestock in Vegetation Management   The use of livestock grazing to meet land management objectives is feasible, but not widely applied for several reasons. First, there is a general mindset on the part of many that livestock grazing—to any degree—will damage natural resources. Second, many livestock managers believe that while the objective may be acceptable, expensive operational needs must be met, such as the right kind of animal to consume the target vegetation, adequate amounts of nutritional forage, necessary watering facilities, and adequate fencing or herding to control the animals on site.

Still, these reservations have been overcome in many instances and animal grazing has helped attain a specific management goal. Livestock have been used in timber plantations to reduce herbaceous and woody competition with the planted trees, often a satisfactory alternative to the use of fire. Spring cattle grazing can be used to reduce herbaceous competition with shrubs to improve deer winter range. Prescribed livestock grazing can also be a biological means of weed control or a method to reduce fine fuels in a fire management program. Periodic grazing by sheep and goats can help keep firebreaks and utility rights-of-way clear of unwanted vegetation (Fig. 11-20). Geese have been used to control weeds on small plots or even vacant lots in urban areas.

FIGURE 11-20
Short-term intensive grazing is one means of effectively controlling undesirable weeds and shrubs. Here, sheep are grazing in a stand of quaking aspen in the Sawtooth National Forest, Idaho.*(Courtesy of USDA Forest Service)*

Signs of acceptance and future application in the use of livestock grazing for vegetation management are found in the increasing numbers of grazing contracts set up to achieve specific objectives, for which the livestock owners receive compensation. Undoubtedly, such uses will expand in the future as ecological and economic restrictions increase for other management tools, such as mechanical and herbicidal control and even on the use of prescribed burning.

SUMMARY

Rangelands have provided vegetation for herbivores for thousands of years, but over the last century, they have been associated primarily with livestock grazing. The rise of the livestock industry created a colorful phase in American history that remains firmly fixed in American culture through literature, movies, and music. These rangelands are located primarily in the South and West in ecoregions of the Atlantic southeast, Great Plains, intermountain basins, southwestern deserts, Pacific southwest, and the Alaskan tundra. Since the 1930s their productivity has been increased and former degraded conditions improved through application of the art and science of range management. Range management itself has evolved toward an integration of disciplines and techniques. Several policy issues affect rangelands, such as undesirable plants, wilderness designation, grazing of public lands, riparian conditions, grazing fees, and the use of livestock in vegetation management.

LITERATURE CITED

Box, T. W. 1979. The American Rangelands: Their condition and policy implications for management. In: *Rangeland Policies for the Future: Proceedings of a Symposium,* January 18–31, 1979, Tuscon, AZ. USDA Forest Service General Technical Report WO-17. Washington, D.C.

Box, T. W. 1993. "On Rewarding Good Stewards: A Viewpoint." *Rangelands* 15:181–183.

Cleary, C. R. 1988. "Coordinated Resource Management: A Planning Process that Works." *J. Soil Water Conservation* 43:138–139.

Dagget, D. 1995. *Beyond the Rangeland Conflict: Toward a West that Works*. The Grand Canyon Trust/Gibbs Smith, Layton, Ut.

Donahue, D. L. 1999. *The Western Range Revisited: Removing Livestock from Public Lands to Conserve Native Biodiversity*. University of Oklahoma Press, Norman.

Elmore, W. 1992. "Riparian Responses to Grazing Practices." In: R. J. Naiman (ed.). *Watershed Management: Balancing Sustainability and Environmental Change*. Springer-Verlag, New York.

Jensen, M. N. 2001. Can Cows and Conservation Mix? *BioScience* 51(2):85–90.

Joyce, L. A. 1989. An analysis of the range forage situation in the United States: 1989–2040. USDA Forest Service General Technical Report RM-180. Rocky Mountain Forest and Range Experimental Station. Fort Collins, Colo.

Mitchell, J. E. 2000. Rangeland resource trends in the United States: A technical document supporting the 2000 USDA Forest Service RPA Assessment. USDA Forest Service Gen. Tech. RMRS-GTR-68. Fort Collins, Colo.

Platts, W. S. 1989. Compatibility of livestock grazing strategies with fisheries. In: R. E. Gresswell, B. A. Barton, and J. L. Kerschner (eds.). *Practical Approaches to Riparian Resource Management: An Educational Workshop.* USDI Bureau of Land Management. Billings, Mont.

Torell, L. A., and J. M. Fowler. 1992. Grazing fees: How much is fair? New Mexico Agricultural Experiment Station Research Report 666.

USDA Economic Research Service. 1997. Agricultural resources and environmental indicators, 1996–97. USDA Agricultural Handbook 712. Washington, D.C.

USDA Forest Service. 2001. Grazing statistical summary: Fiscal year 2000. Washington, D.C.

USDI Bureau of Land Management. 1993. Public land statistics 1992. Washington, D.C.

U.S. Senate. 1936. The Western Range. Senate Doc. 199, 74th Congress, 2nd Session. U.S. Government Print Office. Washington, D.C.

Van Tassell, L. W., E. T. Bartlett, and J. E. Mitchell. 2001. Projected use of grazed forages in the United States: 2000 to 2050: A technical document supporting the 2000 USDA Forest Service RPA Assessment. USDA Forest Service Gen. Tech. Rep. RMRS-GTR-82. Fort Collins, Colo.

Winward, A. H. 2000. Monitoring the vegetation resources in riparian areas. USDA Forest Service Gen. Tech. Rep. RMRS-GTR-47. Ogden, Ut.

Heady, H. F., and D. R. Child. 1994. *Rangeland Ecology and Management.* Westview Press, Inc. Boulder, Colo.

Heitschmidt, R. K., and J. W. Stuth, eds. 1991. *Grazing Management: An Ecological Perspective.* Timber Press. Portland, Oreg.

Holechek, J. L., R. D. Pieper, and C. H. Herbel. 1995. *Range Management Principles and Practices.* (2nd ed.). Prentice-Hall, Inc. Englewood Cliffs, N.J.

Johnson, K. L. 1987. Rangeland through time: A photographic study of vegetation change in Wyoming, 1870–1986. Wyoming Agricultural Experiment Station. Misc. Pub. 50.

McClaren, M. P. 1990. Livestock in Wilderness: A Review and Forecast. *Environmental Law* 20 (4):857–889.

Reid, E. H., C. G. Johnson, Jr., and W. B. Hall. (undated). Green fescue grassland: 50 years of secondary succession under sheep grazing. USDA Forest Service Reg. Pub. R6-F16-SO-0591. Wallowa-Whitman National Forest, Baker City, Ore.

Rogers, G. F. 1982. *Then and Now: A Photographic History of Vegetation Change in the Central Great Basin Desert.* University Utah Press. Salt Lake City, Ut.

Wilkinson, C. F. 1992. *Crossing the Next Meridian: Land, Water, and the Future of the West.* Island Press. Washington, D.C.

Young, J. A., and B. A. Sparks. 1985. *Cattle in the Cold Desert.* Utah State University Press. Logan, Ut.

ADDITIONAL READINGS

See the Society for Range Management Website http://srm.org for a list of reference books and practical handbooks, abstracts of conference proceedings, journals, and brochures.

Burkhardt, J. W. 1996. Herbivory in the Intermountain West: An overview of evolutionary history, historic cultural impacts and lessons from the past. Idaho Forest, Wildlife and Range Exper. Sta. Bull. 58, Moscow.

Chaney, E., W. Elmore, and W. S. Platts. 1993. Managing change: Livestock grazing on western riparian areas. U.S. Environmental Protection Agency. Washington, D.C.

Dagget, D. 1995. *Beyond the Rangeland Conflict: Toward a West that Works.* The Grand Canyon Trust/Gibbs Smith, Publisher. Layton, Ut.

Donahue, D. L. 1999. *The Western Range Revisited: Removing Livestock from Public Lands to Conserve Native Biodiversity.* Univ. Oklahoma Press, Norman.

Ewing, S. 1990. *The Range.* Mountain Press Publishing Co. Missoula, Mont.

WEBSITES

U.S. Department of the Interior, Bureau of Land Management. 2002. *http://www.blm.gov/nhp/index.htm* Washington, D.C. Accessed March 2002.

U.S. Department of the Interior, Bureau of Land Management. 2001. Public land statistics, Vol. 185. *http://www.blm.gov/natacq/pla00* Washington, D.C. Accessed March 2002.

U.S. Department of Agriculture, Forest Service. 2002. *http://www.fs.fed.us* Washington, D.C. Accessed March 2002.

U.S. Department of Agriculture, Forest Service. 2001. 2000 RPA assessment of forest and rangelands. *http://www.fs.fed.us/pl/rpa/rpaasses.pdf* Washington, D.C. Accessed March 2002.

Society for Range Management. 2002. *http://srm.org* Lakewood, Colo. Accessed March 2002.

U.S. Department of Agriculture, NRCS. 2001. The PLANTS database, Version 3.1. *http://plants.usda.gov.* National Plant Data Center, Baton Rouge, La. Accessed March 2002.

STUDY QUESTIONS

1. Give six characteristics of western rangelands.
2. Beyond livestock grazing, what other goods and services do rangelands produce?
3. What factors brought about the end of the "open range"?
4. What are the definitions for an *AUM?*
5. Why did the range associated with the Rocky Mountains, the Great Basin, and the southwestern states not move into private ownership?
6. Name four large herbivores that originally grazed the western ranges.
7. Discuss the riparian vegetation problems of rangelands.
8. What tools do range managers possess to control weeds and undesirable shrubs?
9. Why are some citizens opposed to livestock grazing while others are not?
10. Define *rangeland management* and state its central objective.

Outdoor Recreation and Wilderness Management

The outdoor recreation material was co-authored for the 6th edition by Dr. William Hammitt, Professor of Outdoor Recreation at Clemson University (now ret.) and updated for the 7th edition by Dr. Steve Hollenhorst, Professor and Head, Department of Resource Recreation and Tourism, University of Idaho.

OUTDOOR RECREATION MANAGEMENT

INTRODUCTION

Americans have long enjoyed outdoor activities. Although the original settlers of the colonies may have feared the deep forests, later pioneers recognized the sporting aspects of hunting and fishing, hiked for pleasure to scenic vistas, celebrated with picnics, and found pleasure and relaxation in the out-of-doors. As human habitation has become urbanized for the vast majority of people, outdoor recreation activities have taken on new importance as a release from the intense pressures of city or suburban life.

Today the availability of outdoor recreation opportunities is a feature advertised by states, communities, and individual companies seeking to attract people to their locale. The opportunity to renew one's spirit and productivity through a diversity of outdoor recreation activities and contact with pleasant, natural surroundings is regarded as essential to well-being. Thus, the management of forests and renewable resources to provide such opportunities becomes important. In many areas, recreation opportunities are regarded as having more significance than commodity production.

Chapter 12 concerns itself with natural resource-based outdoor recreation and with wilderness—the values, providers, trends in use, and basic principles and strategies of management of these areas. A spectrum of opportunities is involved, ranging from developed recreation areas in urban parks, greenbelts, and developed campgrounds to dispersed wildland areas and wilderness. The chapter is divided between outdoor recreation and wilderness, since these two important entities have major differences in degree of development, the presence of roads, and requirements for management.

OUTDOOR RECREATION

People recreate in many ways and in many environments, with diverse benefits. For our purposes, *recreation* is defined as outdoor activities that offer a contrast to work and that offer the opportunity for constructive, restorative, and pleasurable experiences.

Recreation takes place in a variety of settings. Urban or suburban parks offer *activity-oriented* recreation, which depends on constructed facilities such as playfields, courts, and swimming pools; serves millions of people; and plays a vital role in our social structure. However, our concern here is with conservation and management of renewable natural resources, so we will consider only *resource-oriented* recreation, defined here as recreation that occurs in and is dependent upon the natural environment. Here management strives to maintain the natural appearance and quality of these environments. Facility developments are designed and located so as to protect the resource conditions that support and enhance recreational opportunities, as well as to provide for visitor comfort and convenience (Fig. 12-1). Although some resource-oriented recreation activities are concentrated in special use areas such as campgrounds, the majority of activities are dispersed over large areas and necessitate fewer restrictions.

Outdoor recreation activities involve a broad spectrum of opportunities ranging from those possible in developed portions of urban parks to those dependent on remote wilderness. Most outdoor recreation takes place on public lands such as national parks, national forests, water-related areas managed by the Army Corps of Engineers or the Bureau of Reclamation, and other federal, state, or county areas. These areas may or may not be designated solely for recreational use. For example, national forests and the national resource lands managed by the Bureau of Land Management are dedicated to multiple use, and recreation is integrated with timber harvest, grazing, production of water, mineral extraction, wildlife, and other uses. National wildlife refuges provide opportunities for wildlife-related recreational activities, but their main function is wildlife protection. Outdoor recreation such as hunting or hiking also occurs on private lands managed primarily for timber or grazing. However, since most resource-based outdoor recreation occurs on public lands, most of the management responsibility falls on those public agencies that deal with natural resources, although some of the essential work or provision of services may be provided by contractors, concessionaire companies, or volunteers.

Outdoor Recreation Values and Benefits

Outdoor recreation forms a part of our history, our culture, and our contemporary lives. People recreate outdoors for diverse reasons—for exercise, challenge, competition, relaxation, a change of pace,

FIGURE 12-1
A resource-oriented campground, designed to concentrate visitors to protect the forest resource yet provide for visitor comfort and convenience. *(Courtesy of Tennessee Valley Authority)*

sightseeing, appreciation of scenery, to be with family and friends, to experience solitude, contemplation, or to affiliate with and study nature (Cordell, 1999). They often engage in a "package" of activities and thus receive multiple benefits. For example, hiking, picnicking, socializing with family or friends, and observing nature commonly occur during a single trip.

Direct participation (including travel to and from the site) is just one phase of the outdoor recreation experience. Outdoor recreation values are often experienced well before people arrive at a recreation site (anticipation), and are enjoyed long after they return home (reflection). The planning and anticipation of a hunting or camping trip, and the recollection of the experience, may be every bit as enjoyable and valuable as the event itself. Some people vicariously enjoy natural resources, such as wilderness, unique parklands, or wildlife, without ever traveling to them. Reading about such areas, reviewing photos or videos taken on the trip, watching professional videos, or

television programs, even seeing a friend's slides photos, or videos, or hearing accounts of his or her trip, can provide *vicarious* but very real enjoyment.

The President's Commission on Americans Outdoors (1987) identified a number of social, economic, and environmental benefits resulting from outdoor recreation including better personal health, social adjustment, heightened environmental awareness for participants, and economic returns to the providers.

Outdoor Recreation Lands

In America, a complex of public (federal, state, local) and commercial and nonprofit private agencies and organizations provides a rich diversity of outdoor recreation opportunities.

Public Lands Nearly one-third of the United States, about 747 million acres (303 million ha), is public land available for outdoor recreation (Zinser, 1995). This includes about 691 million acres (280 million ha) of federal land and 53 million acres (21 million ha) of

TABLE 12-1

THOUSANDS OF VISITS TO FEDERAL RECREATION SITES BY AGENCY AND YEAR, 1988–1996.

Agency	Fiscal Year				
	1988	1990	1992	1994	1996
Tennessee Valley Authority	411.5	447.1	568.4	580.6	603.8
	(100)	(116)	(138)	(141)	(147)
Fish and Wildlife Service	25,307.5	27,878.9	29,964.6	27,091.7	29,468.1
	(100)	(110)	(118)	(107)	(116)
Bureau of Reclamation	41,833.4	40,736.4	38,583.7	38,242.0[b]	38,280.0
	(100)	(97)	(92)	(91)	(92)
Bureau of Land Management	57,841.5	71,821.0	69,418.0	50,743.0	58,922.9
	(100)	(124)	(120)	(88)	(102)
National Park Service	286,160.7	259,767.9	273,298.1	268,636.2	265,796.2
	(100)	(91)	(96)	(94)	(93)
Corps of Engineers	N/A	396,157.9	413,456.2	380,357.7	375,722.3
		(100)	(104)	(96)	(95)
Forest Service	N/A	597,609.9[a]	691,180.5	829,839.8	859,210.0
		(100)	(116)	(139)	(144)
Totals	411,554.6	1,394,419.1	1,516,469.5	1,595,491.0	1,628,003.3
		(100)	(109)	(114)	(117)

[a]FY91 figures.
[b]FY93 figures.

Source: From Cordell (1999). Federal Recreation Fee Reports and the Respective Agencies. BLM changes from FY94 to 96 may not reflect differences in visitation, but new counting methods; BOR statistics for FY94 and 96 are unofficial estimates and exclude BOR units managed by other federal agencies; TVA figures are for fee-charging areas only and are calendar years.

state lands. Local government properties total 2.3 million acres (930,000 ha). Recreation resources are not equally distributed among levels of government, or among regions of the country. Excluding Alaska and Hawaii, the western states contain over 90 percent of the federal recreation lands. The majority of lands in the eastern states are in private ownership.

Federal Lands   Federal agencies manage over 691 million acres (200 million ha) of land, over one-quarter of the U.S. land base. The vast majority of this acreage is located in 11 western states and Alaska. Four federal agencies manage 94 percent of this acreage: the Bureau of Land Management, the Forest Service, the Fish and Wildlife Service, and the National Park Service. In addition, several other federal agencies offer outdoor recreation opportunities, including the Army Corps of Engineers, the Tennessee Valley Authority, the Bureau of Reclamation, and the Bureau of Indian Affairs.

The number of recreation visits and visitation trends from 1988 to 1996 for these jurisdictions are shown in Table 12-1.

State Lands   States manage about 55 million acres (21 million ha) of outdoor recreation lands (National Association of State Park Directors, 2000). State forests tend to be less developed, while state parks typically offer more facilities and easier access. In contrast with the federal recreation lands, more than 40 percent of the 10 million acres (4 million ha) of state parks and reserves, and over 60 percent of the state forest lands, are located in the eastern half of the United States (Cordell, 1999). State lands are often located more conveniently for outdoor recreation uses than are federal lands, because they usually lie closer to populated areas and major travel routes.

All 50 states have a state park system, and some have a state forest system. Some of these state parks and forests are as large, or larger, than some federal recreation areas. They range in size from small areas adjacent to highways, to Alaska's natural areas that average more than 21,000 acres (8,500 ha) (Knudson, 1984). State parks range from the 2.4-million-acre (970,000-ha) wild public lands of Adirondack State Park in New York, to high-upkeep Kentucky resort parks with modern lodges, restaurants, golf courses, and marinas. California, New York, Michigan, Florida, and Pennsylvania have large, long-established, and nationally recognized state park systems.

In addition to state parks and forests, many states manage fish and wildlife areas as well as natural areas and preserves set aside for nature appreciation and recreational use. The number and acreage of these areas vary greatly. Alaska and California have the most acreage in state parks, Washington and Minnesota in state forests, and Mississippi and Pennsylvania in state fish and wildlife areas.

Local Public Lands In acreage, county parks and forests account for less than 5 percent of all public recreation lands. However, another component of local lands, municipal parks, accounts for more than 60 percent of the number of recreation areas nationwide, illustrating their role in providing some types of both activity-oriented and resource-based outdoor recreation opportunities (President's Commission on Americans Outdoors, 1987).

Local outdoor recreation providers include municipal park and recreation agencies, county recreation and park agencies, and special park districts. Types of areas provided include parks, forest preserves, greenways, automobile parkways, beaches, arboreta, zoos, aquariums, and many other types of faculties. Most are day-use oriented. Specific examples of such local outdoor recreation areas are the 640-acre (260-ha) Central Park in New York City; the 64,000-acre (26,000-ha) Cook County Forest Preserve surrounding Chicago; the five-county, 18,000-acre (7,290-ha) Huron-Clinton Metropolitan Park System outside Detroit (Fig. 12-2); and the many outdoor education/recreation areas managed by local school districts. Demonstrating widespread local support, U.S. voters in 2001 approved 137 of the 196 local and state ballot initiatives for open space and outdoor recreation, committing $117 billion in funding. From 1998 to 2001, voters supported 529 referenda totaling $19 billion (Trust for Public Land, 2002).

Private Lands The private sector provides a full spectrum of land resources suitable for recreation. These resources range from the extensive forest holdings of timber companies to highly developed, commercial recreation areas, including about 60 percent of all campgrounds in the United States, to the growing land base owned or controlled by conservation organizations like The Nature Conservancy and local

FIGURE 12-2
An example of a highly developed, intensively used piece of land set aside specifically for recreation, at the local level. Stoney Creek Metropark, near Detroit, Michigan. *(Courtesy of Huron Clinton Metropolitan Authority)*

land trusts. Rural private lands provide an important outdoor recreation resource for many, but only about 23 percent (283 million acres; 115 million ha) of the nonindustrial, privately owned forestland is open to the general public for recreation. Leasing of private land for hunting is very popular in some states, particularly in the South where there is limited public land.

Recreation opportunities provided by the private sector include commercial recreation enterprises such as KOA campgrounds. Private industry and businesses also provide for recreation (Fig. 12-3). Most of the 68 million acres (27 million ha) of private industry lands available for outdoor recreation are located in the southern states, including lands belonging to wood products companies, utility companies, land holding companies, and manufacturing firms such as U.S. Steel and ALCOA.

Private, nonprofit conservation organizations have become a key element in efforts to protect private land for conservation purposes, including public recreation access. For instance, working with various partners (government agencies, corporate landowners, local conservation groups), the Conservation Fund has protected more than 3.2 million acres (1.3 million ha) of private land since 1985 (www. conservationfund.org). Many industries also have donated lands to The Nature Conservancy, which manages the lands or holds them for advantageous trades to federal or state government.

Federal Agencies Providing Outdoor Recreation Opportunities

As noted in Table 12-2, the Federal government provides a vast network of recreation amenities on its diverse lands. An overview of the major participating agencies follows. Recreation use trends at federal recreation sites are managed by seven different agencies are shown later in Table 12-4. Additional information about all of the agencies and organizations mentioned can be found at their Websites on the Internet.

USDA Forest Service National forests, administered by the Forest Service in the Department of Agriculture, are lands managed for multiple benefits, including wood products, water, wildlife, range, recreation, and wilderness. Outdoor recreation is now a leading use on the national forests. The national forests receive more *land-based* outdoor recreational use than any other of the federal lands (see Table 12-1).

A great diversity of recreational activities takes place on national forests. Fishing, hunting, hiking, camping, and picnicking have long been traditional uses, while wilderness, whitewater boating, and ice

FIGURE 12-3
A family enjoys a picnic on a table provided by an industrial timber company on its tree farm in northern Idaho. *(Potlatch Corporation photo by Jack Gruber)*

TABLE 12-2

MANAGING AGENCIES, AREAS, AND ADMINISTRATORS OF OUTDOOR RECREATION OPPORTUNITIES

Agencies	Areas	Administration
Federal		
USDA Forest Service	National forests, national grasslands, wilderness	Department of Agriculture
National Park Service	National parks, monuments, historic areas, wilderness	Department of the Interior
U.S. Fish and Wildlife Service	National wildlife refuges, fish hatcheries, wilderness	Department of the Interior
Bureau of Land Management	National resource lands, wilderness	Department of the Interior
U.S. Army Corps of Engineers	Reservoirs	Department of Defense
Tennessee Valley Authority	Reservoirs	Independent
Bureau of Reclamation	Reserviors	Department of the Interior
State		
State park commissions	Recreation parks, natural areas, some state wilderness and natural areas	State parks departments
State forests, departments of lands and natural resources	Multiple-use forests	Forestry commissions, departments of natural resources
State fish and wildlife departments, fish and game agencies	Fish/wildlife management areas	Game and fish commissions, wildlife resources agencies
Local		
City parks	Parks, recreation areas	City governments
County parks	County and metropolitan parks	County governments
Municipal forests	Forests, preserves	Forest districts
Private		
Industrial forests	Corporate forestlands, company lands	Corporations and companies
Nonindustrial private forests	Private forestlands	Individual private owners
Private enterprise	Private campgrounds, concessions	Private businesses and companies
Nonprofit conservation organizations	Various preserves, protected areas	National organizations (e.g., The Nature Conservancy, Audubon)
	Conservation easements on land held by other private landowners	Local land trusts

and snow activities have increased in recent years. The national forests contain numerous roads or trails suitable for pleasure driving, for off-road vehicles (ORVs), and for mountain bikes (Fig. 12-4). The nation's largest system of hiking trails, as well as many interpretive centers and unlimited opportunities for sightseeing and nature appreciation exist on these lands. Many abandoned logging roads on national forests are being utilized as hiking and mountain bike trails. By 1993, a total of 35.2 million acres (14.3 million ha) in 403 areas of national forest had been designated as wilderness, about 6 million acres (2.4 million ha) of it in Alaska.

Recent federal initiatives have attempted to limit road building in certain tracts of national forestlands to protect the integrity of remaining roadless areas.

National Park Service The National Park System, administered by the National Park Service (NPS) in the Department of the Interior, includes national parks, monuments, historic preserves, recreation areas, seashores, lakeshores, riverways, and urban parks. From the White House (18 acres) (7.3 ha) to Wrangall-St. Elias National Reserve (13 million acres) (5.3 million ha), these areas interpret natural and human history and serve many educational and recreational purposes. Generally, consumptive uses are excluded. For example, hunting is generally not allowed, though fishing is permitted in some areas. However, certain areas purchased and put in protected status in recent years, such as the Big South Fork National River and Recreation Area, administered by the NPS in Tennessee and Kentucky, allow hunting and ORV use.

FIGURE 12-4
National forests provide an extensive system of trails and ORV routes. Here a four-wheel group stops to talk during a drive through the Wenatchee National Forest, Washington. *(Photo by Grant W. Sharpe)*

There are over 385 areas in the National Park System, covering 85 million acres (34.4 million ha) in 49 states, the District of Columbia, American Samoa, Guam, Puerto Rico, Saipan, and the Virgin Islands (Table 12-3). These areas include most of the nation's outstanding natural, historical, and cultural resources, protected and managed for our educational and recreational use and for future generations (Fig. 12-5). By 1996, more than 43 million acres (nearly 18 million ha) in 44 areas of the National Park System had been officially designated as wilderness, 33 million acres (13 million ha) of it in Alaska.

Other NPS Functions Another function of the National Park Service lies in administering the Department of the Interior's special recreation assistance programs for all units of government. These programs help obtain surplus properties and provide planning assistance and matching grant monies to state and local governments. A few are mentioned here.

The Land and Water Conservation Fund (L&WCF) This fund has been a cornerstone for protecting lands for outdoor recreation since its passage in 1964. Areas funded through L&WCF grants must be maintained for recreation use in perpetuity. Through its planning and matching requirements, the fund also

TABLE 12-3

CLASSIFICATION, NUMBER, AND ACREAGE OF NATIONAL PARK SERVICE AREAS

Classification	Number	Acreage
International Historic Site	1	35
National Battlefields	11	13,175
National Battlefield Parks	3	9,674
National Battlefield Site	1	1
National Capital Park	1	6,468
National Historic Sites	77	37,267
National Historic Parks	40	163,082
National Lakeshores	4	228,857
National Mall	1	146
National Memorials	28	8,532
National Military Parks	9	38,798
National Monuments	73	2,706,955
National Parks	57	51,914,773
National Parkways	4	173,865
National Preserves	17	23,709,764
National Recreation Areas	18	3,692,223
National Rivers	5	429,959
National Scenic Trails	3	225,357
National Seashores	10	594,518
National Wild and Scenic Rivers and River Ways	10	308,130
Park (Other)	10	40,120
White House	1	18
Totals	385	84,327,466

Source: U.S. Department of the Interior, National Parks: Land Resources Division, 2001.

FIGURE 12-5

The National Park System contains outstanding natural and cultural resources. Isle Royale National Park, in Lake Superior, is renowned throughout the world as a unique wildlife sanctuary. *(National Park Service photo by Richard Frear)*

aims to stimulate nonfederal and private investments in protection and maintenance of recreation resources, and the development of new recreation opportunities throughout the United States.

To obtain LWCF moneys, states must prepare a State Comprehensive Outdoor Recreation Plan (SCORP). Money for the L&WCF comes from three main sources:

1. Sales of federal surplus real properties
2. A small part of federal motorboat fuel taxes
3. Outer continental shelf revenues derived from leasing of oil and gas sites in coastal waters

By law, 40 percent of the annual appropriation goes toward acquisition of federal recreation and conservation land. The remainder goes to the states to dispense to state and local governments, on a 50 percent matching basis. Authorized at $900 million annually, actual spending has been erratic. Federal grants have averaged only $250 million, and grants to states have been nonexistent since the early 1980s.

In 2001, Congress reauthorized LWCF by passing the Land Conservation, Preservation, and Infrastructure Improvement Act (LCPII), appropriating $1.2 billion for open space and parks in 2001 and $12 billion more by 2005.

The Rivers and Trails Conservation Assistance Program This program operates on the principle of partnerships with government agencies, private organizations, and landowners. The NPS works with these groups toward the use and protection of important land and water resources that lie outside of national parks and forests. Priorities for assistance include projects that combine river and trail "greenways," "rail-to-trail" conversions of abandoned rail lines, or others that have a similar innovative approach to conserving natural resources.

The Federal Lands to Parks Program This program allows states and communities to obtain no longer needed federal lands for use as parks and recreation areas. Typical properties include historical lighthouses, former military installations, abandoned missile sites, surplus research centers, hospital reservations, intracoastal waterways, discontinued coast guard stations, and military bases scheduled for closure.

Urban Park and Recreation Recovery Program (UPARR) Established in 1978, this program provides, through the National Park Service, direct federal assistance to cities and urban counties for reha-

FIGURE 12-6
A visitor watching snow geese on the Chincoteague National Wildlife Refuge, Va.
(U.S. Fish and Wildlife Service photo by John and Karen Hollingsworth)

bilitation of critically needed recreation facilities. Three kinds of grants are available: rehabilitation grants, innovation grants, and planning grants.

Although these NPS functions may not always deal with resource-based recreation lands, they are important in meeting people's recreation needs. The existence of these areas relieves pressure on the more remote and sensitive natural areas, as well as on nearby urban lands.

U.S. Fish and Wildlife Service The Service's major responsibility is management of the National Wildlife Refuge System, a collection of over 500 refuges covering 90 million acres (36 million ha) in 49 states and five trust territories (Cordell, 1999). The Refuge System includes areas located along the major north-south flyways of migratory birds, providing feeding, resting, and breeding areas for migrating ducks, geese, and other birds. Many areas serve as sanctuaries for threatened, endangered, or sensitive species of mammals, birds, reptiles, amphibians, fish, and plants. By 1996, about 21 million acres (8.5 million ha) in 75 areas of these wildlife refuges had been designated as wilderness, about 19 million acres (7.7 million ha) of the total in Alaska.

Although national wildlife refuges and fish hatcheries offer a wide variety of recreational opportunities and draw an estimated 30 million visits annually, these recreational activities are not allowed to interfere with the wildlife protection purposes for which the wildlife refuges were designated.

The National Wildlife Refuge System Improvement Act of 1997 established wildlife-dependent recreation as "a legitimate and appropriate general public use of the system." Specifically, it recognized hunting, fishing, wildlife viewing, environmental education, and interpretation as priority public uses. The legislation states that these uses should be facilitated if they will not materially interfere with the mission of the system (Fig. 12-6).

Bureau of Land Management The Bureau of Land Management (BLM) in the Department of the Interior manages what remains of the nation's once vast land holdings that comprised the *public domain* and are now called *national resource lands*. Many of our national parks, forests, wildlife refuges, and military lands were originally part of the public domain. The BLM today administers about 300 million acres (121.5 million ha) located primarily in the western states and Alaska. BLM management policies are very similar to those of the USDA Forest Service, based on the principles of multiple use and sustained yield of renewable resources. Outdoor recreation resources managed by the BLM are classified into 518 recreation management areas covering 215 million acres (87 million ha); including 134 designated wilderness areas totaling 5.2 million acres (2.1 million ha).

Other Federal Agencies   Three other major agencies, all involved with providing *water-based* (reservoir) recreation, are the U.S. Army Corps of Engineers, the Tennessee Valley Authority (TVA), and the Bureau of Reclamation.

The *Army Corps of Engineers,* located in the Department of Defense, first got involved in outdoor recreation in the late 1800s, when several corps of officers were placed in charge of building roads, bridges, and other engineering works for Yellowstone and Yosemite National Parks. Today, the Army Corps of Engineers manages over 500 impoundments and modified lakes, and with these, the largest water-based recreation program in the United States. Hunting, hiking, camping, picnicking, and nature study are popular activities on lands adjacent to these waters. Several of these areas have visitor centers, and nearly all provide boat ramps, beach facilities, picnic areas, and campgrounds.

The *Tennessee Valley Authority* (TVA), organized as an independent government corporation in 1933, provides a recreation function similar to that of the U.S. Army Corps of Engineers, but is restricted to the Tennessee River and its tributaries. While the main purposes of TVA dams are to regulate flood waters, provide navigation, and generate power, they also attract outdoor recreation, tourism, and vacation home development on the slightly over 1 million acres (.4 million ha) it manages. TVA also manages the Land Between the Lakes, a 170,000-acre (68,850-ha) area in west Tennessee and Kentucky, as a demonstration center for outdoor recreation, environmental education, and natural resource stewardship. Land Between the Lakes provides campgrounds, interpretive centers, roadside exhibits, over 200 miles (322 km) of trails, ORV areas, horse riding facilities, and managed hunting and fishing areas (see Fig. 12-1).

The *Bureau of Reclamation* (BOR), housed in the Department of the Interior, was originally established to help reclaim the arid lands of the western United States for farming and development by providing a year-round supply of water for irrigation. Hoover Dam on the Colorado River and Grand Coulee Dam on the Columbia River are two of the Bureau's best-known projects. Currently, the Bureau of Reclamation has 333 reservoirs and lakes throughout 17 western states that provide millions of visitors with opportunities for boating, swimming, fishing, picnicking, and related activities. However, the majority of recreation areas on BOR lands are not managed by the Bureau, but are turned over to local, county, state, or other federal agencies for management.

OUTDOOR RECREATION PARTICIPATION TRENDS

Predicting outdoor recreation participation is important in meeting human needs because of the demands participants place on resources, the long-term investments and land acquisition required, and the expenditures that must be anticipated.

General Use Trends

With the exception of the World War II era, outdoor recreation on federal lands has increased steadily, and today nearly 95 percent of Americans enjoy outdoor recreation. In recent decades, participation rates increased dramatically for many of the most popular activities. By 1995, 134 million people participated annually in walking (42.8 percent increase), 113 million in sightseeing (39.5 percent increase), 78 million in swimming in a lake or ocean (38.2 percent increase), and 54 million in bird watching (155 percent increase). However, participation in traditional consumptive activities has actually decreased to 18.6 million in hunting (12.3 percent decrease) and 57.8 million in fishing (3.8 decrease) (Cordell, 1999). Trends are apparent in use of federal recreation sites from 1988 to 1996, as shown in Table 12-1. Recreational uses of different categories of outdoor resources and projections by decade from 2000 to 2050 are shown in Table 12-4. Economic and social issues are involved in these changes.

The 2- to 3-week vacation trips that were standard in the past have been replaced by shorter day trips and long weekend trips. Also, more recreation now takes place close to home; the median distance for day trips to federal recreation areas is now 25 miles (40 km), with 130 miles (209 km) for overnight trips.

TABLE 12-4

THE RELATIVE POPULARITY OF KEY OUTDOOR RECREATION ACTIVITIES IN 1995 AND THEIR PROJECTED GROWTH TO 2050

Resource category and activity	Trips in 1995 (millions)	Projected future number of trips as percentage of 1995 demand					
		2000	2010	2020	2030	2040	2050
Dispersed Land							
Camping in primitive campgrounds	146.6	101	101	103	104	105	100
Backpacking	79.2	102	103	108	114	120	130
Nature study	70.8	105	113	120	131	138	
Horseback riding	185.1	101	114	129	146	160	177
Day hiking	557.7	104	112	123	133	139	152
Rock climbing	34.0	102	100	103	109	119	138
Visiting prehistoric sites	16.7	103	107	112	121	128	135
Walking for pleasure	14,381.4	106	117	129	140	150	158
Bicycle riding	1,386.8	108	124	145	167	190	216
Driving vehicles or motorcycles off-road	522.6	98	91	86	82	82	18
Visiting historic sites	482.4	107	126	148	171	193	216
Family gatherings	855.6	98	89	83	79	79	75
Sightseeing	1,209.5	106	120	143	163	172	198
Picnicking	855.6	94	79	70	63	59	55
Camping in developed campgrounds	209.6	103	115	130	146	162	180
Wildlife-related Activities							
Fishing	919.5	103	105	110	112	114	113
Hunting	305.5	102	.99	103	105	105	106
Non-consumptive (bird watching, photography, wildlife viewing)	2,277.1	100	107	115	118	115	108
Water							
Canoeing/kayaking	49.3	102	107	114	122	129	129
Stream/lake/ocean swimming	837.9	101	104	109	114	119	125
Rafting/tubing	61.5	101	103	110	117	123	130
Rowing/paddling/other boating		112	124	136	150	159	
Motor boating	480.4	105	108	116	126	136	148
Visiting beach/waterside	1,667.1	103	110	117	125	131	142
Snow and Ice							
Cross-country skiing	33.5	102	106	113	122	133	149
Downhill skiing	78.9	106	118	136	158	185	222
Snowmobiling	38.6	102	116	135	156	177	210

Source: H. K. Cordell, Outdoor Recreation and American Life: A National Assessment of Demand and Supply, Trends, Sagamore Publishing, 1999.

Recreation managers are dealing with rapid change, as recreation resource issues are highly influenced by a variety of social factors. Researchers examining the more than 30 years of national recreation studies have identified the following 10 factors most likely to shape the future of outdoor recreation participation (Hartmann et al., 1988; O'Leary, 1989):

1. An aging population with earlier retirement
2. A decline in available leisure time
3. Increased population growth, particularly in the South and the West
4. Increasing immigration, probably bringing new patterns of outdoor recreation participation
5. A greater percentage of the work force represented by females, resulting in more dual-income households with increased discretionary income and less family leisure time
6. A changing family structure, including fewer extended families and more single-parent families

7. Higher average education levels
8. Greater physical fitness and health consciousness
9. Baby boomers entering middle age and becoming important consumers
10. Young people delaying marriage and childbearing.

Participation Trends

National surveys of recreation activity participation have been conducted in the United States since the 1960s, and most show that driving or walking for pleasure, sightseeing, family gatherings, and viewing/learning activities are the most popular activities (Cordell, 1999). Cordell (1999) estimates that, of the approximately 4.6 billion outdoor recreation trips Americans took in 1987, about 50 percent were taken to participate in wildland activities such as camping, fishing, hunting, hiking, horseback riding, canoeing and kayaking, bicycling, wildlife observation, winter skiing, and visiting prehistoric sites (Fig. 12-7).

Cordell (1999) also made forecasts of future activity participation based on societal trends, and predicted that the fastest-growing activities will be visits to historic places, downhill skiing/snowboarding, snowmobiling, sightseeing, and nonconsumptive wildlife activities. Most forms of water and snow activities are predicted to grow rapidly. Risk/adventure activities, such as whitewater sports and rock climb-

ing, are also expected to increase rapidly. Participation in more traditional activities like hunting, horseback riding, fishing, picnicking, and driving motor vehicles off-road are expected to grow at a much slower rate during the next 15 to 20 years. Participation in hunting will continue to decrease, as it has since the 1970s.

RECREATION MANAGEMENT PLANNING

Three useful conceptual frameworks for recreation management planning are the recreation opportunity spectrum (ROS), recreational carrying capacity, and limits of acceptable change (LAC).

Recreation Opportunity Spectrum (ROS)

The ROS is a planning framework that recognizes the existence of a spectrum of recreation opportunities ranging from paved urban settings to primeval wilderness, with many points between. ROS recognizes a diverse spectrum of recreation settings where a variety of activities and different experiences can occur. Because recreationists seek activities in different physical, biological, social, and managerial settings in pursuit of various preferred experiences, a diversity of recreational opportunities should be provided.

The U.S. Forest Service and BLM applications of ROS recognize six opportunity classes within this spectrum, ranging from primitive recreation at one end of the spectrum to urban recreation at the other end (Fig. 12-8). The opportunity class settings are further characterized in terms of the following criteria that can then be described or inventoried: access, remoteness from sights and sounds of human activity, nonrecreational resource uses, onsite management, social interaction, acceptability of visitor impacts, and acceptable regulation by management.

By codifying a diverse spectrum of recreational opportunities, the ROS can assist in identifying area capabilities, matching them with visitor needs and preferences and developing recreation management prescriptions and strategies to protect the different kinds of opportunities. For example, each ROS opportunity class features different user experience objectives and may need to be managed in a special manner. Thus, each ROS class (or range of classes) will have limits of acceptable change with respect to the use impacts that can be allowed, and the intensity

FIGURE 12-7
Cross-country skiing is expected to increase in popularity. *(Photo by Alan Ewert)*

and type of management that is appropriate in order to mitigate impacts. For example, in the semiprimitive, motorized, opportunity class, the objective is to provide user experiences that feature contact with nature by small groups of family or friends camping, picnicking, or hiking. Management seeks to protect that opportunity by preserving the natural setting, minimizing development, and preventing crowding—but without intruding on visitor experiences. Thus, there are limits on acceptable change.

Recreational Carrying Capacity

The concept of carrying capacity came from the disciplines of range and wildlife management, referring to the number of animals a given unit of land could

support on a sustained basis without undue deterioration of the resource base. *Recreational carrying capacity* is defined as the maximum amount and type of use that an area can tolerate before the quality of the recreation resource and experience deteriorate beyond an acceptable level. However, for several reasons, determining recreational carrying capacity is not simple. There are several types of recreation carrying capacity, including ecological (e.g., resource limitations), social (e.g., visitor tolerance for congestion), and design capacity (e.g., site, engineering limitations). However, the concept is useful for planning—and sometimes evaluation—for it may be apparent when the capacity for a particular kind of recreation experience has been exceeded.

FIGURE 12-8
The recreation opportunity spectrum (ROS) and the six opportunity classes, from primitive to urban, where a variety of activities, settings, and experiences can occur.

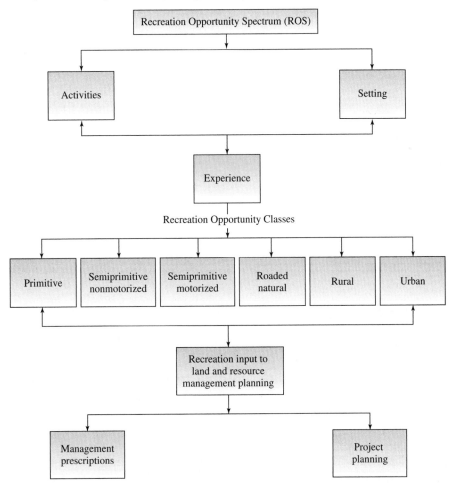

Limits of Acceptable Change (LAC)

The LAC concept emerged when managers came to realize that they should focus on what kinds of resource and social conditions they could hope to maintain, rather than on some magic carrying capacity number. The basic premise of the LAC concept is that change (both environmental and social) is a natural, inevitable consequence of recreation use. This premise redefines the traditional question about carrying capacity from "How much use is too much?" to "How much change is acceptable?" (Hendee & Dawson, 2002). Two important implications for managers emerge. First, LAC focuses their orientation on managing for desired conditions rather than merely controlling numbers of users; and second, it focuses them on the question, *"What is the level of acceptable change?"* The answer to this requires human judgment based on the viewpoints of managers and of citizens as well. Therefore, in the LAC process, user feedback is essential.

FIGURE 12-9

A simple planning framework, illustrating the steps from setting objectives to monitoring conditions. *(Courtesy of D. N. Cole)*

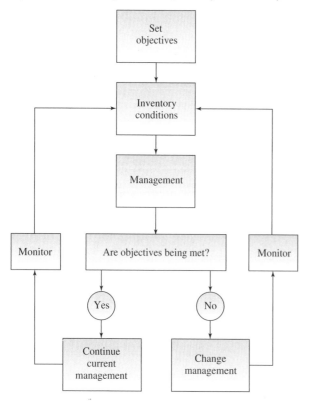

In simple terms, LAC calls for stating the conditions management will provide in the form of clear and measurable objectives, and stating how much impact will be allowed; inventorying conditions to see how they compare to desired conditions as stated in the objectives; and implementing management to maintain conditions, or to achieve or restore conditions where they do not meet objectives. The final step, monitoring, involves continuing evaluation and periodic return to the inventory stage of the process (Fig. 12-9). Monitoring is merely periodically repeating the inventory and comparing current conditions with objectives for the area and the previous inventory.

STRATEGIES AND CONCEPTS FOR RECREATION MANAGEMENT

Certain basic principles and concepts guide outdoor recreation management strategies. Some of these principles will first be outlined; then planning concepts and management strategies for guiding management of recreation visitors and resources will be described.

Key Management Concepts

1. *Quality* is the major goal of all recreation management strategies. Management aims to maintain and increase the quality of recreation experiences that visitors seek, as well as the quality of opportunities (resources) that provide for those experiences, on a sustained basis.

2. *Diversity* is a major characteristic of outdoor recreation. Preferences for outdoor recreation opportunities are varied in the types of activities desired, the way they are carried out, and the types of environments where they might occur. Preferences may change over time, among individuals, and across society, so diversity in the provision of outdoor recreation opportunities and environments is essential to accommodate changing preferences.

3. *Managing change,* that is, planning for desirable change and halting undesirable change in recreation conditions, is key to recreation quality and diversity. Management's role is to manage recreation use skillfully, not to inhibit it. Managing visitor behavior to ensure the protection of natural conditions as well as visitor safety is included in controlling undesirable change.

4. *Clear, measurable objectives* for recreation areas are necessary for high-quality management. Such objectives and their measure can suggest whether or not recreation use has reached an undesirable level. Sensitive areas must be identified, and limits of acceptable change established, to determine what management actions are needed and when to initiate them.

5. *Predictability of demand* can guide management effort. Recreational use tends to be predictable, such as high concentration near access points, special attractions and facilities, and along travel routes. Weekends are busiest, and there are common patterns of activity. The predictability of outdoor recreation use allows managers to schedule staffing and/or restrictions, and strengthen design of sites to accommodate predicted use.

6. *Protecting user freedom of choice* is important, because freedom is characteristic of all forms of recreation and should be respected. Thus, indirect approaches to visitor management that aim to modify user behavior are preferred over direct approaches that regulate and control visitor use and behavior.

The Manager's Role

It would be easy to develop an antiuser bias in recreation management, since protecting areas from serious damage seems paramount. Wouldn't such areas be better protected if recreational use did not occur? Of course they would. But this is unrealistic. We must accept the fact that recreational use is appropriate in forests and almost all natural areas, and that it will result in some impact. Management's role, then, is not to halt use but to manage for acceptable levels of use, associated impacts, and the resultant environmental change. Common sense is a key.

Direct and Indirect Approaches

Indirect approaches to managing visitor use are preferred over direct, regulatory approaches, since freedom and spontaneity are at the core of most outdoor recreation pursuits. Indirect management attempts to influence rather than force user behavior through interpretation, persuasion, or site manipulation.

However, there are situations where direct management and regulation are important and legitimate. Examples include situations where safety is an issue, such as regulations keeping motorboats out of swimming areas; where conflict and interference occur, such as regulations requiring quiet after 10:00 P.M. in campgrounds; and where a few individuals might use more than their share of recreation resources, (e.g., requiring bag limits on numbers of fish and game). Generally, regulations are appropriate where compliance is imperative, where law enforcement is available to back them up, and if users understand the reasons for them. When regulations are instituted, it is important to (1) explain the reasons for regulations, (2) be sure that visitors understand how they are expected to behave, (3) enforce regulations fairly, and (4) regulate at the minimum level.

VISITOR MANAGEMENT

Recreation management is always concerned with enhancing visitor opportunities, experiences, and satisfactions. Concern for resource protection must be tempered with concern for recreation opportunity. As mentioned earlier, visitor management actions that feature indirect or light-handed techniques should be applied before resorting to direct, and more heavy-handed techniques. For example, providing information about the need to protect the natural qualities of a site to persuade visitors to disperse over a wider area should be tried before erecting barriers and limiting or banning use. Figure 12-10 illustrates the line of

FIGURE 12-10
The continuum of indirect to direct approaches in outdoor recreation management.

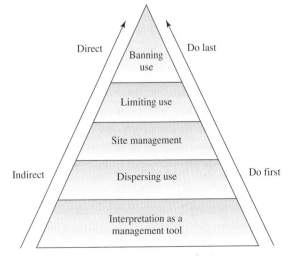

FIGURE 12-11
Through interpretation, visitors develop a keener awareness, appreciation, and understanding of the area they are visiting. The park interpreter's very presence assists in protecting a fragile resource. Mesa Verde National Park, Colorado. *(Photo by Grant W. Sharpe)*

action of moving upward from indirect to direct (first to last) management techniques. Five approaches are discussed here, ranging from indirect to direct.

Interpretation   Visitors often are unaware of their impacts. Interpretation, provided through brochures, exhibits at visitor centers, interpretive talks, and contacts with rangers, is an important means of managing visitors' behavior as well as adding to their enjoyment of the area (Fig. 12-11). By using interpretation as a management tool, it is possible to inform visitors about appropriate ways to appreciate and use recreation resources, and of ways to interact suitably with other users.

Sometimes safety considerations make education necessary before allowing certain forms of recreation. Hunter safety and game identification education, boater safety training, and scuba diving certification are current examples.

Dispersing Use   Many recreation social impact problems, such as crowding and user conflicts, and resource degradation arise from concentrated use in small areas. Although recreationists go to the outdoors to "get away," they may end up in crowded areas at high-use times. Dispersal is a useful management tool in some situations to alleviate these social problems. Visitors can be dispersed within a given area or areas can be zoned for particular types of activity. This can be accomplished by time limits, such as zoning a lake for fishing only before 10:00 A.M. and after 4:00 P.M., and scheduling water skiing during the 10:00 A.M. to 4:00 P.M. period. Encouraging off-season and weekday visitation can also disperse use. Maps, interpretive exhibits, and literature informing visitors of where and when visitor concentrations occur allow them to select less-visited sites and times. However, dispersing visitors can actually increase biophysical impacts. Concentrating visitors in "hardened" areas can alleviate pressure on sensitive areas.

Site Management

Site management can change or modify visitor use of recreation areas. Some site management techniques indirectly manage recreation use and behavior, while others are more direct. For example, addition of facilities and conveniences on durable sites may attract visitors to these areas. The "hardening" of sites, such as surfacing trails, providing tent pads in campgrounds, and paving parking areas, may encourage use while at the same time increasing the ability of the site to withstand this use. Where the level of use must be reduced, not replacing, or even removing bridges and signs may (indirectly) discourage would-be users. Fences and barriers would be a direct method of site management.

Limiting Use

Limiting the amount and type of use to protect recreation sites and experiences sometimes is necessary. But regulating the total amount of use in an area is only a partial solution. Limitations on length of stay, seasonal limits, restricting activity in certain areas (zoning), party size restrictions, and consumptive removal limits (e.g., fish and game bag limits) represent a direct means of managing visitor use. Regulations and accompanying law enforcement are a very direct means of limiting use, but other approaches, in some situations, may be less intrusive. For example, river rafting, backcountry and wilderness camping, or big game hunting can be rationed through request (reservation), by lottery (chance), queuing (first-come, first-served), pricing (fee), and merit (skill and knowledge). Again, the principle of first applying the most indirect or least intrusive approach to limiting use before directly controlling it is recommended.

Banning Use

A total ban on use may sometimes be warranted. When a stream is completely fished out, a campsite worn out, a bear–human interaction problem so severe as to cause safety concerns, or when vandalism is out of control, managers may have little choice but to ban all use of an area. Sometimes natural causes such as extended drought, extreme fire danger, or the presence of an endangered species may call for a periodic ban on use of an area. Again, managers should explore other choices to see if limiting party size or

time or type of use might suffice, before all use is banned.

MONITORING RECREATION IMPACTS

Inventory data provides an empirical basis for monitoring the current condition of recreation resources and opportunities in relation to management objectives, so that problems can be identified. Over time, periodic monitoring will document changes in conditions. These data can be of value in establishing limits of acceptable change, and provide a basis for evaluating the effectiveness of management programs.

Monitoring Systems

Monitoring systems call for periodic measurement of indicators that are assumed to reflect the overall conditions under evaluation. Numerous indicators and techniques are available for monitoring ecological and social conditions in different situations and with different management objectives, but four characteristics are desirable for all monitoring systems.

First, when monitoring campsites and other recreation site conditions, those variables (indicators) most sensitive to use, degradation, and management action should be measured. One cannot measure everything. Only a limited number of meaningful indicators should be selected, such as size of area denuded by trampling.

Second, measurement techniques must be reliable and sufficiently precise to allow the studies to be replicated at a later date. Therefore, if the indicator to be measured is "area denuded," permanent plots should be established and their exact location recorded for later measurement of denuding or recovery.

Third, trade-offs between the reliability, sensitivity, and cost of techniques must be considered. For example, a limited number of plots should be established, yet there must be enough plots to reliably sample the extent of site degradation. The greater the number of plots, the higher the cost.

Fourth, plots should be precisely established so they can be relocated accurately at a later date for remeasurement. If photographs are used to document change over time, the photo point must be related to a permanent feature and clearly described.

In the following section, campsite conditions, two examples of monitoring programs commonly used to assess campsite conditions (ecological impacts) and use crowding (social impacts) are discussed.

Campsite Conditions

Campsite inventory techniques can conveniently be grouped into three classes: measured conditions, estimated conditions, and photographs.

Measured Conditions The best way to get accurate, replicable data is to take careful field measurements. Estimates and photographs are less costly, but the data collected are less precise (Hammitt & Cole, 1987). Cole (1982) sampled campsites in the Eagle Cap Wilderness in Oregon, using a combination of *area, line transect,* and *quadrant* measurement techniques (Fig. 12-12). On each campsite, linear transects were established in 16 directions, radiating from an arbitrarily (but permanently) established center point. Distances were measured from the center point to the first significant vegetation encountered and also to the edge of the disturbed part of the campsite. Thus, the barren central core of the site (bare area) and the entire disturbed area (camp area)

were measured. Tree seedlings and mature trees within the entire area were counted and examined for damage. Furthermore, on each campsite, approximately 15 quadrants, 3 ft by 3 ft, were located along four of the transects. In each quadrant, the total vegetation cover and exposed mineral soil was estimated, and plant species were identified.

Such careful measurement provides good baseline data for future monitoring comparisons, but obtaining such precise information involves time and expense.

Estimated Conditions Where it is not feasible to spend so much time per campsite, estimates must be substituted for precise measurements. This will reduce the accuracy and reliability of monitoring measures.

Two approaches to estimating conditions are available; one utilizes condition class estimates, and the other employs multiple parameter estimates.

In the condition class approach, a series of site condition descriptions are assigned to sites upon vi-

FIGURE 12-12

A measurement system used to determine campsite ecological impacts. Recorded on site are bare area, camp area, number of damaged trees, and number of tree seedlings. *(From D. N. Cole, 1982)*

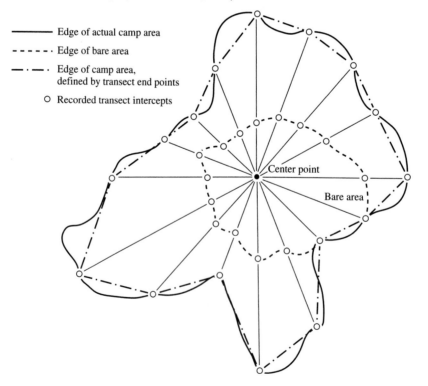

sual inspection. Frissell (1978) suggests the following five classes.

1. Ground vegetation flattened but not permanently injured. Minimal physical change except for possibly a simple rock fireplace.
2. Ground vegetation worn away around fireplace or center of activity.
3. Ground vegetation lost on most of the site, but humus and litter still present in all but a few areas.
4. Most of site worn down to bare mineral soil. Tree roots exposed.
5. Soil erosion obvious. Trees reduced in vigor or dead.

Sites are then rated again in future years to see whether conditions have improved or further deteriorated.

In the multiple parameter approach, information is estimated for a number of separate campsite-impact variables. Each parameter is assigned a rating (i.e., 1 to 5), depending on amount of impact. These ratings are then totaled (scored) to obtain an overall impact rating. For example, sites can be rated according to severity of exposure of mineral soil, vegetation loss, tree damage, or root exposure. Future impact rating scores can be compared to see if conditions have improved or worsened.

Photographs   Photography has frequently been used for monitoring, sometimes systematically and sometimes not, sometimes to enhance field data and sometimes as the only monitoring tool. Photographs are best used as a supplement rather than as a substitute for field measurements. They can be used to validate field measurements and, more importantly, they provide a visual means of conveying information about site conditions quantified in field measurements. Three photographic techniques that have been used as part of campsite monitoring programs are photo points, quadrant photography, and campsite panoramas (Brewer & Berrier, 1984).

Visitor Use and Crowding

Visitor use and its relationship to crowding is the most monitored of social impacts. Too much use, as well as the presence of certain types of use, may influence visitors' perception of crowding, cause user conflicts, and eventually diminish the quality of recreation experiences. To estimate visitor use and determine if intergroup interactions are at acceptable levels, managers must monitor use over time and space as well as visitor reactions to the use. Mechanical use counters, field observations, and surveys are techniques commonly employed to monitor crowding impacts. Use measures include recreation visits (RVs), recreation visitor days (RVDs), or the number of people at one time (PAOT).

Crowding can occur in many different types of recreation areas and activities, such as at campsites, around rivers and lakes, and seasonally in hunting areas. Different variables can affect crowding in each of these situations. However, the monitoring of crowding usually involves at least the following four indicators.

1. *Use level.* Monitoring occurs in an area or site that can be measured by mechanical counters, by hidden cameras, or through use permits. However, use level does not determine crowding per se. Just because many people are in an area does not mean you will come in contact with them. Therefore, it is necessary to monitor the number of encounters, by sight or by sound, that occur among visitors within certain areas.
2. *The number of visual encounters.* This is usually measured by field observations, questionnaire self-reports, or exit surveys.
3. *Visitor expectations.* The number of encounters visitors actually experience in the area must also be determined. For example, those visitors who expected to have several encounters are less likely to react negatively to them than visitors who expected to see only a few other users.
4. *The perceived level of crowding.* How crowded or uncrowded do users perceive an area to be? Data must be gathered on this indicator also.

The relationship of the results of these four indicators to management objectives for an area will need to be reviewed periodically to successfully deal with visitor use and avoid crowding.

Site Hardening and Design

Beyond locating use on impact-resistant sites and providing suitable facilities, site hardening and design can help prevent deterioration of trails, campsites, and other sites. Water bars, steps and culverts in trail construction, surfacing of trails with wood chips or paving materials, and silvicultural treatments, such as thinning the overstory canopy to bring in sunlight to dry the trail tread, are examples of site hardening and design commonly used in outdoor recreation areas.

Site Rehabilitation

In some situations there is no option but to permanently close and then attempt to rehabilitate damaged sites. Common reasons for such action include excessive site damage that cannot be controlled during continued use, a decision to relocate the facility on a more durable or desirable site, or to rehabilitate damage. Many cultural treatments such as watering, fertilizing, planting, seeding, mulching, overstory thinning, and habitat restoration can be used to rehabilitate sites.

SUMMARY—OUTDOOR RECREATION MANAGEMENT

Leisure activity in the outdoors is an American tradition and is provided for on federal, state, and local public and private lands. Federal agencies including the National Park Service, Forest Service, Corps of Engineers, and others have major outdoor recreation management programs, each with a slightly different emphasis fitting resources and user groups. Several strategies and concepts for recreation management have evolved to deal with growing recreation use and impacts. These focus on quality and diversity of recreation experiences, managing change, setting clear and measurable objectives for recreation areas, predicting demand, and protecting freedom of choice for users. Managers seek to balance use against impacts, preferably by employing light-handed, indirect approaches rather than more restrictive strategies. Visitor management concepts and techniques, ranging from indirect to direct, include use of interpretation, dispersing use, site management, limiting, and finally, banning use. Key planning ideas include the recreation opportunity spectrum, recreational carrying capacity, and limits of acceptable change, all approaches to deploying and protecting different kinds of opportunities for anticipated use. To determine the effectiveness of management techniques, recreation impacts are monitored by measuring or estimating conditions and comparing these over a period of time.

WILDERNESS MANAGEMENT

INTRODUCTION

To many people, wilderness is one of the most appealing concepts associated with natural resources. The vision of wild nature left to evolve under its own influences and uncontrolled by humans seems particularly attractive in the midst of the hassles, pressures, and constraints found in almost every aspect of life in the twenty-first century. Wouldn't it be renewing to escape to an area that's still pure and simple—a place governed only by nature's primal forces? Isn't it pleasant just to think about such areas, to know that they exist, and that some places remain where nature, not humankind, is in charge?

This exciting notion of wilderness is true, in part. The United States has set aside 643 areas, nearly 106 million acres (43 million ha), in the National Wilderness Preservation System (Table 12-5) that are still natural, relatively speaking, where visitors have opportunities for primitive types of recreation amid solitude and naturalness, with minimal evidence of human influence (Fig. 12-13). Such experiences are renewing, inspiring, and healing to visitors stressed by modern life. Wilderness is a reality, and remains a possibility on millions of additional acres of wild and roadless public lands. But the naturalness of wilderness cannot be taken for granted. Here is the contradiction at the heart of our dreams. Once such areas are set aside for protection as wilderness, it takes proactive wilderness management and planning to maintain their wildness and to ensure that they are influenced primarily by natural processes.

The recreation opportunity spectrum (ROS), described earlier, illustrates the outdoor experience opportunities that presently exist, from highly developed areas to roadless wilderness. We need this full spectrum of areas on our public lands. These areas have different visitation, and therefore different management goals and requirements, ranging from intense human use and resultant facility development to dispersed use and light-handed stewardship utilizing natural forces. But in all kinds of areas across the recreation opportunity spectrum, management will be required. Even in wilderness, which is committed to naturalness, solitude, and letting natural processes be the principal architect of the landscape, management will be required as "nature's helper" to guide or control recreation use and to overcome human influences that may impinge upon the wilderness environment.

TABLE 12-5

WILDERNESS UNITS AND ACRES IN THE U.S. NATIONAL WILDERNESS PRESERVATION SYSTEM (NWPS) BY FEDERAL MANAGING AGENCY IN 2000

Agency	Units[a]	Federal Acres[b]	Percent of NWPS Acres
NWPS excluding Alaska			
Bureau of Land Management	145	6,226,482	13.1
Forest Service	384	29,043,345	61.1
Fish and Wildlife Service	50	2,009,222	4.2
National Park Service	36	10,295,156	21.6
Subtotal	615	47,574,205	100.0
NWPS in Alaska			
Bureau of Land Management	0	0	0
Forest Service	19	5,752,221	9.9
Fish and Wildlife Service	21	18,676,912	32.1
National Park Service	8	33,753,083	58.0
Subtotal	48	58,182,216	100.0
Entire NWPS			
Bureau of Land Management	145	6,226,482	5.9
Forest Service	403	34,795,566	32.9
Fish and Wildlife Service	71	20,686,134	19.9
National Park Service	44	44,048,239	41.6
Total	643	105,756,421	100.0

[a]The number of units managed by each agency does not equal the total because some areas are managed by more than one agency.
[b]Some acreages are estimates pending final surveying and mapping.

Source: Updated from 1999 data in Landres and Meyer 2000.

FIGURE 12-13
Wilderness experiences provide for enjoyment of solitude and naturalness, with minimal evidence of human influence. Arapaho-Roosevelt National Forest, Colorado. *(Courtesy of USDA Forest Service)*

DESIGNATION OF WILDERNESS

Early in the twentieth century, some farsighted individuals became concerned about the disappearance of areas that were still truly wild—that had no substantial evidence of human activity. George Perkins Marsh, John Muir, Aldo Leopold, Arthur Carhart, and Robert Marshall were among the visionaries who called for protection of our nation's diminishing wilderness resources (Hendee & Dawson, 2002). Out of their concern, and the efforts of many more like-minded individuals, came a process for evaluating such areas and legally setting aside the best of them to be permanently protected as units of the National Wilderness Preservation System (NWPS).

Nationwide, since the 1930s, there has been a continuing review of remaining roadless areas, in an effort to save some land whose conditions of naturalness and solitude still exceed the minimum standard for wilderness. Prior to 1964, such areas were protected only by the policy of the federal agency managing the lands on which they occurred. Then, to provide permanent legislative protection for such areas and create a process for selecting them, the Wilderness Act of 1964 (Public Law 88-577) was passed. The act defined *wilderness* as federal lands that were *untrammeled,* meaning not subject to human controls that hamper the free play of natural forces, and that offered opportunities for solitude or a primitive and unconfined mode of recreation. A total of 54 areas containing 9.1 million acres (3.7 million ha) were designated wilderness by the Act. The Act also called for review of roadless lands in the national forests, national parks, fish and wildlife refuges, and later, lands administered by the Bureau of Land Management, to consider which of these remaining roadless areas should be added to the National Wilderness Preservation System by further acts of Congress. Thus, the addition of areas since 1964 has created a National Wilderness Preservation System of 643 areas totalling nearly 106 million acres (44 million ha) through passage of 131 additional pieces of legislation (Hendee & Dawson, 2002).

Since 1964, resource managers in federal agencies, the American people, and the U.S. Congress by adding areas to the NWPS through legislation, have continued to refine their ideas about what kind of areas could or should be set aside as wilderness, and how they should be managed. At first, only the largest and wildest remaining areas were expected to be designated in a 30–40-million-acre (12–16-million-ha) wilderness system. Then, during the first Forest Service roadless area review and evaluation (RARE) in the early 1970s, and the second roadless area review (RARE II) in the late 1970s, additional refinements of wilderness definitions eased the standards for wilderness qualification on national forests. Furthermore, in 1975, the so-called 'Eastern Wilderness Act' confirmed the intent of Congress to designate additional wilderness areas in the East, nearer population centers, and to admit to the National Wilderness Preservation System areas smaller and showing more evidence of human use than previously acceptable. In 1978 the Endangered American Wilderness Act further confirmed that wilderness areas could be located near cities and could include areas previously modified by human activity, but that were deemed able to recover. Thus, during the 38 years since the 1964 Wilderness Act was passed, the specifications for how large or wild an area must be to qualify for wilderness designation have eased. A much larger wilderness system than originally anticipated has been created, but with some dilution of the original concept of *wilderness* as an inevitable result. For a more detailed review of the evolution of the NWPS, see Hendee and Dawson (2002).

The wilderness preservation idea has also spread internationally. Seven other nations now have legislation protecting wilderness areas (Australia, Canada, Finland, Sri Lanka, Russia, South Africa and the Flathead Indian Nation), and another three countries (Zimbabwe, New Zealand, and Italy) protect wilderness through administrative policy (Martin and Watson, 2002). As with the establishment of national parks, attempts to define and protect wilderness areas originated in and spread from the United States. American natural resource managers increasingly work with international colleagues in expanding application of wilderness management concepts and techniques.

Designated Wilderness Acreage

The National Wilderness Preservation System (NWPS), as of December 30, 2000, contained 643 separate areas covering 105.7 million acres (43.1 million ha) under management of four federal agencies (Table 12-5)—the U.S. Forest Service, the National Park Service, the Fish and Wildlife Service, and the Bureau of Land Management. BLM wilderness is growing as new wilderness bills for various western states are considered by Congress. Beyond this, the Forest Service and National Park Service combined have 30–60 million acres of additional land that may be added to the Wilderness System. Nationwide, about 4 percent of the U.S. land area has been designated as wilderness, and 5 or 6 percent seems a reasonable limit of what might ultimately be included in the NWPS. The Wilderness Society has called for an ultimate NWPS of 200 million acres (83 million ha).

More than half of the current wilderness acreage, 58.2 million acres (23.6 million ha), is in Alaska. Nearly all the rest is located in 11 western states. Less than 5 percent of the nation's wilderness lies east of the Mississippi River, and nearly half of that can be found in a single area, Everglades National Park in Florida. However, only five states lack federal wilderness—Maryland, Rhode Island, Delaware, Kansas, and Iowa. The largest wilderness is 8.7 million acres (3.5 million ha), in Alaska's Wrangell-St. Elias National Park and Preserve. The smallest is 6 acres (2.4 ha) on Pelican Island National Wildlife Refuge in Florida (Landres & Meyer, 2000).

State Wilderness Systems During the era of federal wilderness establishment, several states were also establishing land use systems including wilderness or wilderness-like areas. By 1994, a total of eight state wilderness systems existed, including 58 areas totalling 3.1 million acres (1.3 ha), about 60 percent of it in New York (Hendee & Dawson, 2002; Peterson, 1996). These 58 state-owned areas set aside as wilderness reflect determination at the state level to establish a full spectrum of environments within a state's land use systems.

WILDERNESS PROTECTION

Wilderness is protected from development such as logging roads, dams, or other permanent structures; from timber cutting and the operation of motorized vehicles and mechanized equipment; and, since 1984, from new mining claims and mineral leasing.

Mining operations and livestock grazing are permitted to continue in wilderness if these practices existed prior to an area's designation. Existing operations must be conducted so as to minimize the impact on wilderness and other resource values. In most in-stances, Congress has drawn wilderness boundaries to exclude public lands potentially rich in mineral resources.

Hunting and fishing are permitted in wilderness, although hunting is not allowed in most national park units and some wildlife refuges. In most Alaskan wilderness, float planes are allowed, and in a few large western wilderness areas, primitive airstrips, established prior to wilderness designation, continue to be used for access by small planes.

Some of the same management strategies, concepts, and techniques described earlier as applicable to other outdoor recreation areas, such as carrying capacity and LAC, apply to wilderness, but the framework is a little different since there are no roads and mechanized access is generally not appropriate or allowed.

Information may be provided in brochures or signs at wilderness trail heads urging appropriate behavior and safety, and providing information about the area. Dispersal of use is encouraged by information and by wilderness rangers patrolling the area. Impacts on heavily used campsites are monitored, and sites may be closed and rehabilitation measures imposed where needed. As management plans are developed, the process may be used to reconcile conflicts between uses and to establish use limits. Maintaining the opportunity for a high-quality recreational experience featuring naturalness and solitude receives primary emphasis. These challenges are not easy to meet because protecting naturalness and solitude in the face of increasing wilderness visitation poses growing dilemmas for wilderness management (Cole, 2001).

Principles of Wilderness Management

Hendee and Dawson (2002) describe 13 principles for wilderness management. Five of the most important are given here. These principles provide a framework within which resource management techniques are applied to wilderness needs and problems. They derive from the mandate of the Wilderness Act to "keep wilderness wild."

1. *Maintain wilderness thresholds.* Manage wilderness to maintain its position above the threshold of wildness that separates wilderness from other areas. Wilderness is at the primeval end of the recreation opportunity spectrum and needs to be managed to remain at this end of the continuum. Keeping an area roadless and unmechanized is

assumed in wilderness management. But the naturalness and solitude of an area should also be maintained at least at the minimum threshold that would be considered wilderness, and beyond that if such conditions were present when the area was designated as a wilderness.

2. *Manage wilderness for nondegradation of individual areas.* This concept, long applied in air and water quality management, calls for the maintenance of existing environmental conditions (in this case, naturalness and solitude) if they equal or exceed minimum standards, and for the restoration of those conditions that are below minimum levels.

3. *Manage wilderness as a holistic resource.* The importance of the wilderness resource is found in the whole and not in its separate parts, such as its scenery, vegetation types, or wildlife (Fig. 12-14). The wilderness resource is all those things and more, and it includes the dynamic relationships among them. Thus, in wilderness management are the seeds of the more holistic ecosystem management now being applied on most public lands.

4. *Favor wilderness-dependent activity.* Wilderness hosts diverse activities, but not all of them depend on wilderness conditions for their pursuit. Activities that are not dependent on wilderness can be enjoyed in other settings. Whenever uses conflict, the activity that most depends on the wilderness conditions of naturalness and solitude should be favored. This discriminates the least against user's satisfactions, since those seeking activities that don't *depend* on wilderness can find them elsewhere.

5. *Use minimum tools.* Apply the fewest regulations or tools, or the least amount of force in achieving wilderness area objectives. This so-called *minimum tool rule* may be the most important principle of wilderness management. It says two things: *Do only what is necessary* to meet wilderness management objectives, and *use the minimum-impact method* to get the job done. For example, when wildfire must be controlled in wilderness, the use of bulldozers and chemical slurry airdrops should be avoided if firelines constructed with hand tools will do the job. If educating wilderness visitors on how and where to disperse their campsites will adequately protect wilderness solitude, avoid restricting use or regulating it through permits. If trails can be cleared with cross-cut saws

FIGURE 12-14

Wilderness management includes managing for scenery, vegetation, and wildlife. Here a wildlife professor scans the landscape for sightings of desert bighorn sheep, whose survival depends on large blocks of suitable land free from competition with livestock. Pusch Ridge Wilderness, Coronado National Forest, Arizona. *(Photo by John C. Hendee)*

and hand tools, do not use chain saws. By doing only what is *necessary* to meet area objectives, and using the lightest touch possible, both the wilderness environment and wilderness experiences will be protected.

Wilderness and Naturalness

Many wilderness management discussions center around the issue of what is *natural*. What conditions are we trying to preserve? Is it some vestige of an ancient landscape or conditions such as might have been present just prior to European settlement? Or is it the dynamic, free play of natural processes such as fire, avalanche, wind and weather, and whatever landscape conditions these phenomena produce—knowing that left to natural forces, the landscape will remain within its historic range of variability (HRV), but will appear different from time to time. We favor the latter definition, believing the goal of wilderness management is to facilitate the free play of natural forces. It might be easier to target some preconceived landscapes, and then do what is necessary to produce them. But natural processes are dynamic and unpredictable—the essence of what we are trying to preserve in wilderness. Identifying the natural conditions present in each wilderness, and something about its HRV, so they can be used as a goal for management is the first step in the management of any wilderness (Aplet & Keeton, 1999; Hendee & Dawson, 2002).

The distinction between maintaining natural processes and managing for a particular condition is explicit in the phrase "wilderness managers are guardians and not gardeners." The management challenge lies in preserving wilderness as areas where the free play of natural forces dictates what the area looks like. This challenge is becoming more difficult to meet because the free play of natural forces is constrained by so many factors external to wilderness (e.g., climate change, fire control, etc.) and there is public misunderstanding and confusion about naturalness.

A popular but outmoded idea is embodied in the phrase *the balance of nature*. This suggests there is an equilibrium or static condition toward which, if undisturbed, nature will move. This mythic balance simply does not exist. Any "balance of nature" occurs only over short and constrained periods; the constant in natural systems is *change*. This fact is

fundamental to establishing realistic goals for park and wilderness management (Johnson & Agee, 1988).

Maintaining Naturalness   Contrasting philosophies, called *anthropocentric* and *biocentric,* with perhaps a third philosophy that could be called *ecocentric* (existing somewhere between the other two), illustrate the debate over the retention of naturalness in wilderness recreation and management (Hendee & Dawson, 2002). Under an *anthropocentric philosophy,* wilderness is viewed primarily from a sociological or human-use perspective that sees the naturalness of wilderness as less important than accommodating human use. Programs to alter the physical and biological environment to produce attractive settings would be encouraged, such as selecting for big trees and open vistas, making access easy, or stocking fish and attracting wildlife to viewing places.

The core of a *biocentric philosophy* is to allow natural ecological processes to operate as freely as possible. Adherents of this philosophy feel that, to the extent that naturalness is distorted, the experiential, spiritual, and scientific values of wilderness are lessened. Thus, a biocentric philosophy seems essential because wilderness values depend on naturalness and solitude. Following this line of reasoning, wilderness management, even with respect to recreation use, must focus on maintaining naturalness.

The ecocentric position seeks to mediate between the anthropocentric and biocentric views in an attempt to avoid doctrinaire or purist positions, and suggests that humans, too, are part of nature and, within limits, can be accommodated in wilderness management. However, since maintaining naturalness is mandated in the Wilderness Act (along with primitive and unconfined recreation), and many feel that modern humans and their artifacts are outside this ancient sphere, a biocentric approach seems most appropriate for wilderness. But where do your instincts lie? If interested, you can test your personal philosophy in a quiz available in Chapter 1 of Hendee and Dawson (2002) or in Clark and Kozacek (1997).

Exclusion of Fire and the Naturalness Issue   Of all the external factors affecting naturalness of areas set aside as wilderness, the exclusion of fire may be the most important. As noted in Chapter 8, "Fire

Management," our nation has waged a remarkably successful battle against wildfire. However, the success of the campaign against wildfire has now led to a new kind of problem in wildlands, including wilderness—unnaturally heavy fuel loadings due to the exclusion of fire.

The symptoms and cycles of declining forest health—insects, disease, declining tree growth, fuel buildup, and risk of conflagrations—can be offset somewhat in intensively managed forests, with silvicultural treatments to manage stand density and composition (see Chapter 5, "Silviculture"). Such treatments can include prescribed burning, precommercial and commercial thinning, spraying for brush control and insects, mechanical disturbance, and commercial timber harvest. But in wilderness and wildernesslike areas where there is no access, managers are constrained. Here nature must run its course, so the only option may be allowing fire to resume its accustomed role.

Merely letting natural fires burn may not provide the correction needed, however. Such fires may occur at the wrong time, in the wrong place, and during periods of high fire danger when a disastrous conflagration might result. Such a wildfire might well burn beyond the boundaries of the wilderness. So wilderness management must include prescribed fire deliberately ignited under appropriate conditions to restore the influence of fire in maintaining natural conditions and biodiversity in wilderness ecosystems.

Other issues also challenge the wilderness naturalness ideal. For example, recent studies identify the unnatural impacts of fish stocking in wilderness lakes (Knapp et al., 2001). Questions are being raised about the establishment of guzzler watering facilities in desert wilderness and whether the desert bighorn and other wildlife species they support will be at unnaturally high levels, with accompanying ecosystem impacts. While grazing is legally allowed in many wilderness areas, minimizing its unnatural impacts is an ongoing concern.

WILDERNESS AS AN INSPIRATIONAL RESOURCE

Most of us know people who have spoken with wonder and enthusiasm about a wilderness trip in which they experienced something more than would be expected from a routine outdoor recreation trip. They may have taken a trip with one of the literally hundreds of organizations that offer wilderness experience programs, such as the well-known Outward Bound or the National Outdoor Leadership School (NOLS). More likely, they visited wilderness with a small group of family or close friends, as do the vast majority of wilderness visitors. But their reports of renewal and inspiration may be just as enthusiastic. How do wilderness experiences renew and inspire people?

Wilderness experiences provide *change,* a reprieve from cultural influences and prevailing norms that control behavior and the intensity of daily life. Wilderness experiences offer the opportunity for attunement of oneself to the natural environment. With liberation from daily patterns and pressures, and the chance for such attunement, new perspectives and insights can evolve. Furthermore, wilderness experiences and activities can provide *metaphors* or lessons for application back home in one's daily life. These lessons may result from something as simple as feeling successful in completing the experience; from the teamwork, cooperation, or leadership required; or from observing the complex workings of nature. Such metaphors provide new ways of seeing reality and present an opportunity to reframe old ways of doing things and of thinking about oneself. Wilderness provides a back-to-basics experience that helps restore balance in our lives.

Exposure to primal influences also distinguishes the wilderness as an extraordinary place for personal growth and inspiration. In wilderness, we must pay close attention to what is going on around us and continually adapt and respond. We confront the natural world and sense its indifference to us, regardless of our social status back home. We feel relatively insignificant in the face of nature's awesome power. We must be responsible for ourselves and for each other, in ways that are immediate and direct (Fig. 12-15). The wilderness is not only indifferent to our personal plight, it is unforgiving, and we must pay the price for any mistake.

Entirely new combinations of senses, long repressed or perhaps never fully activated, awaken to help us deal with the intense demands of this environment. We see ourselves more clearly under such conditions and may be both humbled and inspired by

FIGURE 12-15
Wilderness experiences provide a variety of opportunities that are unlike those of most other recreational activities. Being responsible for each other is part of this experience. *(Photo by Alan Ewert)*

the beauty and power of the natural world. In experiencing the world in such a primal, immediate, undistorted way, for a moment we take our rightful place beside the creatures of the wild.

The required "back-to-basics" living and responsibility for self and others, which are enforced by natural consequences in the wilderness environment where you must take care of yourself and get along with others in order to be safe and comfortable, restore healthy functioning (Kaplan & Kaplan, 1989). These and other positive benefits have given rise to a substantial and growing "wilderness experience program" (WEP) industry in which paying clients on organized trips seek personal growth, insight to life's challenges, leadership skills, and for some, therapeutic adjustment (Friese et al., 1998). The WEP industry is proving especially valuable for "at-risk" youth who, regardless of economic or personal circumstances, may be having trouble in their adjustment to adult responsibilities (Russell & Hendee, 2000). Thus, wilderness is proving increasingly valuable to society in generating social benefits, in addition to its substantial economic benefits from recreation visitation and the ecological services that

it provides, such as clean air and water (Loomis & Richardson, 2001).

SUMMARY—WILDERNESS MANAGEMENT

Wilderness management is concerned with the stewardship and use of areas permanently protected in their natural, roadless condition as units of the National Wilderness Preservation System. The goal of wilderness management is maintaining an area's naturalness and solitude—its wildness—beyond a threshold that distinguishes wilderness from other kinds of protected lands.

Wildernesses are not just recreation areas. They provide opportunities for primitive and unconfined types of recreation. The personal growth and healing from wilderness have led to a growing wilderness experience program industry.

Key wilderness management principles call for protecting areas from degradation, focusing on the holistic aspects of the resource, favoring wilderness-dependent activity and using minimum tools in management. Wilderness managers are guardians, not gardeners, in that their function is to strive as much as possible to let the free play of natural forces dictate what an area will look like. Fire, either natural or prescribed, will be one of their tools.

Information about wilderness can be found many places on the Internet, such as www.wilderness.net; on Websites of the federal agencies that are increasingly establishing sites for individual wilderness areas; from The Wilderness Society at www.wilderness.org for wilderness conservation issues; and from the International Journal of Wilderness at www.IJW.org, for a variety of articles on wilderness topics of interest.

LITERATURE CITED

Aplet, Gregory H., and William S. Keeton. 1999. Application of historical range of variability concepts to the conservation of biodiversity. In R. K. Baydack, H. Campa III, and J. B Haufler, eds. Practical approaches to the conservation of biological diversity. pp. 71–86. Island Press, Covelo, Calif.

Brewer, L., and D. Berrier. 1984. Photographic Techniques for Monitoring Resource Change at Backcountry Sites. USDA Forest Service General Technical Report NE-86.

Clark, Kendall, and Susan Kozacek. 1997. How Do Your Wilderness Values Rate? *International Journal of Wilderness* 3(1):12–13.

Cole, David N. 1982. Wilderness Campsite Impacts: Effect of Amount of Use. USDA Forest Service Research Paper INT-284.

Cole, David N. 2001. "Management Dilemmas that Will Shape Wilderness in the 21st Century." *J. Forestry* 99(1):4–8.

Cordell, Ken (ed). 1999. *Outdoor Recreation in American Life: A National Assessment of Demand and Supply Trends.* Sagamore Publ., Champaign, Ill.

Friese, G. T., J. C. Hendee, and M. Kinziger. (1998). The Wilderness Experience Program Industry in the United States: Characteristics and Dynamics. *J. Experiential Education,* May–June, 21(1):40–45.

Frissell, S. S. 1978. "Judging Recreation Impacts on Wilderness Campsites." *J. Forestry* 76:481–483.

Hammitt, W. E., and D. N. Cole. 1987. *Wildland Recreation: Ecology and Management.* John Wiley & Sons, New York.

Hartmann, L. A., H. K. Cordell, and H. R. Freilich. 1988. "The Changing Future of Outdoor Recreation Activities." *Trends* 25(4):19–23.

Hendee, John C., and Chad P. Dawson. 2002. *Wilderness Management: Stewardship and Protection of Resources and Values,* 3rd ed. N. Am. Press of Fulcrum Publ., Golden, Col.

Johnson, D. R., and J. K. Agee. 1988. "Introduction to Ecosystem Management," pp. 3–14, in Agee, J. K. and D. R. Johnson (eds), *Ecosystem Management for Parks and Wildernesses.* University of Washington Press, Seattle.

Kaplan, R., and S. Kaplan. 1989. *Experience of Nature.* New York. Cambridge University Press.

Knapp, Roland A., Paul Stephen Corn, and Daniel E. Schindler. 2001. Fish Stocking Impacts to Mountain Lake Ecosystems. *Ecosystems* 4(4) June, 275–278.

Knudson, D. M. 1984. *Outdoor Recreation,* 2nd ed. Macmillan Publishing Co., New York.

Landres, P. B., and S. Meyer. 2000. National Wilderness Preservation System Database: Key Attributes and Trends, 1964 through 1999. USDA Forest Service General Technical Report RMRS-GTR-18-Revised Edition, 98 pages, Rocky Mountain Research Station, Ogden, Utah.

Loomis, John B., and Robert Richardson. 2001. Economic Values of the U.S. Wilderness System: Research Evidence to Date and Questions for the Future. *International Wilderness,* 7(1):31–34.

Martin, Vance, and Alan Watson. 2002. International Wilderness. Chapter 3 in: Hendee, John C. and Chad P. Dawson, 2002. *Wilderness Management: Stewardship and Protection of Resources and Values,* 3rd ed. Fulcrum Publishing, Golden, Colo.

National Association of State Park Directors. 2000. Annual Information Exchange. http://www.Indiana.edu/~naspd/index.html.

O'Leary, J. T. 1989. Social Factors in Recreation Participation and Demand, in Watson, A. E. (comp.), *Outdoor Recreation Benchmark,* Conference Proceedings, 1988. USDA Forest Service General Technical Report SE-52.

Peterson, M. R. 1996. Wilderness by State Mandate: A Survey of State-designated Wilderness Areas. *Natural Areas Journal* 16(3):192–197.

President's Commission on Americans Outdoors. 1987. *Americans Outdoors: The Legacy, The Challenge,* Island Press, Washington, D.C.

Public Law 88-577. 1964. The Wilderness Act.

Russell, Keith R., and John C. Hendee. 2000. Outdoor Behavioral Healthcare: Definitions, Common Practice, Expected Outcomes, and a Nationwide Survey of Programs. Technical Report 26. Moscow, Id., Idaho Forest, Wildlife, and Range Expt. Station.

Trust for Public Land. 2002. Land Vote 2001: Americans Invest in Parks & Open Space. Available online at http://www.lta.org/publicpolicy/landvote2001.htm.

Zinser, Charles. 1995. *Outdoor Recreation: United States National Parks, Forests and Public Lands*. John Wiley & Sons, New York.

ADDITIONAL READINGS

Cole, David N. 1993. "Wilderness Recreation Management: We Need More Than Bandages and Toothpaste." *J. Forestry* 91(2):22–24.

Gimblett, H. Randy, M. T. Richards, and R. M. Itami. 2001. "RBS,m. Geographic Simulation of Wilderness Recreation Behavior." *J. of Forestry* 99(4):36–42.

Hendee, John C., and Alan Ewert. 1993. "Wilderness Research: Future Needs and Direction," *J. Forestry* 91(2): 18–21.

Nash, Roderick. 1967. *Wilderness and the American Mind*. Rev. Yale University Press, New Haven, Conn.

Peterson, Margaret and Dave Harmon. 1993. "Wilderness Management: The Effect of New Expectations and Technologies," *J. Forestry* 91(2):10–14.

Runte, A. 1986. *National Parks: The American Experience*. 2nd ed. University Press, Corvallis, Oreg.

Sharpe, Grant W., Charles H. Odegaard, and Wenonah F. Sharpe. 1994. *A Comprehensive Introduction to Park Management*. Sagamore Publishing, Champaign, Ill.

STUDY QUESTIONS

1. Name five federal agencies that manage outdoor recreation areas and also five that manage wilderness areas.
2. Describe the *recreation opportunity spectrum* (ROS) and how it is used.
3. What constitutes *monitoring* and what is it used for?
4. What is the difference between *site durability* and *site resiliency?*
5. Define *wilderness*. What are its important characteristics?
6. How is wilderness management different from outdoor recreation management?
7. What is the difference between *anthropocentric* and *biocentric* philosophies of natural resource management? What third possibility exists?
8. What is wrong with the "balance of nature" idea?
9. Why have some states established their own wilderness system?
10. What is the name of the wilderness area nearest you? What agency administers it?
11. What does the wilderness experience program industry do? What social benefits does it provide?

Harvesting Trees

Chapter Outline

INTRODUCTION

Harvesting in forestry usually refers to the cutting and removal of trees as a forest crop. While it is generally true that foresters do not actually harvest the crop themselves, many professional foresters actively work on site during logging operations. They are the managers who, with advice from specialists in fisheries and wildlife, soils, water, and recreation prescribe when, where, and how harvesting will be conducted and the methods to be used to meet ecological and economic criteria. Foresters also check for compliance with contracts and safety and environmental regulations.

The harvesting of trees generates more revenue and has the potential for creating more impact than any other forest activity—and yet this activity is not well understood. It is through harvesting of trees that silvicultural systems are applied, either in removing mature trees to renew the forest or in thinning the forest so the remaining trees can reach their growth potential. Furthermore, the harvesting of trees can enhance certain wildlife habitat and forest aesthetics associated with many forest uses such as recreation and residential construction in the urban/wildland interface.

Chapter 13 describes basic harvesting processes from planning, through felling and bucking, to logging systems to bring logs to a landing for loading on trucks, and then delivery to the mill. Safety and current trends are also discussed.

HARVESTING

Trees, like all other living things, eventually will die. Harvesting represents a human decision about when this will happen, and to some, this decision seems a rude and unwarranted intrusion into the natural course of events. However, it must be remembered that the regulated removal of trees creates space and light for new generations of seedlings, speeding up and in some ways improving upon nature's more leisurely pace. Removing trees before their natural demise allows forest owners and investors to utilize these renewable resources at the desired time and to optimize future growth to meet their plans. In most instances, the harvested mature trees will be replaced with young, vigorous, growing stock. This is the rationale behind the observation that "trees are a renewable resource."

It is also important to point out that, in the ordinary course of events, unmanaged forestlands do not remain static and pristine. Natural events such as fires, landslides, blowdowns, and beetle or fungus infestations effect a harvest of their own.

The loudest objections to timber harvesting are directed at clearcutting, a silvicultural method that is often prescribed for certain sites and certain species, and for economic efficiency. With clearcutting, effects, and possibly, abuses, are readily seen. Also, it will take at least three to five years before the new crop becomes visible at a distance and softens the landscape. In most locations and with most species, several decades must pass before any of the trees are mature enough to remove in a commercial thinning, and more decades before the stand reaches a predetermined *rotation age,* when mature trees are ready to harvest. The practice of clearcutting, now called into question on public lands, is being used with greater discretion even on private lands in response to public concerns and objections. Increasingly, partial cutting approaches are being used to allay these concerns, and to meet ever more stringent environmental regulations.

Silviculture systems prescribe harvesting to meet their goals. Thinnings and selection or partial cuts are also harvest operations. As discussed in Chapter 5 on silviculture, these operations allow forest managers to improve the condition of the remaining stand by lessening competition for moisture, light, nutrients, and space. Logging partial cuts is more difficult, hazardous, and expensive than clearcutting because damage to the remaining trees must be avoided. There is more danger from falling snags or trees that may be caught or deflected by other standing trees, and there will be less volume of wood removed per unit of effort. The use of smaller and specially designed equipment addresses some of these problems. The new forestry systems and ecosystem management techniques that emerged in the 1990s rely to a large extent on variations of partial cutting, and are widely applied on public lands at least. Where clearcutting is practiced, the harvest units are now smaller and designed or located so as to be less intrusive in the landscape (Fig. 13-1).

Logging Plans

Harvesting or logging a forest stand, whether by clearcutting or in a selection or thinning operation,

FIGURE 13-1
Where clearcuts are permitted, they can be designed to blend with natural patterns, textures, and contours of the landscape. *(Courtesy of USDA Forest Service)*

requires careful advance planning. The degree of planning required varies with ownership, geographical region, species, and applicable state or federal regulations.

More than one plan is usually involved, unless the timber to be cut belongs to private industry and the company carries out all aspects of the operations. In that case, the company's foresters, economists, and engineers could develop one comprehensive plan from timber sale layout to marketing of the logs or chips, but even then, in some states with forest practice regulations, the plan may have to be approved and a permit required (see Chapter 18, "Forest Management by the States"). On public lands (federal, state, county, or municipal) detailed analysis of environmental effects may be required. On most small, private ownerships, some parts of the operation may be accomplished by the owner, and others contracted out. A long-range timber-harvesting plan is usually the basis for action on all ownerships.

Typically, once a decision has been made to harvest and a plan has been prepared, a *timber-sale prospectus* describing what is to be harvested, where, and under what conditions will be presented to potential buyers of the timber. This prospectus serves as a database for the buyers preparing a bid to purchase

and remove the timber. Thus, the preparation of such a timber-sale prospectus or sale description is a prerequisite to estimating the value of the timber and the cost of removal for the forest owner, timber seller, or contract logger.

The prospectus defines the boundaries of the sale, indicates the existing roads, and prescribes the locations and standards for new roads, log landing sites, and skid trails to be built. Brief descriptions of the timber and the special operating and environmental protection regulations are also included.

Stumpage refers to the value paid to the landowner for trees to be harvested. It is calculated by taking the market price for the logs at the mill and subtracting the costs required for logging and hauling. A stumpage appraisal must be obtained based on estimates of the volume and value of the standing timber, plus logging and hauling costs. Species, size, quality of timber, and the number and volume of trees per acre are prime considerations. Obviously, dense, pure stands of mature trees on gently sloping ground, with long, clean boles showing few defects, have the highest value. Low-density stands of inferior, crooked trees on steep terrain will result in higher logging costs and, therefore, lower stumpage values.

A vital consideration in stumpage appraisal is accessibility. Difficult terrain, long skidding distances for logs, and a long haul to the mill will increase costs and thus reduce stumpage value, as will the use of logging methods that seek minimum environmental impacts. The effects of all these factors must be considered by landowners in estimating what reasonable price their timber would bring, and by loggers or log purchasers in deciding what stumpage price to offer for the timber.

The logging business is very sensitive to changes in markets, and to the economic climate in general, because of its close ties with housing demands—a key economic indicator. The sale and removal of the timber must be economically attractive. If timber prices are low because of weak housing and construction demand, high logging costs will cut further into the owner's profits. Hence, logging plans must balance economy of operation against silvicultural objectives and environmental and aesthetic requirements.

Contract Logging A small, independent logging contractor (once commonly called a *gyppo logger*) might proceed in the following way in order to secure timber for harvesting. Finding timber for sale that is suitable in age, species, and stocking, that can be cut and removed with available equipment, and that can show a profit for the logger, is the first step. Securing the contract to harvest the timber, usually by submitting the lowest or best bid, is the next step. This requires the logger to plan financing for the operation, having the cash flow to pay the bid deposit and surety bond and to keep the job going by paying wages for labor and providing fuel for equipment until payments begin for work accomplished, or until products are sold.

Logging contracts and timber sales are set up in various ways. Sometimes the logging contractor "buys the sale" and sells the logs. Sometimes the logging contract is for removal and delivery to a processing plant or mill that has purchased the timber. If the contractor buys the timber outright, he or she must contact log buyers and try to estimate prices on the species and grades expected from the harvest, in addition to the logging and hauling costs. Once the contract is secured, the contractor hopes the price will hold or increase until the logs can be delivered. When mill owners buy a sale, they sometimes must provide their own loggers and haulers to deliver the logs, and so they too need a keen eye for estimating logging and hauling costs as well as the value of timber.

Thus, a good logging contractor must be a competent manager, a good judge of timber, and have experience in cost estimation. Good loggers must have knowledge of:

- Type, quality, and size of the timber;
- Logging difficulty, as affected by topography, odd corners, long skidding distances, rock outcrops, and streams;
- Soil type and, thus, the potential for soil erosion;
- Road construction, including culverts to be installed and rock to be hauled and spread to reduce erosion and hold up under heavy logging trucks;
- Maintenance costs for existing roads;
- Fire trail construction; and
- Logging slash disposal.

The logging contractor is foremost a business person—earning a profit is the key to staying in business (Hoffman, 1991).

Environmental protection has become a significant factor. Landowners may have different short- and long-term objectives for their land, and therefore may need to work with different standards or requirements to protect the forest environment. These may require special logging methods using tractors, cable log yarding, or advanced logging systems that process trees into logs in the woods. Some logging systems require greater expense in equipment and operation. There may be requirements for logging slash disposal or for protection of streamside zones or critical habitat for wildlife. If small side streams must be crossed by skid trails, a temporary crossing may need to be constructed. There may be permits to obtain.

There may also be special felling, bucking, and log yarding requirements, particularly in partial cuts, to avoid damage to the remaining trees in the stand. If the timber sale is on public land, all of these requirements will be carefully spelled out in a contract. If the timber is on private land, the understanding between owner and logging contractor may be more informal, but a written agreement is usually signed to ensure compliance. Some states have forest practice regulations that require a harvest permit and inspection of the sale area before, during, and after harvest to see that there is compliance.

Logging contractors generally use their own equipment, so maintenance and scheduling of availability are important when the time comes to move from one job to the next. Sometimes new and improved machines must be purchased or leased to meet the requirements of a logging contract, in which case interest or rental rates and volume to be logged with that system become important.

The road system is often a major cost to be considered in appraisal and planning. Roads must be constructed in specified ways because of environmental considerations and regulations that vary by region, state, and local conditions. For example, instead of side-casting debris along the right-of-way, modern regulations demand that little or no evidence of construction remain beyond the road itself. Debris from road building on steep slopes may have to be hauled away and disposed of elsewhere. Existing roads used in the operations must be maintained.

On public lands, the agency offering timber for sale must often complete an Environmental Assessment, or a full-fledged Environmental Impact Statement, to meet the terms of the National Environmental Policy Act (NEPA), and hold public meetings on the advisability of cutting the timber and the methods to be used. In environmentally sensitive areas, timber harvests may be delayed by claims that environmental standards will be violated. These claims may result in litigation. Environmental concerns are now usually a constraint on both silvicultural and economic objectives.

Logging Camps and Woods Labor

The lumber and logging camps of yesterday were an institution truly North American in character. Typical logging camps varied from a group of crude log buildings with primitive conveniences to groups of portable houses that are still used to some extent in the Northwest, in southeast Alaska, and in British Columbia. Here, such camps are occasionally necessary for the accommodation of employees in out-of-the-way locations if commuting labor is not available. These portable camps are of a size and design adapted to transportation on flatcars, trucks and trailers, or barges.

A good crew is vital to a contractor's success. Dependable, motivated, versatile workers knit into a cooperative crew are one of the small contractor's most important assets. They must be able to comply with state regulations for safety and environmental protection and yet move fast enough to make the operation pay.

Although wages for woods labor are generally good, logging work is seasonal in some areas, and deep snow, wet weather, or spring thaws can stop everything. Mechanization has reduced labor requirements and slowdowns in the economy, resulting in reduced housing starts, have caused unemployment in all phases of the forest industry. Stricter environmental standards for aesthetics and water quality, wilderness set-asides, and requirements for protection of habitat for threatened and endangered fish and wildlife have all decreased timber supply. All of these factors combined have reduced the number of jobs in timber harvesting.

Although logging is a high-risk occupation, it has obvious attractions. A hard worker can make good wages in a job that requires strength, skill, alertness, and endurance, and is conducted out of doors in the company of other hard-working people.

Felling and Bucking

Felling and bucking refers to the cutting down (felling) of a tree and cutting (bucking) it into desired lengths. If felling and bucking are done incorrectly, the losses incurred cannot be recovered in other operations. Studies have shown that significant losses can occur in felling (12 percent), bucking (22 percent), and yarding (2 percent), so quality control in harvesting is extremely important economically (McNeel, 1992).

Unless there is to be a clearcut, the seller, forest manager and, increasingly, the logger after being trained by a forester, choose which trees will be cut and mark them for the fellers (Fig. 13-2). The feller's task is to fell a tree so that it will not lodge against another, will not break because of uneven ground, and will lie so that it can be trimmed of branches, correctly cut or bucked into logs, and skidded or dragged (yarded) to the landing with as little difficulty as possible. The way in which the tree leans, and the side bearing the heaviest part of the crown, influence which way the tree will fall. Accurate felling of the tree depends on a skillfully placed undercut, which produces a sort of hinge (Fig. 13-3). There are other considerations, such as cutting the stump close to the ground and avoiding injury to seedlings by selecting the least damaging place to

When trees are very tall, after felling the tree, buckers must saw the tree trunk into the optimum lengths for the desired end products, such as veneer, lumber, poles, pulp, or furniture stock. Depending on available equipment or the logging plan, this bucking may take place in the woods or at a log landing after whole trees are dragged (yarded) from the woods. In some modern operations, the fellers and buckers may carry hand-held computers that calculate the dimensions of logs to be bucked from trees of various size in order to get the highest value product.

One labor-saving innovation in equipment for felling trees is the use of hydraulic power shears and circular saws mounted on rubber-tired or crawler tractors, or machines that look like hydraulic excavators. Some of this equipment can now work on terrain up to a 50-percent slope. Most of the equipment can handle only tree diameters of less than 24 in (60 cm) in a single pass cut. These *feller-bunchers* shear or cut for trees at the butt and transport them vertically for a short distance, where the whole tree is piled for skidding. Some feller-bunchers can shear several small-diameter trees before having to pile them (Fig. 13-4). *Feller-directors* are smaller machines that can cut and direct the fall of a tree but cannot transport the cut tree. *Harvest processors* that can cut a tree, process it into log lengths, and sort the log lengths into short sawlogs and pulpwood logs at the stump are particularly useful where trees are smaller and the ground is relatively level, and are more widely used in the South.

When whole trees are bunched and skidded to a landing, there is usually a need for some type of mechanized or high-speed process to remove the limbs and cut the tree into log lengths. The mechanized processors, sometimes called *delimbers* and *slashers,* can effectively process trees up to 26 in (68 cm) in diameter.

Log Handing

Primary Transportation (from Stump to Landing)
Teams of oxen were the first kind of power used for skidding logs from stump to log landing in North America. They were used initially in almost every logging region, even in the redwoods of California. Logs 80 to 100 ft (24 to 30 m) long and straight as an arrow are reported to have slid along behind 25 yoke of oxen on the St. Clair River in Michigan, more than 170 years ago. Skidding tongs, resembling gigantic

FIGURE 13-2
A forester marks a tree for cutting in New York State as the landowner watches. *(Courtesy of USDA Forest Service)*

FIGURE 13-3
Felling a large ponderosa pine with a power saw. The undercut is made by removing a wedge-shaped section of the trunk in the direction of the fall. The back cut (seen here) is made on the reverse side. *(Courtesy of the Boise Cascade Corporation and American Forest Institute)*

fell the tree. As in other operations, these environmental precautions may result in higher costs but are a part of doing the job right, and increasingly are required by regulations or concerned landowners.

FIGURE 13-4
The feller-buncher shears the tree at the butt and transports it vertically a short distance where it is piled for skidding. *(Courtesy of Timber/West Publications, Inc., and J. I. Case Company)*

ice tongs, were attached to the log, and two hooks joined by chain were sometimes used to attach other logs to the first one, in tandem. Teams of oxen would then be hitched to the log or crude sleigh, known as a *go-devil,* and cries, goads, curses, and bellows accompanied the slow but steady parade. Work horses gradually took the place of oxen, and tractors and cable yarding equipment eventually replaced horses. Horse and mule skidding is still occasionally used by a few small contractors who work on fragile sites. Today, log yarding or skidding is accomplished by ground-based systems using tractors and/or forwarders, cable systems, or by aerial systems using helicopters.

Tractor Yarding Tractors, in various sizes, have been an integral part of timber harvesting since World War I, and particularly since the early 1930s. Three general types are in use—crawler tractors, rubber-tired articulated skidders, and soft-track, high-speed tractor systems. These machines are well adapted to skidding single logs or trees, "bunched" lots of logs, or whole trees (Fig. 13-5). Tractors are

also used to *cold-deck* logs, which means assembling the logs in piles for future haul to the mill. Rubber-tired skidders similar to the one shown in Fig. 13-6 are highly mobile, can travel at faster speeds than crawler tractors, and cause less soil disturbance.

Forwarding Forwarding systems use a log carrier that is loaded in the woods and towed or driven to a central log landing where the logs are reloaded on highway trucks for transport to a mill (Fig. 13-7). The advantages of forwarding systems lie in reduced costs and reduced environmental impacts from not having to build roads into the woods. Forwarding systems are particularly useful in partial cutting, where damage to remaining trees must be avoided.

Cable Yarding One of the methods used on steep terrain is direct cable yarding to a portable steel tower or spar tree (Fig. 13-8). In such systems, a cable is attached to one or more logs, and yarding is accomplished by pulling the logs in as the cable is rewound on a drum or a winch. The logs may slide flat along the ground, in a cable system called *high*

FIGURE 13-5
A track-type skidder transporting trees to a landing. *(Courtesy of Caterpillar Tractor Company)*

FIGURE 13-6
Rubber-tired skidders are mobile and adaptable to varying soils and topography. They are equipped with either cable or winch, or, as here in South Carolina, with a hydraulic grapple. *(Courtesy of Caterpillar Tractor Company)*

lead or *jammer logging.* When more than one cable is used to support the load, the lifted configuration is called *skyline logging.* The result is that the front ends of the logs are lifted off the ground while the aft ends drag. The entire load may be fully suspended when crossing intervening canyons or low spots. This system allows the suspension of the weight of the log to minimize disturbance to soil and vegetation. Tractor systems are used on slopes of less than 35 percent, and cable systems on steeper slopes.

FIGURE 13-7

The forwarder is a wagonlike carrier that can be loaded in the woods to transport logs to a road for reloading onto highway trucks. *(Photo by Vincent Carrao)*

FIGURE 13-8

A portable steel tower used in high-lead skyline systems of cable yarding in western North America. The yarder or power unit is at the base of the tower. The machine at right, a heel boom loader, loads the logs onto the trucks. Gifford Pinchot National Forest, Washington. *(Photo by Grant W. Sharpe)*

Aerial Yarding   Helicopter logging is used in areas where timber is valuable and access is difficult. The major advantage is that timber can be removed from steep hillsides or inaccessible pockets without having to build a road.

The logging crew and equipment are flown to the logging site by helicopter. A service area and a large, separate log landing area (for logs), adjacent to a road, are necessary.

Cost becomes a major factor. Aerial hauls over 1½ mi (2.4 km) are not usually practical due to high costs, depending on the type of helicopter used.

The ultimate in helicopter logging is the powerful Sikorsky Skycrane, which is capable of lifting 20,000 lb (9,000 kg) per turn (the trip back and forth between the logging site and landing). Here the pilots fly 30- to 40-minute cycles, refuel, and then switch with a fresh pilot. The maintenance crew works through the night, servicing the craft and readying it for lifting 50 to 60 truckloads of logs during its 8- to 10-hour working day. The cutting area and the landing must be relatively close in order to keep the helicopter haul time to less than 4 minutes. The huge machine costs between $4,000 and $5,000 per hour to operate (Fig. 13-9). A smaller helicopter is used to fly the *chokers* (cables to tie around the logs) to the cutting area. These are fastened to logs in preparation for the larger helicopter machine that will transport them to the log landing.

Bucking at the Landing   Whole tree skidding to the landing reduces costs, since there are fewer turns or loads of shorter logs to be hauled and logging slash is concentrated at the landing, where it can be more easily burned or chipped. But in some locations, regulations may require bucking and limbing in the woods to eliminate slash piles at the landings. Concern about whole tree logging centers on the nutrients lost for the next crop in the complete removal of limbs and tops from the forest. Leaving slash scattered in the woods also cushions the ground from the impact of tractors and logging equipment.

Loading at the Site   Loading in the past was back-breaking work, since much of it, with smaller logs at least, was done by lifting or by use of a peavey and rollway. *Parbuckling* was commonly used with larger logs. Parbuckling, or *cross-hauling,* consists of rolling logs up inclined skids from the ground to a truck or railroad flatbed by means of a cable passed

FIGURE 13-9
This Sikorsky S-64 Skycrane lowers its load at a landing on the edge of the Mount St. Helens blast zone. Mount Adams shows in the distance. Gifford Pinchot National Forest, Washington. *(Photo by Scott Shane)*

under the log and doubled back over the vehicle to a power source—animals, a winch, or a tractor. Cross-hauling may still be used in some smaller logging operations.

Stationary Loading   Most log loading today is done mechanically, with diverse methods and equipment. A relatively inexpensive stationary loader, still in use in some places, is the *gin pole,* a small tree held in a leaning position by guy wires so that the loading block is centered over the vehicle to be loaded. The A-frame is similar, but uses two poles separated at the bottom to give greater stability to the operation.

Mobile Loading   Modern logging methods include a great variety of mobile loading devices. In recent years, huge front-end loaders, long in use in warehouses, have become a frequent sight in both the loading and unloading of logs of all sizes. As shown in Fig. 13-10, forklifts operate at right angles to the vehicle being loaded or unloaded, and need a rather large landing area to operate efficiently. They are still used in unloading yards, but are not often found loading logs in the woods. The most common loading method today is the hydraulic loader mounted permanently on either the log-hauling truck or a separate carrier. The boom arm has two segments of similar length and operates much as a person's arm does, with a hydraulic grapple at the end acting like a wrist with fingers (Fig. 13-11). The mobile heel boom loader shown in Fig. 13-8 uses a cable-operated grapple. It is still used where the additional distance provided by the cable is needed to reach and grab the logs. Operation of the cable grapple requires a high degree of skill.

Unloading Logs   Unloading operations utilize most of the loading devices already described, but because the unloading area is a terminus for many loads of logs, the equipment can be considerably larger and unload the entire load of logs at once.

Pulpwood Handling

Pulpwood, which ultimately becomes paper or a wood composite product, is the leading forest product in many sections of the country. Pulpwood handling today is done much like the felling and processing of logs for lumber and board products. The difference is that pulp can be made from smaller-diameter trees cut to shorter lengths. Lengths are usually standardized within regions but vary from one region to another. Most are between 48 and 96 in (1.2 and 2.4 m), the length depending on custom and the requirements of the pulp-mill equipment.

Felling and Loading Pulpwood   There are no standardized methods for felling and loading pulpwood; operators use the method best suited to the species, terrain, and the equipment they can afford. Trees may be felled with power saws, bucked to the appropriate length, and loaded, or hauled tree length or in long logs to the pulp mill for bucking to appropriate lengths. In some areas, whole trees are skidded to the landing for delimbing, topping, and cutting into pulpwood lengths. One- and two-person machines today can fell and process trees by shearing, delimbing, cross-cutting, and stacking the sticks onto a pallet or truck, at the rate of one tree per minute or three to five cords per hour. These machines are usually limited to tree sizes less than 15 to 19 in (38 to 46 cm) in diameter and on slopes of less than 20 percent.

FIGURE 13-10
Unloading a truck by forklift in Alabama. This equipment is available in massive sizes for larger logs. *(Courtesy of Caterpillar Tractor Company)*

FIGURE 13-11
Upper: A truck in Oregon equipped with a self-loader and grapple (seen here in the travel or storage position). Note the piggyback trailer at right, which is lifted on and off with the hydraulic loader. *Lower:* The same truck with the trailer down and the hydraulic loader and grapple being operated by the truck's driver. *(Photos courtesy of United Hydraulics Company)*

Shearing trees at ground level is advantageous in that a greater percentage of the tree is used, and tree-planting or other machines can maneuver more readily in the harvested area. The whole-tree system of harvesting decreases the amount of wood handling required. As already mentioned, whole-tree harvesting raises concerns about whether or not the nutrients present in the limbs and needles or leaves are needed in the forest to help grow subsequent crops of trees.

EVOLVING HARVEST METHODS AND EQUIPMENT

Harvesting, like all elements of natural resource management, has been and still is evolving to meet new environmental and social conditions. Increasingly, the term *forest engineering* is used to refer to the harvest-related actions of road and bridge building and maintenance, harvesting and processing trees into log lengths, and transporting them from the forest to truck, rail, or water loading sites for transport to mills. In advanced logging operations, modern technology and equipment are reducing environmental impacts, and they are also replacing labor, thereby reducing accidents and the high wages and insurance expenses associated with dangerous "on the ground" work by loggers. For example, one study documented a reduction in logging injury rates from 14.2 to 5.2 per 100 workers in 11 companies after they began using feller-buncher machines. Mechanized timber harvesting and processing have evolved rapidly in recent years due to the need for more productivity and profitability, the high cost of workers' compensation insurance, and the transition to smaller trees.

The evolution of tree harvesting and processing techniques has been steady, with changes in equipment and methods documented in trade magazines and journals such as *Timber Harvesting,* published since 1953 (www.timberharvesting.com/); *Loggers World Magazine,* published since 1977 (www.loggers world.com/); and the *International Journal of Forest Engineering,* published at the University of New Brunswick since 1989 (jforeng@unb.ca).

Trees were once harvested and processed with axes and crosscut saws known as *misery whips.* Next came huge chain saws requiring "two men and a boy" to operate. Eventually came the modern chain saw used to fell, limb, top, and buck trees into log lengths, ready for a cable skidder or tractor to pull to a log landing to load on trucks. Then came machines equipped with mechanical shears used to snip trees off at ground level, which were soon surpassed by circular saws to cut trees faster and with less fiber damage. Called *feller-bunchers,* these machines can fell trees and put them in bunches just the right size for grapple skidders to take to a landing to be further processed by a de-limber or chipper. Such equipment completely mechanizes logging operations, removing

labor from the ground. But even these feller-bunchers are now being replaced by "harvesters" that cut trees and process them into log lengths all in one motion. *Forwarders* can then pick up the processed logs in the woods and piggy-back them to the roadside, thereby minimizing the need for roads and landings in the woods (Temperate Forest Foundation, 2002).

Delivery

Log Transportation to Mill or Plant   Several methods of transporting logs, by land, water, and railroad, have been used in North America, with almost every conceivable variation. Today, most logs are promptly hauled by truck to the mill or processing plant. There is often some urgency in moving logs from the woods to the mill, especially in warm spring and summer weather when insects and fungi can damage or stain the wood.

Water Transport   In the past, the presence of rivers and streams in log-producing territory helped establish the custom of floating logs to the mill. Rafting of logs—that is, forming them into large collections called *log booms,* once a common practice, is limited today mostly to tidewater areas of Oregon, Washington, Alaska, British Columbia, and the Maritime provinces. Some rafts used in coastal waters are designed to withstand heavy seas by lacing several layers of logs together with wire rope. In more protected waters, the rafts are flat, long, and narrow, and are made up in sections with the logs lying parallel to each other. As many as a dozen sections are often held end to end and towed by tugboats.

As in the early days of logging, ships and barges are still used for transporting logs to world markets. Several West Coast ports load logs, chips, and other forest products daily for export to Pacific Rim countries. Also, oak and other species are exported to Europe from East Coast ports.

Rail Transport   Railroad logging in the United States dates back to the early 1850s. During its development specially geared locomotives were designed to pull heavy loads up steep grades. They were also equipped with special joints that allowed the shaft to change length when negotiating sharp curves. Although railroad log hauling was still common in the 1930s, it was soon all but replaced by log

FIGURE 13-12
A loaded truck-trailer combination for logs up to 40 ft (12.2 m). *(Courtesy of Kenworth Truck Company)*

trucks. The advent of the crawler tractor for building roads allowed access for logging trucks into the forests. A few logging railroad spurs have persisted in some parts of the country. Since these railroads are standard gauge, cars loaded in the woods can be picked up by common-carrier railroads and transported to the mill.

Trucking   Logging trucks are now the customary means of hauling logs from forest to the mill (Figs. 13-11 and 13-12). The truck or tractor-trailer type of log transportation has greater mobility and, thus, much improved access to logging sites, as compared to rail transport. Trucks are also better suited for use in scattered stands of timber or in certain cutting systems where logs are picked up here and there.

If logs of 32 ft (9.6 m) or longer are being hauled, a truck-trailer combination is required. This truck-trailer combination allows the vehicle to work effectively on roads with tighter turns than would ordinarily be possible for trucks with 40-ft trailers. Tighter turns result in lower costs and less impact on the road and equipment. The bed of the truck supports one end of the load and the trailer the other end (Fig. 13-12). On the return trip from the mill or log dump to the log landing, the empty trailer rides "piggyback" as seen in Fig. 13-11 (upper).

Logging roads are very expensive to construct, and they also cause major scars on the landscape and erode forest soils, with further environmental consequences. The network of forest roads in the United States is huge. Most were built originally to haul logs. Motor graders maintain more miles of forest roads in the United States than exist in the entire interstate highway system (Temperate Forest Foundation, 2002). Thus, there has been research on methods of road construction and maintenance, and experiments with modification of equipment to minimize wear and tear on both roads and equipment. For example, fill slopes may be covered with straw, erosion mats, or mulch to reduce erosion. Road surfaces may be rocked, outsloped, or ditched, depending on moisture and soil conditions. Low tire pressure may be used to reduce wear and tear on both roads and equipment, thus reducing erosion and maintenance costs (Burroughs & King, 1989).

Transporting Pulpwood   A variety of skidding methods is used to transport pulp logs to a central loading point, including the use of horses, mules, and small or large tractors. Pallets are used with truck hauling in some areas. The pallet has a sledlike bottom and may be hauled about the woods with a tractor. Loading is made easier because the pallet sits

FIGURE 13-13
A loaded pallet being pulled aboard a truck in Maine. *(Courtesy of Scott Paper Company and American Forest Institute)*

FIGURE 13-14
The total chip harvester that converts entire trees into chips. *(Courtesy of Morbank Industries, Inc.)*

flush with the ground. The truck reloads with power loaders (Fig. 13-13). While one truck is on its way to the mill with loaded pallets, others are being loaded in the woods. Pulpwood is also hauled by railroad and barge.

Pulp mills can use chips produced from sawmill waste and logging residues. Some chips now come from whole-tree harvesting operations, where everything but the roots goes into chipmaking. Besides the environmental concerns, limitation on the amount of whole chips used arises from the amount of bark a pulp mill can tolerate in its pulping process. When the mill is making fine white paper, the percentage of bark allowed will be very low.

Figs. 13-14 and 13-15 illustrate in-woods chipping. In areas where energy costs are high, these chips can also be burned to make heat, steam, and then electricity. In situations with relatively long hauls where water transport is possible, pulp chips are moved by tug and barge (Fig. 13-16).

FIGURE 13-15
The concept of whole-tree harvesting. (1) The feller-buncher shears the small standing trees, bunches them, and (2) stacks them into appropriately sized piles. The skidder (3) picks up the pile and transports it to the landing (4), where the chipper's (5) hydraulic grapple arm picks up a tree and places it against a self-feeding, rotating disk that converts the tree to chips. The chips come out the other end and are blown into a van (6) for transport. If the chips are destined for a pulp mill, they are screened to remove unusable material. *(Courtesy of Morbank Industries, Inc.)*

FIGURE 13-16
Pulp chips being transported by tug and barge from British Columbia to a pulp mill in Washington State. *(Photo by Grant W. Sharpe)*

SAFETY

The forest products industry is dangerous and has the highest accident fatality rate of any occupation. It is a business concerned with efficient production rates of felling and bucking, yarding, loading, and hauling large, heavy objects such as trees and logs, and this requires the use of heavy equipment. The risk of accidents and injuries to workers is ever present. Logging is the most dangerous part of the forest industry. In spite of tremendous advances in timber harvest technology, the logging industry has seen little improvement in injury statistics (Myers & Fosbroke, 1994). In 1993, 147 logging industry workers died as a result of work-related injuries, with another 13,800 nonfatal injuries, including 5,875 that resulted in time away from work (reported in Fosbroke & Myers, 1996). In 1991, the rate of work-related injuries in the U.S. logging industry was twice the rate for all private industries (Egan, 1996), and from 1980 to 1988 averaged 161 deaths per 100,000 full-time workers compared to 7 deaths per 100,000 in all private sector industry (Myers & Fosbroke, 1994). Fifty-three percent of logging injuries take place at the stump (felling, limbing, topping), with falling objects (limbs, tops, snags, and trees) accounting for most of the deaths (Egan, 1996).

Logging Safety Training

Since logging is the most dangerous part of the forest industry, it has been targeted by the U.S. Department of Labor, Occupational Safety & Health Administration (OSHA) as being an area for specialized entry-level training to reduce the rate of accidents. Program materials, both video and written, have been developed to aid in carrying out this task. Although many major logging companies conduct safety training on a regular basis, smaller companies may not have easy access to training materials. Thus, there is a strong industrywide effort to produce effective training tools for beginning and established loggers. Logging safety is increasingly regarded as an important link in forest management education (Fosbroke & Myers, 1996). Many forestry schools have outstanding logging safety programs that include short courses and workshops. Logger accreditation programs are available in some states, such as Tennessee, with safety a key topic as well as forest management, silviculture, and best management practices (Clatterbuck & Hopper, 1996).

Education and training that aim to reduce injury and fatality rates represent a potential cost-saving edge for the logging industry. For every dollar of direct cost incurred by injury to a worker (medical, workers' compensation, and property damage), there are indirect costs (lost production, lost time, retraining, increased insurance rates, etc.) in addition to the human suffering. In 1996, there were an estimated 36,894 logging firms in 41 states, with 195,650 total employees, so the trend to reduce their risk of injury or death and reduce costs through safety education can pay big dividends (Lewis, 1996).

Safety and job training are also conducted by the industry in conjunction with insurance carriers. Many areas require job training certification as a condition of employment. Lower insurance premiums have been an incentive for industry to become involved. Many states now have insurance premium rates that are more than 40 percent of payroll costs, and a few are approaching 100 percent. Another source reports workers' compensation rates in many states may exceed $50 per $100 of wages (Cubbage et al., 1993).

TRENDS

There has been a change from emphasis on large-tree harvesting to techniques and equipment for small-tree harvesting, since timber sales with large trees are less common and wood is becoming more valuable. In response to these conditions, new technology has been developed, harvesting of smaller trees has become economically feasible, and changes in the economics of log hauling, log yarding, and skidding, as well as new methods of planning, design, and managing forests have evolved.

There have also been changes in the traditional uses of timber. *Dimension lumber* (trees sawed into boards) is still produced, but composite wood products made from wood chips and fiber make up an ever-growing share of the market, and in some places wood is harvested to produce heat and energy. This explains the increasing use in forestry of such terms as *biomass management* and *biomass harvesting techniques*. (In this context, *biomass* refers to the total volume of woody plant material on a given site.) Until these new uses arose, there was no market for the very small pieces, and certain species would not ordinarily be logged. Only selected larger pieces were handled, and equipment was adapted to these sizes. Now, with a wider variety of sizes and a larger proportion of small pieces being used, loggers and manufacturers are changing to equipment and systems that can efficiently handle these smaller dimensions.

Biomass Removal

Because whole-tree harvesting and utilization are relatively new, their impact on forest ecosystems is still being assessed. Research is expanding in response to growing concerns that the removal of branches, foliage, and all biomass, as well as stems, may affect long-term forest productivity through a gradual depletion of essential nutrients, but the extent to which concern is warranted has not yet been quantified (Gessel et al., 1990; Harvey & Neuenschwander, 1991; Perry et al., 1989). At present, there are no standard guidelines for dealing with this issue.

Protecting Water Quality

The yarding of logs and the location, construction, and use of associated skid trails, as well as temporary and permanent roads, can have a direct impact on water quality. Water crossings by any road, whether utilizing fords, culverts, or bridges, is a critical concern. Roads create more pollution in the form of sediment than any other harvest activity, and stream crossings are the most important source of such impacts (Blinn et al., 1998; Taylor et al., 1999). Fords are the least expensive method for stream crossings but create the greatest sediment impacts. Culverts are more expensive to install and maintain than fords but their water quality impacts are less. Bridges are generally the most expensive stream crossings but have the least impact, although substantial impacts can occur during installation of permanent bridges. Portable, temporary bridges are gaining popularity because they can be installed with minimal site disturbance and subsequent water quality impacts (Mason, 1990).

The expert advice on protecting streams and wetlands in logging operations is to avoid crossings. If this is not feasible, there are several alternative approaches to water crossings to be considered, along with evolving best management practices for other phases of the logging operation. Water quality standards are becoming more stringent and meeting them will be a continuing challenge to timber harvesting everywhere.

FIGURE 13-17
A logger contemplates his next move. In the future, fellers and buckers may need training in silvicultural concepts in order to make decisions about which trees to cut and what lengths to buck them. *(Courtesy of USDA Forest Service)*

Silvicultural and Business Decisions by Loggers

Forest management, like all other businesses, is very competitive and is evolving toward the goal of more efficient operations conducted with a reduced labor force. Giving loggers more responsibility for silvicultural and tree-processing decisions on the ground is consistent with this trend (Fig. 13-17). Whereas in earlier times a forest manager might have marked all the trees for cutting, laid out skid trails, and provided detailed oversight, more responsibility may now be given to the logger. The brains as well as the brawn of workers must be enlisted at production levels in order to stay competitive. Thus, loggers may increasingly be more involved in silvicultural decisions as to what trees to cut, as well as in the techniques of removing them from the forest.

SUMMARY

Harvesting the trees and transporting the logs and other usable parts to their destinations at cold decks, mills, or processing plants are integral parts of the forestry enterprise.

As timber supplies change, and environmental standards and costs rise, new machines and methods are being developed to more efficiently cut and transport wood from logging sites to mills. Logging methods and equipment were once designed to harvest large trees for sawmilling into boards. Now small trees are harvested for sawmilling and/or chipping or grinding into pulp for paper or fiber used in composite wood products such as chip and fiberboard. Although logging equipment and methods are changing to adapt to these new conditions, logging is still rough, dangerous work, with physical and economic risks.

Forest managers are usually not directly involved in cutting and hauling logs, but they are the designers, planners, and regulators of these activities, and are concerned with the processes involved and the effects of the various methods on the forest ecosystems. Impacts of log yarding and its associated skid trails and temporary and permanent roads on water quality is a very important issue. The challenge to forest managers to plan environmentally sound harvesting operations has never been greater. The evolving ecosystem management approaches will provide even greater challenges to timber harvesting in the future, expanding the role of loggers in decisions that directly affect the forest environment.

LITERATURE CITED

Blinn, C. R., R. Dahlman, L. Hislop, and M. A. Thompson. 1998. Temporary Stream and Wetland Crossing Options for Forest Management. USDA Forest Service General Tech. Report NC-202, North Central Forest Research Station.

Burroughs, Edward R., and John G. King. 1989. Reduction of Soil Erosion on Forest Roads. USDA Forest Service General Tech. Report INT-264.

Clatterbuck, Wayne K., and George M. Hopper. 1996. Partners in Success: The Tennessee Master Logger Program. *J. Forestry* 94(7):33–35.

Cubbage, Frederick W., Jay O'Laughlin, and Charles S. Bullock III. 1993. *Forest Resource Policy.* John Wiley & Sons, New York.

Egan, Andrew F. 1996. Hazards in the Logging Woods. *J. Forestry* 94(7):16–20.

Fosbroke, David Elton, and John R. Myers. 1996. Logging Safety and Forest Management Education. *J. Forestry* 94(7):21–25.

Gessel, S. P., D. S. Lacate, G. F. Weetman, and R. F. Powers (eds.). 1990. Sustained Productivity of Forest Soils. Proceedings of the Seventh North American Forest Soils Conference, University of British Columbia, Vancouver, B.C., Canada.

Harvey, A. E., and L. F. Neuenschwander (compilers). 1991. Proceedings—Management and Productivity of Western Mountain Forest Soils. USDA Forest Service General Technical Report INT-280.

Hoffman, Benjamin F. 1991. "How To Improve Logging Profits," Northeastern Logging Assoc., Box 69, Old Forge, N.Y.

Lewis, Richard. 1996. Timber! The Cry of the Past: Stewardship! The Voice of the Future. *J. Forestry* 94(7): 8–11.

Mason, L. 1990. Portable Wetland and Stream Crossings. Pub. No. 9024 1203-SDTDC. USDA Forest Service, San Dimas Technology and Development Center, San Dimas, Calif.

McNeel, J. F. 1992. "Value-based Analysis of Timber Harvesting Systems Research," Branch Lines. University of British Columbia Forestry Faculty Newsletter 3(1).

Myers, John R., and D. E. Fosbroke. 1994. Logging Fatalities in the United States by Region, Cause of Death, and Other Factors—1980 through 1988. *J. Safety Research* 25(2):97–105.

Perry, D. A., R. Meurisse, B. Thomas, R. Miller, J. Boyle, J. Means, C. R. Perry, and R. F. Powers (eds.). 1989. *Maintaining the Long-term Productivity of Pacific Northwest Forest Ecosystems.* Timber Press, Portland, Oreg.

Taylor, Steven E., Robert B. Rummer, Kyung H. Yoo, Richard A. Welch, and Jason D. Thompson. 1999. What We Know—and Don't Know—about Water Quality at Stream Crossings. *J. Forestry* 97(8):12–17.

Temperate Forest Foundation. 2002. Forest Engineering eco link. (www.forestinfo.org/organization/index.htm)

ADDITIONAL READINGS

Bell, Jennifer L. 2001. Changes in Logging Injury Rates Associated with the Use of Feller-bunchers in West Virginia. Proceedings of the International Mountain Logging and 11th Pacific Northwest Skyline Symposium (http://depts.Washington.edu/sky2001).

Conway, Steve. 1982. *Logging Practices.* Miller Freeman Publications, San Francisco, Calif.

Dean, William, Terri Richards, and Steven Wilcox. 1984. *Terms of the Trade: Vocabulary of the Forest Industry.* Random Lengths, Eugene, Oreg.

IUFRO. 1985. Proceedings: Symposium on Mountain Logging and Symposium on Skyline Logging. G. V. Wellburn and G. G. Young (eds.), FERIC, Vancouver, B.C., Canada.

Pacific Logging Congress. 1984. *Loggers Handbook,* Volume XLIV (including index to articles in Volume I through XLIII, 1941–1983). Pacific Logging Congress. San Francisco, Calif.

Stenzel, G., T. A. Wallbridge, Jr., and J. K. Pearce. 1985. *Logging and Pulpwood Production,* 2nd ed. John Wiley & Sons, New York.

STUDY QUESTIONS

1. Explain why there is resistance to timber harvesting and how forest managers are responding.
2. How can the harvesting process increase the yield of wood fiber from the forest?
3. What issues must be considered in a timber harvesting plan?
4. Explain what is meant by the following terms:
   a. *stumpage*
   b. *gyppo logger*
   c. *whole-tree harvesting*
   d. *fellers and buckers*
   e. *forwarder*
   f. *biomass*
5. What are some reasons that logging jobs are disappearing?
6. What are some advantages and disadvantages of helicopter logging?
7. Explain the difference between *cable* and *tractor yarding*. Where is each method typically used?
8. Why are smaller trees now being harvested and how is this changing logging?
9. What are the pros and cons of whole-tree harvesting?
10. Why is logging such a hazardous profession? What are some changes that are making it safer?

Forest Products

The substantial assistance in revising this chapter by Dr. Tom Gorman, Professor and Head, Department of Forest Products, University of Idaho, is gratefully acknowledged.

Chapter Outline

INTRODUCTION

Chapter 14 describes the importance and complexity of wood and wood products; their structure, properties, and processing; as well as the end uses of logs, lumber, and plywood. Many engineered and wood fiber commodities are discussed, miscellaneous products are briefly surveyed, and the increasing trend in the industry to utilize waste residues and to recycle is noted.

The wood products industry comprises one of the largest in the nation, and the demand for wood fiber as its raw material provides the principal source of revenue for managing forests (Table 14-1). The U.S. forest products industries employ more than 1.5 million people and annually ship products valued at more than $230 billion, ranking these industries among the largest industrial complexes in the United States.

Without timber harvesting, and the use of revenues derived from harvesting, the scope of forest management would be much reduced. Thus, forest managers must understand wood products manufacturing, both as an integral end product of their forests and as a source of funds to manage for other uses.

Wood constitutes an important part of the lives of the people worldwide, but especially in the United States, where nearly 80 percent of the materials used in construction of most homes is wood based. Wood also figures in the paper we use as newspapers, money, books, packaging, and countless other products we have come to expect as basic human needs.

In the United States, every man, woman, and child utilizes over 2,000 pounds of wood products per year (approximately the equivalent of one mature tree) in 5,000 different products designed for shelter, communications, and packaging. This consumption is projected to increase over the next two decades and beyond (Skog, Ince, & Haynes, 1998).

There are many concerns about overharvesting of forests and the resulting ecological implications. These concerns in part derive from the demand for wood and its many desirable qualities. Wood is superior to its competitive products in its strength-to-weight ratio, economy, versatility, and beauty (Fig. 14-1). Wood is also much superior to its competing substitutes in energy conservation and, unlike most alternative products, is renewable. For example, to manufacture and transport to site 1 ton of 8-foot 2×4 wood studs requires a net energy input of 2.91 million BTUs. If these studs were replaced by steel studs, the net energy required would be 26.67 million BTUs (oil equivalent)—nine times more than wood (Koch, 1991).

A major trend in the marketplace is the demand for "green products," and this trend is being embraced by the wood products industry. Most of the larger companies certify their raw material as harvested from forests managed for sustainability. This movement is in response to consumer demand, leading to several of the major building supply outlets going to green products. Certification is covered in Chapter 19, Forest Management on Private Lands.

TABLE 14-1

WOOD HARVEST AND UTILIZATION IN THE UNITED STATES BY MAJOR CATEGORY, 1997

Category	Volume (million ft³)	Volume (million m³)	Percent of total harvest
Sawlogs	7,120	201.6	43
Veneer logs	1,282	36.3	8
Pulpwood	5,043	142.7	31
Composite products	361	10.2	2
Fuelwood	2,283	64.6	14
Posts, poles, and pilings	175	5.0	1
Miscellaneous products	167	4.7	1
Total wood harvested	16,431	465.1	100

Source: Smith, Brad, John Vissage, David Darr, and Raymond Sheffield. 2001. *Forest Resources of the United States, 1997.* General Technical Report NC-219, USDA Forest Service, North Central Research Station, St. Paul, MN.

FIGURE 14-1
Wood imparts warmth and interest to interiors. Note the flooring, paneling, and furniture—all made from wood. *(Courtesy of Hedrich-Blessing, Chicago, Ill.)*

THE STRUCTURE AND PROPERTIES OF WOOD

Structure

The performance and adaptability of wood depends on its structure and properties. One must know how wood is constituted and how the minute elements fit together to help understand its shrinkage, swelling, strength, elasticity, its reaction to paints and stains, and other properties that are important to end users.

Wood consists of a multitude of minute cellular units that differ from each other in shape, size of opening, thickness of walls, contents, and arrangement. They are more or less firmly grown together, and most of them are slender and drawn out to points on the ends. This type of cell (called a *tracheid*), when isolated and examined, may be as long as ⅓ inch (8 mm) in some species such as redwood or only 1/25 inch (1 mm) in many hardwoods. Cells of this sort in hardwoods are generally referred to as *fibers.* In the softwoods, they conduct water longitudinally in the tree. Other cells of greater diameter, called *vessels,* render this kind of water transport in hardwoods. Water or sap moves in a radial direction in both softwoods and hardwoods by cells that comprise the wood rays. These rays can be seen in a stump cross-section of oak trees as tiny, spokelike marks, and they give oak its well-known "silver grain" when the processing cut parallels the rays or intersects them at a sharp angle.

No two wood species have identical structure. Coniferous trees such as firs, pines, cedars, spruces, and other needle-bearing softwoods have relatively few kinds of cells, whereas the cells of hardwoods, such as oaks, hickories, beeches, and maples, are diverse. Different cells have varying functions, such as conducting water, storing nutrients, and reinforcing structure. The collections of thin-walled, open cells are formed in spring (springwood); the thick-walled ones later in summer (summerwood). The parallel positions of these collections next to each other give the annual- or growth-ring pattern found on stumps and the ends of logs.

Properties

Workability   Sharp tools used skillfully can do wonders with almost any wood. However, wood that is easily shaped and fitted and is still reasonably strong and serviceable is said to be "easily worked." White pine and a number of other softwood species have long been a favorite among carpenters for this reason. Among hardwoods, yellow-poplar exhibits the same qualities. These easily worked woods are used for such things as moldings, trim window sash, and carvings.

Strength   Individual wood species have strength and stiffness characteristics that are frequently the basis for their selection for specific products. For example, ash and hickory excel in toughness or impact bending strength, and for that reason are highly desired for baseball bats, hockey sticks, and other sporting equipment, as well as for machinery parts, hammer and axe handles, and other hand implements.

Combined Strength and Light Weight   The strength of wood per unit weight is the highest of commonly available raw materials. This strength–weight ratio accounts for its wide usage in structures, such as houses and buildings. Aircraft, certain sporting goods, and other products demand a wood that is especially strong and light in weight.

Port-Orford-cedar, and several of the spruces, exhibit a combination of these characteristics.

Hardness   Where abrasion from the walking of people or animals or the moving of heavy objects is likely to wear away floors, pillars, docks, or steps, hardness is imperative. Higher-density species, such as sugar maple and the red and white oaks, possess this property. Wooden cutting boards and counter tops are valued in modern kitchens for their hardness— their ability to take punishment from sharp knives— as well as for their aesthetic qualities.

Resistance to Decay   Natural oils and other materials in the heartwood of redwood, the cedars, bald cypress, locust trees, and other species enable them to resist attack by decay fungi. The heartwood of these species is particularly decay resistant and can be used in contact with soil except in the humid South and tropical climates. Osage-orange, white oaks, red mulberry, catalpa, and black locust are examples of hardwood species with resistant heartwood. Osage-orange and black locust will last many years as fence posts, even in the humid South. No species has resistant sapwood, but virtually all can be treated with preservatives that provide as good or better protection from decay than the naturally resistant heartwood.

Permeability   Woods possessing the type of cellular structure that permits chemicals in solution to enter under reasonable pressure can adapt to treatment to protect the wood from attack by fire, insect, and fungi. The sapwood of red oaks and the southern pines exhibits this characteristic.

Heat and Sound Insulation   We frequently desire protection from heat or cold and extraneous sound. Solid wood or reconstituted wood in engineered products or insulation serves this purpose well. The low density and highly insulative qualities of wood-based fiber insulation boards and cellulose fiber insulation, for example, help maintain houses at desired temperatures and reduce winter heating and summer cooling. Wood serves as an insulator for hot dishes and for handles on kitchenware.

Response to Sound Vibration   The resonance sought in musical instruments comes from the use of wood in sounding boards, woodwinds, and the bodies of large and small stringed instruments. The unique combination of physical and anatomical properties that gives certain woods (such as Sitka spruce) good tone and resonance has defied complete analysis, although high stiffness-to-weight ratio is involved.

Odor   Some woods have a pleasant smell. For example, eastern redcedar adds a delightful fragrance along with its moth-deterrent properties.

Color and Grain   The greatest proof of the "warmth" and beauty of color and grain in wood, aside from the testimony of our own eyes, is the frequent attempt at imitation in metal and other materials. The mellowness and richness of oak, cherry, walnut, and redcedar are particularly evident in antique furniture. Grain, in the sense of figure revealed by various ways of cutting through or across the annual rings, also imparts special beauty and interest to various woods, and is often accentuated by stains, oils, and polishes. *Texture*—a term indicating size and proportional amounts of woody elements—describes wood's appeal to the senses, a quality that no other building material can even approach.

Strength Under Weight and Shock   The ability of wood to withstand various loading conditions without sustaining damage makes it adaptable for many uses, from mallets to bridge timbers or railroad cross ties. The high "fatigue" resistance of wood makes it a desirable material in uses where stresses are repeatedly applied.

Plasticity and Resiliency   In bending wood to permanent forms, such as in chairs and sporting equipment, a considerable volume of wood must be bent or permanently compressed. Beech is a good example of a wood that has this valuable property, called *plasticity*. This plasticity can be enhanced when wood elements are subjected to steam for a period of time. This differs from *resiliency*, which means the ability to spring back to the original shape after deformation.

Chemical Characteristics   Slow combustibility, and resistance to damage by acids and alkalies, constitute two important chemical characteristics of wood. Resistance to volatile and water-soluble chem-

icals and other extractives adds to the useful properties of certain woods. Because of these properties, we have wooden tanks and vats at chemical works, and wood-extracted dyes and tannins for textile and leather industries.

Tyloses are nonsoluble materials that block vessels and prevent bourbon or other liquids from leaking through wood. Thus, the bourbon industry typically uses white oak in aging barrels to take advantage of *tyloses.* Wood also yields food additives, drugs, and adhesives. For example, *taxol,* extracted from the Pacific yew, has become important in the treatment of certain types of cancer. The flavoring *vanillan* can be synthesized from wood-pulping waste liquors.

WOOD PRESERVATION

Rapid decay is a problem in some wood species, reducing their market potential. Preservative treatment may give such woods odors or a soiled appearance, but does make them marketable, giving the forest manager a market for otherwise unwanted species, as well as for small-dimension material from improvement cuttings and thinnings. For example, lodgepole pine fence posts may rot out in four to six years if untreated, but properly dried and treated with preservatives may last 15 years. Furthermore, treatment with preservatives makes it possible for wood to compete with substitutes such as metals and concrete even for such uses as building foundations. Metal and concrete are handicapped by weight, cost, and lack of easy workability, but usually have the advantage of relatively long service. Treatment with preservatives increases the market for wood by making it long-lasting, which reduces building costs and extends wood supply (Preston, 1993).

Drying

In sawmills, now the destination for approximately 43 percent of all wood harvested in the United States, there is a heavy investment in lumber-drying kilns, or else tremendous space for stacking lumber to be air dried. Often both methods of drying are used. Undried or green lumber may contain from 30 to 250 percent water, based on oven-dry weight. Such unseasoned lumber is most commonly used in noncritical applications such as pallets, where checking, cracking, shrinking, and warping can be tolerated.

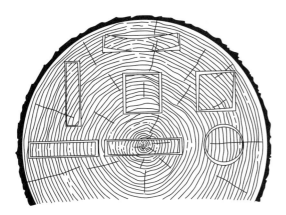

FIGURE 14-2
Characteristic shrinkage of sections from different parts of a log. *(Courtesy of USDA Forest Products Laboratory, Madison, Wis.)*

Free water, held in cell cavities of the wood, is driven off first in the drying process. When free water is gone, the *fiber saturation point* is reached. In other words, wood that holds only the water in its cell walls is at fiber saturation point. This is a good reference point; below it shrinkage occurs, and above it there is no change in size with loss of water. Woods vary in their tendency to shrink, both in total amount and amount per direction. For example, few woods shrink appreciably along the length of a board, but can shrink by as much as 10 to 12 percent across the grain. Figure 14-2 indicates that shrinkage varies with respect to the arrangement, within the piece, of the annual ring pattern.

Wood must contain some moisture, no matter what purpose it is intended for, allowing it to maintain a moisture equilibrium with the surrounding air and be less affected by normal changes in atmospheric moisture. An average moisture content of 8 percent is recommended for interior-finishing woodwork in most parts of the United States.

Air Drying Versus Kiln Drying   Air drying takes a long time and may not dry the wood sufficiently for the intended use. Kiln drying can cut drying time with safety, and can also control the moisture content to levels appropriate for the use of the wood. Shipping and handling costs are reduced in both cases. Both methods reduce the prospect of shrinking, warping, and checking in use, but kiln drying de-

stroys fungi and insects. Careful drying by either method increases the strength of the timber and prepares it for paint and other protective coatings and impregnating materials. Kilns are costly to install and require skill to operate, but they are time savers and make it possible to dry wood to specific moisture contents for specific purposes.

Chemical Preservatives

Wood is commonly kept dry during its use, thus assuring long service. Wood-destroying organisms such as wood fungi require heat, moisture, oxygen, and the presence of a suitable food supply—the wood itself. If wood cannot be kept dry, it can be successfully treated with an appropriate chemical to protect it from fungi, and also from termites, carpenter ants, and shipworms or marine borers. With the problems of decay and attack by fungi, insects, and marine organisms overcome, wood can successfully be used in contact with soil, water, and weather.

The most common method of treating timber involves impregnating the outer portions with coal-tar creosote or pentachlorophenol or solutions of inorganic chemicals. A huge industry has been built up in wood preservation, and many firms purchase or produce wood products for treatment, then sell them as treated stock.

These treatments range from brushing the wood with a preservative to immersion in hot baths of creosote or other preservative for absorption. Pressure treatment involves placing the entire stock into a long cylinder or tank where a vacuum is created, the preservative introduced, and pressure applied (Fig. 14-3). Approximately 40 percent of all southern pine lumber is pressure treated, making it suitable for decks, fences, playground equipment, and other external uses.

Due to severe toxicity, government registration is required when using preservative chemicals, in order to ensure safe application and environmentally acceptable processing technology. While chromated copper arsenate (CCA) has been the leading wood preservative in the United States over the past 70 years, concern over potential arsenic exposure resulted in its withdrawal from most consumer applications in 2002. CCA is still used in applications such as highway guardrail posts, agricultural posts and poles, and some engineered wood products, but several alternative organic preservatives are now used

FIGURE 14-3
A load of wood about to enter a long pressure cylinder for chemical treatment. Such treatment inhibits attack from fungi, insects, and marine borers. *(Courtesy of USDA Forest Products Laboratory, Madison, Wis.)*

for applications such as decks, playground equipment, and outdoor furniture. Impregnation of lumber and structural timber with ammonium and other salt solutions produces fire resistance through release of nonflammable gases when heated.

MAJOR USES—LUMBER AND LOG PRODUCTS

Lumber

Although it is a centuries-old process, the sawing of logs to produce lumber remains the most common method of producing primary wood products. It is a relatively simple process; a saw is passed through a log to cut boards and timbers. However, efficient modern sawmills are technologically sophisticated operations.

A basic commercial sawmill generally consists of a debarker, a log carriage, a head saw or *head rig,* an edger, and a trimmer. Logs are first debarked to remove sand and grit. Next they are passed by the head rig on a carriage or some other feed device and lumber is sawn in the process (Fig. 14-4). The lumber is edged for width and transferred to a trim saw and

trimmed for length. In the edging and trimming process, defects are also removed. In the days of cheap logs and low wages, most operations were labor intensive. Saws with a wide cut, or *kerf,* and other inefficient equipment were often used. Today modern equipment is generally operated by a single operator using buttons or levers to control the saws. Saws now have narrower kerfs and are also far more energy efficient.

Many high-recovery mills have head rigs that are controlled or assisted by computers. Log scanners measure for diameter, length, and form, feeding the data to a computer that determines the best ways to saw the log for maximum yield and value.

Lumber scanners and computers are also used to make decisions on edging and trimming lumber to maximize recovery. Defect detectors can find knots and splits; then computers are used to decide how best to cut up the board for maximum utilization. Technologies such as laser cutting and increased automation are improving lumber-recovery efficiency to allow more of a given tree to be utilized, thereby increasing yields from forest management.

Sawn lumber is used for a wide variety of products, with the greatest volume being consumed for structural use—such as for houses and other light construction. The product is known as *structural dimension lumber* and is manufactured primarily from softwoods. Heavy timbers may be sawn for special uses such as construction of bridges. Other sawn lumber is used for remanufacture (secondary processing). Both hardwoods and softwoods are used in secondary processing to make furniture, molding and millwork, paneling, boxes, pallets, and other materials. Remanufacture or secondary processing of sawn lumber or waste wood adds additional value to the raw material. Thus, *"value added"* or secondary manufacturing generates additional jobs and economic activity and provides opportunities for additional growth in the wood products industry and in communities where wood is processed.

Softwood Lumber Utilization   The term *softwood* does not categorically refer to the hardness of the wood but designates the coniferous, usually evergreen, trees as opposed to *hardwoods,* which are usually deciduous trees. Softwoods account for approximately 80 to 85 percent of the total lumber used in

FIGURE 14-4
These photos show modern head rigs designed to increase lumber recovery.
Upper: The log (the end is shown in the center of the photo) is about to be sawn
into lumber by this bandsaw head rig. The round log is being "squared up" by
means of a mechanical cutting head that chips the excess material for use as pulp.
(Courtesy of Potlatch Corp. Mill, Lewiston, Idaho.) Lower: Small-diameter logs are
increasingly being sawn into lumber by means of curve-sawing equipment such as
that shown here. Lumber is cut along the curve, resulting in substantially increased
yield from logs that contain sweep. *(Photo by Francis Wagner.)*

the United States (Table 14-2). Fluctuations in production are largely attributable to changes in markets. Softwood production and consumption are closely allied to the housing industry and, because of this narrow base, are vulnerable to the cycles of the housing market and associated mortgage rates and financing trends. The U.S. domestic consumption of softwood lumber is about 52 billion board feet and increased more than 15 percent from 1993 to 2003. U.S. consumption now exceeds U.S. production, since more lumber is imported from other countries, especially Canada, than is exported.

TABLE 14-2

U.S. LUMBER CONSUMPTION 1993–2003 (MILLION BOARD FEET)

Year	Softwood consumption (Million bd. ft.)					Hardwood consumption	Total consumption
	Residential construction	Non-residential construction	Repair & remodeling	Material handling & others	Total softwood consumption		
1993	17,379	6,167	14,043	7,793	45,382	10,631	56,013
1995	17,864	6,696	14,296	8,442	47,298	10,928	58,226
1997	19,158	7,541	15,148	9,014	50,861	11,103	61,964
1999	21,384	7,285	16,034	9,128	53,831	11,507	65,338
2001*	19,420	7,194	16,195	9,081	51,890	—	—
2003*	19,657	7,176	16,202	9,169	52,204	—	—

*Estimated

Table prepared by Dr. Tom Gorman from the following sources: U.S. Department of Commerce, Washington, D.C., R. E. Taylor & Associates, Vancouver, B.C., Canada.

Hardwood Lumber Utilization   Total U.S. hardwood lumber consumption remains at about 18 to 19 percent of total lumber consumption and has increased only slightly since 1993, with a total of 11.5 billion board feet consumed in 1999 (Table 14-2). Hardwood flooring is not as common, but hardwood pallets have increased and now account for half of all sawn hardwood use. Hardwoods are also very important for pulp used in paper making.

Trends   The housing and forest products industries were severely depressed early in the 1980s and early 1990s as the U.S. economy experienced a major recession. High interest rates, together with other effects of the recession, combined to reduce housing starts. But in response to low interest rates in 1993, housing starts increased, along with a trend for wood use in home additions and remodeling, which now accounts for more than half of domestic consumption. In 2002, despite recession, housing demand remains high due to low interest rates.

Other economic and industry changes have encouraged more fuel-efficient construction. Changes in structural design can decrease costs, for instance, when walls, floors, and other items are prefabricated. Wood composites provide a less-costly alternative to construction lumber and plywood for some uses.

Other Log Products

Special Grades and Species   From specially graded logs and species come cooperage stock (for barrels), mine timbers, railroad ties, fence posts, handle bolts, shake and shingle bolts, poles, beams, and piling. However, new technology, changes in consumer buying habits, and new packaging techniques have reduced demands for cooperage stock to one-fourth of that in 1950, and demand for posts and mine timbers has similarly decreased.

Whole-Log and Lumber Exports   The volume of logs and lumber, mostly softwoods, and structural panels that are exported from the United States, has stabilized over the past decade, after several years of increased growth. In 2000, the United States exported about 9.4 billion cubic meters of logs, 6.0 billion cubic meters of lumber, and 1.1 billion square feet of structural panels (Table 14-3). Over the past several decades Canada and Japan have been the most important customers for U.S. exports of timber products measured in value terms.

Hardwood log exports, though small in relation to those of softwoods, are largely made up of high-value and relatively scarce species and thus have had some important effects on domestic markets. The export of walnut logs, for example, principally to western Europe, has contributed to large increases in walnut log and stumpage prices.

The export of U.S. logs has been controversial and has centered in the Pacific Northwest where over 85 percent of the softwood log exports originate. About 70 percent of the softwood log exports from this region come from forest industry ownerships and 30 percent from lands managed by the state of Washington. Export of logs cut from timber on federal

TABLE 14-3

U.S. WOOD PRODUCTS EXPORTS

Year	Logs (1,000 m³) Softwood	Hardwood	Lumber (1,000 m³) Softwood	Hardwood	Structural Panels (million ft³—3/8 basis)	Pulp and Paper (short tons) Wood pulp	Paper & paperboard
1996	10,650	1,181	4,387	2,568	1,640	7.2	12.1
1997	9,453	1,462	4,132	2,891	2,056	7.0	13.1
1998	7,436	1,505	2,886	2,501	1,097	6.0	12.2
1999	7,592	1,693	3,224	2,792	1,036	5.9	12.4
2000	7,406	1,981	3,025	2,949	1,136	6.4	11.5

Sources: R. E. Taylor and Associates, Ltd., Vancouver, B.C., Canada; U.S. Census Bureau, Lumber Production and Mill Stocks; USDA Foreign Agricultural Service; and American Forest and Paper Association, Washington, D.C.

lands, or from Oregon and Idaho state lands, is prohibited, with minor exceptions.

Opponents of softwood log exports argue that if these logs were processed domestically, they would offer employment and lower local prices, and the reduced timber harvests would ease environmental impacts. Proponents argue that little of this volume would be processed domestically, that the export market contributes to employment and improved timber management and helps the U.S. balance of payments.

MAJOR USES—PLYWOOD

Plywood is made of veneer layers glued together, with the grain of adjacent layers at right angles to each other. Both softwoods and hardwoods are used in the manufacture of plywood. Originally most softwood plywood was made of Douglas-fir, but western hemlock, larch, white fir, ponderosa pine, redwood, southern pines, and soft hardwoods such as sweetgum and yellow poplar are also used now.

Veneer cutting operations depend on the type of plywood to be produced. Rotary peeling is the most common procedure used in the production of veneer for assembly into construction and industrial-use plywood panels. This operation utilizes a knife, equal in length to the length of the peeler log, to cut a continuous sheet of veneer of uniform thickness from a log's tangential surface (Fig. 14-5, upper).

After peeling, the veneer strip is clipped into various widths up to a maximum size of 4 ft (1.2 m). The veneer then passes through dryers of various designs, which often have multiple tiers. The veneer moves continuously on rolls or belts along each tier, with

the drying time controlled by the speed of the belts and the temperature of the drying air.

After being dried, veneer is sorted into various grades. Because veneer suitable for facing plywood is scarce and consequently valuable, every possible method is used to produce the maximum amount of this grade.

The dried material then moves to the glue spreader. Veneer sheets intended for the core layer are spread evenly with glue on one or both sides and placed at right angles to the face sheets to give the plywood its strength and dimensional stability (Fig. 14-5, lower). From here, the stack of "sandwiched" panels goes to a multiopening press, where each panel is subjected to pressure and heat. Pressing consolidates the veneer layers while curing the thermosetting glue. The resulting bond of glue is stronger than the wood itself. Panels are trimmed, and some may be sent to sanding machines, where both surfaces are sanded to impart smooth surfaces and improve thickness tolerance (uniformity). Before shipment, each panel is inspected and the grade is marked into the end.

Plywood has several advantages compared to solid wood. It has greater resistance to splitting, and its form permits many useful applications where large sheets are desirable. Cross-lamination of veneer sheets in plywood imparts a high degree of dimensional stability. Because in some applications it is permissible to use plywood, which is thinner than normally available sawed lumber, large areas can be covered using less wood fiber.

The plywood industry dates back to the early 1900s, when machinery was developed to cut veneer sheets mechanically, thus increasing production.

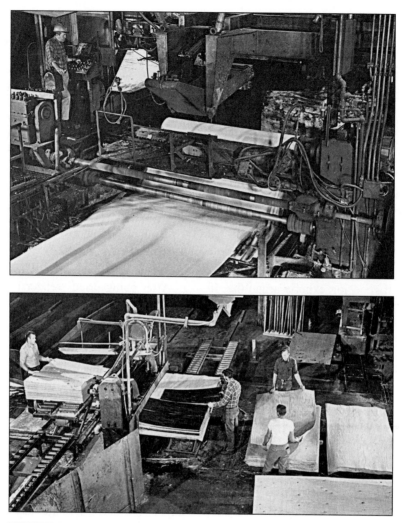

FIGURE 14-5
The manufacturing of plywood from a peeler log. *Upper:* Peeling a Douglas-fir into veneer. The operator at left controls the lathe, rotating the log against the knife, and the result is a continuous sheet of veneer. Next the veneer is clipped, dried, and sorted. *Lower:* Gluing up a plywood panel. The man at left feeds the veneer core sheets through rollers, where glue adheres to both surfaces. The man in the center catches the sheets and places them so that the grain is at right angles to the veneer sheet. The two people at right place two full sheets of veneer on top of the glued core sheets, one sheet for the panel below and one for the next panel to be assembled. The next step is the press. *(Courtesy of American Plywood Association)*

Waterproof glues, developed in the late 1920s, accelerated the use of plywood as a construction material. During and after World War II, plywood assumed an increasing role because of higher lumber prices and its acceptance as a suitable substitute for lumber in housing and light construction. Convenience and speed of handling were also important factors in plywood's use in construction, since more surface area could be covered in a given period of time, thus further reducing labor costs (Fig. 14-6).

In spite of the phenomenal increase in production over the last 50 years, forecasts now point to declining plywood production brought on largely by the introduction of oriented strand board (OSB), which

FIGURE 14-6
Laying plywood sheets for subflooring in building construction. Plywood makes it possible to cover considerable surface area in a short period of time. *(Courtesy of American Wood Council, Washington, D.C.)*

continues to take market share from plywood. OSB utilizes smaller, lower-quality logs and thus is better suited to future raw material supplies.

MAJOR USES—ENGINEERED WOOD PRODUCTS

Engineered wood products use wood fiber, strands of wood or veneer (generally cut from young, fast-growing, or small-diameter trees), or wood residues and, bonded together with adhesive and sometimes other materials, to make timbers and lumber. These engineered products are replacing those originally harvested from large, old-growth trees and greatly extend the wood supply (Balatinecz & Woodhams, 1993; Ryan, 1993; Smulski, 1997).

Glulam Timbers

The term *structural glue-laminated timber* refers to an engineered, stress-rated product often used for structural applications. *Glulam timbers,* as they are commonly known, are comprised of two or more layers of wood glued together with the grain of all layers (called laminations) approximately parallel. Laminations may consist of pieces of lumber end-jointed to form any length of lamination, or of pieces of lumber placed or glued edge to edge to make wider laminations. Lumber used in manufacturing glulam timbers

is usually 1 or 2 in (2.5 or 5.0 cm) thick and must be properly selected and prepared for gluing. Finger joints are the most common way to end-joint lumber to any length of lamination; however, scarf joints, formed by trimming the lumber ends at an angle, are also used for end-jointing. Both finger and scarf joints are designed to facilitate a degree of lateral gluing, because wood cannot effectively be end glued.

After end-jointing, the laminations are stacked, glued, and pressed to form the desired shape and to allow the adhesive to cure. The timbers are then removed from the clamping device, planed, trimmed, and finished according to customer specifications.

The use of glulam timbers offers flexibility in design, great structural strength, and decorative beauty. These timbers may be straight or curved to a shape impossible to attain with solid wood (Fig. 14-7). Curved arches have been used to span structures of more than 300 ft (90 m), while glulam domes have been built to span structures of more than 500 ft (150 m). Straight members with vertical depth of 7 ft (2 m) and spanning over 100 ft (30 m) have been used.

Particleboard

Particleboard is a panel product made of particles of wood bonded together with a binder. The particles are sprayed with an adhesive and bonded together into panels under heat and pressure.

FIGURE 14-7
Laminated beams gave a distinctive dimension to wood utilization. The curved laminated arch provides graceful lines and brings out the natural beauty of wood. *(Courtesy of USDA Forest Products Laboratory, Madison, Wis.)*

Particleboard for floor underlayment and furniture core stock is usually manufactured from wood residues (planer shavings and sawdust). Higher-quality particleboard is made with flakes, using machines with knives to convert roundwood.

The basic steps in the manufacture of particleboard involve reducing the wood particles to the desired size and configuration, drying the particles to reduce their moisture content, and screening the particles to separate those that are either too large or too small. These particles are blended with adhesive and formed into a mattress (mat) to be pressed under heat and pressure (Fig. 14-8). The boards are subsequently trimmed and cut to the desired sizes for shipment.

There are many factors that affect the properties of particleboard. These include particle sizes and shapes, the type of adhesive used, the press temperature and pressure, and the moisture content of the particles. In the manufacturing process, these factors are carefully selected to achieve a certain level of strength, edge tightness, and dimensional stability, as required by the purchaser.

Particleboard has been used as a substitute for lumber in many applications such as furniture core stock, doors, and cabinets. Although production growth continues, that growth is slowing due to competition from other members of the wood composite group, particularly medium-density fiberboard.

Waferboard and Oriented Strand Board

Waferboard began production in 1962; oriented strand board manufacture started in 1981. *Waferboard* is an engineered composite product made with wafers or strands of wood (larger flakes) cut from roundwood, randomly distributed throughout the panel, and bonded together with an exterior type resin.

Oriented strand board (OSB) is made of similar wafers with somewhat rectangular shapes, cut in the same manner as that described for waferboard. However, OSB is manufactured in three layers, with surface and bottom layer wafers aligned along the panel length. The middle or core layer can be either randomly distributed or cross-aligned in relation to the surface layers. This type of construction gives the panels a greater strength along the length of the panel and imparts a better degree of dimensional stability to the board (Fig. 14-9).

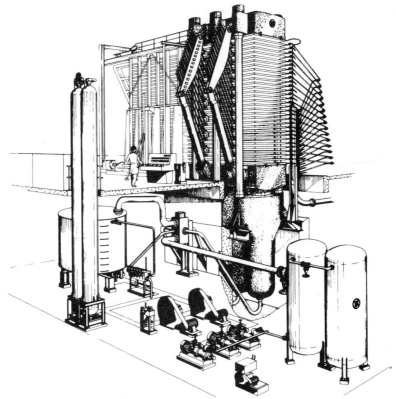

FIGURE 14–8
A typical hydraulic press used for making various types of particleboard and hardboard. This press is equipped with an automatic loader (at left of the main press) and 30 openings where the board is pressed between steam-heated hot plates. *(Drawing by A.B. Motalla Verkstad; courtesy of USDA Southern Forest Experiment Station)*

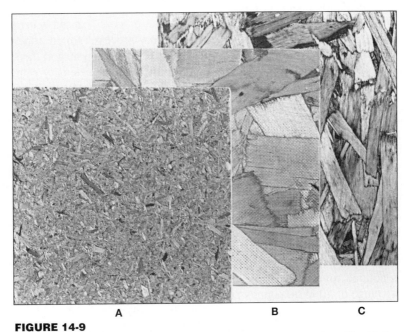

A   B   C

FIGURE 14-9
Three examples of boards made from wood particles, bonded together with a binder, and subjected to heat and pressure (see Fig. 14-8). **A:** particleboard. **B:** oriented strand board. **C:** waferboard.

The manufacturing process for waferboard and OSB is similar to that described for particleboard. However, for OSB the particle alignment requires special technology and less resin than that of particleboard. Waferboard use is gradually declining, while OSB production continues to grow due to the greater level of structural performance that can be achieved with OSB. In addition to sheathing and roof applications, OSB is now also used in house siding and furniture.

Wood Composites—Fiberboards

The term *fiberboard* applies to a family of products made with wood fibers and fiber bundles. The material has been macerated or broken up by a mechanical attrition process. The fibers are held together with an adhesive. Most are sheet products of various thicknesses and densities. Board density is especially important in classifying fiberboards. There are principally three types of fiberboards manufactured; *insulation board, medium-density fiberboard,* and *hardboard.*

Insulation Board   This is made in the density range of 10 to 31 pounds per cubic foot, with thicknesses ranging from ⅜ to ¾ inch. It represents the lowest density class among the three categories of fiberboards noted earlier. Insulation board developed as a by-product of pulp and paper manufacture, which produced large quantities of oversize fiber bundles that had to be removed from groundwood pulp. Insulation board is produced today in manufacturing plants located primarily in the South and in the northern states.

To produce boards of low density, the fiber mat in insulation board is not pressed under heat and pressure, as is the case for other types of fiberboard. Instead, the mat is dried after mild pressing to reduce its moisture content. Insulation board applications include roof insulation and as backer board for aluminum siding. It is also used for ceiling tiles and in the automotive and furniture industries.

Medium-Density Fiberboard (MDF)   This is made of wood fibers bonded together with a synthetic resin to a density range of 40 to 50 pounds per cubic foot, which is similar to that of particleboard. Advantages include tight edges, smooth surfaces, easy machinability, and a wide range of thicknesses. In addition, very little water is used in the manufac-

ture of this product, thus cutting down on pollution. The MDF process offers more product per ton of wood, a desirable density range, high internal bond strength, and board uniformity.

The manufacturing process is accomplished by converting the raw wood to pulp-size chips, followed by a digester-refining process somewhat similar to that used in the pulp and paper industry. The processed material is then dried and blended with a resin prior to forming and pressing. The press consolidates the board and cures the resin. Due to the high bulk density of the fibers, a prepress is used to reduce the thickness of the mat prior to sending it into the hot press. After pressing, boards are cooled and trimmed.

Major uses for the thicker boards (½ to ¾ inch) include core stock in furniture, and in cabinets, replacing either particleboard or lumber.

Hardboard   This development was a result of the need to use sawmill residue from southern pine mills. William H. Mason, who was experimenting with a process to extract rosin and turpentine from wood, developed a device to convert chips into fiber without the loss of lignin. In Mason's digester, which used a wet process, the chips were subjected to high steam pressure, softening the wood and fiberizing the chips. This "wet process" caused water pollution problems. In the "dry process," found in new plants, air is used in place of water to transport the fiber.

Hardboard has a density ranging from 55 to 70 pounds per cubic foot, and it is manufactured in thicknesses from ⅒ to ⁵⁄₁₆ inch. A major use for hardboard has been exterior siding, and it is also used for interior paneling, where finishes simulate natural wood grain. The automotive, construction, and furniture industries are also users of this product, but market areas for hardboard (such as house siding) are being challenged by OSB, leading to a gradual decline in its market position.

Parallam

Parallam, a lumber substitute made of long wood elements, has excellent structural properties. It was developed by MacMillan Bloedel Corporation after nearly two decades of research that sought to provide large structural lumber from veneer elements peeled from either large or small logs. Manufactured lengths up to 66 feet are available.

Wood is broken into long elements, flaws are removed, and then the elements are reconstituted. For the outer layer, currently Douglas-fir or southern pine logs are peeled into veneer. Veneer is dried to 4 percent moisture content, and defects (such as knots) are eliminated through a patented process. The veneer is then clipped into strands $\frac{1}{2}$ inch in width and up to 8 feet in length. Thickness of these wood elements is either $\frac{1}{8}$ or $\frac{1}{10}$ inch.

The veneer elements are coated with resin and oriented along the length of the member being manufactured. This assemblage of coated veneer strips is directed to a press for consolidation and curing. Microwave energy is used to rapidly generate heat for curing. Long, continuous billets emerge from the press and are cut into sizes required by the market.

Parallam has a density close to the wood from which it is made. It can be machined and cut easily. It is used for plated trusses, beams, scaffold planks, doors, windows, and vaulted ceilings.

Parallel Strand Lumber

This is also known as PSL-300 because particle strand or length is up to 12 in (300 mm). The product is an outgrowth of Parallam technology. The first manufacturing plant was located in Minnesota, using aspen wood as the raw material. The use of other light-density species in local areas seems probable. Once the round wood is reduced to thin strands, it is dried to reduce the moisture content, adhesive is blended with the strands, and the blended "furnish" is formed into a mat with the particles oriented along the length of the member. A special press using steam injection technology consolidates and cures the resin. The final billets can be up to $5\frac{1}{2}$ in thick and are cut into the sizes dictated by the market.

Applications for PSL-300 can range from millwork (doors and windows) to direct substitution for lumber in many other applications. This is new technology, and it will be some time before all the uses for this product can be identified.

Laminated Veneer Lumber

Laminated veneer lumber (LVL) is another lumber substitute composed of veneer plies bonded together parallel to the length of the member. Depending on the process and wood species used, a wide range of performance characteristics and products is possible. LVL has been used for many years by the furniture industry to produce curved furniture parts. Over the last two decades, however, increasing quantities of LVL have appeared in the market as high-quality lumber, especially in larger sizes. Major uses are beams consisting of LVL, or LVL as components of glue-laminated members.

The manufacture of LVL is similar to that of plywood, except that veneer sheets are assembled parallel to each other. Veneer is rotary-peeled, dried, spread with adhesive, assembled in the desired configuration, and pressed either in a conventional plywood press or a continuous press. Once the boards emerge from the press, they are cut to desired sizes. LVL is projected to grow rapidly, replacing solid lumber, especially in larger sizes. Further technological development is likely to lead to reduced production costs for LVL, making it possible for this product to compete directly with lumber, but this has not yet happened.

Mineral-Bonded Boards

Mineral-bonded boards combine wood particles or fibers with such minerals as portland cement or gypsum to produce a unique family of products used in construction and industrial markets. This new product is now produced in several countries around the world (Moslemi, 1993).

Cement-bonded particleboard is made by combining wood particles with cement. The resulting product is durable, can be exposed to outdoor elements, is highly fire resistant, and does not rot because of the protection given the wood by the cement binder.

The manufacture of cement-bonded particleboard is similar to that used for resin-bonded particleboard. Wood is reduced to particles of the desired size, but no drying of the particles is necessary. On the contrary, additional water needs to be added to the particles so that adequate moisture is available to the cement for hardening. The cement–wood mixture is formed into a mat, followed by cold clamping for consolidation. The specially designed press clamps 20 to 30 mats at a time. The boards are then cured at elevated temperature for eight hours, trimmed, and placed in the yard for an additional period of two weeks before they are shipped. The product can be prefinished for special applications or shipped without additional finishing. Mineral-bonded boards are used for roof tiles, sheathing, partitions, and siding. Density is in the 75–80 pounds per cubic foot range.

The use of *gypsum* in combination with wood particles and fibers is also growing. Gypsum is combined with recycled wood fibers (obtained from waste paper) to produce an improved product for wall board and other applications. The manufacturing process defiberizes the waste paper and then mixes it with gypsum. Water is added, and the material is pressed. Continuous pressing is possible, due to rapid setting of gypsum. The fireproof product has properties suitable for many structural and industrial uses.

Fiber and Chemical Products

Paper, paperboard, rayon, fiberboard, and insulating materials are all derived from wood or its constituent fibers. The three major chemical constituents are *cellulose, lignin,* and *hemicellulose.* Cellulose is generally regarded as making up the framework of the cell wall. It is colorless and insoluble in ordinary solvents. Lignin, which binds the cells together, is soluble in alkaline solutions at high temperature. Hemicellulose, which contributes to the fiber bond strength of paper, is partially soluble in hot alkalies.

Pulp and Paper   Pulp and paper products consume 68 percent of all wood fiber harvested in the United States. Pulpwood is broken down by either chemical or mechanical processes. Some pulp mills receive chips and sawdust as a by-product of sawmills. The pulpwood is washed, debarked, and chipped into small pieces by forcing the log or bolt against powerful rotating disks equipped with knives (Fig. 14-10). The chips, a considerable amount of which may come from wood-processing residues, go by conveyor to the top of cylindrical digesters. Depending on the process employed and the species of wood used, the chips are cooked in a liquor for a certain number of hours at different pressures and temperatures. There are three common chemical processes: *sulfate* or *kraft* (by far the dominant pulping process), *soda,* and *sulfite.* The first two processes use alkaline cooking liquors and the third uses an acid liquor.

After the lignin and other binding materials have been separated or dissolved, the mass of fibers goes to huge tanks for washing in large quantities of water. Now nearly pure cellulose, the pulp is washed and bleached before going to the beaters where clay, dyes, rosin, and other additives are mixed in. These give the finished product its color, reduce its absorbing qualities, and give it other surface features. Next the pulp flows evenly onto a carrier or screen where the wet sheet loses much of its water before passing over fast-moving rolls, whereby it is pressed and dried. Eventually it comes out in rolls of paper at the other end (Fig. 14-11). Some mills process the wood only through the pulp stage, leaving the papermaking and further manufacturing processes to other plants. The physical characteristics of wood pulp fibers needed in papermaking are fiber flexibility, bonding, and good tensile and compression strength.

Other processes also may be employed to reduce the wood to fibers. One is the mechanical or *groundwood process,* in which wood is forced against a huge grindstone. The rough surface of the rotating wheel tears the fiber from the wood. Water is used to cool the wheel and wash off the pulp. Next the pulp is screened to remove any larger pieces of wood before proceeding to the mill to be made into paper.

A fifth method of making pulp involves a combination of chemical cooking and mechanical action. In this process, the chips are softened with steam or chemical cooking before they are "beaten apart" in an attrition mill. This method is known as the *semichemical process.*

In the chemical process, it takes about 2 tons (1.8 t) of wood to make 1 ton (0.9 t) of paper. The yield is somewhat higher in the semichemical process.

The investment in a paper mill is substantial, ranging from $100 million to a billion dollars or more. A larger paperboard mill can produce up to 2,300 tons (2,090 t) of paper per day.

During the past 25 years, the pulp and paper industry has invested heavily in air- and water-pollution control equipment. These added expenses have resulted in the closure of inefficient plants and contributed to price increases. However, health, safety, and environmental conditions both inside and outside the plants have been improved.

Consumption of pulp and paper products is not as responsive to recessionary cycles as is lumber production. Electronic data systems, business papers, fourfold printout sheets, and other paper-consuming aspects of the information explosion give the pulp and paper industry a more stable outlook than other sectors of the forest products industry.

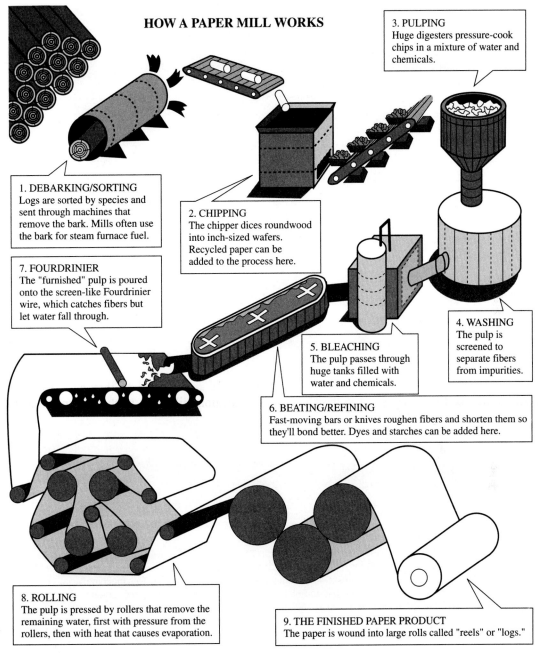

HOW A PAPER MILL WORKS

3. PULPING
Huge digesters pressure-cook chips in a mixture of water and chemicals.

1. DEBARKING/SORTING
Logs are sorted by species and sent through machines that remove the bark. Mills often use the bark for steam furnace fuel.

2. CHIPPING
The chipper dices roundwood into inch-sized wafers. Recycled paper can be added to the process here.

7. FOURDRINIER
The "furnished" pulp is poured onto the screen-like Fourdrinier wire, which catches fibers but let water fall through.

4. WASHING
The pulp is screened to separate fibers from impurities.

5. BLEACHING
The pulp passes through huge tanks filled with water and chemicals.

6. BEATING/REFINING
Fast-moving bars or knives roughen fibers and shorten them so they'll bond better. Dyes and starches can be added here.

8. ROLLING
The pulp is pressed by rollers that remove the remaining water, first with pressure from the rollers, then with heat that causes evaporation.

9. THE FINISHED PAPER PRODUCT
The paper is wound into large rolls called "reels" or "logs."

FIGURE 14-10
The nine steps in paper production. *(Courtesy of David Noonan, Peninsula Daily News, Port Angeles, Wash.)*

FIGURE 14-11
The Fourdrinier paper machine. *Upper:* Step 7 in Fig. 14-10. The wet pulp on the large screen flows to the left into the presses and driers (step 8). *Lower:* The other end of the same machine showing the finished roll of paper (step 9). This machine produces 600 to 650 tons (540 to 585 t) of heavy paper per day. The paper speeds onto the roll at the rate of 800 to 1500 ft (240 to 450 m) per minute, depending on the paper grade. *(Courtesy of Gaylord Container Corporation, Bogalusa, La.)*

Rayon   Previously known as artificial silk, rayon is now produced throughout the world from cotton linters (a short-staple cotton salvaged from the seed after ginning) and from wood pulp. In the viscose process, either wood pulp or cotton is digested by strong alkali to produce alkali cellulose. This product, treated with carbon disulfide, becomes *cellulose xanthate* and dissolves in caustic soda to make *viscose*. This is the clear, gumlike product that is spun into raw cellulose or viscose silk. There are several processes for doing this; all are highly technical, require large investments in plant facilities, and are highly energy intensive.

Other materials (aside from cloth and yarn) obtained from the viscose cellulose solution are photographic films, cellophane, sausage casings, the core layer of shatterproof glass, waterproofing, wallpaper surfacing, paints and lacquers, and viscose rubbers and leathers. From the filament forms come rayon waste and artificial wool, horsehair, and straw.

Plastics   Plastic products combine several ingredients, including dyes, plasticizers, fillers, and binders. When subjected to heat and pressure, they undergo a physical or a chemical change, or both, and may be cast, molded, extruded, or pressed into objects of definite form. Fillers are made principally from wood flour, obtained by disintegrating light-colored, low-density species such as cottonwood, basswood, spruce, balsam fir, or white pine in attrition mills. Cellulose binders are made from chemically modified cotton linters or sulfite pulp. Potentially, fillers and binders could also be made from lignin.

OTHER USES OF WOOD

Fuel-Wood

World Situation   No less than one and a half billion people in developing countries meet at least 90 percent of their energy requirements with wood and charcoal. Another billion people fulfill at least 50 percent of their energy needs this way. It has been estimated that at least half the timber cut in the world still serves in its original role—as fuel for cooking and heating. Fuel-wood shortage is a growing problem assuming a desperate dimension in some countries where wood is poached from forest reserves and land is stripped of shrubbery and other biomass. The challenge lies in the fact that woody plants are renewable resources that, if effectively managed, could alleviate the problem not only for the present, but for decades to come. But tree planting sufficient to keep pace with population growth, decreased consumption of wood for fuel and commercial purposes, reduced loss of forests to other uses such as agriculture, and better designed stoves, kilns, and boilers will be necessary. Otherwise, more millions of people will be without wood fuel for their minimum cooking and heating needs and will be forced to burn dried animal dung and agricultural crop residues, thereby further decreasing food crop yields.

Fuel-Wood Plantations   With practices similar to those used in modern agriculture, intensively cultivated plantations of fast-growing trees can produce as much as 20 green tons per acre (44.8 green tons per ha) per year of wood, bark, and foliage. The possibility of establishing such plantations on a vast scale to provide a steady source of fuel for steam-powered electric utilities, or raw material for chemical conversion to liquid fuels, has received much attention. Plantations of thousands of acres might be required.

U.S. Wood Energy Use   Wood was the major energy source in the United States until the end of the nineteenth century. With the development of low-priced coal, oil, and natural gas, wood energy consumption declined rapidly. This decline continued until the 1970s, when environmental regulations and increased fossil-fuel prices led to a resurgence in the use of wood for energy.

Although a pound of oven-dry wood has about the same energy content regardless of species (8,500 BTUs), some species such as hickory and oak contain less void space within the wood structure and thus provide a greater amount of wood substance (and therefore energy) per unit volume.

As a young nation, we needed iron and steel products. This required wood for production of the charcoal used in blast furnaces. Charcoal was also required for gunpowder. Later, a world war, the airplane, and the automobile created a need for chemical by-products from wood, such as acetone for munitions and acetate for airplane dope and automobile lacquers. Photography and the manufacture of Bakelite (an early plastic) products created a demand for acetate and formaldehyde from wood.

In the mid-1920s, a new synthetic process was commercialized in Germany to produce synthetic *methanol*. This led to the closing of charcoal plants. While some use in metallurgical processes and for filtration continued, charcoal production slowly decreased until the 1950s, when it began to increase again because of the demand for a smokeless recreational cooking fuel. It is now produced as a primary product in concrete or steel "Missouri-type" kilns and large furnaces (Fig. 14-12). Dense hardwoods such as oak, maple, and hickory are preferred for efficient use of the kiln space.

MISCELLANEOUS FOREST PRODUCTS

Christmas Trees

Each year, millions of Christmas trees are purchased in North America. Most of them are carefully raised on Christmas tree plantations in operations that are, in actuality, short-rotation, intensive forestry. Trees require shearing to produce an attractive, well-shaped

FIGURE 14-12
The "Missouri-type" charcoal kiln. The one marked number 5 is undergoing burning. This process is accomplished using as little air as possible. The kiln at far right is loaded and ready for firing. *(Missouri Department of Conservation photo by Robert Massengale)*

product. In addition, the plantations must be protected from unwanted vegetation, insects, and disease, as well as from such animals as mice, gophers, groundhogs, and deer. Douglas-fir, true firs, and pines are most often cultivated for this purpose.

The Christmas tree industry includes thousands of growers who plant many more trees than markets can absorb. A few thousand growers intensively managing their plantations capture a majority of the market. Nevertheless, up to 100,000 people are employed seasonally by the Christmas tree industry. State and federal taxes are paid on the basis of a wholesale crop value of $400 million with a retail value of $800 million. Oregon, North Carolina, Pennsylvania, Michigan, Washington, and Wisconsin are among the largest producers of Christmas trees.

Forest and Ornamental Nursery Crops

One of the fastest-growing miscellaneous forest product sectors comprises nursery crops of ornamental trees and shrubs for urban and community plantings, seedlings for windbreaks, reforestation, and Christmas trees. The burgeoning industry has been spurred by widespread interest in landscape beautification and forest regeneration. Nursery crops total $8.8 billion dollars in sales. Ornamental horticulture operations account for 11 percent of all U.S. farm crop sales (U.S. Dept. of Commerce, 2002). With increasing urbanization, a growing but indeterminate share of these sales consists of tree and shrub seedlings for urban and community use, such as greenbelts and suburban development, plus regeneration forest seedlings and habitat plantings for windbreak and wildlife.

These nursery tree crops bridge the gap between forestry and horticultural plant production but utilize forestry principles. Because of depressed prices for Christmas trees due to overproduction, many Christmas tree growers have switched to ornamental trees, which can bring from $5 to $20 or more per foot for high-quality ornamentals such as carefully shaped Colorado blue spruce.

Naval Stores

The gum that flows from wounding longleaf and slash pine is used in naval stores production to make rosin and turpentine (wood tar and rosin oils). Naval stores are also obtained in the making of paper from resinous conifers, and by extracting materials from resin-soaked old growth stumps of certain species.

Formerly, the raw gum was burned to make the pitch and tar used in shipbuilding; thus we have the name *naval stores*. Today, the principal products are turpentine, which comes from the distillation of the crude gum, and rosin, which is the residue. Turpentine serves as a thinner in paints and varnishes and as a raw material for synthetic camphor, synthetic pine oil, insecticides, perfumes, odorants, and pharmaceuticals. Rosin finds its way into the manufacture of paper, paints, varnishes, lacquers, soaps, greases, printing ink, shoe materials, rubber goods, and polishes. Rosin oils are used for selective flotation in metallurgy and in paints and varnishes. Tar and pitch go into waterproofing or serve as binders in numerous materials and processes. The deep South, especially southern Georgia and northern Florida, accounts for most naval stores production.

At the end of the naval stores cycle, the entire tree can be used for pulpwood, poles, or saw timber, if modern extraction methods are followed. Genetically superior, high-gum-yielding slash pine trees can yield up to twice as much gum and grow 10 to 15 percent more merchantable wood.

Maple Syrup

The making of syrup from maple sap is well adapted to farm woodlots in New England, the Lake states, and eastern Canada. The product is one of our oldest agricultural commodities, one of the few solely North American crops, and the only crop that must be processed on the farm before it is suitable for sale. Sap (2 to 3 percent sugar) is collected in the early spring from taps placed in trees and then boiled down in a nearby "sugarhouse" to 66 percent sugar content. A good woodlot containing sufficient sugar maple trees can produce 40 gallons of syrup per acre. It takes about 30 to 50 gallons (115 to 190 litres) of sap to yield 1 gallon of syrup.

Maple syrup production has declined since the turn of the century because of labor shortages, rising production costs, changing rural customs, and competition from syrup made from less-expensive cane and beet sugars. In the face of declining U.S. production, Canada has continued exporting maple syrup to the United States to supply a U.S. consumption of 2–3 million gallons (7.6 million liters) per year.

Incidental Products

Cascara bark for medicinal purposes; huckleberry branches and fern fronds for floral arrangements; branches of evergreen trees for grave lining; nuts of the pinyon and of certain hardwoods such as walnut, mistletoe, conifer cones, holly leaves and berries, spruce gum, and Canada balsam (an oleoresin); and various drug plants and substances are also classified as forest products. Cellulose compounds add texture and thicken certain foods, and are useful in the manufacture of hygiene products, medicine, and plastics.

INTEGRATION OF WOOD-USING INDUSTRIES

Integration, the production of a full range of products by wood-processing plants, is an important trend in the forest products industries, and most major companies are now fully integrated. For instance, paper companies could obtain higher returns by producing lumber from their larger-diameter logs than in pulping them, so they began building sawmills. Lumber and plywood manufacturers found they could increase their returns by pulping their wastes and small-diameter logs, or by manufacturing some variety of paneling, rather than by burning these materials. So they, too, have diversified, moving outside their traditional fields.

For the market in general, diversification has increased competition and raised barriers to market entry for smaller, nonintegrated companies. Weyerhaeuser Company was once a lumber producer operating primarily in the West, and International Paper Company was historically a paper producer operating primarily in the South, with little competition between them. Today, Weyerhaeuser is still a leading lumber producer, and International Paper is still a leader in paper sales. However, Weyerhaeuser is now also one of the nation's leaders in sales of paper and allied products, and International Paper is a leading lumber producer. Because of integration, they now compete directly with each other. Consumers benefit from this competition, which reduces prices and can lead to new and improved products.

Trends in Use of Residues

In recent decades, there have been continuing and substantial improvements in the utilization of the timber going into domestic, primary timber-processing plants. There is greater utilization of slabs, edgings, sawdust, veneer cores, shavings, and other similar material for woodpulp, particleboard, and fuel. Now 96 percent of the wood fiber going into primary processing plants is utilized for some purpose.

In recent years, stumpage and log prices have been rising rapidly. But over the past 50 years, softwood lumber prices have not increased nearly as much, so this has made it good business to practice fuller utilization in the woods as well as in the plants and construction sites. Opportunities remain for further improvements through increased use of certain kinds of residues, more efficient harvesting and processing techniques, and more efficient construction and manufacturing practices. There is also a need to use more of the wood left behind as unmerchantable species or sizes. However, tree tops and branches, and rough, rotten, and dead trees are now recognized as valuable for nutrient recycling and wildlife habitat, and may be intentionally left on site.

Wood materials generated in urban areas constitute a substantial solid-waste disposal problem and, at the same time, a potential wood source. Such materials include paper, solid wood from building construction and demolition, pallets, crates and dunnage, and urban tree removals. Approximately 105 million tons of wastepaper, 12 million tons of waste solid-wood products, and 3 million tons of urban tree removals are generated annually.

The Use of Residues for Fuel

The paper and allied products industry, and the lumber and wood products industry, historically have used portions of their wood residues for fuel. These industries often use residual wood and waste products to fire boilers for steam, to produce heat, and to generate electricity.

Further Improvements in Utilization

Aside from environmental needs, the principal obstacle to increased use of logging residues, rough and rotten trees, dead trees, and similar kinds of material is economic. In most instances, current market prices for such materials are lower than the costs of harvest and transport to mills. The development of technologies to increase utilization of existing forest resources is particularly important in resolving the conflict between environmental and economic considerations. One answer appears to be mechanized systems that allow rapid collection of whole stems or trees, and, in some instances, onsite chipping for fuel or pulpwood. Another opportunity to reduce waste in timber harvest is through felling and bucking to proper lengths and grades. Such quality control could add several percentage points to sawlog and veneer log output.

Quality control in sawmilling, lumber drying, and remanufacturing could also improve their yield by as much as 10 percent simply through increased attention to equipment maintenance and machine settings.

Small-diameter wood is now being used in innovative ways, including watershed restoration and habitat enhancement. ELWd Systems in Washington state creates structures that mimic the performance of large-diameter, hollow logs. These structures serve as coarse woody debris replacement in streams and other wetlands, cache and nesting space, runways and perches, as well as bulkheads for landscaping. Small logs, 3–5 in (7.6–12.7 cm) in diameter, including fireline and fuel-reduction debris, tops, and thinning poles, can be assembled to form 12–16-in (30.5–40.7-cm) diameter logs of 4–10 ft (1.2–3.0 m) in length. These "hollow-log" structures can be purchased as a kit and built on site by volunteers or conservation field crews using simple hand tools, to produce a strong, all-wood unit. Components can be hand carried to riparian sites to avoid the entrance of heavy machinery (Burks, 2001).

There also are opportunities for shifting a larger part of the wood removed from the forests into higher-value products such as lumber and plywood, especially manufacturing lumber and plywood from small-diameter and short logs. New approaches include high-speed electronic scanning and automated control systems, gluing techniques to produce wide-width or long-length products equivalent to lumber sawn from large logs, and production of construction lumber from hardwoods such as yellow-poplar and aspen.

Continued development of technologies for harvesting and pulping whole-tree chips will greatly increase per-acre harvest and reduce logging residue. Improvements in papermaking technologies could increase the production of pulp from high-yield processes and facilitate the use of fast-growing plantation hardwood.

Reduction of fuel and power costs in forest industries through energy-efficient processing and expanded use of wood and bark fuels would lower the per unit manufacturing costs and thus make products more competitive in price. In a few areas, the distribution through local utilities of surplus steam and electricity from forest products mills could further enhance revenues.

During the 1980s, wood fuel was used in the utility industry in California, New Hampshire, Maine, Vermont, Washington, Montana, and Wisconsin. Fu-

FUTURE PRODUCT POSSIBILITIES

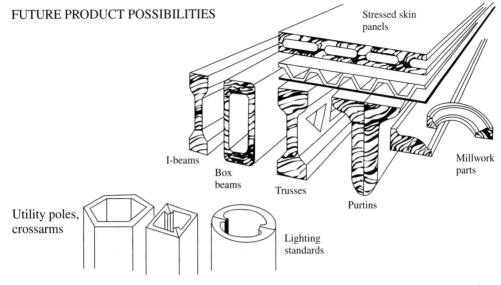

FIGURE 14-13
These drawings demonstrate possibilities for engineered wood products.

ture possibilities may evolve for energy production from biomass fuel where wood waste or fuel-wood plantations exist in the right combination with markets and air-quality conditions.

Improved engineering, design, and construction practices could save 10 to 20 percent of the dimensional lumber required for a conventional house, with no loss in performance (Fig. 14-13). Proper use of preservative-treated products and careful application of water repellents in construction and maintenance could greatly extend the life of structures and thus lower demand for replacement.

RECYCLING OF PAPER AND WOOD

The industrialized world is faced with an ever-increasing solid waste problem consisting of enormous quantities of waste: paper, wood, metals, plastics, glass, yard waste, and various other materials. The rate at which the United States creates solid waste increases because of our growing population, but the average amount discarded per capita each year is also increasing. For example, between 1980 to 2000 the U.S. population grew by 21 percent, but municipal solid waste grew by 46 percent. In 1998, the United States generated 220 million tons of municipal solid waste.

Paper and Paperboard

By weight, paper and paperboard constitute the largest percentage of the solid waste stream, ranging from 30 percent to 55 percent. The landfills are choking with phone books, newspapers, corrugated paperboard, magazines, and "mixed" papers. The last category constitutes the largest single category, amounting to about 22 percent of the total municipal solid waste.

As a result of public pressure, recycling is increasing. More than 47 million tons of paper were recovered in the United States for recycling in 1999, but nearly 58 million tons annually must still be landfilled or incinerated. In a major public commitment to recycling, the U.S. paper industry has set a target of recycling 50 percent of wastepaper, compared to a 45 percent rate in 1999. Clearly, recovering all of the wastepaper and paperboard will be difficult, particularly high-quality magazines and other papers in which difficult-to-remove additives constitute a significant proportion. A key will be new products from recycled paper such as the new cellulose insulation Nature Guard™, made 100 percent from recycled newspaper by Louisiana Pacific Corp. One 50-square-foot bag contains the equivalent of 46 *Wall Street Journals* (Curley, 1993). Approximately 700,000 tons of recycled paper were converted to cellulose insulation in 1999.

Solid Wood Waste

Solid wood makes up only 2 percent to 3 percent of the waste stream, with discarded wood pallets a significant contributor. Treated wood products and wood-containing adhesives or finishes present a problem because they incorporate hazardous chemicals, such as creosote or pentachlorophenol, that complicate disposal. Recycling treated wood is now a technological challenge, but future options include mechanical reduction to particles or fibers for remanufacturing into organic or mineral-bonded composite products.

Sludge and Log Yard Waste

Sludge is a by-product of pulping wood for paper-making. Most sludge is burned for energy, but some is consigned to landfills. Log yard waste is comprised of bark, sawdust, and wood generated by log processing, and it is estimated that about 30 thousand cubic yards of such wastes are generated annually by a typical sawmill in the Pacific Northwest. Left alone, such waste is an environmental hazard. Research is exploring the possibilities of its use for manufacturing fiber, fuel, and additives for agricultural or industrial lands.

FUTURE TRENDS TO CONSERVE ENERGY AND PROTECT THE ENVIRONMENT

The *utilization of wood,* defined as *its collection in the forest, subsequent manufacture, and use,* has evolved and become more efficient during the past 50 years. This trend will continue, but increasingly the focus will be on environmental and energy concerns rather than solely on economic return. Some examples include:

Preservation—Using biological functions rather than the application of unacceptable chemicals to inhibit decay

Innovative drying—Reducing emission of volatiles and reduction of energy consumption

Biological pulping—Substituting biological systems in the pulping process to reduce use of harsh chemicals and to reduce energy use

Wood finishes—Reducing the amount of volatile organic compounds (VOCs)

Integrated building designs—More efficient use of materials in building construction to improve energy efficiency and fire safety

Harvesting—Achieving ecosystem management objectives and improving utilization, rather than simply removing raw material

Recycling—Reducing consumer waste and greater recovery of waste material for reuse

SUMMARY

Wood remains basic to human comfort and survival, and the properties of wood keep it useful and sought after even in our present highly technological society. The wood products industry is one of the major industries in the United States and in North America, ranking among the top 10 largest industrial enterprises. It has seen many changes and continues to evolve through research developments, economic challenges, and environmental concerns.

Wood products come from a renewable resource and are much more energy efficient to produce than competing substitutes. Demand for wood products provides the major revenue supporting forest management. Consumption of wood products in the United States continues to grow and markets for wood exports continue to expand. As the supply of larger trees accessible for harvest decreases, the costs of producing solid wood products, and thus their prices, will increase. Markets for composite products manufactured from smaller trees and recovered waste may increase as a result.

A variety of wood products is produced, including lumber and plywood and engineered products made from reconstituted wood pieces and particles. These include glulam timbers, particleboard, waferboard, and oriented strand board (OSB); wood composites or fiberboards including insulation board, medium-density fiberboard, parallam, parallel strand lumber, laminated-veneer lumber, and mineral-bonded boards. Fiber and chemical products made from wood include pulp and paper, rayon, and plastics. Substantial wood is also used for fuel, and some is still carbonized into charcoal. Miscellaneous forest products include Christmas trees, naval stores, maple syrup, and incidental products.

Wood-using industries are adapting to meet competition and scarcity of resources through integration of wood-processing plants, use of residues for fuel, and recycling wood and paper to supplement raw materials.

Innovative forest management, efficient in both production and utilization of wood fiber, will play an important role in insuring the future of the nation's and the world's forests, while continuing to contribute an array of old and new wood products.

LITERATURE CITED

Balatinecz, John J., and Raymond T. Woodhams. 1993. "Wood Plastic Composites: Doing More with Less," *J. Forestry* 91(11):22–26.

Burks, Jocko. 2001. Correspondence from Jocko Burks, Production Coordinator, Forest Concepts, LLC. Federal Way, Wash.

Curley, Ann. 1993. "More than Recycling: Living in the Future Today," *J. Forestry* 91(11):30–32.

Koch, Peter. 1991. Wood vs. Non-wood Materials in U.S. Residential Construction: Some Energy-related International Implications. Working Paper 36. Center for International Trade in Forest Products, University of Washington, Seattle.

Moslemi, A. A. 1993. "Inorganic-Bonded Wood Composites: From Sludge to Siding," *J. Forestry* 91(11):27–29.

Preston, Alan F. 1993. "Wood Preservation: Extending the Resource," *J. Forestry* 91(11):16–18.

Ryan, Dede. 1993. "Engineered Lumber: An Alternative to Old Growth Resources. *J. Forestry* 91(11):19–21.

Skog, Kenneth, Peter Ince, and Richard Haynes. 1998. Wood fiber supply and demand in the United States. Proceedings of the Forest Products Study Group Workshop www.fpl-fs.fed.us/documents/pdf2000/skog00a.

Smith, Brad, John Vissage, David Darr, and Raymond Sheffield. 2001. *Forest Resources of the United States, 1997.* Gen. Tech Rpt. NC 219. St. Paul, MN. USDA Forest Service. North Central Expt. Sta. 190 P.

Smulski, Stephen. 1997. *Engineered Wood Products.* PFS Research Foundation, Madison, Wis.

Taylor, R. E. and Associates, Ltd., Vancouver, B. C., Canada; U. S. Census Bureau, Lumber Production and Mill Stocks; USDA Foreign Agricultural Service; and American Forest and Paper Association, Washington, D. C.

U.S. Department of Commerce. 2002. Census of Agriculture, 1997, Bureau of Census, Washington, D.C. 1997, www.Nass.usda.gov/census/

ADDITIONAL READINGS

Desch, H. E. 1981. *Timber: Its Structure, Properties and Utilization* (6th ed.). Timber Press, Forest Grove, Oreg.

Haygreen, John G., and J. L. Bowyer. 1989. *Forest Products and Wood Science* (2nd ed.). Iowa State University Press, Ames.

Mullins, E. J., and T. S. McKnight. eds. 1981. *Canadian Woods: Their Properties and Uses.* University of Toronto Press, Toronto, Canada.

O'Neal, Jennifer, Anne Watts, Susan Bolton, and Tom Sibley. 1999. *Comparison of the Hydraulic and Biologic Effects of Large Woody Debris and an Engineered Alternative.* Presented at the American Society of Civil Engineers International Water Resources Engineering Conference, August 11, 1999, Seattle, Wash.

Panshin, Alexis, J., and Carl DeZeeuw. 1980. *Textbook of Wood Technology* (4th ed.). McGraw-Hill Book Co., New York.

STUDY QUESTIONS

1. Describe some of the desirable properties of wood. What are problems associated with wood use?
2. Explain what is meant by the following:
   a. *workability*
   b. *hardness*
   c. *permeability*
   d. *plasticity*
   e. *resiliency*
3. Why is wood usually dried after being processed? What is the *fiber saturation point?*
4. What is *structural dimension lumber,* and what is its major use?
5. What are some of the influences on domestic lumber consumption?
6. What are some pros and cons of log exports?
7. Discuss the reasons why a variety of engineered wood products have been developed and the advantages and uses of different products.
8. Explain the differences among *plywood, oriented strand board, laminated veneer lumber,* and *wood composite fiberboards.*
9. What percent of total U.S. wood fiber goes to pulp and paper use? Where is the paper plant nearest where you live? Where does it get its fiber?
10. Describe some miscellaneous forest products.
11. What are some trends in wood utilization? How will they affect forest management?
12. Explain what is meant by *integration* in the forest products industry.
13. Explain what *value-added wood processing* means.

Economics in Forest and Renewable Resource Management

INTRODUCTION

Economics might be called the science of choice, as it is concerned with the allocation of scarce resources and the effects that follow. Economics tries to answer three questions: (1) *What* goods and services will be produced? (2) *How* will these goals be produced? (3) *Who* will receive the products? Of course, costs of production and market prices are associated with these questions. Chapter 15 describes the focus of forest and renewable resource economics on the efficiency of forest management and shows the necessity of responding to changing markets and economic conditions. The economic importance of forests in the United States is reviewed along with the distribution and ownership of forests and their corresponding productivity. Environmental values receive attention because these are sometimes as important, if not more important, than forest commodity outputs.

Basic concepts of forestland investment, such as accounting for capital and operating costs and capital maintenance, are reviewed along with the basic economic concepts of marginal analysis, amenity values, and opportunity costs. The application of these concepts through land use planning and allocation, as well as forestry incentive and assistance programs aimed at increasing production on private lands, are also described. Finally, historic to current forest economic development is described in terms of settlement, custodial, management, and environmental stages.

FOCUS OF FOREST RESOURCES ECONOMICS

There are two interrelated purposes in applying methods of economics to the management of forests and associated renewable resources. These are (1) production efficiency: managing forests and their related resources for efficiency of outputs; and (2) responding to economic influences: predicting and responding to changes in the economic environment.

1. In evaluating production efficiency, basic economic concepts are used to identify efficient and profitable management programs. Questions focus on the effects of various management investments in terms of increased outputs from the resource base, including all the goods and services for which it is managed.

2. In evaluating response to the economic influences, economic concepts are used to estimate the effect of changes in demand and markets caused by economic factors such as taxes, interest rates, and regulatory activities, as well as to estimate the general level of local, regional, and global economic activity. Based on anticipated demand and markets for goods and services, intensity of forest management levels can be changed, and allocations of resources to different uses can be analyzed and adjusted.

Thus, the economics of forest and renewable resource management is concerned with the production of goods and services that are efficient and profitable to producers, competitive in the marketplace, beneficial to users and consumers, and contribute to the economic productivity and growth of the nation. Economic analyses and methods provide information that helps resource managers and owners to be efficient, profitable, and competitive, as well as responsive to the preferences of consumers.

As the U.S. population continues to increase and our standard of living continues to improve, the demands on forests and associated renewable resources are increasing and becoming more varied. People are demanding more of the traditional forest commodities (items that are bought and sold in markets) such as lumber, paper, and other wood-based consumer products, recreational and year-round home sites, and big game hunting permits; but people also are demanding more nonmarket goods (outputs not bought and sold in markets) such as wilderness, forested landscapes, aesthetics, nongame wildlife habitat, and water quantity and quality. Forests and their associated renewable resources are scarce. There are not enough of them to satisfy all of the demands. Increasing demands, plus scarcity of supply, put more and more pressure on forest resource managers, who consequently must make difficult resource allocation decisions about what products to produce, how to produce them efficiently, and how to distribute these resources while respecting nonmarket demands and other considerations.

The economic concepts and information in this chapter will help readers understand why forests and renewable resources are managed as they are, and why such management is changing in response to changing economic circumstances. Hopefully, the examples used and data presented will demonstrate the importance of these resources to the economic and social well-being of society, and the value of economics in determining and evaluating natural resource management strategies.

THE ECONOMIC IMPORTANCE OF FORESTS IN THE UNITED STATES

Although many forest resources and amenities are hard to measure economically, they are of enormous importance, and their value is increasing as world population and demands on the global environment increase. But the major economic importance of forests to date derives more from their ability to produce renewable supplies of wood commodities, such as lumber, plywood, pulp, and other wood fiber-based building materials and consumer products (Fig. 15-1).

Production of Wood Commodities

The United States is both the world's leading producer and consumer of forest products. In 1990, each American used an average of 681 pounds of paper products and 43 cubic feet of lumber and plywood, more than three times the global average (World Resources Institute, 1992). By contrast, per capita paper consumption in Japan was 28 percent less and that of lumber and plywood was 70 percent less. In China, per capita consumption of paper was only 2.4 percent of that of the United States, and lumber and plywood consumption only 3 percent of the total (American Forest Council, 1991).

Considering the large U.S. wood product consumption and demand, one logically thinks about the U.S. supply of timber—is it sustainable? Recent analysis helps put the situation in perspective, summarized here by Darius Adams (2002a),

FIGURE 15-1
Timber continues to place as one of the top agricultural crops in the United States.
(Photo by Grant W. Sharpe)

Harvest of timber of all types from U.S. forests has grown markedly over the past five decades as have aggregate timber inventories. From 1957 to 1996, total U.S. roundwood harvest from all sources and for all products rose by more than half—a 30 percent increase for softwoods and nearly a doubling for the smaller hardwood component. Over roughly this same interval (1952–97), growth on U.S. forests exceeded harvest, and total inventory rose by about 35 percent—softwoods by 12 percent and hardwoods by 90 percent. On private lands alone, softwood inventories rose by 22 percent and hardwoods increased by more than 80 percent (Adams, 2002a, p. 26).

In fact, net annual growing stock growth exceeded removals by 54 percent in 1976, 41 percent in 1986, and 47 percent in 1996 (Smith et al., 2001). The United States is not running out of timber.

From 1977 to 1990, public lands provided an average of 20 percent of the harvests, but public opinion (and thus, public policies) has led to a reduction of harvests on public lands, which are now predicted to make up only 8 percent of the U.S. timber harvests in the future (Haynes, 2002). So, the level of U.S. timber supply and harvests in the future will depend even more on the 350 million acres (142,000 ha) of private land. Because private lands are expected to decrease in acreage by 3 percent by 2050 due to increasing demand for urban and related uses (Alig et al., 2002), good management of these private lands will be a key to sustaining harvests. Total U.S. timber volume (inventory) will continue to grow in the future, despite growth in consumption and harvests, with the difference between U.S. harvest and consumption made up by imports (Adams, 2002a, b).

Production of Nonwood Commodities and Amenities

Forests are also valuable because of their central role in providing ecological services and amenities, such as clean and stable water flows, replenishing oxygen in the atmosphere, providing habitat for wildlife and fish, providing areas for recreation, and for their aesthetic and scenic contributions to our lives. Other examples are wilderness, wildlife-viewing opportunities, the preservation of threatened and endangered species, and a renewed interest in nontimber forest products such as mushrooms, herbs, and greenery.

Most of these nonwood commodities and amenities are not sold in traditional competitive markets, although they create real benefits for many forest users. The owners and producers of these opportunities usually do not charge full market price for access to them, or charge no user fee at all. As the demand for such opportunities has strengthened, some landowners, both public and private, have introduced fees. However, many such fees remain at a nominal level. Fees for recreation on private lands, for example, are suppressed by the availability of recreation opportunities on public lands at very nominal rates. A long-standing debate over how much to charge for recreation on public land continues. Modest increases have been implemented, but rates are held down by political pressures. The central argument is that opportunities on public land are owned by the public and paid for by taxes, so they should be available free or at nominal rates.

Regardless of the imperfect markets and fee structures for these forest outputs, they are vital for the growth and development of the U.S. economy and for our social well-being. For example, it has been estimated that over 60 percent of households in the United States participated in forest-related outdoor recreation in 1986, and that a significant proportion of the total direct recreational expenditures of $132 billion for that year was related to forest recreation activities (Cordell et al., 1989). More recently, Cordell (1999) estimates that of the approximately 4.6 billion outdoor recreation trips Americans took in 1987, about 50 percent were to participate in wildland activities such as camping, fishing, hunting, hiking, horseback riding, and visits to outdoor attractions and scenery. If the value of these nontimber commodities was set by free market forces, it could conceivably exceed the value of timber. Following is a list of the noncommodity values of forests.

Noncommodity Values of Forests (many of these products could be called *ecological services*):

- Holding soils in place
- Supplying water storage and flows
- Influencing air quality positively
- Ameliorating climate
- Providing habitats for wildlife
- Protecting streamside zones and related fish habitats
- Promoting aesthetic values—scenery
- Storing carbon
- Producing oxygen
- Offering recreational use
- Providing biodiversity and protecting gene pools

The need to protect biodiversity and gene pools of tropical forests, especially for their potentially valuable pharmaceutical uses, is well known. But genetic diversity is important to the adaptation and survival of all life forms. For example, when populations of endangered plants or animals become too small or isolated from each other, their potential to adapt and therefore survive may be threatened by the limited diversity in their available gene pools.

One of the emerging noncommodity values of forests is carbon sequestration. Forest ecosystems serve as greenhouse gas sinks by transforming carbon dioxide (CO_2) into carbon (C) stock components (trees, roots, woody debris, other vegetation, litter, and soils). Although controversial, the Kyoto Protocol to the United Nations Framework Convention on Climate Change provides for greenhouse gas emission targets to be met by reducing emissions *or* by enhancing carbon sinks (Murray et al., 2000). The potential for trading carbon credits and greenhouse debits potentially makes carbon sequestration by forests something more than a noncommodity value (i.e., an ecological service that may be exchanged for value in a market).

Production Management

Production management is the process of combining certain necessary elements, inputs, or actions (labor, capital, raw materials) to produce a desired output. Nature unaided usually provides some timber, forage, wildlife, water, recreation, and other outputs from forests. Production management actions, however, can augment or increase these outputs. Timber stands can be protected from fire and pest losses. They can be fertilized, thinned, and pruned to obtain trees of greater volume and value, and they can be replanted after harvest with genetically improved seedlings that will grow faster. Similarly, the treatment of range allotments to obtain a greater production of forage for livestock or wildlife, and the installation of fencing and water impoundments to control and distribute animal use, can increase forage output and utilization significantly, while reducing environmental damage. Likewise, recreation facilities, such as campgrounds, trails, and specialized facilities for a wide variety of outdoor recreation activities, can increase the recreational values and capacity of a forest for recreation use.

COMMERCIAL FOREST OWNERSHIP AND PRODUCTION

Because forests are important to the economy, we need to be concerned about their productivity, ownership, and management nationwide. The individuals and groups that ultimately make these economic decisions are the landowners. The diversity of forest landowner and, consequently, the diverse management regimes, are a key to current and future productivity. *Forest productivity* is defined as the amount of wood per acre that can be produced in fully stocked natural stands. The "natural potential" is used because measures of the potential are available for most regions of the United States, and they provide a uniform means of describing and comparing productivity. This estimate of biological productivity is also a useful indicator of forest capacity for "other uses" because it is based principally on soil, climate, and topography—factors that affect all uses (Smith et al., 2001).

Ownership

About one-third of America is forested, and over two-thirds of these forests are classified as commercial forest, or *timberlands* as they are called (Fig. 15-2). Commercial timberlands are forests that are capable of producing at least 20 cubic feet of industrial wood per acre per year and have not been reserved by statute or administrative regulation for uses that are not compatible with commercial wood production. Not all commercial timberlands actually provide harvests of timber. Some timberlands are too far from mills or are too poorly stocked with commercial timber to make harvest profitable. Then, too, some owners simply do not want their trees cut or own an insufficient amount of land or timber to make commercial timber harvests feasible. Many private landowners, especially owners of smaller holdings, value their forests more for privacy, scenery, recreation, and wildlife habitat.

Commercial timberlands are owned by the public (29 percent), by the forest industry (13 percent), and by nonindustrial private forestland owners (58 percent) (NIPF; Fig. 15-3). Nationwide, most commercial forestland is located in the northern and southern forest regions, and nonindustrial private owners own

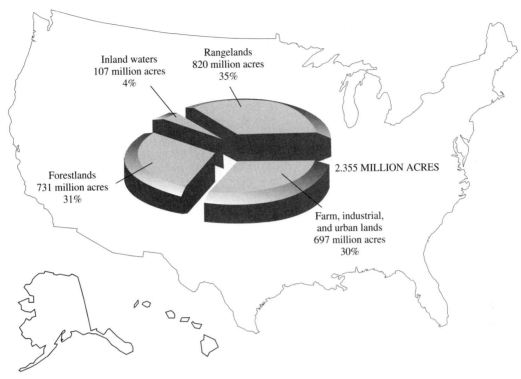

FIGURE 15-2
Area of land and water in the United States, including percentage of rangeland and forestland.

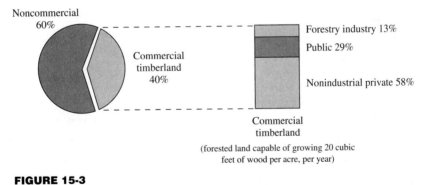

FIGURE 15-3
Percent of commercial forestland in the United States.

the greatest amount (Fig. 15-4). Thus, the United States has substantial commercial timberland, the vast majority of which is in private ownership. This provides a renewable natural resource base for one of the nation's most important industries, while providing environmental benefits at the same time.

Growth and Yield by Ownership and Region

One concern of forest managers is the relationship between the volume of wood harvested and the volume of wood being added to the forest inventory by growth. When removals exceed net growth, wood is being used more rapidly than it is being replaced,

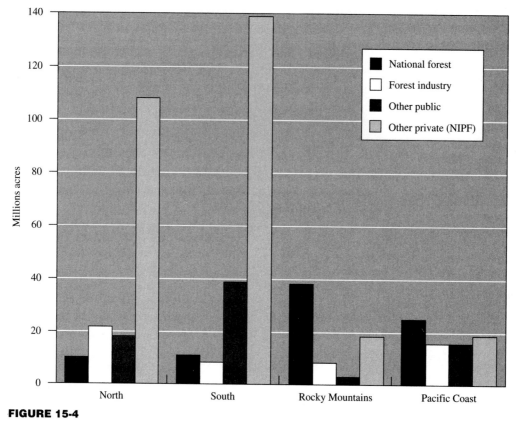

FIGURE 15-4
Area of commercial timberland in the United States, by region and ownership.

and the inventory declines. Such a decline could mean even less growth in the future and a loss of sustainability—that is, the eventual inability of forestlands to sustain harvests at current levels. This, in turn, would impact the economy and employment dependent on wood products. Fortunately, as already mentioned, in the United States, continuing measurement of forest resources shows that, during the period from 1977 to 1987, net forest growth exceeded harvests for decades and continues to do so. This trend is predicted to continue into the future (see Smith et al., 2001).

Nonindustrial Private Forest Ownership, Growth, and Yield

Given the relative importance of nonindustrial private forest owners (NIPFs) in terms of their large acreage of commercial forestland (58 percent, Fig. 15-3), the large volume of growing stock they control (47.4 percent of the U.S. total), and the relatively poor stocking of their timberlands, these owners deserve close attention. The NIPFs have unfulfilled potential for production because they control so much of the total timberland, yet their lands are understocked and a majority of the ownerships are in small parcels.

A distribution of nonindustrial private ownerships by acreage, size, and number of owners (Table 15-1) reveals that there are about 9.89 million NIPFs, and that a mere 6 percent of these owners hold nearly 60 percent of the acreage (Birch, 1996). In other words, there are 627,000 owners that have holdings of 100 acres or larger. The other 94 percent of the owners, more than 9 million of them, all own 100 acres or less, and 8.5 million of them own tracts of 50 acres or less. There is a trend toward large and midsized tracts to be broken up and sold, thereby increasing the number of small tracts and owners (DeCoster, 1998; Mehmood & Zhang, 2000; Sampson & DeCoster, 2000).

An extremely important economic concept called *economics of scale* applies here, because studies have

TABLE 15-1				

DISTRIBUTION OF NONINDUSTRIAL PRIVATE FOREST (NIPF) LANDHOLDINGS BY ACREAGE AND NUMBER OF OWNERSHIPS

Size of forested landholding (acres)	Acres (millions)	Percent of total acres	Thousands of ownerships	Percent of total ownerships
1–9	17	5	5,733	58.0
10–19	16	5	1,243	12.6
20–49	46	15	1,540	15.6
50–99	49	15	744	7.5
100–499	93	30	566	5.7
500–999	23	7	38	0.4
1,000+	71	23	23	0.2
	314	100	9,888	100

Source: Thomas W. Birch (1996). Private Forest Landowners of the United States, 1994. *Resource Bulletin* NE-134. USDA Forest Service, Radnor, Pa.

shown that the per acre costs of forest management rise very rapidly as the size of a forest tract drops below 40 acres. Industrial forests and public forests are usually large tracts, and here managers can take advantage of economies of scale in their operations, which results in lower management costs per acre. Thus, the 10 million NIPF landowners who control 41 percent of the nation's timberland in tracts of 100 acres or less will have higher management costs per acre—unless, of course, they could be organized into cooperative groups to share costs, which is occurring in some locations (Barten et al., 2001).

Furthermore, industrial forests and public forests are generally well-staffed and managed by professional foresters. However, only about 40 percent of the nonindustrial private forests that harvest timber each year employ trained foresters (such as consultants) for assistance. Studies indicate that nonindustrial forest owners that do not use professional assistance in preparing and selling timber get substantially lower prices for their timber than those who do, and that their lands are left in poorer condition for future growth (Fedkiw, 1992). Thus, technical forestry assistance is an important production factor, increasing financial returns from timber harvests—and also increasing production capacity for future harvests.

FOREST PRODUCTIVITY

The productivity of timberlands is the result of many factors. First, not all timberlands have the same climate, topography, and soils. Some areas are simply better suited for growing tree crops. This is perhaps indicated by the relatively high net annual growth of forests in the South and on the Pacific Coast, where milder climates prevail. The lower growth rates on the national forests may reflect not only their slow-growing, old-growth forests, but also the fact that these lands are generally located in high-elevation, mountainous terrain, as growth rates subside with altitude. On a national scale, the economics of timber production is concerned with the allocation of the best tree-growing lands to commercial forest production, because that will most efficiently produce commercial timber from the nation's forests. This, too, explains why the forest industry is concerned when productive timberlands are taken out of production, and why the industry and government would like to see the extensive NIPF lands producing up to their potential.

Improving Forest Productivity

The economics of forest resources deals with the efficiency of management in producing forest outputs, including the allocation of land uses and investments in management and the effects of these decisions on the net growth and yield of forests. In the United States, continuing remeasurement of sample forest plots shows that total volume of forest growth has exceeded removals for several decades (Fig. 15-5) and that total net volume of trees on commercial timberland is increasing (Fig. 15-6). Net growth in 1997 exceeded harvests and mortality nationwide (as previously mentioned). This continued long-term improvement in growth relative to harvest and

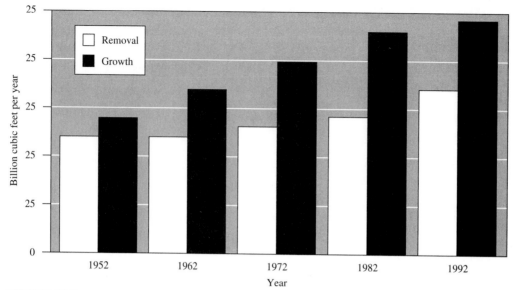

FIGURE 15-5
Timber removals and growth in the United States between 1952 and 1992.

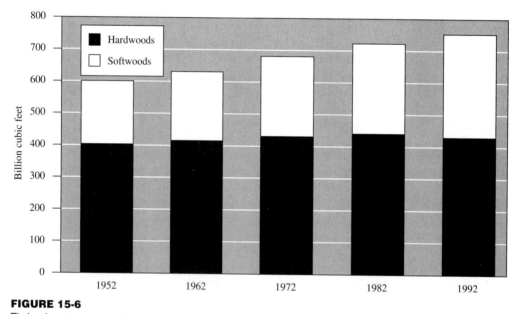

FIGURE 15-6
Timber inventory: net volume of trees on unreserved commercial timberland, between 1952 and 1992.

mortality is largely the result of forest management—protection against wildfire, insects, and disease; reforestation following harvest; forest-stand improvements; and, more recently, fertilization and genetically improved seeds from seed orchards and better seedlings from modern nurseries.

During the 60-year era from 1940 to 2000, professional resource management began in earnest and has intensified. The payoffs have been significant. For example, in the early 1930s the area of forestland annually burned by wildfire regularly exceeded 30 million acres (12,150,000 ha) or 12.150 mi ha. As fed-

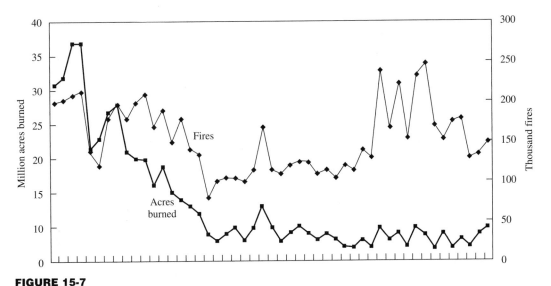

FIGURE 15-7
Number of fires and acres burned on state and private lands and in the national forests between 1946 and 1988.
(Wildfire statistics, USDA Forest Service, in Fedkiw, 1990.)

eral and state fire protection and suppression programs were strengthened through federal–state cooperative programs beginning in the 1930s, the area burned was rapidly reduced to less than 5 million acres (2 million ha) per year by the end of the 1950s, although the number of fires remained about the same (Fig. 15-7). In the 1960s, the annual area burned averaged less than 4 million acres (1.6 million ha), and in the 1970s and 1980s the average annual burn was less than 3 million acres (1.2 million ha) (Fedkiw, 1992). Further reductions in the early 1990s were modestly offset by increased acreage burned in "bad" fire years, aggravated by drought, overstocked stands, due, ironically, to lack of light brush-clearing fires, plus insects and disease (see Chapter 8, Fire Management).

This reduction in wildfires helped to improve the investment climate for plantations and more intensive forest management, since the risk of losing such investments was reduced. Thus, there was a steady increase in acres of trees planted, from a few hundred thousand acres per year from 1940 through the 1950s, to nearly 3 million acres per year in 1989 (Fig. 15-8), utilizing the approximately 2 billion seedlings grown annually in U.S. nurseries, with tree planting subsidized by several federal and state programs for private landowners. During the same pe-

riod, resource management professional expertise and employment steadily increased. For example, membership in the Society of American Foresters (and presumably the employment of professional foresters) grew more than fourfold, from 4,556 in 1940 to 18,000 in 2001 (Table 15-2). A slight decline in memberships in the past two decades may indicate saturation, increased efficiency, fiscal constraints, or all of these factors.

Table 15-2 also shows dramatic increases in the memberships of other professional societies linked to the management of renewable resources—the Wildlife Society, the American Fisheries Society, the Soil & Water Conservation Society, and the Society for Range Management. While the products and environmental benefits from their professional resource management efforts are not so easily measured, it is reasonable to assume that their collective efforts resulted in improved conditions from reduced erosion, better water quality, and improved fish and wildlife habitat.

Thus, the nation's growing concern over its forests and natural resources resulted in investment in federal and state cooperative programs to reduce wildfires (Fig. 15-9). This in turn led to increased tree planting and investments in professional management of forests and their associated renewable resources.

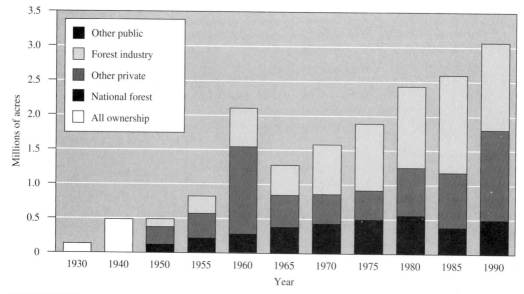

FIGURE 15-8
Acres of trees planted in the United States, by major owner, 1930 to 1990.

TABLE 15-2

MEMBERSHIP IN NATURAL RESOURCE PROFESSIONAL SOCIETIES, 1940–2001

Professional society	Year established	1940[2]	1970[2]	1980[2]	1989[2]	2001
				Number		
American Fisheries Society	1870	848	5,359	8,193	9,011	9,000
Society of American Foresters	1900	4,556	17,227	21,280	19,051	18,000
Wildlife Society	1937	908	5,102	7,785	8,753	8,900
Soil & Water Conser. Society	1945	(1,472)	14,222	13,945	12,855	10,000
Society for Range Management	1948	(634)	3,832	5,907	5,189	5,350
American Water Resources Assoc.	1964	(261)	1,707	2,410	3,421	2,800
Total Membership		6,312[1]	47,449	59,420	58,280	54,050

[1]Total is only for 1940 data. Data in parentheses are for year of establishment of society.
[2]Data for 1940, 1970, 1980, and 1989 are reported in Fedkiw (1992).
See also Hendee and Pitstick (1992) and the appendix for membership.

(Some of the negative effects of successful fire control, and the need for prescribed fire in some stands, are described in Chapter 8.)

Economic and social conditions influence forest values dramatically. For example, a reduction in mortgage rates increases the number of housing units that will be built, and this expands the demand for wood products and timber. Thus, if the price of timber increases, more timber will be harvested in response to the higher prices, and soon investments in forest management become possible. Similarly, growing environmental interest shown in regulations controlling air and water quality has stimulated investment in environmental protection and has affected, among other things, allocation of lands to parks, wilderness, and roadless areas.

FIGURE 15-9
Modern satellite and ground equipment used in fire detection contributes significantly to reduction of wildfires, improving the investment climate for intensive forest management. *Left:* A firefighter pulls data from a Remote Automated Weather Station in Minnesota. *Right:* A ground observer using the satellite communications link at a field location. *(Photos courtesy of National Interagency Fire Center, Bureau of Land Management.)*

FORESTRY INCENTIVES FOR PRIVATE LANDS

To encourage private sector investments in forestlands and their associated natural resources, federal and state government agencies have historically offered a number of incentives. These programs are described in more detail in Chapter 18, "Forest Management by the States" and Chapter 19, "Forest Management on Private Lands," but are mentioned here in their economic context as incentives for private land management. As described previously, there is particular interest in stimulating better management of the NIPF lands because of their yet unfulfilled potential for production.

Forest Taxation Incentives

The way taxes are levied on timber income or forestland has a major effect on the willingness of investors to own forestland and spend on forest management. Thus, forest property tax laws can be an incentive for owning more rather than less forestland, and for managing more rather than less intensively. The reverse is also obviously true.

Simply stated, starting in 1944, income from the sale of timber was treated as a *capital gain* under the federal income tax code. This originally meant that such income was taxed at a lower rate than ordinary income, thus making timber growing a more profitable undertaking than otherwise would have been the case. The opportunity for capital gains on timber provided an important incentive for industrial timberland acquisition and management. While the capital gains rate relative to ordinary income has varied with changes in tax law, it remains a favorable influence on forest ownership and management.

Property taxes on forestlands have been modified by many states. Under these modified property tax laws, the annual property tax may be substantially reduced on qualifying timberlands, and a yield tax on timber harvest income collected instead. This means that the greatest tax is levied at the time timber is sold, when income is available to pay the tax. Tree planting also is given special treatment in the federal tax code. This capital investment may be amortized over an eight-year period, reducing the net after-tax cost by as much as one-half. Thus, tax laws and their variations as applied to forestland and its income are a major influence on forestland ownership and management. Much more could be said on this subject, as it is a complex field.

New ideas about forest management, such as managing for biodiversity and other ecological services, as well as economic productivity, suggest possibilities for new incentives.

For example, as the environmental benefits of forests become recognized, such as greater oxygen production, carbon storage, and more energy-efficient construction compared to substitute materials, economic incentives encouraging sound forest management and use will become possible. Further, creative approaches to economic analysis of noncommercial uses are revealing values and opportunities to be considered in land allocation strategies. For example, a recent analysis suggested that the U.S. National Wilderness Preservation System generates $3–4 billion in value each year (the largest being ecological services) (Loomis & Richardson, 2001).

Forestry Assistance Programs

Federal and state governments cooperate in providing many programs to encourage effective conservation and management on private forestlands. By providing educational, technical, and financial assistance to private landowners, there is an expectation that public benefits in the form of increased wood and amenities will be produced.

Forestry assistance programs fall into three main categories: 1) education, 2) technical assistance, and 3) cost-sharing financial incentives. Additional information on such programs appears in Chapter 18, "Forest Management by the States" and Chapter 19, "Forest Management on Private Lands."

Educational Programs Educational programs for forest landowners are provided primarily by the Cooperative Extension Service, jointly administered by the U.S. Department of Agriculture and the land grant universities of the various states. Cooperative extension agents provide information to landowners through publications, videos, television, short courses, and personal contact.

Technical Assistance Technical assistance to individual timberland owners is provided by service foresters in the respective state departments of natural resources, sometimes by federal foresters employed by the USDA Natural Resource Conservation Service (formerly the Soil Conservation Service) and sometimes by private sector foresters employed by wood products companies (see Chapter 19, "Forest Management on Private Lands"). Wood products companies may provide technical assistance to private landowners in return for rights to buy their timber or pulp at favorable prices or to merely increase the raw material supply from which the company will benefit. Of course, many consulting firms offer technical assistance to forest landowners at a price.

Financial Assistance Financial assistance is available in the form of cost sharing through federal, state, and private programs. For example, cost-share programs may help some private owners to pay for the cost of qualifying forest practices, such as tree planting or thinning, and write forest stewardship management plans for their property. These programs may also serve in cost-share development of ponds and in restoration of wetlands that have wildlife, water fowl, and fisheries habitat benefits. The Conservation Reserve Program (CRP), which pays landowners to retire land for a 10-year period, also provides cost sharing for planting of trees and wildlife habitat improvements on that land.

States may also provide educational, technical, and financial assistance to increase productivity of their private forests, and they may also provide matching costs and cooperatively administer the federal programs.

Tax incentives, coupled with programs of education, technical assistance, and cost sharing, are important incentives for the practice of good forestry in the private sector. Without them, many private (NIPF) owners would not find timber growing to their advantage or would not practice any degree of forest management on their lands.

Thus, federal and state financial and technical assistance programs represent government investment in the nation's forest management enterprise. In the economics of forest management, nationally for the federal government and at state levels for the individual states, investments in financial and technical assistance programs must be weighed against the additional gains in productivity and conservation benefits that will result, effects on land use allocations, and the opportunity costs arising from not using the same money, efforts, and resources on some other use. (These analytical concepts of "marginal analysis" and "opportunity costs" are further explained in the following section, "Forestland as an Investment.")

FORESTLAND AS AN INVESTMENT

ForestLand Investment Accounting

Forestlands and their associated resources are capital assets with enormous (and renewable) capacity for production and use. As *capital assets,* they generate a long-term flow of products, services, use opportunities, and benefits, although for small landowners, the income from timber harvests may be periodic. Three kinds of expenses are assumed by the forestland investor: (1) capital costs, (2) operating costs, and (3) maintenance costs, including replacement and enhancement expenses.

Capital Costs   are the expenses for owning the resources. They represent the largest part of the cost of generating annual flows of products and services. Capital costs include the expense of purchasing the land and recurring costs of ownership such as payment of taxes.

Operating Costs   refer to recurring expenditures associated with the annual harvest and use of the resource, such as timber sales preparation and supervision, services to recreation and other resource users, permit granting and supervision, fee collections, and other related overhead expenses. These are the expenditures associated with capturing the benefits of forest resources for commodity production or human enjoyment and consumption. In normal accounting, these costs are chargeable to the uses and the products produced, rather than to the capital (resource) maintenance account.

Capital Maintenance,   including *replacement* and *enhancement* costs, covers regular forest resource monitoring activities; maintenance of improvements including roads, trails, and facilities; and activities for resource protection from fire, insects, and disease. Reforestation of harvested areas or areas lost to fire, insects, disease, or other perturbations, would be classified as *capital replacement* costs. Also, capital replacement of associated renewable resources might include restocking of fish or wildlife, reconstruction or replacement of depreciated or obsolete recreation facilities, and other restorations.

Generally, both capital maintenance and replacement spending are chargeable to sustaining the current productivity of the natural resources and the existing flow of products, services, uses, and benefits. These costs, like the operating costs, tend to be relatively stable. Some increases would be associated with increased production, such as increased recreation use or timber harvest, where there would be more road and facility maintenance, and replanting on harvested areas.

Capital *enhancement* expenses expand the forest resource capacity to generate more products, services, uses, and benefits in the future. Such expansion can be achieved by investments that improve resource productivity or quality, provide access to underutilized parts of the resource, or actually expand the total resource. Examples would include investing in wildlife and fish habitat improvements; thinning and fertilizing stands of trees to increase growth; building roads and trails to reach inaccessible resources and to reduce losses from fire, insects, and disease; storage of water; afforestation of previously nonstocked lands; planting genetically improved trees or forage; building facilities to increase recreation quality and use; and similar investments that add to the future flow of products, services, uses, and benefits. Thus, capital enhancement costs increase the capital value of the resources. Such investments ordinarily are chargeable to capital accounts and future production and use.

Annual spending for *operating costs* and *capital maintenance,* including *replacement* and resource *enhancement* costs, is usually a small fraction of the capital value of the resources, and this is an important attraction for forestland investors. For example, as compared to investing in apartment buildings, the operating capital maintenance, replacement, and enhancement costs are small, and the forest will literally "grow" in value.

Forestland Investors

Not all landowners have acquired their natural resource capital assets in the same manner or for the same purpose. Generally, nonindustrial private landowners acquire their forestlands as a part of the purchase of a ranch, farm, rural residence, forest estate, or summer home. Many became landowners through inheritance, from real estate speculation, or as an investment in forest property for production purposes. Industry has acquired forest assets largely as strategic investments to assure future wood supplies for their mills. Federal government forests were, for the most part, originally public domain land, with a smaller portion acquired by exchange with other owners and, in some cases, by outright purchase, particularly for lands having a high recreation value. State lands have generally been acquired as federal grants from the public domain, through tax foreclosures, and by direct purchase.

It is not surprising, then, that this diversified group of natural resource asset owners would have a variety of management and investment objectives. For most owners of 100 or more acres (40.5 ha), management and investment for timber growing and harvest is often a leading objective. Owners of smaller tracts typically own land because it is part of their residence and they are more interested in aesthetics and wildlife, but this does not necessarily preclude timber harvest, and such values may even be enhanced by careful harvests.

The larger, private timberland ownerships are held by businesses and individuals as investments. Forestland is an attractive investment in some circumstances because it has an unusual combination of investment characteristics. Forestland, like farmland and some other types of real estate, has proven to be a hedge against inflation. Timberland owners usually have found that the value of their property at least has kept pace with inflation, and in some instances, has appreciated in market value at a substantially higher rate.

As the attractive nature of forestland investments has become better known, significant amounts of capital have been invested in forestland by institutional investors—banks, investment syndicates, insurance companies, pension funds, and foundations, as well as individual businesses and households, both domestic and foreign (Binkley, Roper, & Washburn, 1996). When inflation is substantial, it becomes important to protect capital resources by shifting capital toward those investments that keep pace with inflation. Investors also like the possibility of selling some of their timberland holdings for recreation home sites, housing developments, conversion to farming, and other uses that bring a high price in comparison to the land's original price. Profits on this sort of speculation are not infrequent, making forestland investments even more attractive.

The timber component of a forestland investment grows naturally with time, and so appreciates in value without any effort or additional expenditure. Of course, natural timber growth can be augmented by timely harvest, thinning, and other carefully selected management practices. Also, stumpage prices paid to harvest timber, like land prices, have increased at a rate equal to or greater than the inflation rate in most instances. In some regions, like the Pacific Northwest, where timber harvests on public lands are being reduced for environmental reasons, prices for stumpage have risen dramatically in response to the shrinking supply.

Returns on forestland investments, then, have been quite favorable for many owners. Although such investments are not for everyone, they make good sense for the forest industry in that such investments provide a dependable supply of raw materials for processing mills. Other businesses, institutional investors (such as pension funds), and individuals who have substantial capital resources and a long-term planning horizon can also benefit. Forestland investments serve as a hedge against business cycles because timber continues to grow and add value despite short-term, downward fluctuations in business trends.

Marginal Analysis to Guide Management Investments All production units conform to an important economic principle called *the law of diminishing returns.* This principle states that for each production

---

TABLE 15-3

MARGINAL ANALYSIS SHOWING HYPOTHETICAL REDUCTIONS IN COSTS OF PARK MAINTENANCE AND REPAIR FROM INCREASING INVESTMENT IN COSTS FOR PARK PATROL. THE POINT OF DIMINISHING RETURNS IS LEVEL 4, WHERE THE INCREASE IN INVESTMENT EXCEEDS THE BENEFITS OBTAINED.

Level	Costs of park patrol	Investment in costs for park patrol	Costs of maintenance and repair	Benefits decrease in costs of maintenance and repair
1.	$3,000		$3,800	
2.	$3,700	+$700	$2,800	−$1,000
3.	$4,400	+$700	$2,000	−$800
4.	$5,100	+$700	$1,400	−$600

---

unit, some increase in output is available for only a small additional cost or effort, but a point is soon reached at which additional output is available only at an increasingly large cost. Analysis to determine the point where increased outputs exceed the cost of obtaining them (the point of diminishing economic returns) is called *marginal analysis.*

Consider a hypothetical example. Table 15-3 depicts trade-offs in a park budget between spending for patrolling the park (a cost), and the resulting reduction in the costs of maintenance and repairs (a benefit) due to reduced damage from careless use and vandalism.

By increasing the patrol budget from $3,000 to $3,700, costs of patrolling increase by $700, but the costs of maintenance and repair are reduced by $1,000. Thus, the benefits exceed the costs. By increasing the patrol costs by another $700, from $3,700 to $4,400, the costs of maintenance and repair are reduced by another $800. Again, the benefits exceed the costs. However, an additional $700 increase in the patrol budget to total $5,100 only reduces the costs of maintenance and repair by $600, and thus the costs exceed the benefits. It is apparent that spending $4,400 for the patrol budget is best, since that is the last point where benefits exceed costs. Thus, as stated earlier, marginal analysis is the procedure of looking at successive increments in output compared to the investments required to obtain them, and finding the last increment where the value of added benefits exceeds the added cost. Through marginal analysis, park managers in the previous example determine the optimum intensity of management to apply.

Similarly, the law of diminishing returns and marginal analysis can be applied to a forest management example such as investments in fertilizer. Initially, investments in fertilizing trees may return more value in enhanced growth than the cost of fertilizer. But as additional money is invested in fertilizer, the returns would diminish, and, at some point, the costs would equal or exceed the value of enhanced growth. Because the cost of interest on the money invested in fertilizer must be included in the analysis, it is common to fertilize 5 to 10 years before harvest, lessening the time that interest must be paid and, thus, the cost involved.

Amenity Values

As discussed previously in this chapter, many forests also are managed for *amenity values,* such as wilderness, wildlife-viewing opportunities, the preservation of rare or endangered species of animals and plants, scenery, water quality, and other intangible benefits. These amenities create real benefits for forest users but are intangible in the sense that their economic value is usually not quantified (Fig. 15-10). Nonetheless, a judgment must be reached about how much expense is appropriate in order to secure amenity values. Some of these amenities come from public lands partially or completely dedicated to these ends, such as wilderness areas. But modifications of commercial use to preserve scenic values or water quality, to augment wildlife habitat, or to protect critical habitat for endangered species may take place on public timberland, or on private industrial and nonindustrial lands. All of these exclusions and enhancements have a cost. For the past decade at least, a focal point of controversy in forest policy concerns the degree to which environmental protection costs, such as those for protection of endangered species habitat on private lands

FIGURE 15-10
Scenic amenities such as Sahale Falls create real benefits for forest users, as well as highway travelers. *(Courtesy of Oregon State Highway Commission)*

(e.g., the spotted owl), should be imposed on private landowners by the government without compensation (Hickman & Hickman, 1990).

Opportunity Costs   In some instances, expenses for environmental protection represent an *opportunity cost,* or value foregone, at least in part. For example, when an area receives wilderness designation, its timber cannot be harvested, and the value that is given up or foregone thus creates an opportunity cost. For purposes of economic efficiency, then, this opportunity cost reflects the minimum value that must be associated with the intangible outputs from the area as wilderness. If an area suitable for wilderness has a million dollars worth of timber, then designation of the area as wilderness means foregoing the opportunity for that million dollars of timber income. Such

designation would be a logical (efficient) action only if the area's wilderness values were judged to be worth more than the $1 million foregone.

Thus, the principle of *marginal analysis* can also be helpful in reaching decisions regarding the creation of amenity values, just as it is for the production of market goods and services. The principle suggests that amenity values should be protected whenever the amenity value at stake is more important or valuable than the costs of its creation and/or protection, including the opportunity of costs of other uses foregone. Some amenity values can be partially measured in economic terms, as with the situation wherein improved wildlife habitat enhances wildlife populations for viewing and hunting. Market value here could be partially based on the sale of hunting licenses or entrance fees. Likewise, protection of forest aesthetics might be measured by increased land values. Resource enhancement to improve water quality and increase water yields might be subjected to marginal analysis to see at what point the increases in water quality and yield no longer justify the investment. When markets do not, or only partially, reflect the values and opportunity costs at stake, such decisions may be made politically, perhaps through legislation designating parks, wilderness, or some other form of protection. The marginal value of the resources at stake is determined through political debate.

User Fees

As mentioned earlier under Production of Nonwood Commodities, user fees are one way that landowners may capture these values—for example, for wildlife viewing, hunting, fishing, or other recreation uses such as camping or hiking. However, the level of user fees is generally below their market value because of the U.S. tradition of free or low-cost access to public lands where such amenities are available.

For example, the fees charged for camping in national forests and national parks are below what they would be in a free market where prices are used to balance supply and demand. Thus, in popular national parks, campers may wait in line every morning to claim the first available site vacated by another party, after having paid a nominal fee for a camping permit. Or they may have made reservations months earlier to ensure they would have a camp site. In a free market, prices would be allowed to rise so that demand by campers willing to pay such fees would equal the supply of campsites. Perhaps in the future, the fee for a site with a water view will be commensurate with the cost of a desirable hotel or motel room.

Since there is a large supply of nonwood commodities available on public lands at nominal prices, it is difficult for private landowners to charge market-level prices for private outdoor opportunities, except where there are limited public opportunities nearby or where the private supply is of superior quality. Thus, in Texas, where most of the land is privately owned, fee hunting on large ranches is an attractive enterprise for landowners to supplement livestock production. Private upland game bird and duck hunting clubs are popular in many places where habitat and food supply is maintained to attract and hold birds, thereby providing hunting opportunities superior to those available on nearby public land and thus allowing handsome user fees to be charged. Likewise, in the eastern United States, where public land is limited, more private camping opportunities exist. This is because realistic user fees, reflecting the cost of doing business, can be charged.

Some argue that we should have public policy of charging higher recreation user fees on federal lands, not only to pay the costs of providing such opportunities, but because it would also stimulate more private recreational opportunities. The issue of user fees for recreational use of public lands, even wilderness, is a topic of continuing debate, and trial programs are common, such as the Fee Demonstration Program applied in selected areas by federal agencies. Fees at federal recreation sites now generate $100 million per year, but are a small fraction of the several hundred billion dollars spent by recreationists to visit them (Crandall, 2001).

LAND USE PLANNING AND ALLOCATION

This is a process for reaching decisions about how a forest will be managed. Some parts of the forest may be best adapted for timber growing, and others, particularly public land, for recreation, wilderness, watershed, or wildlife habitat. In the process of assigning various parts of a forest to their best, or highest-valued, uses, it should be stressed that under the principle of multiple use, one use does not necessarily preclude another. For example, timber stands provide wildlife habitat for various species during different stages of forest growth and development. Water is a product from almost all forested land, and with the current focus on water, watersheds, and wetlands, additional emphasis on water is appearing in management plans. Thus, most forests provide multiple benefits, and land use allocation is an important decision forest managers must make. They do this by considering both the natural characteristics and capabilities of each part of the property and the kinds of outputs or uses in greatest demand.

The *economic principle of best* or *highest use* suggests that the choice should be the use, or combination of uses, that provides the greatest total value of output. This highest or best use does not necessarily provide the greatest monetary income, as many of the highest valued uses are based on amenities not bought and sold in traditional economic markets—for example, wilderness, recreation use, scenery, wildlife habitat, and watershed values.

In this way, the forest properties are subdivided into tracts, areas, compartments, allotments, wildlife habitats, watersheds, and other units, each devoted primarily to the production of a particular output, or multiple use combination.

FORESTRY AND ECONOMIC DEVELOPMENT

The use of forestlands has followed a common pattern in the economic development of many places in the world, including the United States. Some nations and regions are at one stage, others at another, but a similar pattern of development is apparent.

Settlement Stage

The first, or *settlement, stage* occurs when the commercial timber resources of a region are initially harvested. Large quantities of old-growth timber are extracted and transported to mills for conversion to lumber and other wood products. This creates jobs and income in the region, while at the same time it clears the land, a first step in conversion to farming uses. All too frequently in the past, this stage was characterized by destructive logging, followed by uncontrolled fires, the conversion to "stump farms" with grazing or marginal crop production around the stumps, clearing and plowing of land poorly suited for agriculture, and a failure to establish new crops of trees on those lands best suited for growing trees.

However, it should not be overlooked that this development stage, even when poorly accomplished, does make possible the settlement process. The ax precedes and facilitates the plow. In many countries, creation of agricultural land from forests is promoted to help feed growing populations. Such was the case in the 1800s in the United States, when millions of acres of forest were converted to farmland.

There is currently great concern about settlement stage deforestation in tropical developing countries where slash and burn practices are employed to create new agricultural land and are thus destroying tropical forests (see Fig. 21-5, page 000). Today, with greater knowledge of the global effects of such practices, people are asking whether the world can afford to allow them. Can we afford the global impacts from such activity as the depletion of tropical forests that produce substantial oxygen, the biomass burning that contributes directly to atmospheric CO_2 build-up, and the loss of biodiversity on which we may ultimately depend? Partly in response to such concerns and world pressures, nations with expanding populations and tropical forests are beginning to modify their development policies. But it is not clear whether such changes will do enough to preserve remaining forests and their beneficial effects on the global climate and biological diversity. The costs of settlement-stage activity are now borne on a worldwide scale, and thus there are attempts to influence this stage of the traditional economic pattern in developing countries.

Custodial Stage

A protective era commonly follows settlement. This *custodial stage* is a constructive response to the preceding destruction and loss of forests. This occurs when people of a region understand that their timber resources are not limitless, there are other values from the forest, and that it is possible and desirable to control the logging, stop the fires, and regenerate the woods. During this stage, logging and forest-based industry decline as readily accessible supplies of commercial timber are exhausted, and forest industry becomes much less important in the region's total economy. In this stage, forestland is placed under professional management, and a rebuilding process begins. This was the situation in the United States during the early to mid-1900s, when the national forests were created. Millions of acres were re-

planted in large public works programs such as the Civilian Conservation Corps (see Chapter 1), and federal programs began to assist the states in controlling fires and encouraging good forest management. Thus, a common focus of international forestry programs in developing countries (see Chapter 21) is education and training of resource managers and field workers, and financial assistance to stop the exploitation, establish management structure, and implement tree planting and conservation measures.

Management Stage

The custodial stage leads to a *management stage* as forests are regrown and are capable of supporting commercial harvest once again. This stage of development seeks to increase the value and productivity of the forest and its related resources through better management; to encourage, yet control, commercial use at sustainable levels; and to develop the recreation, wildlife, and other uses of forests for the benefit of a broader segment of society. In this stage, forest resources begin to make more stable, sustainable, and diverse contributions to a region's economy.

The forested regions of the United States have experienced the first two stages of this sequence of economic development and are well into the management stage—some would say we are into the next or "environmental" stage.

Environmental Stage

Since the widespread awakening of public interest in our natural resources in the 1960s, there has been a struggle to bring forest management into balance with environmental protection. For example, national policy has mandated protection of habitat for threatened and endangered species, such as the northern spotted owl, the salmon and the marbled murrelet in the Northwest, wolves and grizzly bear in the Rocky Mountains, and the red cockaded woodpecker in the South (see Fig. 9-6, page 195). More stringent standards for air and water quality, aesthetics, and protection of biodiversity have been established. The management stage has not concluded, but it is constrained by public input. Thus, there have been major changes in forest management direction and practices on public and private land in response to increased awareness of how forest management affects the lives of people and other living things.

SUMMARY

The forestlands of the United States have a very substantial and growing economic importance. They constitute about one-third of the American landscape and are owned by the government, wood-using industries, and many private individuals. They provide for widespread outdoor recreation activities, habitat for fish and wildlife, and watershed protection and aesthetics. The timber harvested from forestlands each year is the largest commercial product of forests.

Economic principles influence the management of forests in a number of ways. Land use planning seeks the allocation of various parts of a public or private forest to their most valuable or highest use. The intensity of management practiced at any time or place responds to the possibility for increased values of outputs. The more valuable the outputs of timber and other products or services, singly or in combination, the more intensively forests will be managed. Forestland also has proved to be a good investment because it has tended to increase in value more rapidly than inflation, due to growth of the trees and the increasing scarcity of timber supply. Tax laws also have an important influence on the profitability of private forest management.

In public forests, politically expressed demands for more forest recreation facilities, for more environmental protection, for wilderness and roadless areas, for critical habitat to protect endangered wildlife, and for protection of water quality result in a mix of products and services for which these forests are managed.

Because forest protection and management are considered desirable, the federal government has established educational, technical, and financial assistance programs for forest landowners, as have some state governments. Private companies may also provide landowner assistance in return for timber purchase rights or simply to increase the local supply of raw material available to operate their mills.

It is clear that demands will continue to increase for forest products, goods, and services as the nation grows and as the value of forests for amenities and quality of life is recognized. This anticipated demand for all the forest resources will increase the complexity of forest management in the future. But it will also make forest management economically more attractive as an investment for private entities seeking financial returns from commercial forest products, and for public entities seeking multiple products emphasizing those values contributing to quality of life.

We thank Dr. Brett Butler, National Woodland Owner Survey Coordinator for the USDA Forest Service, Northeastern Research Station, for his excellent review, suggestions, and help providing new information sources for this chapter.

LITERATURE CITED

Adams, Darius M. 2002a. Harvest, Inventory and Stumpage Prices: Consumption Outpaces Harvest, Prices Rise Slowly. *J. Forestry* 100(2):26–31.

Adams, Darius M. 2002b. Solid Wood Products: Rising Consumption & Imports, Modest Price Growth. *J. Forestry* 100(2):14–19.

Alig, Ralph, John Mills, and Brett Butler. 2002. Private Timberlands: Growing Demands, Shrinking Landbase. *J. Forestry* 100(2):32–37.

American Forest Council. 1991. The American Forest: Facts & Figures, Washington, D.C.

Barten, Paul K., David Damery, Paul Catanyaro, Jennifer Fish, Susan Campbell, Adrian Fabos, and Lincoln Fish. 2001. Massachusetts Family Forests: Birth of a Landowner Cooperative *J. Forestry* 99(3):23–30.

Binkley, Clark S., Charles F. Roper, and Courtland L. Washburn. 1996. Institutional Ownership of U.S. Timberlands: History, Rationale and Implications for Forest Management. *J. Forestry* 98(9):21–28.

Birch, Thomas W. 1996. Private Forest Landowners of the United States, 1994. *Resource Bulletin* NE-134. USDA Forest Service, NE Forest Expt. Station, Radnor, Pa.

Cordell, H. Ken, John C. Bergstrom, Lawrence A. Hartman, and Donald B.K. English. 1989. *An Analysis of the Recreation and Wilderness Situation in the United States: 1989–2040,* Publication no. F5-345, Washington, D.C.

Cordell, Ken (ed.). 1999. *Outdoor Recreation in American Life: A National Assessment of Demand and Supply Trends.* Sagamore Publications, Champaign, Ill.

Crandall, Derrick. 2001. Recreation Fees at Federal Sites. *Intl. Journal of Wilderness* 7(3):19.

DeCoster, Lester A. 1998. The Boom in Forest Owners—A Bust for Forestry. *J. Forestry* 96(5):25–28.

Fedkiw, John. 1992. Natural Resources: Federal Spending and Resource Performance. 1940–1989. US Dept. Ag. Office of Budget and Program Analysis.

Haynes, Richard W. 2002. Forest Management in the 21st Century: Changing Numbers, Changing Context. *J. Forestry* 100(2):38–43.

Haynes, Richard W., and Kenneth Skog. 2002. The Fifth Resources Planning Act Timber Assessment: A Critical Tool for Resource Stewardship. *J. Forestry* 100(2):6–12.

Hickman, Clifford A., and Maribeth R. Hickman. 1990. Legal Limitations on Governmental Regulation of Private Forestry in the United States. International Union of Forest Research Organizations (IUFRO) Working Party 54.08-3.

Loomis, John B., and Robert Richardson. 2001. Economic Values of the U.S. Wilderness System: Research Evidence to Date and Questions for the Future. *Intl. Journal of Wilderness* 7(1):31–34.

Mehmood, Sayeed R., and Daowei Zhang. 2001. Forest Parcelization in the United States: A Study of Contributing Factors. *J. Forestry* 99(4):30–34.

Murray, Brian C., Stephen P. Prisley, Richard A. Birdsey, and R. Neil Sampson. 2000. Carbon Sinks in the Kyoto Protocol: Potential Relevance for U.S. Forests. *J. Forestry* 98(9):6–11.

Sampson, R. Neil, and Lester DeCoster. 2000. Forest Fragmentation: Implications for Sustainable Private Forests. *J. Forestry* 98(3):4–8.

Smith, Brad W., John S. Vissage, David R. Darr, & Raymond M. Sheffield. 2001. Forest Resources of the United States, 1997. USDA Forest Service. NC Expt. Station, St. Paul, Minn.

World Resource Institute. 1992. World Resources Institute 1992–93, Oxford University Press. New York.

ADDITIONAL READINGS

Best, C., and L. A. Wayburn. 2001. *America's Private Forests: Status and Stewardship*. Island Press, Washington, D.C.

Curtis, Robert O., and Andrew B. Carey. 1996. Timber Supply in the Pacific Northwest: Managing for Economic and Ecological Values in Douglas-Fir Forests. *J. Forestry* 94(9)4–7; 35–37.

Dixon, John A., and Paul B. Sherman. 1990. *Economics of Protected Areas: A New Look at Benefits and Costs*. Island Press, Washington, D.C.

Ellefson, Paul V. (ed.). 1989. *Forest Resource Economics and Policy Research: Strategic Directions for the Future*. Westview Press, Boulder, Colo.

Frederick, Kenneth D., and Roger Sedjo (eds.). 1991. *America's Renewable Resources: Historical Trends & Current Resources for the Future*. Washington, D.C.

Klemperer, D. 1995. *Forest Resource Economics and Finance*. McGraw-Hill, New York.

MacCleery, Douglas W. 1992. American Forests: *A History of Resiliency and Recovery*. Forest History Society, Durham, N.C.

National Research Council. 1998. *Forested Landscapes in Perspective: Prospects and Opportunities for Sustainable Management of America's Nonfederal Forests*. National Academy Press, Washington D.C.

WEB SITES

www.Timbertax.org Timber Tax Website

www.fia.fs.fed.us USDA Forest Service, Forest Inventory & Analysis Website

www.fs.fed.us./woodlandowners USDA Forest Service, National Woodland Owner Survey Website

STUDY QUESTIONS

1. What are two interrelated purposes of applying the methods of economics to forest resource management?

2. Describe the extent and ownership of commercial forestland in the United States.

3. Discuss nonindustrial private ownership of forestlands and why the government would want to invest in improving forest management on these lands.

4. Do you know any NIPF owners? How much acreage do they own? What forest management practices do they use?

5. Describe some of the investments in forest resource management during the past 50 years, and the economic effects of these investments.

6. Discuss three kinds of costs borne by forest landowners and how they might compare to expenses involved in owning some other kind of capital asset.

7. Using an example pertinent to your own experience, explain the economic principle called *the law of diminishing returns* and how this is applied in *marginal analysis.*

8. What are *opportunity costs?* How do they apply to the protection of forest amenity values?

9. Through newspapers or other media, identify a current economic problem based on natural resource issues. (Such problems may relate to import or export policies.)

10. Identify the four stages in development and use of forestlands, and comment briefly on each. What stage do you think we are now in, and why?

CHAPTER 16

Measuring and Analyzing Forests and Renewable Resources

This chapter was prepared in part by Dr. J. Michael Vasievich, Natural Resource Information System, USDA Forest Service, East Lansing, Michigan.

INTRODUCTION

Forest mensuration is the measurement of forest resource conditions and use. This has traditionally included measurement of forest stands and wood products, and also survey of the land on which they grow. Forest mensuration now includes measurement of nearly all characteristics of forest resources. These include forest attributes such as understory vegetation, landscape patterns, wildlife abundance and habitat quality, aesthetic and recreational uses and values, aspects of environmental quality, and other intangible values. Measurement of forests is a rapidly evolving discipline that relies heavily on statistical theory.

This chapter describes many of the ways forest resources are classified and measured, and some of the specialized tools used by resource managers (Fig. 16-1). Development of innovative methods to measure forest resources and estimate the use and value of forests has expanded because of the many new demands placed on forests and their amenities. Computers are used extensively to analyze forest resources and support forest management decisions.

Today, forest managers must consider more biological, economic, and social factors than in the past, when most forest management decisions were aimed at producing commodities. Chief among these new considerations is information on the condition, productivity,

FIGURE 16-1

Left: Tree-measuring equipment and materials. (1) Tele-Relaskop, (2) Relaskop, (3) increment borer, (4) diameter tape, (5) bark gauge, (6) increment hammer, (7) timber cruising book, (8) Cruz-All angle gauge, (9) Panama angle gauge, (10) Suunto compass, (11) Haga altimeter, (12) pocket calculator, (13) wedge prism and case, (14) Suunto clinometer, (15) Wheeler Pentaprism caliper, (16) microcomputer program diskettes, *Right:* (17) log rule, (18) Biltmore stick, (19) tree/log scale books (three), (20) tally book, (21) tree marking paint gun, (22) Spencer logger's tape, (23) crayon and holder, (24) tally meter, (25) tree/log scale stick, (26) caliper. Land-measuring equipment and materials, (27) 100-foot steel tape, (28) transit, (29) hip chain, (30) flagging ribbon, (31) chain tape, (32) map measurer, (33) altimeter, (34) map scales (two), (35) stake wire flags, (36) hand level, (37) staff compass, (38) tally board, (39) plat sheets, (40) grids, (41) aerial photographs with stereoscope, (42) Abney level, (43) Silva ranger compass. *(Courtesy of Forestry Suppliers, Inc., Jackson, Miss.; photos by John Gwaltney)*

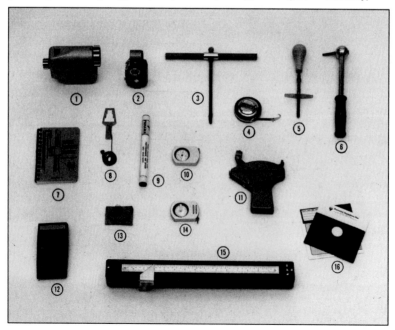

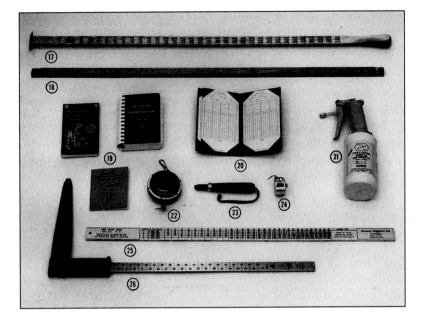

FIGURE 16-1
Continued

health, and diversity of the forest; information on the many different, and often competing, values and uses for forest resources; and linkages to surrounding forest resources and communities.

Methods for measuring timber remain very important and continue to evolve because of high values and increasing scarcity of timber, expansion of intensive timber management, needs to improve timber recovery, use of specialized mill equipment, and new electronic measurement technologies. Careful measurement of most nontimber characteristics has a much shorter history, but interest in this area is intense.

Some areas where significant changes are continuing to take place include the following:

- Improved estimates of forest yields and responses to management, particularly for nontimber resources
- Monitoring of changes in forest health and productivity
- Use of remotely sensed data; integration of computer-based mapping and geospatial information systems to support decision making
- Measurement of ecosystem and landscape characteristics such as biological diversity and spatial patterns
- Development of new ways to classify forest resource conditions
- Measurement of and predicting the ways people value and use forests and respond to proposed management actions
- Estimation of biomass and multiple products from trees
- Measurement of urban forest resources and their effects

CLASSIFYING FOREST RESOURCES

Various schemes have been developed to classify forestlands and the biological and physical resources that occur on them. The purpose of any such system is to group similar areas to simplify and improve planning and management. Each forest area can be classified by any number of systems to recognize different resource characteristics and patterns. For example, areas may be classified on the basis of timber or vegetation type, ecosystem, watershed, wildlife habitat, soils, soil moisture conditions, or other characteristics. Each classification scheme rec-

TABLE 16-1	
THE COMMON FOREST TYPES OF EASTERN AND WESTERN UNITED STATES. THESE CLASSES EXCLUDE MAJOR FOREST TYPES COMMON IN TROPICAL AREAS AND INTRODUCED SPECIES.	
Eastern United States Broad Forest Types	**Western United States Broad Forest Types**
White-red-jack pine	Douglas-fir
Spruce-fir	Ponderosa pine
Longleaf-slash pine	Western white pine
Loblolly-shortleaf pine	Fir-spruce
Oak-pine	Hemlock-sitka spruce
Oak-hickory	Larch
Tupelo (gum)-cypress	Lodgepole pine
Elm-ash-cottonwood	Redwood
Maple-beech-birch	Hardwoods
Aspen-birch	Pinyon-juniper
Nonstocked	Chaparral
	Nonstocked

ognizes different forest attributes. Classification schemes that recognize the ecological relationships within a landscape are relatively new. These classification systems are based on the structural and functional relationships between biological and physical components.

Timber Type Classification

Forest areas can be categorized by the predominant tree species, tree size class (seedling, sapling, poletimber, or sawtimber), stand age class, stand origin (natural regeneration or plantation), site productivity class, or some other category. The most common forest types are shown in Table 16-1. In many instances, forest managers can subdivide these into many, more detailed classes for management purposes.

Ecological Classification Systems

One of the newest systems for classifying forest areas is based on similarities in ecological characteristics such as the occurrence of interdependent communities of plants and animals and their associated physical resources such as soil, water, landscape features, and climate, which combine to create *ecosystems*. They range from small areas to large regional landscapes. Ecological differences in forests have been recognized for a very long time, but many new ecological classification systems now deal with the emerging focus on *ecosystem management*. The new ecological classification systems offer managers more effective ways to understand and thus manage forests and renewable resources in order to conserve and enhance biological diversity and to sustain ecological structure and functions. Maps can now be produced showing ecosystem types. One very practical classification system identifies land by dedicated use or landowner objective.

Measuring the Human Dimension

People shape and transform landscapes. Landscapes also shape the lives of people. Measuring the interactions between people and forests and understanding how to use this information effectively to craft management choices is one of the most challenging tasks faced by forest managers. Failure to effectively address the desires of people and the social acceptability of proposed actions may challenge management, lead to appeals and legal confrontations, or, at worst, cause significant human suffering.

Social scientists are especially interested in understanding many aspects of the relationships among individuals, communities, and forests in order to make better decisions. These interactions include forest uses; values, attitudes, and beliefs about forests; demands and needs; behaviors; and effects of forests on people and people on forests. The types of data collected may, for example, include counts of people, patterns of economic or social activity, behaviors, and laws and policies of social organizations and institutions. People are especially interested in the sustainability of resources that they find valuable; however, the concept of social sustainability is particularly difficult to define or measure.

Although some of the same principles apply, measuring human dimensions usually requires different approaches than those used to measure physical or biological resources. Collecting and analyzing information about human interactions requires counting or observing people, assessing their actions and the effects of their actions, or interviewing them using structured questionnaires to determine their views. While some data regarding people are quantitative (such as counts), many are qualitative and require significant skill and expertise to interpret.

Sampling people to determine some measures of their actions or beliefs is difficult at best. In each case, questions arise regarding whom to sample, what to measure, and how to measure it.

Devising an appropriate sampling strategy is complex. People can be selected randomly for sampling from some list (called a *list frame*), or every person can be sampled at a particular location for some time period. Also, all people within a specified land area (an *area frame*) can be sampled. Some individuals may respond to the questions, others may not. Development of the questions to ask and the way to ask them is difficult and requires careful choices of wording and presentation. People can be interviewed in person, over the phone, over the Internet, or often by mail.

Other methods are also used to elicit information about people. Specific focus groups can be formed to respond to a particular situation or set of questions. The Delphi method uses experts to develop needed information and often involves a series of responses intended to refine that information. The choice of methods can greatly affect the results.

MEASURING AND MAPPING FORESTLAND

Systems of Land Survey

As any property owner knows, adequate land surveys are important to define ownership boundaries, to prepare maps, and to generally define locations of physical features on the surface of the earth. Two surveying systems are common in the United States—*metes and bounds* and *rectangular surveys*.

Metes and Bounds Surveying In the New England and Atlantic coastal states (excluding Florida) and in Pennsylvania, West Virginia, Kentucky, Tennessee, parts of Georgia, and Texas, a method of subdividing land called *metes and bounds* evolved from pioneer settlement and is still in use today. Under this confusing system, property lines follow physical features such as streams, lakeshores, or ridge tops, or artificial features such as fences and roads. Thus, locating legal property lines is now a common problem, and surveyors must often spend days seeking information as to "who owns what." Because land lines followed convenient land features, which may change over time, this system presents an often confusing and uncertain method of defining boundaries. Although property corners may have once been marked with monuments, their descriptions are often vague and many corners have long been lost.

The Rectangular System of Surveys   The more practical rectangular system of surveys applies to the rest of the United States. In this system, land is generally mapped and subdivided in a rectangular grid. From this system, established in 1784, familiar terms such as *township, sections,* and the *lower forty* derive. This system uses established *meridians* and *baselines* as references from which land surveys are made. All states not on the metes and bounds system use it. Each state or small group of states has a common east-west baseline that parallels (but does not necessarily coincide with) the major parallels of latitude. Similarly, one or more states has a north to south principal meridian. The starting point of the survey, where the baseline intersects the principal meridian, is called the *initial point.* The systematic numbering of square townships north and south of the baseline begins at the initial point. Township distances east and west of the principal meridian are measured by ranges.

This numbering system is illustrated in Fig. 16-2a. Township 1 North, Range 3 West, is usually stated T 1 N, R 3 W. This and other areas have been located as examples on the grid. Each township is theoretically 6 miles square (9.66 km), contains 23,040 acres (9,324 hectares), and is divided into 36 1-mile-square sections of 640 acres (259 ha) each, as shown in Fig. 16-2b. Section number 1 is in the northeast corner of the township.

Each section is 80 chains[1] (1 mile) square and contains 640 acres (259 ha). Sections are further divided into quarter sections of 160 acres (65 ha) each, which in turn are divided into four 40-acre (16.2 ha) squares. This last subdivision of 40-acre squares is termed a 1/4–1/4 (quarter-quarter) section, or more simply a "forty." The legal descriptions of the single forties are designated in Fig. 16-2c. Irregular areas within a section, for example, caused by lakes and surveying errors, become lots. The numbering of lots is left to the judgment of the surveyor, although by convention they are usually numbered from east to west and north to south.

To partially compensate for convergence of the meridians in the northern hemisphere due to curvature of the earth, the northern boundary of townships is usually somewhat shorter than the southern boundary. Especially in the southern tier of states involved with the Louisiana Purchase, the rectangular surveys are severely impacted by former Spanish land grants and other odd legal configurations brought about by heavy dependence on river transportation. The nearer to the north pole the township lies, the greater the convergence. Some early surveyors were not particularly careful, and in some instances field measurements may not closely follow this theoretical pattern.

The actual laying out of section corners started in the southeast corner of the township so that human survey errors and discrepancies due to convergence fall in the northern and the western tier of forties, which often makes forties lying along these boundaries more or less than 40 acres.

For each 24 miles east or west of the principal meridian, a guide meridian is established; similarly, a standard parallel is established each 24 miles north and south of the baseline, as shown in Fig. 16-2a. These lines are termed correction lines and are established to prevent further extension of any errors. In many areas, roads follow section lines. Offsets or corrections are often indicated by sharp bends in a road where section lines do not converge.

Marking Corners   Corners of boundaries in either system are marked in many ways, and unfortunately, many corners are not very permanent. Corners were often indicated by stone monuments, piles of rock, or wooden posts and blazed trees. Many of these markers have since disintegrated. Notes of early surveyors in Western and Prairie states make for colorful reading. Often surveyors were paid by the monument, and they would leave town to mark corners only when they ran out of money. However, many early surveyors made careful notes that described the areas they surveyed, especially trees that were present at corners. These Government Land Office (GLO) notes have been used to reconstruct maps of forest types that were common before new settlers arrived. Maps of the original forest made from GLO notes are not precise, but often show that forest conditions have changed significantly over time, especially in the eastern United States.

The Bureau of Land Management, charged with locating corners on their public lands, uses iron or concrete posts with stamped brass caps. Whatever material is used, field notes should describe corner

---

[1]A surveyor's chain is 100 links, 66 feet, or 20.1 meters.

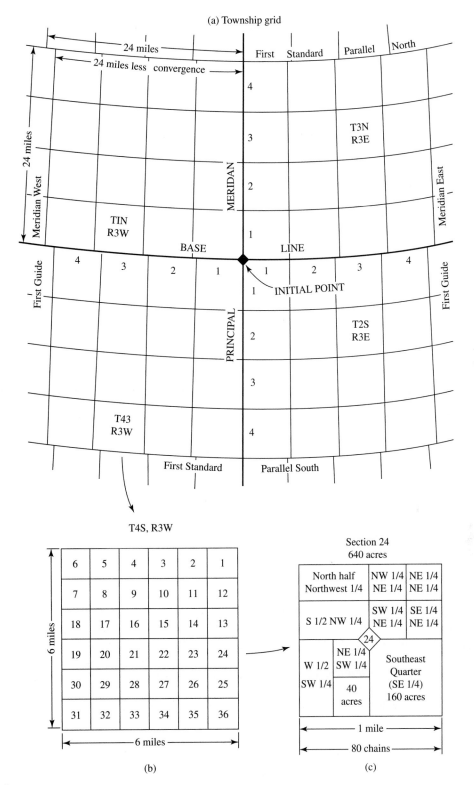

(a) Township grid

T4S, R3W

Section 24
640 acres

(b)

(c)

FIGURE 16-2
Generalized diagram of the Rectangular System of Surveys, adopted from the Bureau of Land Management. (a) Township grid showing initial point, baseline, principal meridian, standard parallels and guide meridians, and examples of township and range designations. (b) Subdivision of township into 1-mile sections and the system of numbering sections from 1 to 36. (c) Subdivision of section into half sections, quarter sections, and forties.

markers and surrounding land features so that they can be found again or relocated in the future. Current technology provides more accurate ways to relocate permanent corners, such as the global positioning systems.

Aerial Photography and Remote Sensing

Aerial photographs are valuable tools for land management. They enable managers to view large areas and save time and money when preparing maps, conducting inventories, monitoring vegetation changes, and planning roads, harvest systems, and recreational uses of forests.

Most areas in the United States and Canada have been aerially photographed, often several times. Periodic as well as special-purpose flights over forestlands are made by companies and government agencies. Most aerial photography done by public agencies is available through state and federal data centers or from private firms.

Aerial photographs are only one type of remotely sensed imagery available. Scanned electronic data from satellites are also used, but satellite imagery cannot provide the detail available in low-level photographs. However, the availability of images for very large areas, in electronic form and for modest cost, makes such imagery valuable for inventories of extensive land areas. Satellite imagery is most often analyzed by using computers to process electronic data files produced by scanning devices that generate data records of the image as they scan a grid of ground points on the earth's surface. Each point may represent the average characteristics of a few to many acres as seen from space. The use and interpretation of satellite imagery is a highly technical field, driven by military intelligence, but from which useful resource management applications continue to evolve (Greer, 1993).

The availability of high-resolution satellite imagery to the general public from both government and commercial sources at low cost has greatly expanded its use. These images, in both digital and photographic forms, are commonly used to map even small areas. For an example of available imagery, see http://www.terraserver.com.

Types of Photography   The scale of an aerial photograph is perhaps its most important characteristic. The scale is the ratio between distance measured on the photograph and distance measured on the ground. A ratio of 1:5,000 means that 1 foot or meter measured on the photograph equals 5,000 feet or meters on the ground. Objects appears twice as large on a 1:5,000 photograph as on one with a scale of 1:10,000, but it takes four times as many photographs to cover the same area. The increase in detail for large scales must be weighed against the increased costs for acquisition and interpretation. Most aerial photography in the United States comes in 9-in by 9-in (23-cm by 23-cm) contact prints, with the scale indicated in inches, a common scale being 1 in equals 1,320 feet (2.56 cm equals 33.51 meters).

The type of film also determines what information can be interpreted from a photograph. Black-and-white aerial photography is the most common. Color film is used to distinguish between vegetation types but is limited to low altitudes because of atmospheric haze. Color infrared is most often used for high-altitude photography because it penetrates haze well. Live vegetation appears bright red on infrared film, which is excellent for identifying vegetation types and locating stressed or dying trees.

Forestry Applications   Two of the primary forest applications for aerial photographs are map making and measurement. Overlapping photographs of the same area, when viewed through a stereoscope, produce a three-dimensional image of the terrain (Fig. 16-1, No. 41). This capability has long been used to make topographic measurements and maps. Most U.S. Geological Survey topographic maps are made from aerial photographs.

Foresters and other resource managers use aerial photography to measure tree height, crown closure, and stand density; to lay out roads and harvest systems; to measure forest areas; to identify vegetation types and make type maps; to asses damage from fire, disease, and insects; to classify landforms and soil types; to locate sample plots, property boundaries, and roads; and to evaluate wildlife habitat for certain species.

Aerial photography has some limitations. Accurate measurement of tree diameters and understory vegetation is possible only from the ground. But combinations of technologies are useful. Aerial photographs are often used to design multistage samples of forest resources. For example, satellite or high-altitude imagery can be used to map similar forest units. Low-altitude photographs of sample strips within units are then used to measure factors such as tree heights and stand density. Representative plots within these strips are then sampled on the ground to measure basal area, volume, species composition, and other characteristics. Aerial photography and remote sensing thus serve to complement ground measurement, reducing time and extent of field work rather than replacing it.

Global Positioning Systems

The newest technology for collecting spatial information is the *global positioning system* (GPS). The system utilizes an array of satellites that transmit radio signals to the earth based on extremely precise atomic clocks. Thus, with the latest equipment, a surveyor or forest manager can record the coordinates of a point in the field with accuracy of 6 to 30 feet (2 to 5 meters). GPS receivers can store thousands of coordinates, which can represent lines or polygons. To record the coordinates of a road or trail for accurate mapping, it is technically possible now to simply drive a road, or bike or walk a trail, and use a GPS receiver to locate precise coordinates every second (Fig. 16-3). GPS technology can also be used to record and locate section corners, research plots, or other features that previously had to be laboriously surveyed to mark or locate (Lance, 1993). Costs of GPS receivers have declined substantially and features have improved in the past 10 years. GPS technology has continually improved with reduced size and much greater integration with computer mapping systems. The most significant limitation that remains is interference with satellite signals caused by tree canopies and mountainous terrain.

Geographic Information Systems

A *geographic information system* (GIS) is a computer-based tool kit for analyzing various data from maps. Data are commonly stored in a GIS as points (e.g., permanent plot locations, fire tower locations, campsite locations), lines or arcs (e.g., hiking trails, streams, roads), and polygons (e.g., forest stands, lakes, soil mapping units).

Data are most commonly represented in a GIS in either a *grid cell* (*raster*) or *vector (arc) format* (Fig. 16-4). Typically, each major information class is stored as a point, arc, or polygon layer (also called a *theme* or *coverage*). Because each point,

FIGURE 16-3

A field operator demonstrates the use of a global positioning system (GPS) receiver. The antenna unit receives signals from a network of earth-orbiting satellites, and through triangulation the operator can precisely locate the position of the antenna wherever it is on the earth's surface. Field notes can be entered on the keyboard of the six-channel receiver. This equipment maps locations for fire control, land survey, forest inventory, endangered species habitat, archeological excavations, recreation facilities (roads, trails, campgrounds), or any other situation requiring precise location. *(Trimble Navigation photo by Fred Schuller)*

line, or polygon has a unique identification code, a database management program can be used to efficiently store and analyze information on map features stored in attribute tables.

Data Input The most common method of entering data into a GIS is by manual electronic digitizing from maps (Fig. 16-5). Spatial data are usually organized as separate layers, each representing one group of map features. Typically, each GIS layer is digitized by electronic tracing over points, lines, or polygons on existing maps to record their spatial coordinates, although geo-data files are becoming available from U.S. Geological Survey and some commercial sites. Scanning devices can also be used if a map is in a machine-readable, black-line format.

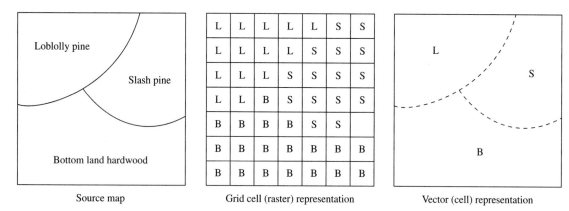

FIGURE 16-4
A GIS representation of a forest stand. The source map at left is portrayed by a grid cell *(center)* and a vector *(right)*.

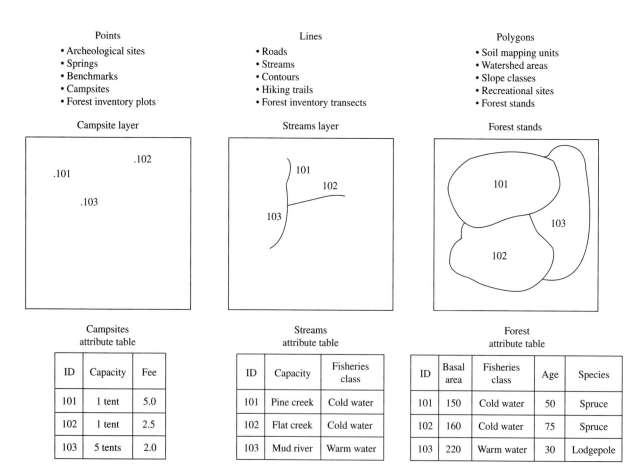

FIGURE 16-5
Examples of GIS layers and attribute tables.

Tables describing the attributes of features in each layer can be updated using the computer keyboard or by joining an already existing database. Engineering survey notes, such as distance and bearings, can also provide input into a GIS if they are recorded in a nonlocal coordinate system such as universal transmercator (UTM), state plane, or longitude/latitude. Another electronic means of locating land features uses GPS to map coordinates automatically.

Remotely sensed data from satellites provide a valuable source of spatial information that has been used to describe many vegetation and landform features. Satellite data can be obtained in grid cell form and can be analyzed by computer image processing techniques to develop broad vegetation layers for large areas. Digital elevation data can also be obtained in grid cell form and used to develop GIS layers describing attributes such as slope aspect, slope gradient, elevation zones, sunlight intensity, watershed areas, and viewshed areas.

GIS Tools   Geographic information systems have many tools that resource managers can use to solve problems. Although GIS have powerful capabilities, these features are of little use unless data on the attributes of interest are carefully collected and entered with acceptable accuracy. Specific tools vary by system; at the minimum, most GISs have the following general functions.

Attribute Query   Questions about points, arcs, or polygons can be answered by using a database management system that stores information associated with each layer—for example, "How many acres (hectares) of ponderosa pine were precommercially thinned or prescribe-burned within a watershed during the last two years?"

Spatial Query   Questions about spatial relationships can be answered because points, arcs, and polygons are stored as coordinates. For example: How many primitive campsites had at least 50 percent occupancy last year? How many are within 2 miles (kilometers) of a parking area? What are the locations of all plantations that had less than 60 percent survival, were planted to Douglas-fir or ponderosa pine, and were planted with bare root seedlings?

Overlay   GIS layers can be overlaid to answer questions concerning combinations of layers. For example, we might have one polygon layer of land ownership, another polygon layer of vegetation types, and another polygon layer of slope class. For example: Where are cedar stands on public land, at least 1 acre (hectare) in size and on a slope of less than 15 percent? Or, the overlay technique can be used as a prediction tool. For example, broad site classes can be predicted by overlaying an elevation layer, a soils layer, and a vegetation layer. Overlay tools can be used to map broad habitat classes such as elk winter range, spotted owl habitat, old-growth forest, and high-biodiversity areas.

Buffer   A *buffer* is a zone of some specific width around a point, line, or polygon boundary. For example, buffer zones along streams or lakeshores generally receive special protection during timber harvest. Thus, buffering questions might concern the extent of harvestable stand areas after buffer zones around streams have been excluded.

Network Functions   A *network* is a set of interconnected arcs such as a stream drainage or road network. Networks can be used for prediction of network loading, such as transport of sediment and water in a drainage system; route optimization, such as scheduling of log trucks on a road, or timing of canoe departures on a river; and resource allocation, such as assigning timber supply to mills or recreation visitors to campgrounds.

Neighborhood Operation   Neighborhood operations answer questions concerning values surrounding a particular location. For example: What are the maximum and minimum stand site indexes within a certain soil type? What is the average size of visible clearcuts within 2 miles (kilometers) of a lookout vista?

Topographic Functions   One of the most powerful features of a GIS comes from the new layers of information that can be generated from one or more source layers. For example, given a layer of elevation points, a GIS can be used to generate contours, slope class, slope aspect, sunlight intensity, watershed areas, and viewshed areas.

GIS Applications in Forestry and Conservation

By using GIS tools, new forest management strategies that focus on landscape and ecosystem qualities and spatial relationships can be considered. Forest management systems demanded by a critical public and by globally competitive industry require that complex information be considered; thus, GIS tools have become essential to successful forest management.

To this end, GISs expand the use of data by using the conceptual notion of space and location as a means to integrate knowledge. If data can be described with a map, it can be put in a GIS set-up and integrated with other knowledge (Convis, 1993). With GIS we can go beyond traditional combinations of soils, climate, wildlife, and plant data and add such things as demographic data—for example, population, income, taxes, and health statistics. GIS expand manager's capabilities, so they can go beyond uses of maps that display timber volumes and habitat types to more complex information. This might include baseline ecological conditions; species occurrence or richness; pollution sources and impacts; current and proposed land uses, classification, and zoning; and simulation of effects of proposed policies such as establishing wildlife travel corridors or protecting old-growth resources. Furthermore, the integrated data can be visually described in a variety of maps so decision makers can see potential effects of alternatives and thus fine-tune decisions so they will be more effective. GIS have become more powerful over time but use of many spatial analysis features requires advanced training.

MEASURING FOREST RESOURCES

Continuous Forest Inventories

Long-range information on forest growth and mortality is obtained by continuous forest inventory. With this method, commonly referred to as CFI, permanent sample plots are located systematically in forested areas and resampled every few years. Data from these remeasured plots, such as species, diameter breast height (DBH), height, form class, understory, wildlife habitat, and other forest characteristics are thus collected periodically and analyzed to monitor forest conditions and determine trends. The USDA Forest Service, in cooperation with state forestry organizations, carries out a continuous forest inventory by measuring several thousand plots in each state on a recurring cycle. The forest inventory has changed from measurement of all plots every 7 to 15 years to an annual sampling scheme that measures some plots every year. This change also integrates remotely sensed data and is designed to track dynamic forest processes, such as harvest rates, more effectively. These data provide estimates of timber growth, harvests, forest conditions and, thus, potential yields for each state or substate region (called *forest survey units*). CFI has done a good job on this kind of broad information, but improved growth and yield models (described later) are replacing CFI on private ownerships.

The availability of tree and plot measurements collected in a forest inventory enable many kinds of analyses such as determining tree distributions, measuring tree biodiversity, estimating growth and yields, determining forest health and risks, and documenting long-term changes in forest composition.

Forest Sampling Methods

Timber cruising is the process of determining timber volume in a forest stand, but the same sampling methods are often applied to measuring nontimber vegetation. Because forests usually cover a large area, the cost of measuring every tree or plant is prohibitive. Thus, timber volumes and other vegetation measurements are usually estimated by measuring plots and then multiplying by the appropriate expansion factor. The sampling can be done several ways.

One of the first tasks of timber cruising is to obtain or prepare a simple map of the forest area. The location of lakes, streams, meadows, roads, changes of timber type, understory vegetation, and topography are some of the features that must be recorded, whether logging operations, wildlife habitat improvements, or further road locations are planned. Using such a map, various forest sampling procedures can be designed, based on the precision required.

The process of establishing and measuring a sample is termed a *protocol*. Protocols provide a precise recipe for data collection, analysis, and interpretation. They help assure that data are collected consistently and produce results with known precision.

Several computer innovations improve the efficiency of forest sampling or timber cruising. One of the most important improvements is the use of handheld field data recorders to tally plot information as trees are measured. These devices can also transfer cruise data directly to office computers to calculate timber stocking, volumes, and other measures of forest conditions.

Random Sampling   Random sampling is a system that gives each part of a forest an equal and independent chance to be sampled. The forest area is subdivided on a map into many areas of equal size. Then, plots equaling an acceptable proportion, say 5, 10, or 20 percent, of these areas are chosen at random and measured.

This method has several disadvantages. Some areas are measured completely and others not at all. Uniform coverage of the forest tract is left up to chance, and some areas may be missed altogether. Foresters may want to sample some portions of a forest more intensively than others because they contain more valuable or more variable timber. Also, completely random sampling does not provide an easy or

effective method for locating sample plots or for mapping land features while doing the sampling (Fig. 16-6a).

Stratified Random Sampling   It is desirable to design a sampling procedure that minimizes costs yet provides estimates of acceptable statistical precision. Improvements in sample design and efficiency can be achieved if some additional information is available. Aerial photographs can be used to subdivide the area into distinct forest cover types for separate sampling in each cover type. The result is a stratified random sample. Aerial photographs also simplify the task of locating sample plots and features on the ground. Topographic and cultural features needed for planning can be mapped directly from photographs, and the area of timber types determined. Overall variation is reduced by sampling within each cover type, so the total number of plots

FIGURE 16-6
(a) Random sampling method wherein plots are selected by drawing numbers from a hat. As indicated in this example, the plots are seldom evenly distributed. Although the cruise covers 10 percent of the total, it is not representative of the true cover because no plots fell in the swamp or grassland. (b) Systematic sampling method showing 16 quarter-acre plots, each with a radius of 58.9 feet. The cruise is 10 percent of the "forty," or 4 acres. The arrows indicate the direction of the cruise. The plot centers are 5 chains apart. The cruiser started in the southeast corner and offset 2½ chains before heading north 2½ chains and locating the first plot.

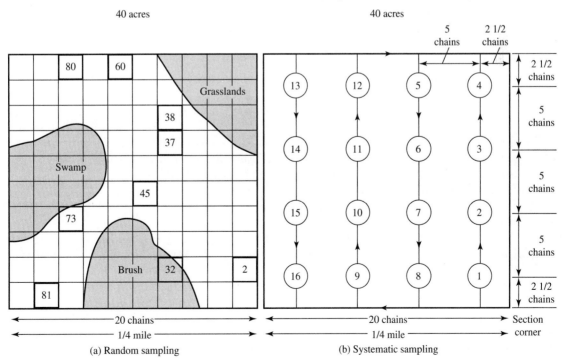

(a) Random sampling   (b) Systematic sampling

TABLE 16-2

COMMONLY USED PLOT DIMENSIONS FOR CIRCULAR AND SQUARE PLOTS, AND ITEMS MOST EFFICIENTLY MEASURED

Plot size acres	Plot radius feet	Square dimension feet	Commonly used to measure
1/1,000	3.7	6.6	Regeneration (small seedlings)
1/100	11.8	20.9	Young stands (saplings)—understory vegetation
1/20	26.3	46.7	Young stands (poles)
1/10	37.2	66.0	Timber
1/5	52.7	93.3	Timber
1/4	58.9	104.4	Timber
1	117.8	208.7	Vegetation type—large, high-value timber

required for a given degree of precision is reduced in stratified random sampling.

Systematic Sampling   This method of forest sampling assures an even distribution of plots and thus lends itself to map making (Fig. 16-6b). Plots are located along predetermined lines, and the timber cruiser follows these lines, usually with a compass, establishing measurement plots at given intervals.

Sampling with plots located at equal distances on several parallel lines is the most common sampling method for measurements of timber, understory vegetation, seedling survival, or other vegetation characteristics. Plots are usually circular, but square plots may also be used. The radius of the plots may vary with the sampling intensity. Plots ¼, ⅕, or 1⁄10 acre in size are usually used for timber cruising. Commonly used plot sizes are shown in Table 16-2.

Strip cruising is another systematic sampling method in which all trees or vegetation are measured for a given distance on each side of a compass line. The cruise strips are laid out parallel to each other, and spaced so that the area measured provides a sample of the desired intensity. Strips are usually 1 chain (66 feet or 20.12 meters) wide and spaced 5 chains apart (a 20 percent sample) or 10 chains apart (a 10 percent sample). With 1-chain strips, the cruiser measures all trees within 33 feet (10.1 meters) on each side of the compass line. Strips ½ chain wide may be used in dense, young stands, or 2-chain strips may be used in open, larger timber.

Variable Plot Cruising   Another type of systematic sampling is variable plot cruising. Variable plot cruising is also known by several other names, including

prism cruising, angle count, point sampling, plotless cruising, and *variable radius plot sampling.* Samples are taken at regular intervals along cruise lines as described, but fixed radius plots are not used. At each sample point, the cruiser counts all trees whose diameters exceed a constant angle when viewed from the sample point. Therefore, the chance that a tree is sampled, or counted, depends on its size and its distance from the sample point.

Several devices called *angle gauges* may be used to determine if each tree is either "in" or "out" of the sample; a common gauge is the *wedge prism,* a piece of glass that shifts the image of the tree to indicate whether the tree falls within the angle or not (Fig. 16-7). Other types of angle gauges are also available.

The basal area per acre estimate at each sample point is computed by multiplying the tree count by the basal area factor of the prism or angle gauge. The timber cruiser is really measuring several plots because each tree generates its own plot proportional to its square feet of basal area. By measuring the diameter and height of the counted trees, volume per acre can be calculated from the sample.

Measuring Tree and Stand Characteristics

Tree Diameter   Tree diameters are most commonly measured at a standard height, called *breast height,* 4.5 feet (137 cm) above ground level, because most tree species have a lot of taper just above their base and by 4.5 feet this taper has subsided. Adjustments can be made for leaning trees, abnormal growth, or trees on steep slopes. The diameter at breast height, or DBH, of a tree is usually measured outside the

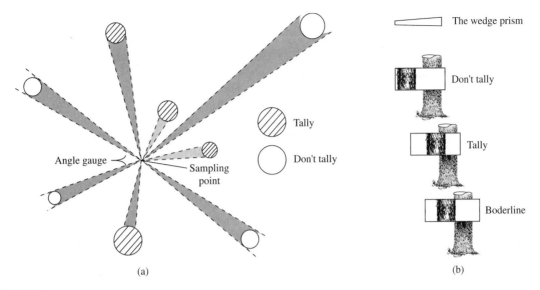

FIGURE 16-7
How variable plot cruising works. (a) The sampling point is the position of the observer's eye. The circles represent the DBH of trees. The shaded area shows the angle of projection defined by the edges of the angle gauge. Only those trees that intercept a greater angle are tallied. (b) The wedge prism and how it tells which trees to count. If the displacement of the tree, as seen through the prism, is beyond that part of the tree not seen through the prism, the tree is not counted. If the displacement is less, the tree is tallied. If the displacement coincides with the tree's edge, it is borderline and is counted as half a tree.

bark to the nearest 0.1 inch, 1 inch, or 2 inches (or 1 cm), depending on the accuracy desired. Measurements are usually made with a diameter tape measure, tree calipers, or Biltmore stick (Fig. 16-8).

Tree Height To determine tree volume and site index (see page 375), the height of trees must be ascertained. Several measurements are of particular interest to foresters—total height to the top of the crown and merchantable height to any of several possible top diameters. Tree heights may be measured in feet (meters) or as the number of merchantable logs or pulpwood sticks. Skilled observers can quickly and accurately estimate tree heights, especially the number of merchantable pieces.

It is usually quite difficult to measure the height of a tree with a tape measure, so hand-held optical instruments such as the Abney level (Fig. 16-1, no. 42), Relaskop (Fig. 16-1, no. 2), or clinometer (Fig. 16-9) are available. These devices contain a sight, a level, and an angle scale. The angle scale shows the height of a tree when viewed from a set distance. The tree height is obtained by sighting at the top and base of the tree from a known distance. The two readings are

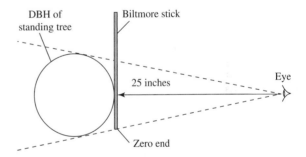

FIGURE 16-8
Measuring the DBH of a standing tree with the Biltmore stick. The zero end is held to the left; the tree's diameter is read at the point where the line of sight, from the observer's eyes to the right of the tree, intersects the stick.

then added or subtracted, depending on the position of the observer. Such readings can be quite accurate, but skill and careful measurement of the distance from the base of the tree are required.

Tree and Stand Age Foresters often want to know the age of individual trees or wish to examine tree rings to determine growth rate and estimate site

FIGURE 16-9
The forester at right is using a clinometer to determine the height of a tree. At left, the forester is demonstrating the use of the hand-held "ranger" compass. *(Photo by Grant W. Sharpe)*

FIGURE 16-10
Using the increment borer to determine the age of a tree. Here, the borer has been twisted into the tree and the core extracted to examine the growth rings. *(Photo courtesy of the USDA Forest Service)*

index. In temperate climates, trees form distinct growth rings each year. These rings occur because trees grow faster in the spring than in the summer or winter. When a tree is cut, its age can be determined by counting these rings. To get the total age of a tree, it is also necessary to estimate the number of years it took for the young tree to reach the height of the cut.

Tree ages can also be determined without felling them by using a hollow, augerlike tool called an *increment borer,* which is twisted gradually into the trunk. As the increment borer is forced toward the center of the tree, it cuts a core of wood that can be removed so that the rings can be measured and counted (Fig. 16-10). Another instrument, an *increment hammer,* can be used if only the outside inch of wood is needed to assess recent growth.

Tree rings also tell an experienced observer about the tree's past growth and the growing conditions such as moisture, nutrition, and competition that influenced it. The science of understanding the relationships between past tree growth and environmental conditions is called *dendrochronology.*

Stocking and Stand Density These represent two important measures of forest condition. *Stocking* is a measure of the number and size of trees growing in a stand relative to an optimum number of trees. The

concept of *optimal stocking* is not an exact one. Forests are managed for different products and goals, so the best level of stocking may vary, depending on the circumstances. Stands may be understocked, overstocked, or fully stocked if they have too few, too many, or just the right number of trees. Stocking of a stand can be represented as a percentage of full stocking, such as 80 percent stocked. In general, diameter growth in overstocked stands is slower, and mortality is higher, than in understocked or fully stocked stands.

The old concept of a "normal" forest is related to full stocking. Such a forest is theoretically composed of the number of trees to fully occupy the site in order to achieve maximum volume growth. This outdated model of a perfect forest is not used much now because more useful models of forest growth are available. It does, however, reflect the idea that forest stands, if undisturbed, tend to approach some "normal" condition and grow naturally toward full stocking as they get older.

A simple measure of *stand density* is the number of trees per acre. This measure is not very useful without more information about the size or spacing of the trees. A better measure of stand density is basal area expressed as square feet per acre. This is the sum of the cross-sectional areas of all trees measured at breast height (4.5 feet or 137 cm). Basal area is a useful indicator of the extent to which trees occupy the site. It is directly correlated with total stand volume and is an important variable used in all modern growth and yield models. Basal area is usually determined by individually measuring trees on plots or with a prism or angle gauge. The same basal area may exist as many small trees in a young stand or fewer large trees in an older stand. Models have been developed to indicate the relationship between basal area and percent stocking for different forest types. The percent stocking varies with average tree size and the number of trees per acre. The basal area for full stocking increases as stands get older and average tree size increases.

Site Productivity   Foresters are concerned with the productivity of forestland—the capacity to grow trees. One of the most basic measures of site productivity is *site index*. Site index is expressed in feet of total tree height at a certain base age. The most commonly used base age is 50 years, but 25 years is also often used, especially for fast-growing species. Site index measures the relationship between the height of dominant trees in a stand and their age, but it is also generally correlated with potential stand volume growth. Site index can be measured on any site with large trees. Accurate measurements of tree age and total tree height are needed for several dominant trees in a stand to estimate site index. This usually requires very careful measurements of the best trees on a site. To determine site index, the measure values are compared with site index curves or equations, which are generally available for most species and locations throughout the United States. This method is not very reliable for use on young trees because early tree growth is variable (see page 000).

Some large public and private landowners use other specialized systems to classify forest sites and estimate productivity. One common site productivity classification system used by the USDA Forest Service and others identifies seven site productivity classes (by species) based on the potential of sites to grow timber in fully stocked stands (Table 16-3). The actual productivity of many stands is less than the potential amount indicated because they are less than fully stocked or because off-site species are growing there.

Many factors such as soil structure, fertility, and water abundance affect the productivity of forestland.

TABLE 16-3

TIMBER SITE PRODUCTIVITY CLASSES AND THEIR PRODUCTIVITY LEVEL

Timber site productivity class	Potential annual productivity in tree growth
Class 1	225 + cubic feet/acre/year
Class 2	165 to 224 cubic feet/acre/year
Class 3	120 to 164 cubic feet/acre/year
Class 4	85 to 119 cubic feet/acre/year
Class 5	50 to 84 cubic feet/acre/year
Class 6	20 to 49 cubic feet/acre/year
Class 7	Less than 20 cubic feet/acre/year

While some limitations occur naturally, some result from past land use. In many cases, these limitations can be reduced and site productivity increased through management actions. For example, drainage can improve growing conditions on waterlogged sites, nutrient deficiencies can be improved by adding fertilizer, and growth on areas with compacted soil can be increased by subsoil ripping. Furthermore, actual productivity can be increased substantially through intensive management and by planting genetically improved trees or species better adapted to sites where less desirable and less productive species are now growing (see Chapter 19, "Forest Management on Private Lands").

Form Class   *Form class* is a measure of the amount of taper in the lower section of a standing tree. It is usually expressed as the ratio of the DBH and an upper stem diameter—usually the diameter inside bark at the top of the first 16-foot log. A form class of 78 indicates that the inside bark diameter at the top of the first 16-foot log is 78 percent of the DBH. Form class usually varies from 76 to 84 percent, but some species may fall outside this range.

Tree and Stand Volume   *Tree volume* is an estimate of the useful volume of wood found in a tree. Volume estimates can be expressed in terms of board feet, cubic feet, cords, total biomass tons, or other pertinent measures. Volume tables or equations take into account variation in volume due to species, tree diameter and height, form class, and common manufacturing processes. Tree volume tables or equations are developed by actually measuring the product yields from many trees.

Three types of volume tables or equations are in general use. *Form class volume tables* are based on species, DBH, merchantable or total height, and form class. *Standard volume tables* are based on species, DBH, and merchantable or total height. *Local volume tables* are based on species and DBH only and are usually developed for localized conditions from form class or standard volume tables.

Measuring Logs

Foresters involved with timber production need to predict the amount of wood products that can be produced from harvested trees, as do the mill owners or loggers who may purchase timber.

Board-Foot Log Rules   A *log rule* is a formula or table that gives the volume of a log in merchantable logs products, based on its diameter and length. *Board-foot log rules* predict the lumber that could be sawn from logs. Buyers of sawlogs know that not all of a log can be used for lumber. Logs are not square, straight, or uniformly round. The first cuts, which square the log, remove *slabs*. Each pass of the saw through the log produces sawdust, and the width of this cut is the saw *kerf*. Side trimming of each board or plank removes *edging*. Log buyers use log rules to estimate what can be cut from logs so they can estimate their value and volume before they are sawn. Log rules account for wood lost to saw kerf, slabs, and trimmings, although considerable overrun is common, especially in recent years with the development of computer-guided saws of much narrower width. There is considerable variation in the design of log rules, although each shows the estimated usable volume based on log measurements. A few of the more common board-foot rules are described here.

Scribner Log Rule   This is the oldest rule in general use and is the official rule in many parts of the United States. This rule is based on a simple diagram of boards drawn in a circle that represents the small end of a log. The volume represented by these boards was computed for various log lengths and put into table form. The Scribner rule underestimates the actual amount of lumber that can be recovered from logs, particularly small-diameter logs sawn with modern sawmill techniques. Because this rule does not account for log taper, it also underestimates the lumber contained in long logs.

International Log Rule   The International log rule is in common use, especially in the eastern United States. This log rule is based on a series of formulas and accounts for both taper and saw kerf in its various forms.

Doyle Log Rule   This is another very old log rule derived from a simple mathematical formula based on the diameter and length of logs. Although it is still widely used, it is very inaccurate and underestimates the volume of logs up to 30 inches and overestimates the volume of large logs.

Cubic-Foot Log Rules   The cubic foot is being used increasingly to measure logs and trees because of greater accuracy and usefulness for conversion to a wide variety of forest products. Cubic-foot volumes are often expressed in *cunits,* or hundreds of cubic feet (e.g., 600 cubic feet = 6 cunits). Estimated volumes are based on mathematical formulas that represent the size and shape of logs.

Table 16-4 shows estimated volumes of 16-foot logs using several different log rules and based on inside bark diameter at the small end of the log.

The term *overrun* is used when the volume of lumber actually sawn from a group of logs is greater than the volume estimated by one of the log rules. The term *underrun* indicates that the sawn volume was less than the estimated log volume.

TABLE 16-4

A COMPARISON OF COMMON LOG RULES FOR 16-FOOT LOGS

Log diameter inches	Scribner board feet	International ¼-inch board feet	Doyle[1] board feet	Cubic foot/ cubic feet
10	54	65	36	11
12	79	95	64	15
14	114	135	100	20
16	159	180	144	25
18	213	230	196	32
20	280	290	256	39
22	334	355	324	46
24	404	425	400	55
26	500	500	484	64
28	582	585	578	73
30	657	675	676	84

[1]The Doyle rule underestimates volume for logs under 30 inches and overestimates volume for logs larger than 30 inches.

Metric Volume Measurement   Unlike the United States, many countries use the metric system to measure timber and wood products. Foresters in the United States will eventually need to use metric measures if, for no other reason, than to participate effectively in global trade. Many publishers of scientific forestry literature now require authors to report their findings in metric equivalents so that foreign readers can easily compare U.S. results with those in their countries. All distance, area, and volume measurements can be easily converted to metric units.

Log Scaling   A person who uses log rules to determine the volume of logs is called a *scaler*. Log-rule values are available in booklet form or may be printed on scale sticks. These are strong, flat, wooden measuring sticks. The scaler places the stick against the small end of the log to measure the diameter and, knowing the log length, the volume in board feet (or other measures) can be read directly from the stick (Fig. 16-11). Adjustments are then made to account for log defects. Decay is the most common form of defect, but splits, cracks, breaks, fire scars, and *log sweep* (crookedness) also affect net volume and quality.

FIGURE 16-11
A log scaler reading the log volume directly from numbers on a scale stick. The scaler must recognize log species and know how to deduct for defect.
(Courtesy of USDA Forest Service)

Logs can be scaled in the woods, on trucks, in mill yards or sorting ponds, or as they are entering the mill on a conveyor, depending on how they are purchased, transported, or processed. There is considerably more to log scaling than reading numbers off a scale stick. Scalers must be able to recognize species from the wood or bark and must know the rules and regulations on log specifications, how to recognize log defects, and how to apply this knowledge. Gross volume is determined from log dimensions using a log rule, but net volume is calculated by deducting an estimated volume of defect (called *cull*) from gross volume.

Cordwood Measurement   A measure of stacked round or split wood for fuel or pulpwood is a *cord*, which somewhat loosely occupies 128 cubic feet of stack space. The *standard cord* is a stacked pile 4 feet by 8 feet by 4 feet high (1.22 meters by 2.44 meters by 1.22 meters). Solid wood or wood and bark occupy 75 to 100 cubic feet, depending on the size of the sticks, their straightness, smoothness, and how well they are limbed and piled. Another measure is the *short cord, rick,* or *face cord*. These terms are used for short-length fuelwood. The pieces in a face cord are less than 4 feet long, so this unit is less than a standard cord.

The number of cords in a stack of roundwood is determined by measuring the length, height, and depth of a pile, calculating the volume and dividing by 128, the volume of a standard cord. When the length of sticks exceeds 4 feet, the measure may be referred to as *long cords*. When wood is not stacked but thrown into the back of a truck, it is guesswork as to how much is there.

Weight Scaling   Many mills purchase logs and pulpwood by weight. This method has several advantages—it is easy, accurate, and expedient. A cord of green wood weighs approximately 5,400 to 5,600 pounds (2,430 to 2,520 kg). A thousand board feet weighs from 9,000 to 13,000 pounds (4,050 to 5,850 kg). With weight scaling, loggers have an incentive to deliver logs soon after they are cut. Log owners who are paid by weight make considerable effort to get logs to the mill before they dry out, as they will then weigh less. In addition, logs left in the woods too long may become infested with blue stain fungi and insects.

Product Recovery Factors

Harvested timber is made into many products. It is often inaccurate to measure tree and log volumes in traditional units such as board feet, because there are so many end products and manufacturing processes. Some log rules are based on outdated technologies, as much wood is not sawn at all but is peeled, sliced, chipped, split, ground, pulped, or used for nonlumber products.

A more useful approach to predicting the amount of end products that can be made from logs or trees is to measure wood volume in cubic feet and apply recovery factors for the desired product. Product recovery factors, when multiplied by the cubic volume of raw material input, estimate the amount of product expected to be manufactured from the input. For example, a lumber recovery factor of 7.3 indicates that 7.3 board feet of lumber can be produced from each cubic foot of logs used.

Recovery factors are most accurate when based on the particular equipment, product mix, and log characteristics of a specific production facility. In many cases, equations or tables have been developed to relate the amount of wood input to the production of end products. Owners and operators of wood-processing plants generally know their conversion factors, and this guides them in bidding for timber. By using cubic-foot units to measure input volume and appropriate product recovery factors, more accurate estimates of output quantities can be made.

Biodiversity

Managing forests to sustain and enhance biodiversity has become a key objective, especially on public lands under ecosystem management approaches. The concept of biodiversity recognizes all of nature's variety, including ecosystems, species, and genes, in a variety of scales. Thus, some measure of biodiversity in forests helps ascertain whether ecological processes are being maintained or enhanced by management.

Several biodiversity indices have been developed and are most useful for comparing two or more areas. The most simple measures of biodiversity, which reflect only the number of different species in an area, are called *species richness*. Perhaps the most commonly used measure of biodiversity is Shannon's Index, which describes the uncertainty in predicting the species of an individual selected at random from a community. In a community with low diversity, one would be more likely to predict the species of an individual chosen at random.

Biodiversity indexes can be calculated for many different characteristics of a forest such as tree species richness, the diversity of understory plants, the diversity of bird species present, diversity in the sizes of trees, or the canopy structure of a forest. However, there is no single measure that combines the many dimensions of biodiversity. That we should even contemplate such a measure shows that humans in their management mode look for simplicity and uniformity even though nature radiates diversity.

Ecologists are concerned with forest fragmentation and patterns in forest and rangeland landscapes for many reasons. For example, some species need a large and unbroken habitat area to maintain a viable population. Also, cumulative effects of human activities and management across landscapes are often revealed in patterns of change. Aerial photos and satellite images can be used to analyze and compare the size, boundary, and configuration of forest patches in large landscapes. A technique called GAP analysis uses satellite imagery to analyze broad patterns of viable species richness to detect areas vulnerable to loss of other species dependent on ecosystem diversity (Scott et al., 1993; Pennisi, 1993). GAP analysis is now being applied in several states.

MEASURING URBAN FORESTS

Forests are crucial to enhancing the quality of life in cities throughout the world. Many cities actively manage urban trees and forests to increase benefits and minimize costs. Street trees are inventoried and mapped to document the type, size, health, and location of trees. Such inventories are used to track the spread of disease or other problems, plan maintenance operations, route utility lines, and determine the need for tree replacements.

Residents and city planners also have a substantial interest in the extent and distribution of trees and green space in cities. Their interests range from the vegetation growing around buildings and in residential yards to street corridors, parks, and undeveloped open spaces such as farms, forests, river corridors, and wetlands. These areas make up the "green infrastructure" of a city and provide wildlife and bird habitat, areas for water runoff, infiltration and flood prevention, and recreation opportunities. Measurements and maps of these areas are critical to effective urban planning to meet the utilitarian needs of people.

An important area of urban forestry involves measurement of the value of urban trees and forests and their effects on people. Most often, these are special studies that estimate effects such as the contribution of trees and green space to property values, effects on flood prevention and other weather hazards, mitigation of local climate, and effects on heating and cooling costs.

FOREST GROWTH AND YIELD MODELS

Timber production is often a major goal of forest management, so predicting forest growth and yield is important. Growth can be measured as the change over time in the size or volume of an individual tree, such as height growth, diameter growth, basal area growth, or volume growth. Foresters are often interested in the growth of a forest stand—a collection of trees—or in the growth of an entire forest—a collection of stands.

Stand or forest growth is usually expressed per unit area for a specific time period—for example, annual volume growth per acre, or basal area growth per acre, over a 1-year or sometimes for a 10-year period. Stand growth is the accumulation of volume on all *growing stock trees,* meaning live trees larger than a minimum size and of acceptable quality (i.e., not cull trees).

For volume estimation purposes, forest or stand growth consists of three components that reflect the net change in volume during a particular period. *Volume increments* is the increase in volume of all growing-stock trees during the growth period. *Ingrowth* is the volume of trees that grow large enough to be counted as growing stock trees during the growth period—trees that enter the smallest measurable diameter class during the growth period but were too small to measure at the beginning. The third component is *mortality* or *loss of growing stock* trees to death or damaging agents that severely reduce their vigor.

Increment and ingrowth increase stand volume. Mortality decreases volume. It is also important to define the time period to which forest growth estimates apply. *Current annual increment* (CAI) refers to net growth for a specific, current year. *Periodic annual increment* (PAI) is the average net growth per year over a short period of time, say, 2 to 10 years. *Mean annual increment* (MAI) is the average annual growth accumulated over the life of the stand at a specific age.

Yield Tables

The concept of forest yield differs from that of forest growth. *Yield* is the accumulation of growth less mortality over time. Yield is the total amount of timber volume or stocking as measured by basal area or number of trees expected at a given age, for a particular site index, and may be represented as total or merchantable volume. Yield estimates are generally organized in yield tables or as mathematical models; several types are used, including normal (based on natural, fully stocked stands), empirical (based on average stands), variable density (based on stand density, age, and site index), and managed stand yield tables (which incorporate such things as genetically improved trees and fertilization).

Growth and Yield Models

Mathematical models of timber growth and yield have evolved from early yield tables and are more precise and applicable to specific conditions. Modern models generally include sets of equations from regression analysis that estimate height and diameter growth, mortality, timber volume, and sometimes stand structure such as diameter distributions. The models are based on data from remeasured, long-term research plots or from continuous forest inventory records. The most useful models estimate the effects of common management treatments such as site preparation, fertilization, and growth responses to thinning. Volume equations are based on the diameter and height estimates and the number of surviving trees. Several types of models exist, including whole stand models (to project development of entire stands), diameter distribution models (to project size classes), and individual tree models (to project development of individual, representative trees).

COMPUTERS AS RESOURCE MANAGEMENT TOOLS

The growth and availability of computer technology has driven advances in forest measurement and analysis from the continuous forest inventory (CFI) approach of the 1950s, to linear programming and simulation of the 1960s, to remote sensing and strategic planning with computer models in the 1970s, GIS in the 1980s, and expert systems GAP analysis, artificial intelligence and integrated information systems in the 1990s and beyond. The trend will continue, making computer literacy essential for all resource managers. New hardware, software, and information management methods are constantly being developed. Microcomputers have increased substantially in power and are lower in cost than ever before. Disks, CD-ROMs, and small tapes can now store massive amounts of information compactly for low cost and thus expand the opportunities for sharing data electronically. Advances in electronic networks, especially the Internet, have made information sharing commonplace.

Demand and competition have increased the need for highly specialized information and analysis among landowners, forest products companies, and public forestry agencies. To manage forests and rangelands for more diverse objectives, managers must now consider a much broader list of biological, economic, and social factors.

Effects of Computer Use

Computers are used as information management and decision tools to improve the organization of information, expand access to databases, and apply more powerful analytical procedures and decision aids. Computers are also used to automate many repetitive tasks once done manually. Tasks that once required substantial computations are done more quickly, and often more accurately, than before. Many problems too complex for "hand methods" are routinely analyzed with computer programs.

One of the most significant effects of computers has been the rapid extension of new technology to practitioners. In many cases, new scientific findings are made available as analytical models implemented on a computer. Also, computer programs have allowed access to more technical information such as large bibliographic databases that quickly reveal key published references on highly technical subjects. Remotely sensed satellite imagery and access to large databases and maps that describe forest conditions and trends are commonplace. These applications are extremely useful for evaluating management strategies, and to support the landscape-level analyses that are now being demanded. Computers

are making possible the network sharing of data between computers, and thus between users, at instantaneous speeds over thousands of miles.

Computers also allow resource managers to use problem-solving mathematical techniques such as linear programming, simulation models, statistical procedures, numerical methods, and system dynamics that were out of reach before. Properly applied, computers help resource managers identify and evaluate more options and project their consequences more broadly than ever before. Although computers and networks expand access to data, the sharing of information and knowledge is more problematic. This step requires that data be made useful to the receiver.

INTEGRATED INFORMATION SYSTEMS

Forest management organizations, especially public agencies, are often spread over large areas with decentralized management structures. In the past, this organizational structure often led to disparate systems for measuring and managing forest inventories. Local managers decided what data to measure, how to measure the data, and how to store the information. There was little need to gather inventory information from many places. The result often was that information collected at one time and place could not be easily joined and compared with information collected elsewhere. This situation is now viewed as a management obstacle, especially for public lands because of the need to address resource conditions across large ecological regions.

The USDA Forest Service has developed an innovative approach—the Natural Resource Information System (NRIS). This collection of databases and analytical tools supports standard protocols for measuring forest resource conditions. Data are collected once and can be used many times because they are stored in the same database structures for all national forests. This approach to standardization means that information is collected in a consistent way, meets established measurement quality objectives, and is stored in common formats. Employees who move from one place to another do not have to learn new systems for collecting and managing forest data. Better yet, information from many places can be combined much more readily to draw a picture of conditions across whole ecosystems.

Effects of the Internet

The rapid expansion of electronic communications has had a dramatic influence on public perceptions and participation in forest management decisions. Residents are no longer isolated when they can surf the Internet and sample hundreds of television channels from a satellite dish or high-speed cable. These information channels provide new ways for people to become informed and engaged in forest resource issues and provide methods for special interests to target large audiences. The effects have been to heighten awareness and contention over most social issues, including those that involve forests.

The effects on forest management organizations have been substantial. The public demands more information about forest conditions and trends and often wants it available electronically. Many public forest management organizations routinely distribute planning documents and many have online systems for delivering interpreted data and maps over the Internet. This has significantly increased demand for employees who know about forests and can effectively communicate using the Internet and other electronic media.

The great expansion of information available electronically has presented some additional opportunities to better understand the public as well. For example, systems are available to automatically scan electronic copies of news stories to track the development of issues related to forests (or any other topic). This process, called *content analysis,* can quickly filter large volumes of published news articles to determine patterns of concerns considered important enough to be reported by the press.

Expert Systems

One especially important use of computers is to reproduce the combined knowledge of many experts, with programs that consider information and draw conclusions very much like humans would. Users interact with the program through ideas stored and expressed in words. These programs, called *expert systems,* capture specialized problem-solving expertise that is scarce and make it available in a consistent and inexpensive way.

Expert systems are now used to help forest managers diagnose tree diseases, recommend silvicultural strategies, determine fire management strategies, provide advice to landowners, help people interpret complex natural resource models, and integrate, search, or manage databases. For example, a call from a citizen reporting symptoms of a sick plant might be "keyed out" on an expert system to diagnose the affliction and suggest treatment.

Expert systems are a recent outgrowth of a larger area of computer science called *artificial intelligence,* or *AI.* Artificial intelligence involves making computers behave in ways that people recognize as intelligent. Expert systems should be able to communicate intelligently with the person using the system. These systems demonstrate their reasoning by clearly explaining to the user how a conclusion was reached. They are able to formulate conclusions with incomplete, ambiguous, or uncertain information. If the person using the computer doesn't know a particular bit of information or is unsure, the expert system should be able to continue by asking other questions or providing answers without that information. The more effectively a computer program can do these things, the better it reflects human intelligence. Expert systems represent a major step forward by making human expertise available to aid decision making by managers who have neither the time nor training to learn and use complex computer programs. This will enhance the power and precision of management decisions.

A particularly promising area of development involves combining expert systems with GIS. Systems such as the Ecosystem Management Decision Support (EMDS) system allow users to test complex management strategies in mapped landscapes. The EMDS system provides an expert framework to evaluate and prioritize projects and helps users understand and estimate their effects on landscape processes such as forest sustainability.

Analyses to Support Resource Management Decisions

Forest and Range Planning Forest companies and public agencies commonly use large analytical models to evaluate various management options and identify preferred strategies based on analysis of outputs under various constraints. Linear programming and other optimization methods have been key planning tools, a notable example being the Forest Service's FOR-PLAN model, on which their land management plans in the 1980s were based. The forest products industry uses linear programming methods to prepare harvesting schedules, optimize budgets, determine the optimum mix of products to produce, and evaluate long-term management policies and the cost of supplying wood to mills.

However, such methods have several disadvantages in that they require large amounts of data and generate tables that may have thousands of rows and columns. They are also computationally intensive and take a long time for runs, even with very powerful computers. Most importantly, they reflect a simplified, linear view of the processes being analyzed. Adequately accounting for the many social, biological, and spatial relationships involved in planning and managing forests and rangelands calls for skills beyond the purview of computers. Linear programs can only maximize or minimize the value of an objective function such as timber or forage production, costs of producing a certain amount of wildlife, or any similar outcome. While solutions can suggest tradeoffs between functions, human judgment is needed to make the wisest choice.

Simulation   Computers can also be used to estimate the effects of proposed management actions before they are implemented. Complex natural and human systems have often been simulated because they are too difficult, expensive, or time-consuming to experiment with directly. For example, simulation has been used to estimate effects of forest pest management strategies, initial attack and suppression effort on wildfires, watershed and streamflow response to management alternatives, and the forest tree development and growth outlined earlier. Simulation models have also been applied to forest product trade, timber supply, demand, and market processes. Of course, answers produced by such models are limited by how realistically the system can be represented by the data and quantitative relationships within the computer. Because they are time-consuming and expensive to develop and use, simulation models are most useful for large or complex problems.

VISUALIZATION SOFTWARE

How forests appear is very important to most people. Technological advances in computer hardware and software have enabled systems to visualize whole landscapes and to immediately "see" the effects of management treatments such as harvesting activity,

road placement, or other developments. With computerized visualization tools, managers can examine or "fly over" a forested landscape to see how it would be viewed from any vantage point. Sharing images of proposed actions with interested communities can avoid many conflicts and improve understanding of how the spatial arrangement of proposed actions affects scenic quality.

Database Management Systems   Resource managers often deal with large amounts of information in the form of lists, tables of data, observations, field plot records, periodic repetitive reports, and inventories. Computer programs, called *database management systems,* are very useful for entering, storing, changing, managing, and reporting almost any kind of information. Database systems also provide methods to sort, find, and subset records so that reports can be produced for selected groups of data. Applications include bibliographies, fire-weather data, reports of wildfires, timber sale records, physical and biological inventories, stand and compartment records, grazing allotments, recreation use, and mailing lists. One common application deals with the timber received at wood yards. A computer is connected directly to weight scales and the operator enters the identification of the supplier as loaded trucks arrive. The computer automatically records the weight and value of timber entering the wood yard, updates the inventory, and uses the information to pay suppliers for delivered wood and inform mill operators of changes in inventories.

A *management information system* (MIS) is a broad extension of database systems that combines databases, analyses, and reports needed for decision making and strategic planning in a single format for use by managers. These systems can display information in many formats—graphs, charts, tables, and text—for routine use by managers who monitor and guide production processes.

Estimating Values   Computers facilitate determination of various types of values, including commodities such as timber and forage, opportunities for outdoor recreation, and qualities of wildlands such as scenic beauty and relatively unmodified ecosystems. Measurement of forest and rangeland values is an important part of management but is also very imprecise. The value of any particular resource quality or commodity is related to how scarce it is and how much people want it, but the term *value* can have different meanings when used in different contexts. Many resource values are hard to measure and are not traded in markets where dollar values are generated, yet estimates of the relative values or worth of resources and uses are essential to comparing management options.

The most straightforward concept of value is the amount of money needed to purchase goods or services. This is sometimes called *exchange value* or *fair market price.* Where active and competitive markets exist, values for resources can be based on the prices paid in established markets by examining actual market transactions for similar resources. The price paid for standing mature timber or acres of forestland is a good example. Values can also be determined by calculating how much it costs to produce a resource, by figuring how much it would cost to replace it, or by estimating future costs and revenues. Resource managers use cash flow analysis at prevailing interest rates to estimate market value.

Some of the greatest difficulties come in valuing nonmarket natural resources or *use values,* such as recreation experiences (see Chapter 12, "Outdoor Recreation and Wilderness Management"). Because landowners do not usually charge people for using forests, or charge only nominal fees, value must be imputed from other evidence. Value here represents a broader concept of the relative worth or benefit people get from the resource and its use.

An even more extended concept of the value of forest and rangeland resources may reflect the intangible benefits from just knowing that they exist. For example, people greatly value wildlife, endangered species, scenic beauty, wilderness, and the spiritual qualities of forests. These *existence values* are easy to recognize but quite difficult to quantify for an individual or for society.

Some indirect methods such as *willingness to pay* or *contingent valuation* can be used to estimate nonmarket or intangible values. Value, in these instances, is determined by what people have to give up or are willing to give up to have the resource. These values are imprecise and do not represent market values, but they can be used to broadly estimate the relative importance or worth of the resource. Significant difficulties emerge when people try to compare values derived by using different techniques. In many cases,

market and nonmarket resources cannot and should not be compared (see Chapter 15, "Economics in Forest and Renewable Resource Management").

SUMMARY

Measurement and analysis of forest resources is rapidly changing, driven by the need to solve broader and more complex resource problems and the availability of computers, new technology, and analytical methods. There is increasing emphasis on measurement of nonmarket and human dimensions of natural resource use—from recreation trends to rural economic development, to demands for conserving biodiversity and endangered species. Emerging demands to manage for sustainable ecosystems will require more detailed information on forest and range conditions than has been collected in the past, greater emphasis on the spatial relationships among physical and biological components of ecosystems, new information about the distribution of many species, and models to better describe, understand, and manage them.

Resource managers often need to perform specialized, and usually computer-based, analyses to support planning and management decisions such as: forest or range investment analysis; design and analysis of field samples and inventories; growth and yield estimation; appraisals of timberlands or rangelands; statistical procedures such as regression, analysis of variance, and analyses of field data and inventories; analysis of experimental results; and trend projections. Because software programs are available to carry out such procedures, managers can use them without detailed knowledge of complex computations if they understand proper applications and interpretations.

LITERATURE CITED

Convis, Charles. 1993. *The Appropriate Use of GIS and Its Relation to Conservation.* ARC News. Summer 1993. Env. Systems Research Institute, Inc., Redlands, Calif.

Greer, Jerry D. 1993. "The View From Above: An Overview of GPS and Remote Sensing Options." *J. Forestry* 91(8):10–14.

Lance, Kate. 1993. "Bringing Technology Down to Earth: A GPS Consumers Guide." *J. Forestry* 91(8):17–19.

Pennisi, Elizabeth. 1993. Filling in the Gaps: Computer Mapping Finds Unprotected Species. *Science News* Vol. 144, 248–249, 251.

Scott, Michael, Frank Davis, Csuti Blair, Reed Noss, Bart Butterfield, Craig Groves, Hal Anderson, Steve Coicco, Frank Derchia, Thomas C. Edwards Jr., Joe Ulliman, and Gerald Wright. 1993. Gap Analysis: A Geographic Approach to Protection of Biological Diversity. *Wildlife Monographs.* No. 123, January 1993.

ADDITIONAL READINGS

Avery, Thomas E., and Harold E. Burkhart. 1994. *Forest Measurements* (4th ed.). McGraw Hill Book Co., New York.

Berry, Joseph K., and Joyce K. Berry. 1988. "Assessing Spatial Impacts of Land Use Plans." *J. Environmental Management* 27:1–9.

Brown, D. G. 1997. Mapping historical forest types in Baraqa County, Michigan, as Fuzzy Sets. *Plant Ecology* 134(1):97–111.

Consolletii, W. L. 1986. "GIS in Industrial Forest Management." *J. Forestry* 84(9):37–38.

Haight, Robert G. 1993. "Efficient Management of Forest-Dependent Wildlife: A Review of Decision Models." Proceedings, Society of American Foresters, National Convention, October, 1992, Richmond, Va.

Marcot, B. G., R. S. McNay, and R. E. Page. 1988. "Use of Microcomputers for Planning and Managing Silviculture—Habitat Relationships." USDA Forest Service General Technical Report. PNW-GTR-228.

Reynolds, K. M. 2001. EMDS: Using a logic framework to assess forest ecosystem sustainability. *J. Forestry* 99(6):26–30.

Rodcay, G. 1990. "Tonto Fires Tests GIS/GPS." *GIS World* 3(5):53–55.

Schuster, Ervin G., Larry A. Leefers, and Joyce E. Thompson. 1993. "A Guide to Computer-based Analytical Tools for Implementing American Forest Plans." USDA-Forest Service, Intermountain Forest Experiment Station, General Technical Report INT-296.

STUDY QUESTIONS

1. Define *forest mensuration*. In what ways has its scope changed in the past decades?
2. Describe a nontimber characteristic now subject to measurement and analysis. Why is its measurement now important?
3. Identify an ecological area, of either large or small dimensions, that you are aware of. What larger area is it part of?
4. What are the problems of the *metes and bounds* system of land survey? What is the newer system called?
5. Which gives more detail—a map with a ratio of 1:5,000 or of 1:10,000? Name six natural resource applications for aerial photos.
6. Discuss three management uses of GIS and GPS.
7. What advantages does *weight scaling* have over *log scaling? Cubic foot volume measure* over *board feet measure?*
8. What data can growth and yield models provide?
9. In what ways do urban planners make use of forest mensuration techniques?
10. How has the expansion of information through the Internet affected public forest policy?

CHAPTER 17

Forest and Renewable Resource Management by the Federal Government

INTRODUCTION

This chapter describes responsibilities for the principal agencies of the federal government that manage public lands or provide technical or financial assistance to other landowners. Because the information is descriptive, and some programs evolve and change, we primarily cite agency Websites as sources of additional information.

The United States encompasses vast expanses of forests and associated rangelands that are diverse in composition and in ownership. Differences in climate, and other site factors, further complicate management. The federal government is a major influence on forest and rangeland management through ownership and management of about one-third of all wildland, and through cooperative programs and regulations that influence management of lands under other ownerships. For example, although it has direct ownership responsibility for only about one-fourth of the commercial forestland of the United States, through research and state and private land cooperative programs providing technical and financial assistance, the federal government extends its influence to almost all of the nation's commercial forestland. The federal government also owns a large share of the noncommercial wildlands, including wilderness, national parks and monuments, wildlife refuges, and other reserves dedicated to noncommodity uses.

More than 15 federal agencies practice forestry or renewable resource management or give important aid or regulation to its practice. Besides the agencies described more fully in the remainder of this chapter, the Department of Commerce reports information through its economic and statistical studies of commodity production and distribution, such as log and lumber exports and imports. The National Oceanic and Atmospheric Administration (NOAA), in the Department of Commerce, supplies vital fire-weather data in certain areas. The Internal Revenue Service in the Department of the Treasury should be mentioned, since it evaluates timber and rangeland properties and taxes outputs from their management. The Office of Management and Budget has a major influence on all federal programs as the central agency planning and managing the federal budget.

The major practitioners of forestry, range, and wildlife management, however, are found in the Departments of Agriculture and Interior, and, to a much lesser extent, in the Tennessee Valley Authority and the Department of Defense.

THE DEPARTMENT OF AGRICULTURE

The U.S. Department of Agriculture is a cabinet-level department of the federal government with several agencies that manage three principal renewable resources or provide related services (http://www. USDA/gov). These three are the Forest Service (USFS), the Cooperative States Research, Education, and Extension Service (CSREES), and the Natural Resource Conservation Service (NRCS).

The Forest Service

The United States Forest Service is the largest forestry organization in the world. Its mission is to provide national leadership in forest management and protection, and in utilization, research, and international forestry. The Forest Service is involved in designating national priorities for land use, formulation of programs to meet national objectives, and establishment of federal forest and rangeland policies to assure maximum contribution of environmental, social, and economic benefits to present and future generations. The Forest Service mission includes four major areas of operation: 1) *The National Forests:* management, protection, and development of the 191-million-acre (77.3-million-ha) National Forest System; 2) *State and Private Forestry:* cooperation with state foresters in providing technical and financial assistance to private owners of forestlands, wood processors, and public and private agencies; 3) *Research:* conducting research that directly or indirectly supports the Forest Service mission and benefits all forest and rangeland-related natural resources; and 4) *International Forestry:* the planning and implementing of forestry programs worldwide to meet U.S. objectives, exchanging information with other countries, and working toward environmental health worldwide.

1. The National Forest System The greatest physical undertaking of the Forest Service is the management of the National Forest System. Units in the system (national forests and national grasslands) are shown on the map in Fig. 17-1.

The national forests, first called forest reserves, were originated in the Forest Reserve Act of 1891

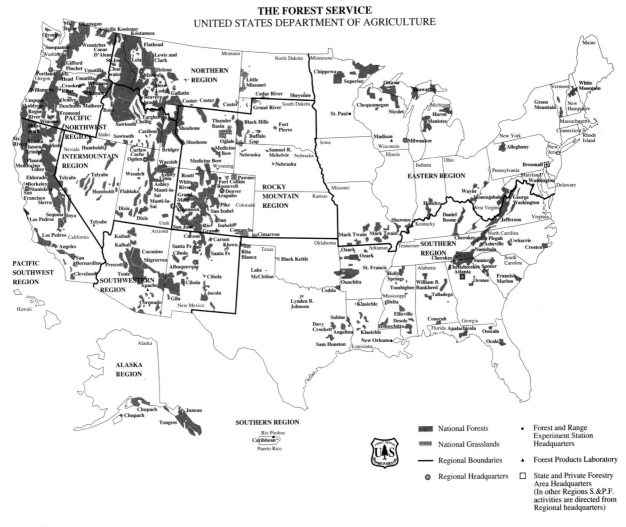

FIGURE 17-1
Map showing regional headquarters and boundaries, national forests, national grasslands, and experiment station headquarters.
(Courtesy of USDA Forest Service)

following two decades of congressional debate over the nation's forests (Steen, 1991). Subsequently, on June 4, 1897, President McKinley signed the Forest Management Act, the organic act that determined the purposes for which national forests could be created, primarily to ensure predictable supplies of water and timber. The 1897 Act provided direction to national forest management until 1960, when the Multiple Use and Sustained Yield Act was passed. Subsequently, in 1976, the National Forest Management Act was passed, giving further direction to the na-

tional forests (see Chapter 2 for a more complete discussion of the creation of the national forests).

Today, the 191 million acres (77.35 million ha) in the national forest system are located in 44 states, Puerto Rico, and the Virgin Islands. They are divided for administration purposes into 9 regions, 156 national forests, and 19 national grasslands (http://www.fs.fed.us 5/8/02).

From their inception, the national forests were dedicated to use. The first set of regulations to guide management of the national forests was called the

Use Book, and the tradition continues today in the "multiple use" doctrine that, along with new concepts and constraints, guides national forest management. A discussion of some of the major activities of the U.S. Forest Service follows.

Forest Fire Management National forests are found from the northeastern United States to California and from Alaska to Texas and Florida. Somewhere in this vast area there is fire danger in every season, and multiple risks and hazards threatening the resource. There will be snow on many of the high western forests while the severe spring fire season is on in the Carolinas. During the summer fire danger period in the Rocky Mountain lodgepole pine forests, moist green cover still exists in the hardwood forests of the Appalachians.

In addition to the control of wildfires, there are other aspects to fire management. For example, prescribed fire is used to improve wildlife habitat, enhance natural cover, control certain forest diseases, decrease slash and other fire hazards, and increase water yields. In wilderness, fire management plans may allow for naturally caused fires to burn, with only a watchful eye in case the fire threatens to escape outside the wilderness boundary, at which point it will be attacked.

Today, after years of effective fire control, many stands are overgrown, and forest health impacts from insects, disease, and drought have further increased fuel loads, leading to recent years with large fires and high fire danger. A national fire plan is being implemented to address this problem by thinning dense forests, restoring burned areas, and helping communities strengthen their fire prevention and suppression capabilities. This national fire plan embraces a large and important agenda for the agency (http://www.fs.fed.us).

Timber Growing and Selling Forests in all regions have stands of commercial forest where timber is harvested judiciously with full provision for regeneration and improving the quantity and quality of the next crop. The Forest Service does not cut and manufacture its own timber, but it does sell planned amounts of standing timber at auction to the highest bidder for harvest under supervised conditions that are designed to safeguard the ecosystem and other values.

In selling timber commercially, the Forest Service estimates the timber volumes on a proposed timber sale area in accordance with the management plan for that national forest. It then appraises the value of the timber, sets minimum prices for the various species, and advertises the total amount for sale, awarding the contract to the highest bidder capable of fulfilling it. The timber harvest by the successful bidder is then administered to see that provisions of the sales contract are met and payments are received from the buyer. Under some contracts, the purchaser deposits money to be used by the Forest Service for disposing of the logging slash and for covering the costs of planting or other forest improvements. Otherwise the purchaser of the timber sale is responsible for these activities.

In 1992, the Forest Service adopted an ecosystem management approach to all its land management activities. Under ecosystem management, everything affected by a practice is considered. Thus, in selecting a silvicultural method for timber harvest on national forests, regeneration of seedlings for the next crop is an important consideration but so are the effects on wildlife, water quality, and landscape aesthetics. The prescribed method of harvest depends on the tree species, associated wildlife, terrain, and the characteristics of the forest stand, such as whether it is even aged or uneven aged. For example, in even-aged Douglas-fir or lodgepole pine stands, small clearcuts were once used with regeneration dependent on seeding from adjacent timber or planting. Now, Forest Service policy is to not practice clearcutting unless it is needed to accomplish some other objective, such as improving forage for wildlife or enhancing forest diversity (Fig. 17-2). In uneven-age forests, partial cuts may be utilized to allow the younger trees to keep growing while still retaining some older trees.

Range Management Mountain meadows, open rangeland, foothill grass and brush areas, sagebrush, and scattered forest openings in the national forests produce a vast amount of forage that is made available through grazing permits and fees for use by local owners to graze their livestock (Fig. 17-3). Such areas are also highly favored by wildlife.

A predetermined grazing fee per head, for cattle, horses, sheep, goats, or hogs, is charged for grazing the animals on national forest ranges. Grazing fees

FIGURE 17-2
The clearcuts in the foreground and beyond were once common practice in even-aged stands of pioneer species such as Douglas-fir and lodgepole pine. Due to public concerns, clearcutting has been discontinued on national forests, unless the objective is something other than harvesting timber, such as for wildlife, regeneration, or forest health. *(Courtesy of USDA Forest Service)*

FIGURE 17-3
Sheep grazing under permit in the Challis National Forest, Idaho. *(Courtesy of USDA Forest Service)*

are based on a charge per animal unit month (AUM). A system of permits, allotting permission to use designated areas of range and limiting the numbers of stock allowed in any one area, is central to managing and using the rangeland while perpetuating the forage resources. Rangeland managers work to prevent overgrazing of the range, to eradicate poisonous plants, to develop water supplies, and exclude stock from sensitive areas. See Chapter 11 for further discussion of this topic.

Watershed Management Much of the nation's water supply flows from the national forests. Use of these lands and their associated water resources is constantly expanding and intensifying. This has led to increasing emphasis on broad interdisciplinary planning and ecosystem management for protection and enhancement of water resources, including methods for increasing streamflow and improving water quality. Watersheds are revegetated quickly after wildfires, and in other instances, to correct or avoid erosion.

Wildlife and Fish Management   These are shared resources in which society has obvious rights and interests. They can be affected by forestry practices. Although the states are traditionally responsible for managing fish and wildlife populations, these populations depend on the considerable habitat found on the forests and rangelands managed by the federal agencies. Obviously, a high degree of cooperation is required between federal agencies and state fish and wildlife agencies to coordinate their respective habitat and population management roles. Lately, wildlife and fish management has used partnerships of federal, state, and private organizations to pool resources and jointly improve wildlife and fish resources on federal lands.

Recreation and Wilderness Management   The concept of multiple-use management that guides the Forest Service provides for recreational use of those lands, where possible and desirable. As the pressures of population and the demand for outdoor recreation use have increased, so has the recreation emphasis in national forest management. In fiscal year 2000 there were more than 250 million visits to recreation sites on the 191 million acres (77,355 million ha) of the National Forest System, and another 250 million people viewed national forest scenery from roads and waterways. Beyond these locations accessible by car or boat, a total of 34.2 million acres (13.9 million ha) of the national forests have been classified as wilderness in the National Wilderness Preservation System, where purposes include recreation and most of the multiple uses except motorized travel and timber harvest (Hendee & Dawson, 2002; www.wilderness.net). Recreation and wilderness on the national forests and other wildlands is considered in greater detail in Chapter 12, "Outdoor Recreation and Wilderness Management."

Engineering and Development   The protection of natural resources and the associated amenities and products available to the public involves careful planning. The execution of these plans requires personnel trained in the physical sciences and engineers to put in place and keep in repair the facilities and developments called for. Initially, these facilities were telephone lines, roads and trails, bridges, cabins, barns, and corrals. By 1911, hydroelectric engineers were needed as dams were built on Forest Service lands. With the advent of the Civilian Conservation Corps (CCC) in 1933, demands for engineering skills were greatly increased as the Forest Service was called on to design and supervise the construction of roads, trails, buildings, campgrounds, and watershed improvements. During and after WWII, Forest Service engineers expedited cutting huge amounts of timber on government lands, as well as working in flood control and photogrammetry.

Today, there is a different emphasis in national forest management, with fewer funds for roads and timber sales. Major road and bridge construction is usually handled cooperatively between the Forest Service and the Federal Highway Administration in the Department of Transportation (Fig. 17-4). But there are an increasing number of other facilities and special uses on the National Forest System, including electronic sites such as microwave relay stations and transmission lines, recreation residences, ski areas, resorts, private road rights-of-ways, and various types of pipelines for which engineering design or monitoring is needed.

Human Resource Programs   The Forest Service was a lead agency in implementing the Civilian Conservation Corps (see Chapter 2). In the period following World War II, there was public support for a similar program, which led to the Job Corps. This program was designed to focus on training and educating youths 16 to 22 years of age from economically disadvantaged circumstances. The Job Corps Program is funded by the U.S. Department of Labor and, as of 2001, there are Job Corps centers nationwide. Thirty-seven centers are in rural locations designated as conservation centers and managed by federal natural resource agencies, 19 of them by the Forest Service and the others by the Bureau of Reclamation and the Bureau of Land Management. The remaining centers are managed by private corporations under contract to the federal government.

The Forest Service and other land management agencies have also participated in many other human

FIGURE 17-4
This national forest road, once used for logging, has been closed and seeded to grass. *(Courtesy of USDA Forest Service)*

resource programs over the years, such as the Youth Conservation Corps (YCC), Young Adult Conservation Corps (YACC), Senior Community Service Employment Program, and others. These programs vary based on available funding and perception of needs by Congress. But one program that has continued strong for years is Volunteers in the National Forests (and in other land management agencies). In this program, groups and individuals donate their time, talents, and knowledge to productive work in all kinds of forest service activities, ranging from serving as campground hosts and wilderness rangers to trail crews and other work, expanding the capability of the agency. Volunteers are sometimes paid expenses, but the real value to participants is the sense of service and connection with the natural heritage of our country.

The general public views conservation work as providing excellent opportunities for the employment, training, and inspiration of people including young professionals, the elderly, the disabled, and disadvantaged and poor youth. These programs that invest in human resources and also develop natural resources make the national forests and other public lands places for growing people as well as trees.

2. State and Private Cooperation   This second major branch of the Forest Service facilitates cooperation with the states through technical and financial advice and assistance on privately owned lands, and through assistance in forest pest control. For example, Forest Service personnel in this work include experts in fire equipment, organization, tactics, planning, and finance, and in their duties they cooperate with state and private organizations in forest-fire fighting, equipment warehouses, fire-research stations, fire-control training conferences, and conferences for planning and reviewing the failures and successes of prevention and suppression efforts.

Similar relations are maintained with state authorities through growing and distributing seedlings for farm and private land reforestation in nurseries partially financed by federal funds, and in supplying instruction and advice to small, nonindustrial private forest owners (see Chapters 18 and 19).

3. Research   The Forest Service operates the largest forest research program in the world for the benefit of all forests and rangelands—federal, state, and private. Plans for research by the Forest Service

FIGURE 17-5
The modern headquarters building for the Southeastern Forest Experiment Station, Asheville, N.C.; one of eight regional stations. *(Courtesy of USDA Forest Service)*

are coordinated with other federal programs, and with research at land-grant universities and other forestry schools. Forest Service scientists include foresters, engineers, botanists, physicists, economists, chemists, entomologists, pathologists, and sociologists, who are supported by technicians, clerks, and administrative personnel. Forest Service research takes place in the Forest Products Laboratory at Madison, Wisconsin, and in nearly every state, under its regional Forest Experiment Stations and outlying field laboratories (Fig. 17-5). The research program produces approximately 2,500 publications each year that describe research findings. Forest and environmental protection as well as commodity production problems are major elements of the research program. A few examples include determining the best means of controlling undesirable forest vegetation; finding the minimum size of log that can be sawed profitably; developing remote-sensing techniques for identifying root rot and other forest diseases; developing better fire behavior simulation models; determining habitat requirements of endangered species such as the spotted owl in Pacific Northwest forests and red cockaded woodpecker in southeastern forests; identifying values and benefits sought by wilderness and recreation visitors and how to increase them through management; and how to improve water quality and quantity through different management regimes. The scope of Forest Service research may be more fully appreciated by browsing its list of publications (www:fs.fed.us/research/publications).

4. International Forestry The topic of international forestry is covered in detail in Chapter 21. The Forest Service is the lead agency in establishing direction for international forestry and cooperating with other agencies and departments of government.

Other Department of Agriculture Agencies

Two other Department of Agriculture agencies have natural resource programs that are covered here. Both of these agencies, Cooperative State Research Education and Extension (CSREES) and the Natural Resource Conservation Service (NRCS), reflect reorganization in the USDA to consolidate natural resource program responsibilities in fewer agencies. The programs administered by these agencies are a principal means of providing technical, educational, and financial assistance to improve renewable resource management on private lands.

Cooperative State Research Education and Extension Service (CSREES) The CSREES administers federal grant funds in agriculture, agricultural marketing, and rural life, and for cooperative forestry research and research facilities. Funds are made available to the State Agricultural Experiment Stations, state land-grant universities, and other institutions in the 50 states and Puerto Rico, Guam, and the Virgin Islands. The CSREES also administers a grant program for basic scientific research, and the McIntire-Stennis research program, which provides money for forestry research at land-grant universities, with allocations based on

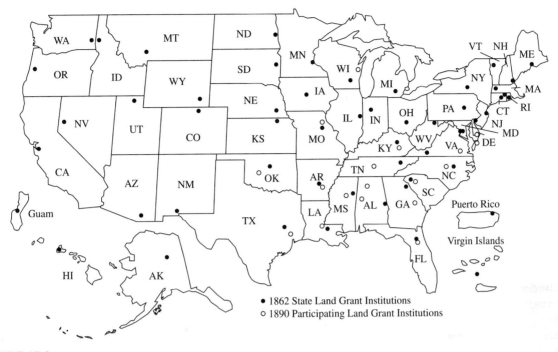

FIGURE 17-6
The location of land-grant institutions in the United States. *(Courtesy of USDA Extension Service)*

the standing timber volumes and timber harvests in each state.

The agency's technical staff reviews research proposals received from the institutions, conducts on-site reviews of the research completed and in progress, and gives leadership in planning and coordinating the funded research. It also participates in coordinating research between the state institutions and the U.S. Department of Agriculture.

The state land-grant universities and other institutions include two types. The so-called *1862 state land-grant institutions* were established under the Morrill Act of 1862 and include land-grant universities in each state. The so-called *1890 participating land-grant institutions* were established by legislation that year and are often predominantly black schools (Fig. 17-6).

CSREES also hosts the Cooperative Extension Service, established by the Smith-Lever Act of 1914, and based on the belief that human welfare can be enhanced if new technical information from research can be translated to individuals and then to the land. The success of this system in developing the world's most productive agricultural system has been inter-nationally recognized. Extension activity focused on natural resources was enhanced by the passage of the Renewable Resources Extension Act of 1978, under which funds are provided for expanded extension programs for forest and rangeland renewable resources.

There are Cooperative Extension programs in each of the 50 states, the District of Columbia, Puerto Rico, the Virgin Islands, Guam, American Samoa, Micronesia, and northern Marianas. Extension programs operate in 3,150 counties as well as in many cities, strongly keyed to local areas. Typically, states and counties supplement the funding provided by the federal government.

Extension programs focus on broad categories of agriculture, forestry, wildlife, fisheries, and other natural resource areas, including the environment, home economics, 4-H clubs, and other youth and community resource development efforts. Extension staff members live and work among the people they serve, and help them plan ahead and deal intelligently with their resource problems.

Spearheading Extension work are the county extension agents. They work with farmers, private forest owners, agricultural industries, homemakers, youth, and community leaders. Most of their assistance is provided through meetings, demonstrations, workshops, short courses, practical how-to-do-it publications, mass media, and personal consultations.

The efforts of extension foresters are concentrated on providing assistance in four major areas: 1) forest resources management; 2) forest products processing, marketing, and distribution; 3) wildlife and fisheries; and 4) urban forestry.

Extension foresters prepare publications aimed at many levels of users—radio, television, and other media programs adapted to current needs; short courses, workshops, conferences, meetings, field tours, and demonstrations to explain current forestry problems; and new techniques of management and details about numerous federal, state, and private agency programs. Because of the close cooperation among teaching, research, and extension, the land-grant universities are in a unique position to transmit forestry education and research findings to local landowners and other users through cooperative extension programs.

To see the full scope of CSREES organization and natural resource programs, visit its Website at www.USDA.gov/CSREES. For several articles and notable forestry extension programs and initiatives see the special issue of *Journal of Forestry,* March 2001, featuring "Forestry Extension: Reaching out to Private Landowners." Some good case studies of extension forestry are described in this special issue.

The Natural Resource Conservation Service (NRCS)

The Natural Resource Conservation Service, formerly called the Soil Conservation Service, was renamed to reflect its broadening responsibilities for conservation programs on private lands. NRCS works closely, often side-by-side out of the same office, with personnel in the 3,000 Soil Conservation Districts, which are organized under state laws and are subdivisions of state government. Conservation districts are similar in some respects to drainage, irrigation, or sanitary districts formed under state laws. In some states, conservation district boundaries coincide with county boundaries; however, some districts cover two or more counties, whereas others contain only a part of a county and yet others are formed along watershed boundaries.

The NRCS assists conservation districts in preparing conservation agreements with landowners and in implementing planned conservation work on individual landownerships. Where those ownerships contain forestland, or land that would benefit from planting trees, NRCS foresters assist the owners accordingly and provide technical forestry services that are not otherwise available. Among the most well-known efforts of NRCS are the windbreak plantings in the Great Plains states, back when the agency was called the Soil Conservation Service. These plantings may be for erosion control, for protection of farmsteads, for the protection and encouragement of wildlife, or for a combination of these uses (Fig. 17-7).

Today, the NRCS helps the sponsors of resource conservation and development projects and watershed projects, as well as other governmental units, to recognize and encourage improved forest and natural resource management. The NRCS provides technical advice for tree planting, forest management, and wildlife habitat enhancement, wetland restoration, grazing initiatives, and other conservation efforts. For example, the Forestry Incentives Program (FIP) provides cost-share reimbursements to small landowners for tree planting and timber stand improvements.

The Conservation Reserve Program (CRP) is designed to take marginally productive land out of production in order to reduce soil erosion, paying landowners a fixed rate per acre for a period of 10 years if they maintain permanent vegetation cover and do not harvest crops. Cost sharing for tree plantings of ultimate timber crops, and wildlife habitat enhancements including ponds, are available. For a complete description of NRCS organizations and programs see their Website at www.USDA.gov.NRCS.

THE DEPARTMENT OF THE INTERIOR

The U.S. Department of the Interior (USDI) is a cabinet level department of the federal government with five agencies that have renewable resource management programs: the Bureau of Indian Affairs (BIA), the Bureau of Land Management (BLM), the National Park Service (NPS), the Fish and Wildlife Service, and the Bureau of Reclamation. Some other USDI agencies provide information in support of renewable resource management, such as the Geological Survey.

FIGURE 17-7
This farmstead is well protected from wind and snow by a windbreak of conifers, fruit trees, and shrubs. Near Clifford, N.D. *(Courtesy of The Natural Resource Conservation Service)*

Bureau of Indian Affairs (BIA)

The Bureau of Indian Affairs Division of Forestry provides policy direction and staff assistance to tribal groups facing forest management and improvement issues. The principal activity is managing and protecting the Indian-owned forest and natural resource lands held under federal trust. In the continental United States 193 Indian reservations in 33 states contain a total of 16.8 million acres (6.8 million ha) of forestland and 9.3 million acres (3.77 million ha) of woodlands (less than 5 percent crown cover). The 5.7 million acres (2.3 million ha) of Indian forests managed for timber production contain 44 billion board feet and support an annual harvest of 850 million board feed (Morishima, 1997). Washington State, plus the states of Alaska, Arizona (Fig. 17-8), California, Montana, New Mexico, and Oregon, contain the vast majority of the forest resources in Native American ownership, and revenues from its management and other forest uses contribute to tribal economies.

Direction and policy for management of these forestlands are provided in the National Indian Forest Resource Management Act (NIFRMA) of 1990. The intertribal timber council, with representatives from tribes with timber resources, was an important influence in the legislation and helps provide coordination for the BIA forestry program. Forestry operations performed on these reservation lands include forest management inventory and planning, fire management, forest pest management, and forest products business management and development. A special woodland management program provides support for noncommercial forest such as pinyon–juniper in the Southwest and sabal palm in Florida. The complexities of landownership presents special problems. Tribes may own land as a group, or individuals may own allotted land through various patents. These patents may be held by an individual owner or may be in an undivided and complicated heirship status.

On forested reservations where the volume of timber available for cutting is in excess of that being utilized by the Indians, open market sales are authorized, provided consent is given by the tribal governing body for tribal timber or by the majority Indian interest of individual allotments. Some reservations have mills owned by tribal entities. Mills

FIGURE 17-8
White Mountain Apache forest development crew taking time out from reforestation work. Fort Apache Indian Reservation, Arizona. *(Bureau of Indian Affairs photo by Jay R. West)*

are usually managed by a board of directors, with Indians filling positions of importance. Grazing on more than 25 million acres (10 million ha) of Indian lands is administered by the Division of Water and Land Resources in the Office of Trust Responsibilities. Many benefits accrue to the tribes through these operations.

More information on forest and renewable resource management on Indian lands is found in a special issue of the *Journal of Forestry* (1997) devoted to forestry on tribal lands.

Bureau of Land Management (BLM)

The BLM administers 262 million acres (106 million ha) of America's public lands, located primarily in 12 western states, with 84 million acres (33.6 million ha) of forest and woodland. Most of the remaining BLM land is rangeland, with grazing, wildlife, or recreational values. The BLM also administers subservice mineral rights on about 300 million acres (121.5 million ha) of additional lands and is responsible for wildfire management and suppression on a total of 388 million acres (157 million ha).

This bureau was established in 1946 by combining the General Land Office with the U.S. Grazing Service, which had been established to handle the graz-

ing lands in the public domain under the Taylor Grazing Act of 1934 (http://www.blm.gov).

The BLM was redirected as a multiple use agency under the Federal Land Policy Management Act of 1976 (FLPMA). An important responsibility is managing public rangelands in the West for livestock forage through issuing grazing permits; wildlife, watershed, and recreation values are also of concern.

Much of the BLM's forestry work consists of managing the Douglas-fir forests of the revested Oregon and California (O&C) Railroad grant lands in the Pacific Northwest, but it is also active on public domain lands in the intermountain states (Fig. 17-9) and the interior forests of Alaska.

Primarily during the past decade, Congress designated several National Conservation Areas on BLM lands, such as the 485,000-acre (196,000 ha) Snake River Birds of Prey Area in Idaho (1993) and the 425,000-acre (172,000 ha) Steens Mountain Cooperative Management and Protection Area in Oregon (2000). This prompted the agency to establish a National Landscape Conservation System (NLCS) to administer these lands that have exceptional natural, recreational, cultural, wildlife, and other outstanding features. Among its NLCS lands, BLM manages 145 designated units in the National Wilderness

FIGURE 17-9
Ponderosa pine from BLM lands in western Montana. *(Courtesy of Bureau of Land Management)*

Preservation System totalling 6.2 million acres (2.5 million ha). About 700 additional areas totalling 25 million acres (10 million ha) in 12 western states are being studied for potential wilderness designation. Ultimately, the BLM may have 15–20 million acres (6–8 million ha) of wilderness to manage, which, along with its other national conservation area responsibilities, will significantly expand its stewardship responsibilities (http://www.blm.gov).

One of the most underused BLM activities is the wild horse and burro adoption program, established to remove surplus animals when they threaten to overgraze rangelands. In fiscal year 2000, there were nearly 50,000 wild horses and burros on western rangelands, with 8,631 of them removed to protect the threatened ecosystems, 6,192 of which were adopted by members of the public (http://www.wildhorse&burro.blm.gov).

Why does the federal government maintain two bureaus doing similar work, such as the forest and range activities of the Forest Service and of the Bureau of Land Management? The answer is that the lands concerned are managed under different laws and long-established practices of administration. The history of the development of these two bu-

reaus, as outlined in Chapter 2, helps explain their differing mandates (see Culhane, 1981; Robinson, 1975; Steen, 1991b). As suggested in the case of rivalry in National Park Service and Forest Service recreational responsibilities, interservice competition between agencies in forest management can also stimulate creative thinking and work against bureaucratic stagnation.

National Park Service (NPS)

The National Park Service, in the Department of the Interior, has as its primary mission the preservation and use of the National Park System comprising the national parks, monuments, historic sites, seashores, recreation areas, lakeshores, battlefields, preserves, and riverways. The total area encompasses 83.6 million acres (33.8 million ha), which includes 4.3 million acres (1.74 million ha) that remain in private ownership (http://www.nps.gov). Each of these park system units may contain various mixes of three different management zones—natural, historic, or recreational—depending on the primary national significance of the area involved and the congressional constraints inherent in the enabling legislation that created the area.

National park resource managers have the difficult challenge of preserving natural resources for both the use and enjoyment of the American people—no small task, considering the popularity of the park system units and their extensive visitation. A major emphasis is the interpretation of natural and cultural features and processes for the enjoyment and education of visitors. The National Park System also contains 44 wilderness areas totaling more than 44 million acres (17.8 million ha), with more than 30 million acres (12 million ha) of additional NPS lands either under wilderness study or awaiting congressional action, having been recommended for wilderness designation (Hendee & Dawson, 2002).

Resource management practices in the NPS units are primarily concerned with protection, the perpetuation of native species, and the continuance of natural ecological processes. These are important natural resource management challenges. When exotic species threaten indigenous resources, prudent and feasible eradication efforts may be undertaken (Fig. 17-10).

Each park prepares a fire management plan that reflects NPS policy and the specific characteristics, legislative obligations, and environmental and social considerations of the area. Fire is used as a management tool in NPS areas; therefore, fire management in the national parks embraces all wildland fire-related activity, including suppression and the use of prescribed fire (see Chapter 8, Fire Management). Unfortunately, the historical exclusion of fire from NPS areas has resulted in some dangerous and unnatural fuel accumulations, such as those that led to the

FIGURE 17-10
National park biologists push the limits of "prudent and feasible" in attempting to eradicate the exotic mountain goats from Olympic National Park through the removal program. *Left:* Live capture by helicopter. The goats are transferred to other ownerships within their natural range. *Right:* Implanting a contraceptive hormone in a female goat as part of the research into using sterilization to control the park's unwanted goat population. *(Courtesy of Olympic National Park)*

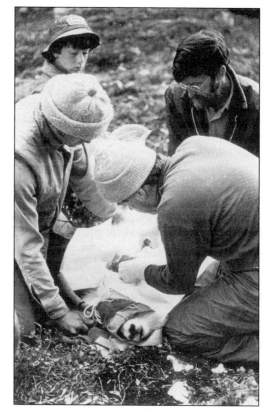

Yellowstone fires of 1988, and pose a continuing risk of conflagrations in other areas. Issues of forest health and biological diversity are important in NPS areas as well as in other natural resource jurisdictions.

Commercial harvesting of trees is not permitted in NPS areas. Moreover, even naturally occurring hazards such as dead and downed timber may not be removed in national parks except to promote or protect park uses and resources. For example, it may be desirable in recreational or historic areas to remove certain trees that have been killed or downed by winds or attacked by insects or disease when these pose a threat to personal safety or become a reservoir for undesirable insect infestations. Removal of trees in these zones may also occur in compliance with specific management objectives, such as the restoration of a historical scene or maintenance of a viewpoint. Another important function of the National Park Service recognizes and assists in the protection of natural and cultural properties outside the system.

Fish and Wildlife Service

The U.S. Fish and Wildlife Service (FWS) is "the principal agency through which the U.S. Government carries out its responsibilities to conserve, protect, and enhance the nation's fish and wildlife and their habitats for the continuing benefit of people (http://www.fws.gov. Specifically, the Service manages migratory birds, endangered species, certain marine mammals, and freshwater and anadromous fish."

Today, the Service employs 7,500 people located in a headquarters office in Washington, D.C., seven regional offices, and 700 field units, including national wildlife refuges, national fish hatcheries, and management assistance offices of law enforcement and ecological services field stations. The FWS has major land management responsibilities in administering the 91.3 million acres (37 million ha) in 454 national wildlife refuges that include 71 wilderness areas containing about 20.7 million acres (8.38 million ha) (Hendee & Dawson, 2002).

Bureau of Reclamation

The Bureau of Reclamation is charged with maintaining water supplies in the arid West (http://www.usbr.gov, 5/5/02). Its reservoirs, used for irrigation, hydroelectric power, municipal and industrial water, and flood control, also attract people in search of recreation. If the area has national significance, the recreation administration is turned over to the National Park Service (or the Forest Service if the project area is within or adjacent to a national forest). In areas of less than national significance, the recreation opportunities are administered by a local or state agency. With few additional dams likely to be built, the Bureau's mission has evolved to include water conservation and management; and enhancing and preserving wetlands, fish wildlife habitat, endangered species, instream flows, and water quality. Millions of acres of western landscape are involved.

OTHER FEDERAL AGENCIES

Tennessee Valley Authority (TVA)

The TVA was created in 1933 to conserve, rehabilitate, and develop the Tennessee River Watershed. TVA operates an extensive reservoir system for flood control, navigation, and hydropower production in seven states, while also maintaining and enhancing water quality. It is the largest public power producer in the United States and also hosts 100 million visitors each year at its recreation sites (http://www.tva.gov). As an independent corporate agency concerned with all natural resources, the Tennessee Valley Authority has a separate Office of Natural Resources (ONR) aimed at stimulating improved management and productivity of the land and water resources in the 26-million-acre (10.4-million-ha) Tennessee River Watershed.

Nearly 80 percent of the Tennessee River watershed is forested, a total area of 21 million acres (8.5 million ha). Here, the Office of Natural Resources seeks to improve management on privately owned lands, maintain capacity of forests to protect soil and water resources, and help forest industries—from logger to manufacturer—in the development of new products and production methods. Regeneration of forests for the future, reclamation of land and water resources damaged by mining and pollution, enhancing wildlife habitat, and boosting environmental awareness through educational programs are equally important goals.

The ONR of TVA is responsible as well for management and operation of Land Between the Lakes, a 170,000-acre (68,000-ha) area between Kentucky

FIGURE 17-11
The area known as Land Between the Lakes provides boating and fishing at
Kentucky Lake and Lake Barkley. The TVA concerns itself with the management of
all natural resources in the 26-million-acre Tennessee River Watershed.
(Tennessee Valley Authority, photo by Jeff Brown)

Lake and Lake Barkley in western Kentucky and Tennessee. Since its inception in 1964, Land Between the Lakes has been managed under a multiple-use concept to increase public recreation and education opportunities, and to enhance wildlife habitat and the quality and appearance of the area's forest and open lands. Camping, fishing, hunting, hiking, boating, and environmental education activities are featured (Fig. 17-11).

Department of Defense (DOD)

More than 30 million acres (12 million ha) of federally owned lands are managed by the Department of Defense through the three military departments: the Army, Navy, and Air Force. While these lands are primarily managed for defense purposes such as military testing and training, activities such as grazing, timber production, and wildlife management are permitted in some areas. The Army Corps of Engineers has separate responsibility, through the Department of the Army Civil Works Program, for the administration of many federally owned areas. Its activities include waterway improvement, flood control, river-flow regulation, and recreation development in civil works projects. It regulates the use of navigable waters of the United States. The areas of federally

	TABLE 17-1	

TOTAL ACREAGE OF FEDERALLY OWNED LAND BY DEPARTMENTS OF AGRICULTURE, INTERIOR, AND DEFENSE.

Agency	Acres millions	Hectares millions
Department of Agriculture		
Forest Service	191.0	77.3
Department of the Interior		
Bureau of Land Management	262	106
National Park Service	83.6	33.8
Fish and Wildlife Service	91.3	36.6
Bureau of Indian Affairs	52	21
Bureau of Reclamation	—	—
Department of Defense		
Department of the Army	11.9	4.8
Department of the Air Force	9.2	3.7
Department of the Navy	3.6	1.5
Corp of Engineers (Civil Works)	7.0	2.8
TOTAL	712	287.5

owned land managed by these agencies are given in Table 17-1.

The Department of Defense develops policies and establishes an integrated, multiple-use program for the renewable natural resources in these forests and woodlands. Fish and wildlife, soil, water, grasslands, outdoor recreation, and natural beauty are all managed so as to be compatible with the military mission.

Multiple-use management of the Department of Defense lands under these policies has resulted in many benefits to local people and the public in general. Visitors seeking outdoor recreation, including hunting and fishing, have especially benefitted. In addition, there has been extensive forest planning and timber stand improvement. Timber sales, where appropriate, have resulted in forest improvements. Grazing has also been allowed in some areas. Such land management policies for military lands are believed to have resulted in improved public relations, in addition to enhanced productivity of the natural resource base and social and economic benefits to local communities.

Environmental Protection Agency (EPA)

The Environmental Protection Agency does not practice forestry but is responsible for administrating many environmental standards and regulations that influence forest practices. The EPA was created in 1970 to address widespread public concerns about the deterioration of the environment, especially air and water quality.

The EPA is charged by law with responsibility for coordinated and effective government action on behalf of the environment, and with the abatement and systematic control of pollution by proper integration of research, monitoring, standard setting, and enforcement activities.

It also has the responsibility for coordinating and supporting research and antipollution activities by states and local governments and private and public groups. In addition, EPA must reinforce the efforts of other federal agencies with respect to abating the impact of their operations on the environment. To accomplish their mission, 10 regional field offices and several national research laboratories have been established in addition to a Washington, D.C. office.

The passage of the National Environmental Policy Act (NEPA) in 1970 has had a profound effect on federal, state, and private forestland management. Through NEPA processes, including public involvement, questions have been raised about the impacts of forest harvesting, especially clearcutting, and road construction on water quality, and of

activities such as slash burning and the use of insecticides and herbicides. Air-quality standards, to be met by certain dates, have been proposed for certain areas designated for Class I air quality, including several national parks and certain wilderness areas on the national forests. The influence of the NEPA environmental impact assessment process on resource management programs is described in Chapter 2.

INFLUENCES ON MANAGEMENT OF FEDERAL LANDS

Landownership Patterns

Landownership patterns influence forestry practices. Federal, state, and private owners must determine and agree on landownership boundaries, rights-of-way, and other management concerns. They have mutual problems in fire, insect, and disease protection. In recent years, it has become obvious that in land areas of mixed ownership, all owners must coordinate their management activity such as timber harvest, road building, and other development so their cumulative effects on water quality and the environment will not exceed tolerable limits. Thus, cooperative management plans are increasingly being developed by all adjacent landowners, public and private, in areas of common influence such as watersheds. An example of an area of diverse ownerships is given in Fig. 17-12.

Public Involvement

Of all the influences on federal forest and rangeland management, none has been more important or far-reaching than the trend toward increased citizen involvement in public land management. Through public involvement, the growing environmental awareness and concern of the U.S. population have greatly influenced resource management on federal lands. Inevitably, public involvement strengthens the environmental considerations of proposed projects. Decentralization has been another result of citizen involvement as local voices are now heard appealing for local considerations in federal decisions. Public involvement in natural resources also includes participation in the work of management through volunteers and partnerships with user groups and organizations. This helps get work done, saves money, and develops stakeholder interest.

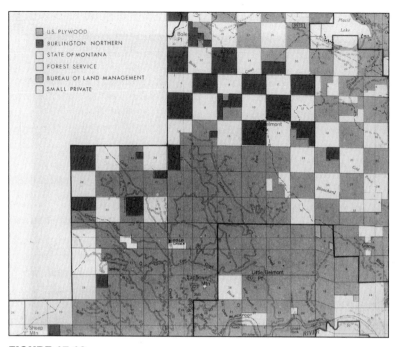

FIGURE 17-12
The need for cooperation between public and private landowners becomes apparent under this diverse ownership in Montana. The checkerboard pattern of ownership originated when grants of alternate sections of land were given to the Northern Pacific Railroad by the federal government in the 1870s and 1880s. *(Courtesy of USDA Extension Service)*

Competition for Natural Resources

When the U.S. Forest Service was created in 1905 to manage the national forests, there were limited demands on forest resources. Today there is intense competition for all natural resources, and still the demand continues to grow. For example, there is always demand by forest industries and local communities to increase timber harvests on the national forests to ease the pressures to overcut private lands and enhance community stability. Yet, despite these pressures, federal timber harvests have been greatly decreased in response to public demands for other uses, including higher standards of environmental quality. Continuing initiatives to ban road building in yet unwooded areas of the national forests are being debated at both the national and local levels. Such pressures are also felt in the public involvement in forest plans mandated by the National Forest Management Act, in proposals for more wilderness and parks, and in many laws, such as those requiring pro-

tection of threatened and endangered species, biodiversity, and air and water quality. Increasingly, federal land management is focusing on environmental protection and restoration in response to public pressures and concern for the nation's natural heritage and environmental quality.

Forest Policy

Many resource analysts believe our forestland has the potential to meet more of our domestic needs while also providing wood products for export. Thus, U.S. policy continues to encourage forestry on the small, nonindustrial private forestlands (NIPF) with cooperative and incentive programs, and to continue federal investments on the national forests for timber reproduction, where appropriate, but with other uses such as wildlife, fisheries, and recreation receiving more emphasis than in the past. Reflecting a trend toward decentralization, even the forestry cost-sharing incentive programs are being implemented with more delegation of programs to the states and local jurisdictions. While the United States has the potential to meet all of its wood products needs, it is predicted that the United States will continue to import wood products, the volume depending on the success of the cooperative and incentive programs in stimulating production of wood products from NIPF sectors and other ownerships. Canada is expected to continue to be the largest U.S. wood products importer, but several other countries are gradually assuming a larger market share of our wood imports.

SUMMARY

At the federal level, forest and renewable resource management is characterized by coordination, leadership, and financial assistance through many federal agency programs, and by the setting of standards through protection, management, and development of federal forestland.

The U.S. Forest Service has the largest and most comprehensive forest and renewable resource programs, including managing the national forests, conducting federal forest research, and providing financial and technical assistance to state and private landowners and for international forestry. The Bureau of Land Management, National Park Service, Bureau of Indian Affairs, Fish and Wildlife Service, and Department of Defense also have significant land holdings on which they practice forestry and other resource management.

LITERATURE CITED

Culhane, Paul J. 1981. *Public Lands Politics. Interest Group Influence on the Forest Service and the Bureau of Land Management,* published for Resources for the Future by Johns Hopkins University Press, Baltimore, Md.

Hendee, John C., and Chad Dawson. 2002. *Wilderness Management: Stewardship & Protection of Resources and Values* (3rd ed.), North American Press of Fulcrum Publ. Golden, Colo.

Journal of Forestry. 1997. Forestry on Tribal Land. Special issue with eight articles on tribal forestry. 95(11).

Journal of Forestry. 2001, March. Forestry Extension: Reaching Out to Private Landowners. Special Issue: 99(3).

Morishima, Gary S. 1999. Indian Forestry: From Paternalism to Self Determination. *J. Forestry* 95(11):4–9.

Robinson, Glen O. 1975. *The Forest Service. A Study in Public Land Management,* published for Resources for the Future, Inc. by Johns Hopkins University Press, Baltimore, Md.

Steen, Harold K. 1991a. The Beginning of the National Forest System. USDA Forest Service Publ. FS-488.

Steen, Harold K. 1991b. *The U.S. Forest Service: A History,* University of Washington Press, Seattle.

Szymanski, Marcella, Joe Colletti, and Lisa Whitewing. 1998. Meeting the Winnebago Tribe's Needs through Agroforestry. *J. Forestry* 96(12):34–38.

Yazzie-Durglo, Victoria. 1998. The Right to Change Tribal Forest Management. *J. Forestry* 96(11):33–35.

ADDITIONAL READINGS

Everhart, William C. 1983. *The National Park Service* (2nd ed.), Westview Press, Boulder, Colo.

Muhn, James, and Stuart R. Hanson. 1988. *Opportunity and Challenges: The Story of the BLM.* U.S. Government Printing Office, Washington, D.C.

Steen, Harold K. (ed.). 1992. *Origins of the National Forests: A Centennial Symposium.* (Twenty-one contributions by numerous authors marking the 100 years of the National Forest System.) Forest History Society, Durham, N.C.

STUDY QUESTIONS

1. Fill out these acronyms that represent public agencies involved in land management and related services: *USFS, NRCS,* and *CSREES.*
2. List the five major resource management agencies in the USDI.
3. What are the four major functions or activities of the U.S. Forest Service?
4. What is the closest national forest to you? Where is the forest supervisor's office? Where is the nearest ranger district office?
5. What are some issues surrounding forest management on tribal lands?
6. What does an *extension forester* do? Where is this office located, nearest to you?
7. What special challenge do national park resource managers face? How has the exclusion of fire complicated their task?
8. Discuss some aspects of the NEPA as it has impacted forestland management by both the USFS and the NPS.
9. Why should the Department of Defense have renewable resource management responsibilities?

Forest Management by the States

INTRODUCTION

This chapter focuses on forest management by the states, although there are other renewable resources like parks and recreation, range, wildlife, and fish for which states also have programs. But forests have been a long-standing concern of the states, and due to their economic importance to the nation, there are also many well-developed programs of federal financial and technical assistance to the states to encourage good forestry on their private lands.

From our beginnings, we Americans have been concerned about our forests. For example, in 1681, William Penn, in his charter of rights to his colonists, directed that, in land clearing, "care be taken to leave one acre of trees for every five acres cleared" (National Association of State Foresters, 1982). Similar provisions were included in the charters or laws of many of the other original colonies. That these well-intentioned efforts were sometimes in vain because of ignorance or lack of any effective means of enforcement in no way minimizes the early recognition of the value of forests by a few of our farsighted antecedants.

There was limited forest regulatory activity in several states in the 1860s and 1870s. In 1885, several states moved for the first time to create a state forestry agency. From this date progress in establishing state forestry agencies came more rapidly, and, in fact, until 1900 most forestry leadership came from state capitals. After 1900, the federal government assumed a leadership role and in the ensuing decades shaped the development of forestry programs nationwide. This unique relationship between federal and state governments in forestry matters makes the history of U.S. forest regulation important to the evolution of the federal–state relationships in general.

The separate states, in many instances, faced problems relating to their own state and privately owned forestlands, more or less by accident. Each state had acquired lands through educational grant lands from the federal government and by other means. These included lands to help support public schools and agricultural colleges as well as tax-delinquent cutover lands and swamp and overflow grants. These land grants contained considerable timber. Obviously, these tracts, even without their stands of timber, were of some value, but they were not easy to sell, and the better of them were excellent targets for trespassers. Although the states shared the easygoing attitude of the federal government about any management obligations, they did see the need for fire control. Moreover, the forestry idea had made some headway before 1900, as outlined in Chapter 2. Much was heard near the turn of the century about public forests for the nation, culminating in the 1908 White House Conference from which many governors returned so enthusiastic that several state forestry departments were promptly modified, and a number of new ones organized.

The importance of forest management by the states is underscored by the fact that a total of 23.4 million acres (9.4 million ha) of commercial forestland is owned by the states. This ownership ranges from a few acres in some states to 3.8 million acres of state forest in Michigan. Furthermore, the states are deeply involved in overseeing activity on privately owned forestlands and in coordinating state laws and responsibilities with all forestland ownerships, public and private, within their state boundaries. Finally, as outlined in this chapter, state forestry organizations provide technical and financial assistance to private landowners to encourage good forestry and to provide and coordinate fire protection for these lands.

STATE FORESTRY DEPARTMENTS

"The powers not delegated to the United States by the Constitution, nor prohibited by it to the states, are reserved to the states, respectively, or to the people." This clause in the U.S. Constitution means the states have the job of managing their own forestlands, of convincing their citizens that they must not start wildfires, of reclaiming land by tree planting, and of making it possible to harvest forest products, manage forest influences, and provide for enjoyment. Often federal help is available if a state so desires. State laws can be passed that enable the federal government to establish national forests within the state, or state forestry activity may merit federal aid or cooperative funding for forestry action programs. State pride, concern, and interest in forests fostered the organization of state forestry agencies or the undertaking of forestry work under existing state agencies. Thus, state forestry programs were formed between 1885 (New York, Ohio, California) and 1966 (Arizona). All

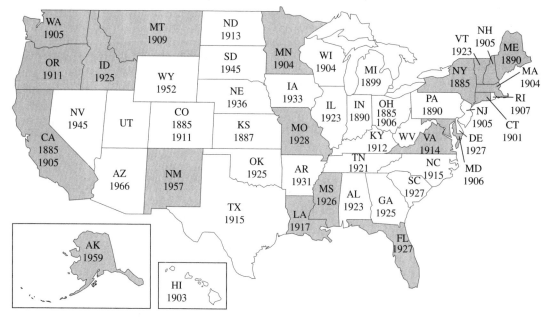

FIGURE 18-1
The initiation date of each state forestry organization is shown. Shaded states have state forest practice acts. *(Courtesy of National Association of State Foresters)*

50 states now have official forestry programs. Figure 18-1 shows the year each state forestry program was initiated (National Association of State Foresters, 1982).

The National Association of State Foresters (NASF)

As more states formed forestry organizations, and as individual state programs grew, there developed a need for organizing in order to share ideas and to enable action as a group on matters affecting forestry in all states. Interest accelerated with the passage of the Weeks Law of 1911, which authorized cooperation with the states in forest fire control on private and state lands, and appropriated $1 million for state forest fire cooperation in 1920. The need for a nationwide association that could represent the interests of the states in forestry matters became urgent.

Gifford Pinchot, then chief of the Pennsylvania Department of Forestry, recognized the need and the opportunity and invited all state foresters, along with U.S. Forest Service chief W. B. Greeley and others, to a meeting in Harrisburg, Pennsylvania, in 1920. During this meeting the State Foresters Association was formally organized. Today, NASF is a nonprofit organization that represents the directors of state forestry agencies in each state and U.S. territory (www.stateforesters.org). These agencies provide management assistance and protection services for over two-thirds of the nation's forests, including private lands that contain 70 percent of the nation's wood supply. From its office in Washington, D.C., NASF staff is actively involved in forestry affairs and keeps the state foresters up to date on policy and legislative and legal issues affecting forest management. NASF publishes a monthly newsletter, the *NASF Washington Update,* and provides many statistics about state forests and their forestry programs (www.stateforesters.org/statistical).

Types of State Forestry Organizations

There are several ways in which forestry organizations are set up within the individual states. These are best described by noting the type of authority under which they operate. Their location in state government is also an indication of the state's approach to managing its forestlands. Four types will be delineated here, with a representative state example for each. In the first example, Oregon, a little more detail is given to illustrate the kinds of activities that state forestry departments typically carry out.

Independent Forestry Boards   In this particular grouping, the office is usually at the state capital. The location of the state forester's headquarters is sometimes an indication of the character of that officer's duties. Nine states have organizations of this type, and the duties of the state forester in each instance are technical, administrative, and, to some extent, involved in policy formulation. The states are Arkansas, Florida, Georgia, Idaho, Louisiana, Maine, Mississippi, New Mexico, and Oregon. In all except Maine, the state forester is the executive of a board or commission of usually 5 to 12 members. The Commissioner of Forestry in Maine is an independent executive, reporting directly to the governor. Some of the boards in the eight other states have administrative duties and all are a part of the policy-making process. In Idaho, the governor chairs the Land Board of Commissioners that oversees forestry activity by the Idaho Department of Lands, the state's forestry agency.

Example: Oregon   The seven-member Oregon State Board of Forestry is appointed by the governor. Appointees are citizens representing forest regions and the public at large (www.oregonforestry.org). The board provides direction for the Oregon Department of Forestry's programs through its strategic plan, the "Forestry Program for Oregon."

The Oregon Department of Forestry (ODF) is the administrative arm of the Oregon State Board of Forestry, operating in seven major areas: forest protection; insect and disease management; state forestland management; forest nursery/orchard; service forestry for private, nonindustrial landowners; forest resource planning; and forest practices.

Oregon is the nation's leading timber state (27.5 million acres) (11 million ha) and contains about one-fifth of the nation's softwood timber supply. Of the 61,356,000 gross acres (24,850,000 ha) of land in Oregon, almost 45 percent is forested, with 57 percent of the forestland in federal agency ownerships and 37 percent privately owned. Including approximately 900,000 acres (365,000 ha) of state forests, the ODF protects almost 19 million acres (7.7 million ha) of private and public forests from wildfires with an average of more than 1,000 wildfires burning

more than 20,000 acres (8,000 ha) annually (much more in bad fire years). For example, in 2002, the Biscuit fire in Oregon burned nearly 500,000 acres (202,326 ha). Strong cooperative programs exist between the ODF and other fire-control agencies, including the U.S. Forest Service, Bureau of Land Management, and Bureau of Indian Affairs. These agencies share resources and cooperate to use the closest forces in initial attacks on wildfires. The forest industry also cooperates and provides equipment and crews for wildfire control.

Insect and disease management advice is available from the ODF for the 11 million acres (4,455 million ha) of state and private forests. As in other western states, pest problems have become serious. For example, in the 1970s, the Douglas-fir tussock moth was a major problem in northeastern Oregon. In the 1980s, mountain pine beetles and western spruce budworms inflicted major damage in forests of eastern Oregon. These losses exceeded 200 million board feet a year, requiring cooperative insect control efforts by federal, state, and private forest landowners, creating a major forest health issue in northeastern Oregon that continues today.

Management of the 800,000 acres (324,000 ha) of state forestlands in Oregon is intensive and includes replanting with genetically improved seedlings, management of competing vegetation, precommercial thinning, second-growth thinning, and fertilization. Timber sale income from these lands has averaged about $40 million a year and is split among the ODF, counties, local taxing districts, and school districts. A principal accomplishment of the ODF has been the reforestation and rehabilitation of the former Tillamook Burn, now the Tillamook State Forest (Oregon Department of Forestry, 1993). In August 1933, the forest east of Tillamook, Oregon, caught fire from a logging operation. Within 20 hours, 240,000 acres (97,000 ha) were burned, consuming 12 billion board feet—enough to build 1 million five-room houses. The forest burned again in 1939, 1945, and 1951 for a total of 355,000 acres (144,000 ha) (Fig. 18-2).

Oregon has about 25,000 private nonindustrial forest landowners (NIPFs) who own 5,000 acres (2,000 ha) or less—16 percent of all the forestland in Oregon. The productivity and level of management has been increased on these lands through on-the-ground assistance provided the ODF service foresters. Service foresters assist in the development

FIGURE 18-2
Starting in 1933 a series of wildfires swept through Oregon's coast range. This area
was known for many years as the Tillamook Burn. *Upper:* The Wilson River bridge
and freshly burned snags, part of the 180,000 acres (73,000 ha) that burned in
1945. *Lower:* Snags left behind act as stark reminders of the devasting fires. Note
the new growth. The entire burn area now comprises the Tillamook State Forest.
(Courtesy of Oregon Department of Forestry)

of forest management plans to encourage increased growth and harvest of forest products while enhancing and protecting related resources.

The Oregon State Board of Forestry administers the Oregon Forest Practices Act. This Act, passed in 1971, was the first of its kind in the nation and has been updated and expanded to meet new concerns. It is a broad program regulating all forest practices on all forestlands in the state. Some of its requirements are successful reforestation of forestlands, careful road construction and maintenance, safe application of chemicals, good harvesting practices, and slash disposal to reduce fire hazard. It also requires protection of associated forest resources, such as air quality, water resources, soil productivity, fish and wildlife, scenic areas, and special use areas. Foresters employed by the ODF inspect field operations for compliance with these regulations, issue citations for noncompliance, and seek correction of the situation—such as for failure to initiate reforestation within 12 months of harvest with established seedlings "free to grow" at the end of six years (Rose & Coate, 2001).

The "Forestry Program for Oregon" is a plan for all 27.5 million acres (11 million ha) of forestland in the state. First published in 1991 after three years of research and public discussion of Oregon's critical forestry issue, a new strategic program is scheduled for completion in 2003 (www.oregonforestry. org/fpfo/2003/). The program is a plan that describes the board of forestry's guidance to the state forester, legislature, governor, and citizens on matters the board deems important. The guidance is provided in terms of a mission statement, objectives, and an action plan containing policies. It addresses all ownerships (federal, state, and private), and all forest resources including fish and wildlife, soil, air, water, recreation, grazing, and timber. This extensive planning effort reflects the importance of forestry and forest products as Oregon's largest industry and is key to its quality of life. The program promotes enlightened management on all of Oregon's forestlands: federal, state, and private.

Nonforestry Administrative Boards   These may or may not have advisory groups, and the headquarters are usually elsewhere than at the state capital. For some peculiar reason, forestry in certain states is not considered comparable with other state functions, even though forestry activity is financed and administered as an independent state function. In such cases, the state forester is usually located at a state educational institution. Such foresters may even give a portion of their time to teaching. Examples include Texas, where the state forester is located at College Station and is responsible to the Board of Regents of the Texas A&M University System. Nebraska, although it has a "Game, Forestation and Parks Commission," turns its forestry work over to an extension forester working out of the Agricultural College at Lincoln. In most instances, forestry functions have apparently been thought of as more closely allied to agriculture than to any other industry. Forestry work in Kansas for many years has been under the aegis of the State College of Agriculture at Manhattan. In 1955, Colorado shifted the responsibility of its state forester from the Board of Land Commissioners to the State Board of Agriculture, with headquarters at the Colorado State University in Fort Collins.

There is, of course, a wide variation in the volume of forestry activity among the states with this type of organization, and some of the most progressive forestry departments are governed in this manner.

Example: Texas   The Texas Forest Service (TFS) was created in 1915 as part of the Texas A&M University System. TFS is mandated by law to "assume direction of all forest interests and all matters pertaining to forestry within the jurisdiction of the state" (txforestservice.tamu.edu). Today, the TFS provides statewide leadership in all areas of forestry through programs in wildland and rural community fire protection, forest resources management, economic and industrial development, insect and disease management, urban and community forestry, and public education. Texas is big and has 22 million acres (8.9 million ha) of forest, including 7.2 million acres (2.9 million ha) of NIPF and 3.8 million acres of industry lands, as well as 103 million acres (42 million ha) of land classified as pasture range. All of this converts to responsibility for wildfire protection on 148 million acres (59.9 million ha) of non-industrial land and 50 million acres (20.25 million ha) of industry-owned land (http://www.stateforesters.org/statistics/).

In the 12-million-acre (4.86 million ha) East Texas southern pine forest, the agency provides professional technical assistance to 150,000 nonindustrial

FIGURE 18-3
The forest industry in Texas employs 60,000 people.
Left: A forest worker collects cones from genetically
improved trees in a state seed orchard. *Upper:*
Employees weeding in the East Texas tree nursery.
This nursery produces 25 million pine seedlings each
year to sell to landowners at cost. *(Courtesy of Texas
Forest Service)*

private forestland owners by developing forest stewardship management plans, administering both federal and state cost-share programs, and providing management services such as resource inventory, pest suppression, prescribed burning, and fire break construction. The East Texas tree nursery annually produces 25 million pine trees for reforestation (Fig. 18-3). In arid West Texas, the TFS provides a windbreak program that includes nursery facilities for the production of 400,000 windbreak trees.

In the state's urban areas, the agency provides technical assistance and coordination for the protection and enhancement of urban forest resources, and coordinates the efforts of one statewide Urban Council and 12 regional Forestry Councils.

The agency maintains a ready fire-fighting force for a 22-million-acre (8.9 million ha) area in East Texas and, in addition, provides support to volunteer fire companies statewide through training and equipment programs to support rural fire protection needs of 800 small communities with whom they have formal agreements.

The TFS also supports applied research programs in forest genetics, wood products, fire technology, silviculture, and pest control.

Despite its rank as the nation's third most populous state, with 80 percent of its population living in large metropolitan areas, the forest industry is the number one or two manufacturing industry in 31 of 43 East Texas counties, employing nearly 60,000 workers.

Departments of Natural Resources (With Boards)
Under this type of organization, broad natural resource management responsibilities are vested in commissions or boards, and these have forestry as a major division. Many states have combined the administration of a number of natural resources, such as forests, parks, other state lands, waters, minerals, and wildlife and fish (the last two frequently, but not always, are combined). In some states, environmental protection was added as a division, such as in the Wisconsin Department of Natural Resources. In 1970, New York established a Department of Environmental Conservation.

FIGURE 18-4
These posts and poles were harvested from the Mohican State Forest in north-central Ohio, in a thinning operation of pine plantings 25 years earlier. The thinning not only produced valuable products but also improved the growth and vigor of the remaining stand. *(Courtesy of Ohio Department of Natural Resources)*

Michigan is an example of an almost inclusive department of natural resources, as it deals with everything but agricultural lands. The state forester is responsible to the director of natural resources, who is in turn responsible to the conservation commission of seven members with staggered terms of office. Such terms are designed to control partisan political domination that might evolve if terms did not overlap periods covered by state elections.

In all, 33 states recognize forestry as being on a par with other natural resources and have placed this function in departments similar to that found in Michigan, and answerable to a policy-making board or commission. Ohio will serve as our detailed example.

Example: Ohio Ohio's forestry work is centered in the Division of Forestry (ODF), one of 13 divisions of a conservation department known as the Department of Natural Resources, and which also includes divisions of Natural Areas and Reserves, Wildlife, Parks and Recreation, and Soil and Water Conservation, among others (www.dnr.state.oh.us/divisions).

ODF's mission is to "promote and apply management for the sustainable use and protection of Ohio's private and public forestlands." Ohio has almost 7 million acres (2.8 million ha) of NIPF lands, 170,000 acres (68,850 million ha) of forest industry lands, 330,000 acres (133,650 million ha) of state lands, and 9.3 million acres (3.8 million ha) to pro-

tect from wildfire (www.stateforesters.org/statistics) (Fig. 18-4). Two state tree nurseries distribute seedlings to various landowners, including farmers, industrial concerns, schools, and other organizations. A considerable amount of this nursery stock is hardwood, and efforts are underway to provide several species of genetically improved stock. Special planting stock is grown for use in windbreak plantings.

Ohio is known for producing some of the highest quality hardwoods in the world, particularly oak and walnut. The forest industry in Ohio grows, processes, and manufactures wood products valued in excess of $7 billion, and its wood manufacturing firms employ 70,000 people with a payroll of about $1 billion. The typical Ohio forest products industry firm employs less than 20 people, so the forestry division plays an important role in supporting a diverse grass-roots enterprise with its service forestry and fire protection.

Under the Ohio Forest Tax Law, hundreds of thousands of acres of privately owned land dedicated to forest management have been certified by division foresters for reduction of taxes. Technical assistance is provided to owners of private forestlands and to the forest products industry under the division's service forestry program. In addition, urban foresters employed by the state provide urban forestry assistance to municipal governments, contractors, developers, and other agencies.

Departments of Natural Resources (Without Boards) Minnesota and Kentucky have Divisions of Forestry within Departments of Natural Resources that are not answerable to boards or commissions.

Example: Minnesota The operating divisions within the Minnesota Department of Natural Resources (DNR) include forestry, conservation law enforcement, fish and wildlife, minerals, parks and recreation, waters, and trails and waterways. The administrative organization of the DNR consists of the commissioner of natural resources, a deputy commissioner, assistant commissioners, and the various division heads. In 1982, the Minnesota Division of Forestry (MDF) was mandated by the Minnesota legislature to conduct a comprehensive assessment of forestry and related resources, thus implementing a planning document that is to be updated every 10 years. The planning approach has evolved to use ecological rather than administrative units as its basis and to include substantial citizen input.

Minnesota has 32.6 million acres (13.2 million ha) of forestland, half of it in NIPF lands, and more than 10 percent each in national and state forests (www.stateforestry.org/statistics). Its federal land includes important assets such as Voyageurs National Park and Boundary Waters Canoe Area Wilderness in the Superior National Forest, as well as other reserved areas in state and local parks.

Minnesota has a sizable forest industry with 596 sawmills and 16 pulp mills processing 264 million cubic feet of roundwood, three-fourths of it by the pulp mills (Hackett & Dahlman, 1997). The industry is growing as jack pine is more often used for lumber, red pine for poles, and aspen for pulp.

Minnesota's Division of Forestry has a comprehensive wildfire and pest management program for private, state, county, and municipal lands. It also actively promotes forest management on private forestlands, and state nurseries produce seedlings for reforestation of private and state lands. The Division of Forestry, in addition to its other forestry duties, administers 55 state forests, which include 64 primitive campgrounds, over 1,250 miles of trail, and 172 public water access sites (Fig. 18-5).

FIGURE 18-5
Visitors enjoy the sun, water, and land in Minnesota's Mississippi Headwaters State Forest. *(Courtesy of Minnesota Division of Forestry)*

STATE FOREST PLANNING APPROACHES

In recent years, almost every state has experienced intensive pressures on forest resources and all states have some kind of forest planning process. State plans are the basis for coordinating with the federal government's nationwide assessment and proposed program prepared by the Forest Service every 10 years under the Forest and Rangeland Renewable Resources Planning Act (RPA) of 1974. The RPA assessment and proposed program are keys to the investment of federal funds for state and private forestry assistance, much of which is administered by the state foresters under their plans. State forest plans also chart a path for the future and are a basis for annual budget requests to state legislatures. In addition to analyzing the state forestry situation and proposing future direction and actions, the planning process provides the chance to involve the public and forestry constituents in charting the state program.

The following state forest plan examples for Georgia and Illinois should give readers an idea of the scope of state forestry planning.

Example: Georgia   Georgia's plan for 1991–95 included renewed emphasis on reforestation to close a gap between acres harvested and acres regenerated, a problem stemming in part from taxes imposed on standing timber as well as a timber harvest tax. Other topics featured in the plan were smoke management, expanded tree planting on agriculture lands qualifying for the federal Conservation Reserve Program, expanding woodland management through the Georgia Stewardship Program to help and encourage landowners in preparing multiple-use management plans, applied research to promote international trade and marketing of wood energy, and information and education to inform citizens about Georgia's working forests.

Georgia's forest planning has undoubtedly been a key to the growth of support for forestry in the state, the diverse program of the Georgia Forestry Commission, and significant growth in the wood-using industries. At the time of this plan's implementation, Georgia's wood products had increased 250 percent in the prior 25 years to more than 3 billion board feet annually, from 521 primary manufacturers and 1,250 secondary manufacturers, with growing stock reserves of 138 billion board feet.

Forestry is Georgia's number one industry, contributing $10 billion to the state's economy and employing 81,000 people, about half in metropolitan areas. Georgia has 17.2 million acres (7 million ha) of NIPF land, more than any other state, plus 4.9 million acres (1.98 million ha) of forest industry lands that generated about 1.7 billion cubic feet of pulpwood (48 percent), sawlogs (41 percent), composite panel, veneer, and other industrial output in 1999 (Johnson & Wells, 2002). Georgia's economic dependence and leadership in forest products rest on an aggressive program of regenerating, developing, and managing the 23.6 million acres (9.558 million ha) of mostly private forest that is the industry's economic base. Thus, it is not surprising that the Georgia Forestry Commission has a well-developed planning system to guide the forestry activities so important to the state.

The Georgia Forestry Commission developed a Forest Resource Plan for 1991–95 as a five-year guide to the state forestry program, building on progress in its previous plan for the years 1986–90 (Georgia Forestry Commission, 1991a). The plan identifies state foresters' goals and objectives for 15 program areas, and gives background and highlights of the proposed activity and the estimated additional expenditures needed for each area. The program areas include forest protection, rural fire defense, forest management assistance for nonindustrial private forestlands, forest pest management, urban forestry assistance, and reforestation (Fig. 18-6). These subject areas coincide with the headings under which progress is reported each year in the Annual Report of the Commission. Thus, each year progress on the strategic plan can be assessed.

Example: Illinois   In 1983, the Illinois Forestry Development Act created what became the Illinois Council on Forestry Development, a diverse group representing citizens, forest landowners, federal and state agencies, organizations, and industries related to the forest. First the council assessed the state's forests and presented findings in a 1986 report entitled "Forestry in Illinois: Opportunities for Action." The council then stimulated preparation of "A Long-Range Plan for Illinois Forest Resources" (Illinois Council on Forestry Development, n.d.). The plan presents an assessment of the state's forestland resources including rural and urban forests, soil and water, recreation, and forest habitat, and discusses

FIGURE 18-6
A young seed orchard of grafted slash pine. *(Courtesy of Georgia Forestry Commission)*

the Illinois forest industry and its key concerns. Five goals, each with several objectives and specific actions toward their achievement, are set forth.

The five goals establish direction to 1) maintain and 2) expand the state's forestland base and 3) stimulate the wood industry to improve and expand its capacity. Community forestry programs 4) are encouraged to make urban areas attractive and liveable, and conservation education 5) is to be strengthened and expanded to encourage citizen support for forest conservation and development.

The direction of Illinois's forest plan reflects the history and current status of the state's forest resources. Since the time of the first land surveys in 1820, forest acreage in Illinois has been drastically reduced, and today only 12 percent of the original forestland remains. In 1820, forests occupied 13.8 million acres (5.59 million ha), which were reduced by 1923, through agricultural conversion and urbanization, to only 22,000 acres (8.9 thousand ha). Since then, forest acreage increased, both naturally and through planting, to 4.26 million acres (1.73 million ha) in 1985; 90 percent of this land is privately owned.

More than 300,000 people settled on the Illinois prairies in the 1830s, creating great demand for housing material, fuel, and fence posts. Local timber was in such great demand that timberland was worth $35 per acre while prairie land sold for $5 per acre. About 1860, the timber industry began to flourish, and by 1870, 92 of the 102 counties had wood-based manufacturing establishments. During the 1880s, more than 350 million board feet were produced annually, production that was sustained until 1900 when supplies were depleted. Slowly, the forests began the recovery that continues today. By 1983, annual growth was estimated to be 437 million board feet, with only 309 million board feet harvested. Illinois ranks fifth among the states in wood consumption because of its large population, but ranks 32nd in production. However, due to extensive secondary manufacturing, the forest-related industries in Illinois employ about 60,000 people with an average payroll of $1 billion (Fig. 18-7). The state's urban forests are also valuable and include more than 100,000 acres (40,500 ha) of urban forest and 140,000 acres (56,700 ha) of urban areas with trees, including 6.5 million street trees. Urban forestry, therefore, is a $300 million industry, employing 3,000 people and supporting 500 tree-care businesses.

FIGURE 18-7
Though only 12 percent of Illinois's original forest remains, through wood brought in
from other states, it has a $1 billion forest products industry, with 60,000 people
employed. These maple furnishings exemplify the thriving furniture industry.
(Courtesy of John Boos & Company, Effingham, Ill.)

FEDERAL ASSISTANCE FOR STATE FORESTRY PROGRAMS

The U.S. Congress, recognizing the national interest in promoting management and protection of forest resources in the states, has enacted federal laws to assist state forestry programs. These laws establish a series of cooperative programs to provide federal financial support and coordination to assist local leadership of the state forestry agencies. From 1911 to the 1970s, several laws were passed authorizing and funding cooperative fire protection, pest (insects and disease) management, forest management, forest incentives, urban and community forestry, and other cooperative programs. The Cooperative Forestry Assistance Act of 1978 encompassed all of the past laws, and brought them under one legislative authority. More recently, the Food, Agriculture, Conservation and Trade Act of 1990 (known as the 1990 Farm Bill) amended the 1978 law, authorizing both a new Forest Stewardship Program (FSP) to assist private landowners in planning and implementing multiple

use activities on their lands, and the Stewardship Incentive Program (SIP) to stimulate enhanced management through cost sharing of approved practices.

Recently, a new farm bill (the Agriculture, Conservation and Rural Enhancement Act of 2002) was signed into law. This massive document has a forestry title that strengthens financial and technical assistance to NIPF landowners. Perhaps most noteworthy is the new Forestland Enhancement Program that replaces the current Stewardship and Forestry Incentive Programs. It will be a few years before the implementation and integration of new and existing forestry assistance programs are routine and well understood, but the intent of Congress to provide forestry assistance to NIPF landowners, much of it through programs administered by the states, is clear.

These cooperative programs are administered by the state forestry agencies. Federal responsibilities are administered by the State and Private Forestry Branch of the Forest Service, in the U.S. Department of Agriculture, and the Natural Resource Conservation Service (mentioned later). The Forest Service allocates federal funds to several states, conducts fiscal and program audits, and provides technical specialists to assist the states. Major programs administered by the Forest Service are listed under the following three major categories: Cooperative Resource Protection, Forest Management and Utilization, and Other Cooperative Programs. Because the organization of new and old programs under the 2002 Farm Bill is not fully developed at this time, the programs described should be regarded as illustrative of federal–state cooperative programs for forestry assistance. Details for current programs are being developed.

Cooperative Resource Protection

Rural Fire Prevention and Control Cooperative Fire Protection improves the efficiency and effectiveness of state and local firefighting agencies by providing matching federal funds for state equipment, training, information gathering and distribution, technology transfer, and technical assistance and coordination. The dramatic reductions in wildland acreage burned annually since cooperative fire assistance programs began (from 30 million to over 2.5 million acres annually) are discussed later.

Forest Pest Management (FPM) Seeking to protect forest resources on lands of all ownerships from insects and diseases, FPM provides help in surveying, prevention, and suppression of forest insects and diseases, and supplies technical as well as financial assistance. Pest management projects save millions of dollars worth of timber and helps protect recreation areas, wildlife habitats, and watersheds. For example, detection and evaluation surveys, along with recommendations and advice about suppression needs and available alternatives, are provided to managers of the affected lands. *Integrated pest management* approaches are used to protect the forest resources, including timber, watersheds, recreation, wildlife, and aesthetics. A variety of tactics is used, including silvicultural, biological, chemical, and mechanical means. Major pest control projects have focused on the gypsy moth in the Northeast, the southern pine beetle in the South, and dwarf mistletoe, mountain pine beetle, and western spruce budworm in the West.

Forest Management and Utilization

Federal forestry assistance to the states in this category supports and supplements programs in forest resource management, ranging from help in management planning to tree improvement.

Forest Resource Management The Forest Service cooperates with state forestry agencies to provide technical assistance to nonindustrial, private forest landowners for managing their forestlands by helping NIPF landowners prepare multiresource plans, plant trees, and make timber stand improvements.

Forest Stewardship (FS) This includes a broad range of forest practices including wildlife, soil, and water conservation, in addition to timber management. Under the Forest Stewardship Program, implemented in 1990, landowner plans are developed that identify actions to protect soil, water, range, aesthetic quality, recreation, timber, or fish and wildlife resources on NIPF lands. This is a popular and effective program based on at least one study (Egan et al., 2001) reporting high landowner satisfaction and a strong association between items included in Stewardship plan and those subsequently accomplished.

Thus, in addition to forest practices such as tree planting and stand improvements, cost sharing is available for enhancement of soil and water resources, riparian areas and wetlands, fish and wildlife habitat, and forest recreation. The Stewardship program, authorized in the 1990 Farm Bill, grew out of an initiative sponsored by the U.S. Forest Service and National Association of State Foresters, and is closely coordinated with the Tree Farm program sponsored by the forest industry, and plans under either program can meet each others' requirements. The new Forestland Enhancement Program in the 2002 Farm Bill will provide new additional cost-share assistance.

Other Forest Service, State, and Private Programs

Seedling, Nursery, and Tree Improvement This program provides high-quality, genetically improved tree seed and planting stock for reforestation to protect soil and water resources and improve productivity.

The Forest Service also provides *taxation* information to forest landowners to assist them in planning for their financial future. Forest Service tax coordinators write articles for forestry magazines and conduct or participate in numerous forest taxation workshops in timber dependent communities.

All forest-related activities for commodity and amenity resources influence *rural development.* By wisely using timber, minerals, wildlife, fishing, recreation, and other forest-based resources, rural economies can be diversified and strengthened. Many opportunities exist for developing joint projects between the Forest Service and other public, state, and private entities concerned about sustainable development through resource conservation and management. These strategies include helping communities organize for resource-based economic development, targeting value-added processing, promoting alternative goods and services, enhancing productivity, strengthening marketing, promoting technology transfer, and improving local human capital through training.

Forest Products Conservation and Recycling

This broad-based program focuses federal, state, and private utilization and marketing expertise and leadership on three major national issues—economic development, recycling, and forest products conservation. Technical assistance, training, demonstration projects, and technology transfer are the mechanisms by which this program seeks to retain and expand forest-based businesses to enhance local economies in rural and urban areas; to deliver new technologies and provide technical assistance in recycling wood fiber-based materials into usable value-added products; and to train forestry service providers to better help their client forest owners understand utilization options, values, and ecologically sound harvesting practices.

Nearly all state forestry agencies develop *state forest resources plans* utilizing data from Forest Service inventory and research. Besides helping states better plan their forestry programs, the cooperative approach generates resources data for strategic federal planning efforts at the national level, including the 10-year assessments and programs developed by the Forest Service, as required by the Forest and Rangeland Renewable Resources Planning Act (RPA) of 1974.

FIGURE 18-8
The Maryland Forest and Park Services spans the spectrum of forestry options from rural forests to urban trees. Here two state foresters are seen taking a street tree survey. *(Maryland Forest and Park Services photo by Jerry Koch)*

Urban and Community Forestry   This program promotes and improves the economic, environmental, and social well-being of communities through the planting and management of trees, shrubs, and other vegetation. These efforts seek to enhance urban environments; improve soil, water, and air quality; and help conserve energy and reduce carbon dioxide in the atmosphere. Increasingly, state forestry programs and federal assistance are involved in urban areas, which now cover 3.5 percent of the United States, contain 3.8 billion trees, have an average tree canopy cover of 27 percent, and contain 75 percent of the population (Nowak et al., 2001) (Fig. 18-8).

Conservation Reserve Program (CRP)   This program was established under the 1985 Food Security Act to remove highly erodible cropland from production. Farmers are paid competitive annual fees per acre for 10 years if they remove land from production and maintain permanent vegetation cover. Congress established a goal of 12.5 percent for tree planting on the 40 to 50 million acres (16 to 20 million ha) targeted for protection in the reserve.

The USDA Natural Resources Conservation Service administers several *cooperative watershed protection projects,* drawing on expertise from the Forest Service for forestry aspects of small watershed protection projects. Both agencies provide assistance on both public and private lands to reduce threats to life and property following natural disasters.

Resource Conservation and Development (RC&D)

This program is also administered by the NRCS with forestry expertise from the Forest Service. RC&D funds are cost-shared with state forestry organizations that provide technical assistance to develop natural resources and improve economic, social, and environmental conditions in qualifying areas, as called for in local RC&D area plans.

Cooperative Forestry Extension

The CFE, described in Chapter 17, is an important program because county extension agents are often the first point of contact for NIPFs who need information. Forestry extension involves federal support (as well as state and county support) through the USDA Cooperative State Research, Education and Extension Service (CSREES), which responds to state needs and plans (Journal of Forestry, 2001).

Financing of Cooperative Programs

Annual congressional appropriations for federal, forestry-related cooperative assistance programs have ranged from $66.6 million in fiscal year 1987 to $182.4 million in fiscal year 1991 to $165+ million in fiscal year 2002, depending on what is included. The programs generally allow for 50–50 cost sharing, but in most instances, to stretch scarce dollars, the federal funds available to match state contributions have varied in programs from 10 to 50 percent. Funds are allocated to the states on a formula basis that considers needs and other factors on a national basis.

The proper role of the federal government versus that of the state governments in financing these cooperative forestry programs has been debated for years. Many believe that the federal government should give strong leadership, including financing, while others claim that the states should have more control and a greater share of the financial burden. The philosophy of the political party in power influences this ongoing debate. The cooperative role of federal and state governments in forestry will continue to be debated based on prevailing philosophies of federal versus state responsibilities, competition for scarce federal and state dollars, the effectiveness of the various programs, national needs, and other contemporary issues in forest policy.

Impacts of Forestry Assistance Programs

Without question, cooperative forestry programs have provided important benefits to the nation and will be even more important to forest productivity in the future. About 60 percent of the nation's commercial forestland is nonindustrial private land and is expected to provide most of the increases in timber supplies in the future. These small ownerships cannot afford, nor do they have the expertise, to carry out all the forest management activities that are important to protect their lands from fire, insects, and disease, and to enhance productivity through planning and investing in treatments such as planting, thinning, fertilization, and regularly scheduled harvests (Fig. 18-9).

The evidence of the benefits from these programs is compelling. For example, from the turn of the twentieth century to 1930, the area of forestland burned annually by wildfires exceeded 30 million acres (12.2 million ha), but with the strengthening of cooperative fire protection programs, the average acreage burned fell from 5 million acres (2 million ha) in the 1950s to 4 million acres (1.6 million ha) in the 1960s and to less than 3 million acres (1.22 million ha) in the 1970s, 1980s, and 1990s—with the exception of a few "bad" fire years. These reductions were achieved despite the fact that fire-frequency totals remain between 100,000 and 200,000 fires per year (Fedkiw, 1992). On the other hand, introduction of exotic pests like gypsy moth, chestnut blight, and Dutch elm disease, along with exposure to air pollutants and changing ecological conditions, have impacted many forests. Nevertheless, cooperative pest management programs have reduced losses from what they would have been without such activity. Recent economic analyses of pest management programs demonstrate benefit/cost ratios greater than 5 to 1. Recent studies document the benefits to landowners from participating in programs that provide professional resource management assistance. One study showed that landowners using professional forestry assistance received higher stumpage values and left their land in better economic and environmental condition.

FIGURE 18-9
A state forester with the New York Department of Environmental Conservation assists a farm forestry owner with a firewood thinning project. Diseased ash is removed, leaving behind healthy maple and hickory. *(Courtesy of USDA Forest Service)*

Additional evidence of the positive effects of assistance for private landowners is found in the fact that more than a dozen states have their own programs over and above the federal cooperative assistance programs (Cubbage et al., 1992). The importance of federal, state, and private cooperative assistance programs will be further evident in Chapter 19, "Forest Management on Private Lands."

FOREST MANAGEMENT REGULATION— STATE AND LOCAL

State Forest Practice Laws

Government regulation of forest practices on private land has been debated for more than 100 years. Proposals for federal regulation of forest practices have always been defeated, but state and local regulation of forest practices now prevail throughout the United States. Twenty states (see Fig. 18-1) have forest practice laws that regulate forestry activities on lands within their boundaries (Cubbage et al., 1992). A few other states, such as Ohio, do not have forest practice laws but may have other state legislation regulating the use of soil erosion practices on agricultural land

and silvicultural operations. Although most of the states with forest practice laws enacted them before 1950, during the rising environmental concern of the 1970s many states strengthened their laws, and three additional states added laws. For example, Oregon passed a conservation act in 1941 requiring seed trees to be left after harvest. In 1971, that law was replaced by the Oregon Forest Practice Act to comprehensively regulate forestry. It was further amended in 1987 (Cubbage et al., 1992). During the past two decades, the current federal emphasis on environmental quality and monitoring has heightened concern by the states for the regulation of their forest practices (Ellefson et al., 1995; Ellefson & Cheng, 1994).

The emphasis of these state forest practice acts has been on protecting environmental quality and ensuring continuous productivity of forestland. Most laws regulate forestry activities so as to protect water, wildlife, fisheries, soil productivity, recreation, and aesthetics, and a few laws recognize the economic contribution of timber harvesting to the economy. The laws range in stringency from merely requiring the leaving of seed trees or replanting of harvested acres, to prescribing slash disposal and road construction practices, to submission of detailed timber harvesting plans prepared by a registered forester (New Hampshire and California) and the issuing of permits to harvest timber. State laws generally provide a statement of forest policy and call for some kind of rule making by the state forestry board to regulate specific practices.

Best Management Practices (BMP)

One of the common approaches to ensuring that forestry activities are acceptable is through the concept of Best Management Practices, or BMPs. These are required, recommended, or voluntary management practices by private landowners and timber harvesters, depending on the state (National Research Council, 1998). In 1996, Ellefson et al. (2001) reported that 47 states had programs to promote BMPs and, in 1997, 34 states monitored BMP compliance. This is a strong, positive trend because BMPs generally evolve in an iterative manner wherein acceptable forest practices are specified, applied, monitored, and then subsequently changed to be more effective in the future (Ellefson et al., 2001). This field of forestry will continue to evolve as will the need for rigorous regulations.

Local Ordinances to Regulate Forestry

In addition to regulation of forest practices by the state, local government units are increasingly regulating what happens to forest cover and trees on private property. A survey conducted during the latter half of 1990 identified 359 local ordinances that regulate forestry in the eastern United States (Hickman & Martus, 1991). These ordinances, enacted by county, townships, or municipal governments, are a relatively recent phenomenon, with 72 percent of them passed within the past decade. They are of four general types: general environmental protection (32 percent); public property and safety (14 percent); special feature habitat protection, such as wetlands or shorelines (12 percent); and forestland preservation (2 percent). Local regulation in the West is also emerging in certain situations (Salazar, 1989).

The proliferation of local regulation of forest practices in the eastern United States was attributed by authors of the study to such things as the shift in population from urban to rural settings. Residents seek a high-amenity lifestyle, and many may continue to work in the city. They respond quickly when they think their amenities are threatened by unregulated forestry or agriculture practices. The widespread general concern for the environment has also contributed to local regulation, with pressure on elected officials to take whatever steps are needed to stop practices perceived to threaten environmental quality. In the East, many woodlots have nearby home developments. There are strong markets for roundwood products, including firewood, from these mature stands, but associated harvest activity in the midst of growing rural populations has raised concerns.

In some states, other laws have encouraged local forest regulation. For example, Maryland's Forest Conservation Act of 1991 requires all municipalities with planning and zoning authority to implement forest conservation programs. The programs will require any person applying for subdivision, grading, or sediment control permits on 40,000 square feet or more of land to submit a forest stand delineation and forest conservation plan. Unavoidable tree loss must be mitigated by either onsite or offsite planting and, where little or no tree cover exists, minimum afforestation levels must be met to create a new forest. Long-term protective agreements for these forest conservation areas are required (Maryland Forest, Park and Wildlife Service, 1991). Elsewhere around the country, city ordinances may regulate what happens to trees on private property during any proposed residential construction, calling for an arborist's report to identify any impacts to trees and/or to prohibit impacts on large trees of certain species.

Impacts of Forest Practice Regulations

A series of events occurring since the enactment of the National Environmental Policy Act of 1969 (NEPA) has resulted in extensive public involvement in land management decisions, not just on public lands but in rule making and forest policy applied to private lands. Forest harvesting methods, the use of herbicides, insecticides, and chemical fertilizer, and many other forest practices are being closely monitored throughout the nation. This overlap between various state and local forest practice regulations where they exist, and the federal Environmental Protection Agency (EPA) and U.S. Fish and Wildlife Service, over common resources such as air, water, soil, fish, wildlife, and endangered species habitat has resulted, in certain instances, in a struggle for jurisdiction. Recent polls indicate strong public support for continued high standards for air and water quality and all the associated forest environment amenities, even at the expense of jobs and economic activity, if necessary. All these influences have led and are leading to more regulation of forestry activity.

Regulatory activity, however, has also resulted in a great amount of monitoring, permits and paperwork for landowners, forest managers, state and federal agencies, and others. This has slowed production and raised the costs of doing business (Cubbage, 1991). For example, in California, which has the most rigorous forest practices regulations in the nation, a book of forest practice rules spells out where, when, and how forestry operations can be conducted. A timber harvest plan must be prepared by a registered professional forester and reviewed by the California Department of Forestry and Fire Protection, with input from agencies responsible for fish and game, water quality, and others, if applicable. Following this review, a preharvest inspection is made in the field, and in some parts of the state a public meeting is held, the proposed sale area is posted, and adjacent landowners notified. Decisions to proceed with the harvest can be appealed to the California Board of Forestry, and its ruling can be challenged in state court. Harvests can be conducted only by licensed timber operators using best management practices and following other guidelines (California Department of Forestry and Fire Protection, 1991). California's first forest practices act was passed in 1945 and revised in 1973, becoming the most comprehensive regulation process in the nation. Each year CDF reviews 1,500 timber harvest plans, performs 7,500 inspections, and issues 50–75 corrective actions and citations.

All eyes are on California and its rigorous forest practice regulations, and it is the hope of some and the fear of others that California may be setting a standard toward which other states will be forced to move.

In other states with forest practices acts (see Fig. 18-1), the requirements and procedures may be less stringent, but they too reflect the public's great concern over the management of forests—public and private. Balanced approaches are needed that protect the environment while allowing sound forest management programs and practices to be implemented in an economically feasible manner. The administration of state and local forest practice regulations lies at the heart of achieving such balance and of earning public support for forestry programs in the states.

The public regulation of private forest practices will continue to be a key forest policy issue in the future. Driven by public concern for air and water quality, aesthetics, public access, wildlife, fish, and endangered species, we expect to see many more state laws and local ordinances regulating forest practices and also some creative self-regulation efforts by the forest industry and other forest landowners and their associations.

SUMMARY

Many state forestry programs and organizations evolved before or near the turn of the twentieth century in response to state acquisition of lands from federal grants, and also in response to problems associated with forestlands such as wildfire control. In 1920, the National Association of State Foresters was formed, although it did not formally acquire that name until 1964. Over time all states have developed forestry organizations to manage state lands and implement cooperative programs providing financial and technical assistance to private landowners, and to carry out forestry activities important to all landowners such as protection from fire, insects, and disease. State forestry organizations operate under four general types of oversight and authority.

Many federal programs provide cooperative technical and/or financial assistance to the states and private landowners in matters concerning fire, insect, and disease prevention and control, wood utilization, forest planning and management, tree planting, stewardship, and erosion control. The benefits of these programs have been impressive, leading to important gains in forest productivity as well as a tenfold reduction in the average annual acreage of forestland burned since 1930, despite approximately the same number of fires.

Many states have forest practice regulations that are administered by state forestry organizations. Regulation of forestry activity by local jurisdictions has increased rapidly, especially in the East but nationwide as well. There is a continuing debate over the relative share of financial responsibility and authority to be assumed by the states versus the federal government in cooperative forestry programs. This debate ebbs and flows with other states rights issues and the philosophy of the political party in power.

LITERATURE CITED

California Department of Forestry and Fire Protection. 1991. California Forest Practice Rules. Sacramento, Calif.

Cubbage, Frederick W. 1991. Public Regulation of Private Forestry: Proactive Policy Responses. *J. Forestry* 89(12):31–35.

Cubbage, Frederick W., Jay O'Laughlin, and Charles Bullock, III. 1992. *Forest Policy.* Chapter 16 Regulation of Forest Practices; Chapter 17 Public Assistance for Private Forest Owners. John Wiley and Sons, New York.

Egan, Andrew, David Gibson, and Robert Whiskey. 2001. Evaluating the Effectiveness of the Forest Stewardship Program in West Virginia, *J. Forestry* 99(3):31–36.

Ellefson, P. V., and A. S. Cheng. 1994. State Forest Practice Programs: Regulation of Private Forestry Comes of Age. *J. Forestry* 92(5):34–37.

Ellefson, P. V., A. S. Cheng, and R. J. Moulton. 1995. Regulation of Private Forestry Practices by State Government. *Station Bulletin* 605–1995. Agricultural Experiment Station, St. Paul, Minn.

Ellefson, Paul V., Michael A. Kilgore, and Michael Phillips. 2001. Monitoring Compliance with BMPs: The Experience of State Agencies. *J. Forestry* 99(1):11–17.

Fedkiw, John. 1992. The Evolving Use and Management of the Nation's Forests, Grasslands, Croplands and Related Resources. USDA Forest Service General Tech. Rpt. RM-175.

Georgia Forestry Commission. 1991a. Forest Resource Plan: A Five-year Guide for the State Forestry Program. Macon, Ga.

Hackett, Ronald L., and Richard Dahlman. 1997. Minnesota Timber Industry—An Assessment of Timber Product Output and Use, 1992. St. Paul, Minn. USDA Forest Service, N.C. Forest Experimental Station.

Hickman, Clifford, and Christopher Martus. 1991. Local Regulation of Private Forestry in the Eastern United States. Proc. S. Forest Econ. Workshop. SOFEW, North Carolina State University, Department of Forestry, Raleigh, N.C.

Illinois Council on Forest Development. (n.d.) A Long-Range Plan for Illinois Forest Resources. Gary Rolfe, Chairman, University of Illinois, Department of Forestry, Urbana, Ill.

Johnson, Tony G., and John L. Wells. 2002. Georgia's Timber Industry—An Assessment of Timber Product Output and Use. USDA Forest Service, Southern Research Station, Asheville, N.C.

Journal of Forestry. 2001, March. Forestry Extension: Reaching Out to Private Landowners. Special issue *J. Forestry* 99(3).

Maryland Forest, Park and Wildlife Service. 1991. The Forest Conservation Act of 1991 Summary. Tawes State Office Bldg., Annapolis, Md.

National Association of State Foresters. 1982. "History of State and Private Forestry," U.S. Department of Agriculture Forest Service, Washington, D.C.

National Research Council. 1998. Forested Landscapes in Perspective: Prospects and Opportunities for Sustainable Management of America's Nonfederal Forests. National Academy Press, Washington, D.C.

Nowak, David J., Mary H. Noble, Susan M. Sisinni, and John F. Durger. 2001. Assessing the U.S. Urban Forest Resource. *J. Forestry* 99(3):37–42.

Oregon Department of Forestry. 1993. Tillamook State Forest—Tillamook Burn to Tillamook State Forest. Oregon Department of Forestry. Salem, Oreg.

Rose, Robin, and Jeremy Coate. 2001. Reforestation Rules in Oregon: Lessons Learned from Strict Enforcement. *J. Forestry* 98(5):24–28.

Salazar, Debra J. 1989. Counties, States and Regulation of Forest Practices on Private Lands in Robert G. Lee, William R. Burch, and Donald R. Field (eds.), *Communities and Forestry: Continuities in the Sociology of Natural Resources.* Westview Press, Boulder, Colo.

Wigley, T. Bently, and M. Anthony Melchiors. 1987. "State Wildlife Management Programs for Private Lands." *Wildl. Soc. Bull* 15(4):580–584.

ADDITIONAL READINGS

Egan, A., and J. Rowe. 1997. Compliance with West Virginia's Best Management Practices, 1995–1996. Department of Forestry, Charleston, W.V.

Southern Environmental Law Center. n.d. North Carolina Office, Chapel Hill, N.C. Model Forest Practices Act.

STUDY QUESTIONS

1. What kind of land ownerships are the focus of state forestry organizations?
2. Explain why there is a need for a National Association of State Foresters.
3. What are the four types of guiding authority under which state forestry organizations operate?
4. Why are state forest plans important to federal forest planning?

5. Explain why it is in the national interest for the federal government to provide technical and financial assistance for state and private forestry programs.

6. Describe four federal programs that provide cooperative assistance for state and private forestry.

7. What are some benefits that have resulted from cooperative technical and financial assistance for private lands?

8. Why should the states have forest practice regulations that apply even on private lands?

9. What type of organization operates in your state to regulate forest practice concerns? How are members of this organization chosen? What issues are currently before it?

10. Look up your state forestry organization on the Internet. Identify some of its responsibilities that you didn't know about.

Forest Management on Private Lands

INTRODUCTION

This chapter describes forestry on private, as contrasted with publically owned, lands, including forest industry lands, which are usually large ownerships that are managed intensively. Nonindustrial private forestlands owned by individuals or families (NIPFs) are also described (see additional information in Chapter 15, "Economics in Forest and Renewable Resource Management").

Differing economic considerations and other motivations affect the various types of landowners and their land management decisions. Federal, state, and private cooperative programs that influence management on nonindustrial lands are described, as are the intensive forestry practices used on forest industry lands, including some kinds of research carried out to further increase productivity. Industrial and privately sponsored forestry programs such as Project Learning Tree and the American Tree Farm Program are described. Also described are the green product certification initiatives that have emerged during the past decade and are now embraced by the industry with leadership efforts by many companies.

PRODUCTIVITY OF PRIVATE LANDS AND THEIR IMPORTANCE

Private forestlands, both industrial and nonindustrial ownerships, in general occupy the better timber-growing sites. One of the major reasons for this is that early federal land disposal laws encouraged sale and settlement of the better and more accessible timbered lands. Thus, private lands, in general, are either producing above the average or have the potential to do so. Because of their more intensive management, forest industry lands have the highest rates of production (see Table 19-1). But because they contain 58 percent of the nation's timberland, these nonindustrial, private forestlands (NIPFs) are extremely important (Fig. 19-1). They produce 59 percent of the nation's annual timber harvest (Smith et al., 2001). Timber removals from NIPF lands increased 17 percent between 1986 and 1996, no doubt in response to reduced harvests on public lands. Thus, policies to increase the intensity of management and, thus, the future yields from these smaller, NIPF holdings are a high priority. On the other hand, industrial forest companies own 13 percent of U.S. timberland

TABLE 19-1

TIMBERLAND OWNERSHIP AND REMOVALS (HARVEST) IN THE UNITED STATES

Ownership Category	Acres	(hectares)	%	Removals (%)
Private, nonindustrial	291	(118 ha)	58	59
Forest industry	67	(27 ha)	13	30
Public—National forests	50	(20 ha)	19	5
Other Public—Federal, state, and local governments	96	(39 ha)	10	6
	504		100	100

(*Source:* Smith et al., 2001).

(67 million acres; 27 million ha) but they account for 30 percent of the U.S. harvests in 1996.

Seventy-one percent of the nation's timberland is privately owned, and these private lands (combined) provide 89 percent of the timber harvested in 1997 (Smith et al., 2001). Despite the fact that the federal government, primarily the USDA Forest Service, owns and manages 29 percent of the nation's timber volume, these lands produced only 11 percent of the timber harvest in 1996. Harvests on the national forests declined 62 percent between 1986 and 1996 when it was only 5 percent of total harvests, and are being reduced by a number of factors, including environmental constraints, proposals to set aside more wilderness, to safeguard ancient forests to protect remaining roadless areas from development, proposals to protect endangered species, and growing pressures to limit harvests and modify harvesting methods on public lands. Thus, private forestlands, including those owned by the forest industry, and especially the smaller NIPF lands, are extremely important to the nation's future timber supply. That is why intensive management of all private forests is such a high priority.

But there are also pressures on private forestlands that will impact their future production potential. The Endangered Species Act applies some of its restrictions to private lands and could impact future timber production. The wetlands provision of the Clean Water Act could restrain the use and management of private forestland such as draining wetlands to increase productivity. State and local forest management regulations are becoming more numerous and

FIGURE 19-1
Commercial forestland in North America. Note the large amount (71 percent) of lands held in private ownership in the United States.

demanding, and timber harvesting has been banned or limited near some urban areas. Fear of development has prompted public acquisition of forestland in some heavily populated regions and areas. Private forestlands are becoming more fragmented due to subdivision and sales. In some instances, when their location makes them too valuable for just growing timber, forest industry companies may offer these prime lands for sale for suburban or vacation homes.

All of these trends are impacting the ability of the United States to produce wood products, pulp, and paper. Why should that be a concern? It is because wood is a renewable resource and far more efficient in the energy required for its extraction, manufacture, use, and disposal than competing wood substitutes—

such as aluminum, plastic, steel, and cement (Alexander & Greber, 1991). Also, on a tonnage basis, wood products constitute about 45 percent of U.S. consumption of all materials including plastics, concrete, steel, aluminum, and other metals (Smith et al., 2001). Wood products manufacture may also be less polluting than these substitutes. Furthermore, forests store carbon, consume carbon dioxide, and produce oxygen, with fast-growing young forests being especially efficient in these functions. If wood production in the United States is restricted unnecessarily, wood imports will increase, in many cases stimulating exploitation of world forest resources in countries without adequate environmental protection and reforestation requirements. Thus, U.S. wood production under sustainable, environmentally sound management is an important conservation strategy. In order to meet the rising demand for wood products and construction in the United States and abroad, forest resources must be allocated, managed, and used carefully. Public education about the goals of forestry and the national and global—as well as local—consequences of forest management strategies provides important support for this direction.

U.S. FOREST OWNERSHIP AND PRODUCTIVITY

Lands that grow 20 cubic feet of wood per acre per year are considered to be of commercial quality and are officially known as *timberlands.* The North American continent has 14 percent of the world's timberland, and nearly one-third of this (about 4 percent of the world's timberland) is in the United States. In the United States 71 percent of the timberlands are in private ownership, with the remainder held by the federal government (22 percent) and the rest (7 percent) by state and local government (Fig. 19-1). In Canada, most of the forests from Quebec west are owned by the provinces.

Private owners of various kinds own and manage most timberlands in the United States. Until the last three decades, under our market-driven economy, U.S. private forest owners managed their land with relatively few government restrictions or controls. Instead, education, financial incentives, and demonstrations were widely used at all levels of government to influence private owners to practice sound forestry and to coordinate their efforts with those of other landowners. This situation has changed as environmental protection laws and standards have come to bear on private forestland. State forest practice acts have been expanded, strengthened, and more intensively applied to regulate forest management on private lands.

In the United States, private owners have the freedom to produce timber products in response to market demands, whereas in some wood-producing nations, such as Canada, a quota system is imposed. A paradox arises here, for despite the fact that private owners in the United States enjoy the advantage of good land and freedom to operate, the United States has imported some of its forest products annually for

many years, with most of these products coming from Canada. It is true that production by U.S. private ownerships, industrial and nonindustrial, meets the major part of our domestic requirements for forest products, but the U.S. forestland base has the potential to supply all or most domestic needs and also produce enough for export. For example, net annual growing stock growth exceeded removals by 47 percent in 1996 (Smith et al., 2001). This includes hardwoods and softwoods—not all of it equally desirable or yet saleable under today's processing technology, markets, and environmental requirements. But it is good to know that the United States is only harvesting about half of the net growth of its commercial forests each year.

U.S. private forests are diverse in size, varying from only a few acres owned by many nonindustrial, private forest owners to millions of acres owned by some corporations. The forests are also ecologically diverse. They range from second, third, and even fourth rotation northeastern forests with low volume and slow growth rates, to the large-volume, fast-growing Douglas-fir forests of the Pacific Coast, to the even faster-growing third and fourth rotation pine plantations of the South. Most of this privately owned timberland is in the eastern half of the country, with 36 percent in the North and 50 percent in the South. The Rocky Mountains and Great Plains have only 6 percent and the Pacific Coast 8 percent of the private timberland.

There are 9.9 million private forestland ownerships with 358 million acres (145 million ha) of timberland in the United States. Forty-five percent of that private forestland is in ownerships larger than 500 acres (200 ha) and is owned by less than 1 percent of the total ownership units, mostly forest industry but some private individuals, families, or investment organizations. There are about 9.9 million NIPFs, but about 6 percent of these owners hold over two-thirds of the acreage. In other words, there are 627,000 owners who have holdings of 100 acres or larger. The other 94 percent of the owners, more than 9 million of them, all own 100 acres or less, with 8.6 million of them owning tracts of 50 acres or less. Of current concern is a process called *parcelization,* in which larger ownerships are being sold off into smaller tracts, further fragmenting the NIPF ownerships (De Coster & Sampson, 2000).

FORESTRY ON NONINDUSTRIAL PRIVATE FORESTLAND (NIPF)

Timberland held by farmers and other nonindustrial private owners has been termed *nonindustrial private forestland* (NIPF) and is an important and large segment of private land holdings, making up 58 percent of all U.S. commercial timberland (Fig. 19-1; see Table 19-1). These lands are important, not only for wood fiber, but to meet other needs as well such as providing wildlife habitat and forest cover protecting watersheds.

These ownerships represent a wide range of financial needs and capabilities: diverse species compositions, forest age classes, land capabilities, and geographical variations. NIPF owners may be business and professional people, wage earners, salaried workers, or homemakers, and an increasing number are retirees. Some NIPF holdings include highly productive forest sites close to markets for wood products and are an important source of timber for wood-using industries.

Yet, all too often these NIPF landowners primarily harvest trees only when they need ready cash or prices are high, even though the trees may be growing at their optimum, and, if pruned and thinned and allowed to continue growing for several more years, would yield even more wood fiber and cash (Fig. 19-2). Conversely, many NIPFs harvest their trees much later than financial maturity. Many NIPF owners do not live on the land and are absentee owners who make their living elsewhere in various occupations. They may hold their land only for the satisfaction of being a landowner or having a place to go for recreation and relaxation. Although many such owners practice forestry, many do not. Possibly they acquired the land through inheritance and may never have visited it. Some NIPF landowners are retirees who own and live on small tracts of rural, forested land.

FIGURE 19-2
A farm woodland owner pruning his trees to develop knotfree wood in the first log. Note also the thinning of less desirable trees in this stand of ponderosa pine.
(Courtesy of the American Forest Institute)

If the owner has a forest management plan and is practicing sound forestry, validated by a field inspection by a professional forester, the property may qualify as a "certified tree farm" under the American Tree Farm System of the American Forest Foundation. Owners may also qualify for various federal and state technical and financial assistance programs to further improve their management. The Tree Farm System is described later in this chapter.

With respect to private forestland, the principal role of government is to provide assistance and coordination in protection from fire, insects, mammals, and diseases—matters that cannot be accomplished successfully or economically on an individual basis. Federal programs also provide cooperative technical and financial assistance, education, and incentive programs to encourage good forestry and natural resource management. Some of these programs are described in Chapter 15, "Economics in Forest and Renewable Resource Management," and Chapter 18, "Forest Management by the States." Gradually, there has been a shift to include more state and local funding for all these programs. Jurisdictional boundaries become less important in addressing all forms of fire, insect, and disease control as coordinated plans are developed. Insects, disease, and fires don't care whose land they impact.

The government role in private woodland management in the foreseeable future will continue to emphasize the historic roles of fire protection, educational advice, and financial incentives; however, controls are increasing. State and local forest practice regulations, controls for water quality protection, air-pollution regulations, coastal zone management, and guidelines for protection of critical habitat for endangered and threatened species limit management options even on private land, but they are far from imposing total control.

PRIVATE COOPERATIVE FOREST MANAGEMENT

With federal and state assistance shifting toward education and incentives, on-the-ground professional forestry advice is coming more often from private sources, including consulting foresters and service foresters employed by forest industry corporations. The number of consulting foresters nationwide has steadily increased, providing federal, state, and local government agencies, as well as private landowners, an available source of forestry expertise and services. To secure and expand their raw material supplies, forest industry firms may lease NIPF land for timber production and then manage it, or they may help landowners directly through landowner assistance programs, often called *cooperative forest management* (CFM), either in return for future harvesting privileges, or without any formal agreement but with the knowledge that such help will lead to good public relations and future harvesting opportunities in their area.

One positive example of a private cooperative forest management (CFM) program is operated in the Southeast United States by Mead/Westvaco Corporation, a leading global producer of high-quality paper, industrial packaging, office products, and chemicals. Before merging with Mead, and continuing today, the former Westvaco Corporation operated four mills and managed 1.3 million acres (52,650 ha) of company-owned land. To make up the deficit between what it grows on its own lands and the needs of its mills, Mead/Westvaco helps small landowners in the region manage another 1.4 million acres (56,700 ha) under its CFM program (Fig. 19–3). In return for a written letter of agreement giving Mead/Westvaco the opportunity to bid for any forest products the landowner might decide to sell, the company provides important management assistance. If Mead/Westvaco and the landowner are unable to agree on a sales price, the landowners can sell the forest products by whatever method they choose.

Mead/Westvaco provides the following management assistance:

- Serves as forest management advisor
- Prepares written forest management plan complete with forest type map and stand recommendations for management that considers landowner objectives
- Assists with implementation of forest management plans that may include:
  Locates contractors to perform activities
  Provides onsite review and advice on progress and quality of work by contractors
  Designs logging plans to achieve good soil and water quality management
  Estimates total timber volumes

FIGURE 19-3
Through Mead/Westvaco's cooperative forest management program, small landowners receive assistance with a range of technical forestry services, including the availability of seedlings for reforestation. *(Courtesy of Mead/Westvaco Corporation)*

Marks timber for thinnings and harvest cuts
Provides tree seedlings for reforestation (seedling programs vary by region from provisional free seedlings to matching programs)
Sets up timber sales
• Provides other services:
Obtains American Tree Farm certification upon qualification
Assistance in qualifying for the Forest Stewardship program and other assistance programs.

The Mead/Westvaco Cooperative Forest Management (CFM) program is just one example of forest industry assistance for NIPF landowners. Several other companies have comparable arrangements to fit their circumstances.

Factors Influencing Private Forest Management

Studies identify taxes in all forms as inhibiting small forest landowners from making investments in forest management. On the other hand, a variety of cost-sharing programs, both federal and state, provide incentives for good forest practices, reforestation, and environmental protection. Most states also offer technical and financial assistance for wildlife habitat improvement and forest management on private land. Incentives such as further simplification of forest taxation will encourage private landowners to be good forest managers. A major step in this direction is the Forest Stewardship program, which involves state forestry organizations in a nationwide federal–state cooperative effort to provide NIPF landowners with technical assistance in forest planning and cost sharing for reforestation and other forest practices. Several studies suggest that only about 50 percent of NIPF landowners seek advice from professional foresters during harvest or reforestation operations, so inspiring and educating NIPF landowners about these new programs will be a key to their success.

There will be continued attention to small, private holdings by government and the forest industry in order to increase productivity and regulate harvests (see Chapter 15, "Economics in Forest and Renewable Resource Management"). Studies show that landowners using technical forest management assistance make more money from timber harvests and that their land is left in better condition. Landowner cooperatives have and must become more common as a means of sharing specialized equipment and improving marketing opportunities.

The role of private forestlands in the nation's economy will become even more important as yields from public lands decline due to further restrictions to safeguard environmental quality and as more public land is set aside in wilderness, parks, and other reserves to meet demands for competing uses. The small, nonindustrial, private forest ownerships will contribute greatly to the future needs of people, not only for wood products, but in all goods and values that these lands can provide such as habitat for wildlife, aesthetics, water, environmental protection, and abatement of air and noise pollution. While harvesting timber for profit is one reason that NIPF landowners own land, it may not be the most important (Argow, 1993). Wildlife-related motives are very important, and technical assistance for wildlife habitat protection and management is growing.

Many small landowners just like having a place to go to enjoy privacy and be close to nature. But these NIPF owners' forests grow, can benefit from management and generate potential wood products. Thus, one challenge and opportunity for foresters and resource managers is to teach these landowners how their goals can be even better served through good forest management, and that some timber can be harvested as a by-product, which may also provide a little income.

There are many challenges facing private forests now and in the future, such as parcelization and fragmentation into smaller ownerships, conversion to other land uses, and sprawl, all of which may lead to diminished ecological function (Best, 2002). Fourteen states lost 2 percent or more of their forestland between 1982 and 1997 (USDA-NRCS, 1999). It will take new, creative approaches to stop these losses, conservation easements with flexibility being one of them (Society of American Foresters, 2002). People and forests must be able to co-exist sustainably.

FORESTRY ON INDUSTRIAL FORESTLAND

Forest industries own only 13 percent of U.S. commercial timberland but account for 30 percent of the annual timber harvests (see Table 19-1 and Fig. 19-1). Although some companies utilize cooperative forest management and long-term options for harvesting wood products from other lands, and many buy timber from the federal government, companies have more flexibility if they own some of their timberlands to supply at least part of their wood fiber needs. This advantage has increased over the past decade as harvests from public lands have been reduced, especially in the West. This continues a shift in overall private forest production to the South, where public lands now account for 64 percent of U.S. removals in 1996, up from 51 percent in 1986 (Smith et al., 2001). The larger wood-processing companies may grow half or more of their raw material supplies on their own lands. Pulp-and-paper and wood-processing plants represent huge investments and because they are not mobile, it makes sense to have a secure, nearby timber supply. The owners of large holdings, as well as some with medium-sized holdings, therefore also see the need for intensive management of their lands to increase yields. As previously mentioned, many also help increase the productivity of nearby smaller, private holdings through cooperative forest management in return for opportunities to purchase the timber. This pattern is especially pronounced in the southern and eastern states.

Documenting the claim that forest management enhances production is the fact that since 1953, about the time that forest management became intensive where feasible, the net volume per acre of U.S. timberlands has increased by 37 percent (Smith et al., 2001). In 1997, about 54 million acres (21.9 million ha) (11 percent of U.S. timberland) was of planted origin, two thirds of it in the South.

The intensity of industrial forest management continues to increase because yields of wood fiber from intensively managed lands are higher. Especially on the lands they own, timber companies want to grow as much wood fiber as possible on a sustainable basis, because this safeguards their future raw material supply. While the biggest investment in forestry is ownership of the land, intensive management practices represent substantial additional investments for which returns will not occur until the next harvest. These practices aim at increasing future yields and include such things as protection against fire, insects, disease, and damage from mammals; clean logging practices such as slash disposal and often prescribed burning or disking and scarification of the ground, all as site preparation for seeding or planting the next crop, probably with genetically improved seed or seedlings; noncommercial thinnings to increase growing space and thus the growth and health of the trees that are left; applying fertilizer to increase growth; and sometimes pruning to increase the quality of future crop trees.

As noted in Chapter 15, "Economics in Forest and Renewable Resources," calculating the financial returns from various management practices is a key to the success of intensive forestry investments. Financial returns may vary with many factors, including site productivity, markets, interest rates, and the length of time until the increased values can be realized. But intensive management can increase biomass yields from the forest by 50 percent or more. This is why private owners and corporations invest in intensive forest management.

For example, one international forest corporation, Weyerhaeuser, has long been a leader in intensive forest management. In 1967, the company created a program called *high-yield forestry* to increase biomass and financial yields from their forests by planting improved seedlings, fertilization, suppressing competing vegetation, and pruning (Weyerhaeuser, 2002; www.weyerhaeuser.com). Nationwide, Weyerhaeuser owns and operates 5.7 million acres (2.3 million ha) of privately managed forests in 10 states, harvesting about 2 percent of its volume per year in the West and 3 percent in the South where growth rates are faster, and planting 100 million seedlings per year. Weyerhaeuser was founded in 1900 when Frederick Weyerhaeuser and 15 partners bought 900,000 acres (364,500 ha) of timberland in the Pacific Northwest and has long been a forestry leader. The company founded its first tree farm in 1941 (more on that later), made intensive forest management visible with its high-yield forestry program, and became a leader in the Sustainable Forestry Initiative (SFI) described later in this chapter. Weyerhaeuser was also the first wood products company to establish an environmental director to oversee protection of fish and wildlife habitat and water quality on its lands and operations.

Like several of the largest forestry corporations, Weyerhaeuser produces a variety of wood products, including lumber, veneer and plywood, OSB, engineered wood products, and raw materials such as logs, chips, lumber, pulp, paper, paperboard, containerboard, and packaging and other products, generating $16 billion in sales in 2001, including $1.5 billion in exports (Weyerhaeuser Co., 2002). Truly a global forestry leader, Weyerhaeuser employs 47,000 people in 17 countries, primarily in the United States

and Canada, where the company owns 664,000 acres (268,920 ha) and has long-term leases on 32.6 million acres (13.203 million ha), to support its diversified wood product line and timber supply. And yet Weyerhaeuser remains globally competitive, in recent years acquiring other companies such as Trees Joist, maker of engineered beams, and MacMillan Bloedel, a large Canadian timberland owner and wood processor. In 2002, the company acquired Willamette Industries, boosting its projected annual sales to nearly $20 billion and putting Weyerhaeuser among the top three producers in all of its product lines.

Weyerhaeuser is not alone in the forest industry in seeking competitive advantage and responding to changing economic conditions thru acquisitions and mergers to expand timber supply and/or product lines. In 2001, Plum Creek Timber Company merged with Georgia Pacific's The Timber Company, making Plum Creek the United State's second-largest holder of timberland. Mead and Westvaco merged in 2001, creating Mead/Westvaco, a leading global producer of packaging, coated and specialty papers, and other products, with annual sales of $8 billion and 30,000 employees operating in 33 countries and owning 3.5 million acres (1.410 million ha) of sustainably managed forests (www.Meadwestvaco.com). These are several examples of change occurring in the industry.

So, while fragmentation and parcelization are reducing the size of forest ownerships on NIPF lands, mergers and acquisitions are concentrating timberland ownership among fewer industrial companies. The potential effects on intensive management contrast, as it is difficult on the smaller NIPF lands and easier on larger ownerships to apply intensive forest management.

Intensive Management

Table 19-2 shows four phases of intensive management practices that the forest industry may implement during the life of a stand

If interest rates are high, the carrying costs of borrowing capital to invest in intensive forestry may preclude certain intensive management practices. During such periods, *extensive* rather than *intensive* forest management may be practiced. Extensive management includes only basic practices necessary to protect the forest, such as fire protection, and insect and disease management. Sometimes intensive practices (such as fertilization) are applied only 5 to 10 years

TABLE 19-2

INTENSIVE FOREST MANAGEMENT PRACTICES THAT MAY BE APPLIED DURING VARIOUS STAGES IN THE LIFE OF A STAND

Site Prepartion Phase	• Clean logging (not necessarily clearcutting) to remove all economically usable materials, reduce slash, and to make room for seedlings. • Logging slash disposal, usually by piling and burning, broadcast burning, or occasionally, by chipping. • Scarification of soil or other seedbed preparation as necessary. • Control of brush and competing vegetation, by spraying if necessary.
Regeneration Phase	• Planting—with seedlings that are often genetically improved, and matched to the site. • Control of competing vegetation by spraying as necessary. • Control of animal damage (rodents, deer, and elk). • Fertilization to get seedlings established and assure rapid growth.
Intermediate Stand Management	• Precommercial thinning of trees (not yet merchantable) to provide more growing space for those left. • Brush control and animal damage control as necessary and justified. • Commercial thinning to provide additional growing space for crop trees and for initial financial yields. • Fertilization—sometimes to intermediate stands, but more often to mature stands 5–10 years before harvest (to reduce interest charges).
Mature Stand Management	• Salvage of diseased, dead, or dying trees. • Fertilization. • Harvest crop trees crop at rotation age. • Begin the growing and management cycle again, or continue it in the case of uneven-aged systems.

before anticipated harvest, when the end result of the rotation is in better focus. In this way, the carrying costs of the investment (interest) will be less, risk of losing the intensive management investment (from fire, insects, or disease) is reduced, and returns can be more easily estimated, based on projected markets and the species and sizes of trees that will be harvested.

Levels of Management Intensity

Growing of trees for wood and fiber production is usually the objective of forest management on large holdings of private lands, less so on smaller (NIPF) ownerships. The intensity of management practices (or systems) will vary under different forest ownerships and often with the productive potential of the land. The most productive land is usually managed more intensively, since the payoffs from investments in management will be greater.

For convenience, we can divide the intensity of timber management practices into increasing levels of intensity, I through V. The amount of investment increases with each level of management. If there is no protection from fire, insects, or disease, and no

deliberate reforestation, for practical purposes there is no forest management being practiced, and this would be referred to as *zero level of intensive management*. This situation might arise where absentee owners live far away.

- Level I—Low-investment, custodial management, which means basic protection from wildfire, perhaps through a cooperative agreement with other landowners or the state. Reforestation, if it happens, is by natural regeneration only. Owners do only what is necessary to keep the stand in place and harvest opportunistically.
- Level II—Basic forest management would include further wildfire protection, perhaps some prevention such as slash disposal, controlling access, and possibly other presuppression techniques such as maintaining fire access roads. Management considerations at this level might also include assessment of the effects of insects and diseases on the stand and strategies for dealing with these. Some kind of management plan is in place, guiding silvicultural treatments, such as a regeneration program

that assures prompt establishment of seedlings on harvested acres either through artificial means, such as planting or seeding, or by leaving some standing trees to serve as a seed source. Level II also includes site preparation, such as getting rid of logging slash and removing undesirable vegetation before planting. This entails greater investment than Level I, and should result in a considerable improvement in productivity.

- Level III—Moderately intensive forest management with a more complex management plan and growing stock inventory to guide additional practices and investments. At this level, precommercial thinning and fertilization to improve productivity might be implemented strategically in certain areas. Commercial thinning is practiced where possible, with tighter control on access and better-maintained interior roads.
- Level IV—Intensive forest management is driven by a complex management plan and inventory to optimize tree growth and financial yields from investments in stand treatments. This level features a stocking control program with precommercial thinning to concentrate growth on the best trees optimally spaced, commercial thinning to maintain optimal spacing and pay for management, fertilizing where it can pay for itself with increased yields, site preparation, and regeneration with a tree-improvement program planting genetically superior seedlings to produce better trees and reduce the time to grow them to harvest age.

Costs rise at each level and thus economic considerations are usually the deciding factor. Most small, private, nonindustrial ownerships (NIPFs) are at Level I, because such landowners are not willing or able to invest capital in a higher level of forest management. As already mentioned, many federal and state, as well as private industry, cooperative programs are focused on encouraging NIPFs to practice some degree of management, since this will increase the health and growth of forests. But for some of these owners, timber harvesting is frequently based on financial need and market prices rather than on optimum growth rates and timber stand maturity (i.e., opportunistic harvesting). Such landowners may be unwilling to pay any costs for intensifying management, even though much help is available.

The majority of publically owned timberlands are managed at Level II or III. Some critics would like to

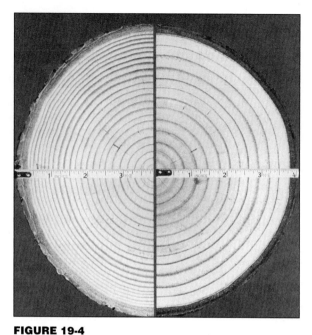

FIGURE 19-4
Managed trees grow faster. *Left:* Cross-section of a 24-year-old tree in an unmanaged forest. *Right:* Cross-section of a 14-year-old tree in a managed forest. *(Courtesy of Weyerhaeuser Corporation)*

increase the level of management on public forests, but Congress provides few funds for intensive management. It is the timber companies that own their own lands—the privately owned industrial forests—that manage at Level IV.

Productivity in Florida's third forest illustrates the response to intensive management. Yields have been increased several times over the first forest harvests in the early part of the century. The second forest grew back from the 1930s until the 1960s, partly from planting and some from natural regeneration. Wood for the 1980s, 1990s and early 2000s is coming from a third—and in some places a fourth—forest, a planted forest managed intensively like an agricultural crop. Fig. 19-4 illustrates the comparison of managed and unmanaged practices on a tree cross-section.

Research in Intensive Forestry

Knowledge created by research is vital for increasing forest yields and minimizing impacts of intensive forestry on other resources and values. Many of the larger forest industry companies have established their own research programs to improve operations on their lands, as well as to solve specific problems. Companies may also use research results from federal agencies and universities and often support such research politically, contribute to it financially, and even participate in studies. Whatever the source, application of research findings has resulted in more intensive forestry on industrial lands. For example, research has led to more complete utilization of harvested material, leaving less slash and waste in the woods, and the burning of wood processing waste at mills to generate energy for the wood processing plant. In some instances there is surplus energy to sell.

Following are a few examples of research topics that have helped advance forest management and yields on private lands, and on some of the most productive public lands where funds for intensive management have been made available.

Control of Undesirable Vegetation   Growing conditions favorable to forest seedlings are often equally favorable to the growth of undesirable brush that competes with and suppresses the young trees. Chemical control of the unwanted vegetation with herbicides helps give seedlings a better start. There have been increasing restrictions on the use of certain chemicals used in herbicides as well as in pesticides used to control insects and disease. Thus, the development of management practices to reduce the need for herbicides, and the development and testing of safer and more acceptable herbicides, are important, ongoing research topics.

The concept of *integrated vegetation management,* meaning the combined use of several techniques for vegetation control in addition to chemicals, is being widely studied and used. For example, prescribed burning or scarification of the site with crawler tractors or bulldozers may be used to prepare the seedbed for natural regeneration and to reduce slash and brush to facilitate planting. These practices also allow sunlight to reach the young seedlings. In the South, prescribed burning has been found to help control brown spot needle disease. By integrating a variety of such practices, competing vegetation may be controlled and the need for use of chemicals reduced to a minimum.

Intensive site preparation is becoming a standard practice in order to give seedlings a headstart on competing vegetation. Site-preparation techniques for planting in the South may include disking and chopping with large, tractor-drawn roller choppers, with or without prescribed burning, followed by hand or machine planting to immediately establish a growing crop of trees. Tractors have also been used to pile logging slash for burning in the West, the Lake states, and the South, sometimes in combination with broadcast burning. Live brush, where present, may be sprayed to kill it for a better burn. Such methods in various combinations continue to be studied and tested to find the best combinations for different sites and to respond to increasing restrictions on use of chemicals, and on burning because of air quality. There is also growing concern that sufficient waste wood be left on the ground to provide nutrients for the next crop of trees, and this is also the focus of current research, but without conclusive findings so far.

Forest Soils and Site Identification   Intensive forest management depends on a good understanding of soils and their productive capacity. Many states have federal–state cooperative soil survey organizations that provide information to landowners about their soils. Soil classification relates soil features to productivity, giving many clues for improved management on different soil types. Usually, the better the site, the more intensive the forest management that can be justified financially. The best-suited species to plant may be indicated by the soil type, thus reducing planting failures. The chemical properties of the soil, and nutrients present or absent, may indicate the kinds of fertilizer that should be used.

Other soil information may identify soil-related hazards, such as the potential for erosion or compaction, so steps can be taken to compensate by using specialized equipment or timing operations to avoid excess moisture and reduce compaction. A soil pit of the type shown in Fig. 19-5 reveals the soil's physical properties. Chemical analysis of soil samples is provided free or at nominal cost by many states in order to encourage good soil conservation and management.

Forest Tree Improvement   Some of the most important developments for intensive forestry have been in tree selection and breeding techniques. Seed from wild trees that have one or more highly desirable qualities, such as good form, rapid growth, straight limbs, favorable crown-to-trunk ratio, windfirmness,

FIGURE 19-5
The physical properties of soil revealed in a soil pit in North Carolina. The profile shows roots and root holes, as well as the difference in texture of the loose topsoil and the compact subsoil. After chemical analysis in the laboratory, a decision can be made about the appropriate species for planting in this soil. *(Courtesy of USDA Forest Service)*

good seed-germinating ability, desirable wood density, and perhaps even a resistance to certain insects and diseases, is collected and given special care. Seed orchards, which produce seed with such highly desirable genetic characteristics in a relatively short number of years, are then developed. Another technique involves grafting branch tips from selected trees onto young orchard stock of the same species, thus producing a crop of superior seeds several years before the normal cone-producing age is reached. Several generations of seed orchards have been established, and research has developed new strategies for the mass production of seeds.

Forest geneticists need to control the haphazard process of tree pollination. Lightweight metal ladders may be strapped to a tree to allow access to the tree crowns in a seed orchard, or, where possible, a ladder truck may be used. Once in the tree, pollen from selected male cones, collected in sacks, can be injected into sacks that isolate the growing female cones. Controlled pollination is the result, and, with success, the geneticist can improve on nature by de-

veloping strains and hybrids with superior characteristics (Fig. 19-6). Once highly desirable trees have been developed, in some species tissue culture techniques allow genetically identical seedlings to be created in the laboratory, bypassing the need for seeds.

Tree breeding has been a widespread practice since the 1970s. Research in tree improvement and testing of improved progeny have been financed by forest industry and government, with cooperation from universities. Regional "tree improvement cooperatives" have been established whereby several companies and agencies contribute funds (often to a university) to hire genetic scientists and technicians to carry out studies and field tests of improved seedlings. The huge success of tree improvement may be partly due to this cooperative approach, which facilitates sharing of research knowledge and widespread professional involvement. Genetic improvement activities, when integrated with other management practices, have resulted in productivity gains of up to 40 percent.

FIGURE 19-6
A forest geneticist using a hypodermic needle to inject selected pollen into a plastic bag. The female flowers of this loblolly pine are bagged to protect them from unwanted pollen. Seed orchard near Plymouth, N.C. *(Courtesy of Weyerhaeuser Company)*

Hybrid Trees Artificially induced crosses between species of certain genera have produced fast-growing hybrid trees with useful characteristics, particularly in certain pine species and members of the poplar family. The formation of hybrids capable of rapid establishment and growth on hostile sites makes such hybrids ideal for screening, for covering coal spoil banks and strip mines, or for planting of sanitary land fills. Not only does the rapid tree growth improve site aesthetics, but wildlife cover is improved and erosion reduced. Many forest industry companies have established intensively managed forests of hybrids for pulp or fuel in short rotations (Fig. 19-7). In addition to rapid growth, other advantages sought in hybridization are disease resistance, good form, and frost or drought hardiness.

Of course, all this progress in tree breeding, tissue culture, and hybridization is not without criticism. Some people object to tinkering with natural selection. Others fear the risks of reduced genetic variability that might make the trees more vulnerable to insects and disease.

Wood Utilization Important research conducted by forest industries, and related industries that produce equipment and supplies for the forest industry, has been directed toward improved wood utilization. This research has helped the industry adapt to the reduced quality of wood supplies, such as smaller-sized trees and species not before utilized, to expand markets for pulp and reconstituted wood products, and to increase efficiency in converting wood fiber into wood products. The huge Forest Products Laboratory of the USDA Forest Service in Madison, Wisconsin, and company and university wood products research programs, have also contributed to wood utilization advances. By increasing the value and utility of trees, this knowledge increases the value of forestland, thereby making intensive management more feasible and increasing U.S. competitiveness in international trade.

Important recent advances in wood utilization research led to development of a curve-sawing headrig that actually follows the natural curve of logs being sown into lumber. This improves utilization of logs with sweep, and small-diameter logs that might otherwise be unmerchantable for lumber (see Chapter 14, "Forest Products"; see also Fig. 14-11).

Molecular Biology Advances in technology have created the ability to look at living organisms at the molecular level, opening up new opportunities for understanding life processes and genetically manipulating these processes through biotechnology to produce desired results. Admittedly, trees are much tougher to work with than agricultural plants such as corn, because trees take so much longer to grow to maturity. However, breakthroughs gained in the engineering of agricultural plants will eventually lead to major genetic improvements in forest species.

Humans have produced new genetic combinations in plant and animal breeding for thousands of years. We select and favor desirable traits already existing in species, or bring them in from other species by interbreeding. In the past, if two species couldn't be

FIGURE 19-7
A 4-year-old hybrid poplar energy plantation in Pennsylvania. *(Courtesy of Miles W. Fry)*

hybridized, traits couldn't be transferred—pine and fir can't interbreed, so disease-resistance genes from one couldn't be moved to the other. Working at the molecular level, scientists can engineer genetic combinations that would have been impossible to produce in the past. Like changes in computer technology, new molecular techniques are rapidly being developed, constantly changing what is possible in biotechnology. For example, the *polymerase chain reaction* (PCR) allows us to make numerous copies of any specific region of the DNA blueprint, theoretically allowing any gene to be studied, changed, or transferred to another species—for example, a rapid growth or disease-resistant gene.

Some potential applications include stopping trees from flowering so that reproduction energy will be available for improved growth; the transfer of insect, disease, herbicide, or fire resistance traits to another species where they are needed; or perhaps even moving the ability to fix nitrogen in root nodules from legumes or alder, which also has that ability, to other species such as pine.

Obviously, the forest industry is keenly interested in the possibilities for enhancing productivity and forest health through molecular biology, and some applications may also enhance environmental protection. Molecular methods such as PCR are also being used to learn more about the natural genetic variation in forest trees, shrubs, and grasses. This knowledge will permit more efficient management of wild species, and is essential to the conservation and restoration of natural systems.

Advances in molecular biology have made possible a shift from forest biology research using experimental methods (such as studies comparing fertilized and unfertilized trees) to a focus on life processes to determine why a certain response occurs. It is no longer enough to just determine the results of different treatments. It is necessary to understand how changes occur so the results can be more widely applied and expanded through an understanding of the basic principles and processes involved. For example, how do different growth rates relate to photosynthesis under different environmental conditions, and why do different results come about?

The shift from mechanistic, experiment-based research to determine what happens under various treatments, to molecular biology studies to determine why responses occur, is especially significant now. For example, we know that the conditions under which growth takes place are changing. Scientists agree that CO_2 concentrations have increased in the atmosphere, and that ultraviolet radiation is increasing from dissipation of the ozone layer. If conditions for growth are changing, then it is more important than ever to understand underlying life processes in order to predict how they may be affected.

Other Research and Development   A variety of other research goes forward, including work on topics such as fertilization and site preparation techniques and their effect on seedling growth; seed storage, and seed sowing from aircraft; animal damage and control (Fig. 19-8); and water quality relationships to harvest practices.

Research and associated development, often called *R&D,* is the combination of research to create knowledge and the application of research findings to create new products and practices through engineering or other applications. R&D is vital to large wood products companies, which must stay competitive for long-term survival. Thus, research and development are frequently linked. Development seeks to utilize new research findings. Some examples might include increasing integration of operations, development of new and improved equipment to implement new harvesting methods, or the use of improved regeneration

FIGURE 19-8
An industrial forester inspects a young Douglas-fir that has escaped damage from browsing deer. A paper cup protected the leader of the seedling in its more vulnerable period of growth. *(Photo by Grant W. Sharpe)*

systems through seed orchards and nurseries; site preparation; and introducing the use of improved genetic stock or planting methods (Fig. 19-9). Computerization has allowed the simulation of forest management practices in various combinations with one another, and this capability can suggest priorities for additional research and development, management strategies, and other long-range planning directions.

Environmental research is also an emerging focus of the forest industry because individual companies and the industry at large must meet increasingly strict environmental standards. In addition, the consumer-driven demand for "green products" generated by sustainable forest management has prompted virtually all companies to develop data about their operations that are linked to sustainability as well as environmental implications of their practices. Many forest companies aspire to be industry leaders in

FIGURE 19-9
A nurseryman tending redwood seedlings in an industrial forest tree nursery in Fort Bragg, Calif. The moisture level, nutrients in the soil, and general health of the seedlings receive constant checking. *(Courtesy of Georgia-Pacific Corporation)*

FIGURE 19-10
Some of the larger forest industry companies have research programs focused on improving the productivity of their intensive management programs. Here we see the Forest Sciences Laboratory of Mead/Westvaco Corporation at Summerville, N.C., one of the company's several research facilities. *(Courtesy of Mead/Westvaco Corporation)*

environmental and sustainability issues and go beyond what is required to meet prevailing requirements or standards.

Applied Research

Many of the larger forest industry companies have formal research programs with laboratories (see Fig. 19-10), experimental sites and employ their own research scientists and technicians who are dedicated to particular projects and topics. Beyond this, the large, and many smaller, companies may conduct more applied studies, gathering field data and experience as to best practices to increase seedling survival, improve utilization, reduce erosion, meet environmental standards, and operate efficiently and safely. The practices of yesterday are inadequate today, and today's standards will be unacceptable tomorrow. Forestry is evolving, improving, and responding to needs and demands by searching for better ways to operate both in the laboratory and in the field.

INDUSTRIAL AND OTHER PRIVATELY SPONSORED FORESTRY PROGRAMS

The American Forest and Paper Association (AFPA)

The company forest is an institution that has its origins in the period before the Civil War, and the stories of the lumberjacks and river drives, and the Paul Bunyan legends that came out of these days are familiar to most of us. Colorful though they might be, they are stories of forest exploitation. Yet, although these owners treated the land harshly, they should not be judged too strictly. For the most part, they were ignorant of even the basic ideas of good forest practice and unaware of its eventual rewards, and, the forests were considered to be unending. The early land laws encouraged forest exploitation. Then, too, working as they were in seemingly limitless old-growth forests, there was little incentive for forest conservation. As explained in Chapters 1 and 2, the early lumber and logging companies did what was dictated by circumstances. Wood was needed to build the country and settle the advancing western frontier, and they supplied it, albeit often in a wasteful and destructive manner.

After the turn of the century, the threat of government or state regulation was always on the horizon (see Chapter 2). Forest industries had to establish and maintain a good reputation. They needed to publicize their intentions, their growing sense of permanence, and their increasing commitment to good forest practices. They also needed to forge relationships with nonindustrial, private forest owners who increasingly controlled wood supply and political action regarding forestry. In 1941, the American Forest Products Industries, Inc., was established to help with this task. It was renamed the American Forest Institute (AFI) in 1968, and then the American Forest Council (AFC) in 1986, when it became jointly sponsored by the National Forest Products Association (NFPA) and the American Paper Institute (API), thereby connecting the solid wood and paper sides of the industry. In 1993, the name changed again to the American Forest and Paper Association (AFPA), as a result of additional mergers, consolidation, and reorganization.

Today, AFPA (www.afandpa.org) is the national trade association organization for the forest, paper, and wood products industry, leaning heavily toward educational programs on the merits of growing trees and the importance of wood in the U.S. economy and lifestyles (AFPA members include manufacturers of over 80 percent of the paper, wood, and forest products produced in the United States). Through its national organization, under all of these names, the forest industry has supported sound forestry by means of programs such as fire prevention and control, public education, and working for policies that are good for forestry and for the industry. AFPA, through its nonprofit arm, the American Forest Foundation (www.affoundation.org) sponsors the American Tree Farm System (www.treefarmsy tem.org), Project Learning Tree, and is also a direct sponsor of a variety of educational programs for school children. It provides teacher tools for programs on trees, forestry, and recycling (www.afandpa. org/kids_educators).

The American Forest Foundation

Project Learning Tree (PLT)   PLT was launched in 1973 to help young people better understand forests and forestry (www.plt.org). Using materials developed by educators, forestry experts, and extension agents, and with 25,000-plus teachers attending workshops each year, PLT has become one of the nation's most outstanding environmental education programs, and is the most widely used public forestry education program in the nation. Now approaching its 30th year, PLT has a national network of 3,000 active volunteers and state coordinators who have trained more than 300,000 educators in an expanding menu of PLT modules. It is now available in 49 states as well as in Canada, Sweden, Finland, and Mexico. PLT materials provide teaching techniques for several curriculums emphasizing understanding of forests and forestry. Several major textbook publishers have correlated their textbooks to the PLT curriculum, making it easier for teachers to use it in the classroom. PLT has been so successful that similar programs have been developed for wildlife and water.

The American Tree Farm System   A *tree farm* is defined as an area of privately owned forestland managed to produce continuous crops of trees while protecting or enhancing the soil, water, range, aesthetics, recreation, wood, fish, and wildlife resources. Generally, 10 acres or more are required. Today the term *tree farm* is a registered trademark owned by the AFPA. An area designated as an official *Tree Farm* has been certified by a professional forester as meeting standards of good forestry, with the owner agreeing to adhere to such standards and be recertified every five years. Today, such certification has additional meaning in the context of green products coming from sustainably managed forests (Simpson, 2001).

About 7,000 professional foresters—from industry, government, and consulting firms—volunteer their time in the Tree Farm program, inspecting and reviewing tree farms and their management plans. State committees and state co-sponsors, generally forestry or landowner associations, operate the Tree Farm System locally, using materials and systems developed nationally by the American Forest Foundation. A bi-monthly publication, *Tree Farmer: The Practical Guide to Sustainable Forestry,* is the official magazine of the 65,000-member American Tree Farm System (www.treefarmsystem.org).

The tree farming idea is a major departure from previous ways that people dealt with forests. A turning point came in 1940–41. Americans had been cutting forests and not replanting them since the early 1600s. In addition, wildfires burned 30–50 million acres (12–20 million ha) of U.S. forests most years. In 1940, commercial forestland in America was at a record low of about 460 million acres (186 million ha). Net annual growth of trees was about 11 billion

cubic feet (.33 billion m^3), and 168 million acres (68 million ha) of forests were unavailable for forestry, either because they were set aside in parks or other reserves, or they were unsuitable for renewable forest management because of other conditions.

Foresters had been preaching fire control and forest management since 1900, but forestry remained something considered only by big industry or big government. Most forest-owning individuals in 1940 allowed trees to grow on their land only when they couldn't find another use for the land. The woods they owned provided shade for cows on land too rough to farm. The trees were cut when they needed money or wood, and it was up to nature to regenerate the forest and take care of what grew back.

Meanwhile, large forest products companies were concerned and frustrated; so was the federal government, because forestland was not a safe investment, since it was likely to be lost to fire before a crop matured. Forests were being burned, cleared, degraded, wasted, and not renewed. Future wood supply seemed to be at risk. Even where the large owners tried to manage large acreages, fires were a constant risk and fire control systems were nonexistent or ineffective. Fuel management was not yet even considered.

The Weyerhaeuser Company, tired of having its woods burned, became concerned about the future of its forests and the nation's forests. On June 12, 1941, the company invited people to come to see forestry in action on one of its properties, a 120,000-acre (48,600-ha) forest near Montesano, Washington. People came, looked around, and talked; some trees were planted, speeches were made, lunches were eaten, and people went home. But the new idea, *Tree Farm,* stuck in their minds. Weyerhaeuser proposed to keep land for growing trees rather than selling it off once it was logged, and use Tree Farm management to maintain and renew those forests. Furthermore, the company urged others to become tree farmers too.

The tree farm idea took root. Today we have 483 million acres (195,000 ha) of commercial timberland (23 million acres, 9.3 million ha, more than in 1940) capable of growing trees renewably. Net annual growth of trees is now more than 22 billion cubic feet (.66 million m^3), *about twice the rate of 1940.* We hold fire losses to about 3 million acres (1.2 million ha) most years, a tenfold reduction. We also have 247 million acres (100,000 million ha) of

FIGURE 19-11
Wild turkey, one of the numerous wildlife species finding food and habitat on tree farms. *(Courtesy of National Wildlife Federation)*

noncommercial forest, much of it set aside from harvest by law in parks, wilderness, and other special designations.

And from that first tree farm, dedicated in 1941, has grown a tree farm system stretching from Maine to Alaska, with 65,000 owners and totaling over 26 million acres (10.5 million ha) of certified tree farms. To our wood supply concerns we have added wildlife, soil and water protection, wetlands, biodiversity, global warming, endangered species, aesthetics, recreation, and other forest values. Tree farmers provide food and habitat for many species of forest-dwelling wildlife. Deer, turkey, grouse, woodcock, black bear, elk, and wolves find shelter on these tree farms (Fig. 19-11). Soil has been stabilized in the windy plains and on previously eroding hills.

American forests and American lands are in better shape today than they were in 1941, in part because of the growth of the Tree Farm idea—an idea that has changed the conventional way forestland owners view their relationship with forests.

Tree Farms in the Future   Because nonindustrial forestlands contain 58 percent of the commercial forestland in the nation, they have a significant impact on the country's timber supply and other forest amenities. Thus, spreading the tree farming idea is important to the forest industry and to everyone else. The American Forest Foundation, through the American Tree Farm System, sponsors a Tree Farmer of the Year award to recognize the most outstanding tree farmer at state, regional, and national levels. These tree farm winners, and the stories of their successes, are good examples of forestry and conservation in action. Their stories, and other tree farmer success stories, are well covered in *Tree Farmer* magazine and other publications such as *American Forests* (www.americanforests.org) and *National Woodlands* (www.woodlandowners.org).

Public Relations

Even the best forestry efforts will go unrecognized if there is no effort to make them visible to the public (Fig. 19-12).

Through the public relations work of the AFPA, positive stories about tree farming and forestry appear in news media, reaching tens of millions of people each year. The AFPA and its predecessors have also run a variety of national advertising programs over the years, backed up with publicity and press briefing packets of information. However, communications problem that AFPA, the forest industry, and forestry in general face is that Americans have moved from an agrarian, rural society to an urbanized, service-industry society. Most Americans now live in cities or urban areas (80 percent). They work at jobs unrelated to growing crops and making products, and thus, as urban-service people, they lack understanding of the opportunities nature presents for renewable, sustainable production of the forests' bounty—wood, water, recreation, and wildlife. Yet, even today's urban dwellers are drawn to nature—it's in the genes of all of us. Thus, the tree farming story for many of these people is best presented as an environmental story, with timber harvests explained as a means to improve bird and wildlife habitat, water yields, and recreation opportunities—as well as generating some income to support such improvements. An emerging problem for the tree farm movement is that in the move of urban dwellers "back to nature," they are taking up permanent or seasonal residence on increasingly small tracts of forestland as larger ownerships are subdivided into smaller lots. The resulting forest fragmentation and larger number of NIPF landowners is a major issue for everyone

FIGURE 19-12
A forester explains a wood processing operation to members of the public and press during a show-me trip to a pole-sorting yard. Communications skills are valuable in the practice of forestry. *(Photo by Grant W. Sharpe)*

seeking to bring these forests under sustainable management (De Coster & Sampson, 2000).

Other Organizations

States with extensive forests may have state-level forestry associations also, whose membership includes persons with forestry interests. There are three basic types of associations that support forestry: (1) state forestry associations that are really forest industry-supported trade associations—such as the Alabama Forestry Association; (2) state woodland owner associations of NIPF landowners; and (3) composites of (1) and (2), usually an industry trade association with an active NIPF landowner division. The Small Woodland Owners Association of Maine is an example of a state-level association (type 2), formed to help promote citizen ownership of woodlands and to promote management in accord with sound silvicultural principles.

A National Woodland Owners Association (NWOA) was founded in 1983 and links 32 statewide organizations in a national alliance of forest and woodland owners (www.woodlandowners.org). The association works cooperatively with state woodland owner associations and agencies, and through its *National Woodlands* magazine, to exchange information and promote the best interests of nonindustrial private forest owners.

Other organizations giving impetus to the promotion of forestry include the Society of American Foresters, the professional organization to which most professional foresters belong; American Forest and Paper Association (already mentioned); American Forests, the nation's oldest conservation organization promoting forestry; National Association of State Foresters, Future Farmers of America, and the Association of Consulting Foresters, as well as many other organizations interested in educating their own members and the American public about forestry, and in recognizing and encouraging sound forestry practices.

Looking at renewable resources on a broader scale, the Renewable Natural Resources Foundation (www.RNRF.org), a consortium of 15 professional, scientific, and educational organizations, is located in a 35-acre office-park complex called Wild Acres in Bethesda, Maryland (Fig. 19-13). Three RNRF buildings house the national offices of organizations including the Wildlife Society, American Fisheries So-

FIGURE 19-13
The site plan model for the Renewable Natural Resources Center, housing
17 member organizations in Bethesda, Md. *(Courtesy of Renewable Natural
Resources Foundation)*

ciety, American Society for Photogrammetry and Remote Sensing, the Nature Conservancy, and the Fish and Wildlife Reference Service. The national offices of the Society of American Foresters is also located on Wild Acres, and has its own two buildings.

CERTIFICATION OF SUSTAINABLE MANAGEMENT

One of the most important developments of the past decade in forestry, and especially on private lands, has been the movement to certify forests as being managed for their sustained productivity. The current movement began in Europe when environmental groups came home from the 1992 World Summit on Sustainability (known as the Earth Summit) in Rio, concerned about the loss of tropical forests (Simpson, 2001). They wanted to create a movement that would verify to consumers that wood harvested from tropical sources was being harvested and managed in a sustainable way, so as to stop the loss of tropical forests. So, they promoted the idea of setting standards of sustainability in order to "certify" that wood harvested from forests meeting the standard was from "sustainable sources." This "green certification" movement quickly caught on with environmentally concious consumers—and the

retail outlets they patronized—and rapidly spread to North America.

But it was not so easy to define sustainable forest management, and the debate continues in forestry circles as to what factors should be considered (Journal of Forestry, 1998; Lober & Eisen, 1995; Sample, 2000; Vogt et al., 1999). By 2001, there were already 60 forest certification systems around the world, presented by various groups with different ideas about what should be included in the standards to be met to earn certification. For example, in Europe the forest owner associations of the European Union came together to create their own certification system—Pan European Forest Certification (PEFC). In the United States, there was already the American Tree Farm System, which for decades had been inspecting and then "certifying" that its member tree farms were properly planned and managed (Simpson, 2001). In 1993, to help resolve confusion and differences among certification systems, the World Wide Fund for Nature, based in Switzerland, brought together representatives of forest industry and environmental organizations to form the Forest Stewardship Council (FSC). The FSC then adopted a set of principles and criteria to apply to the management of tropical, temperate, and boreal forests worldwide, and established a process for third-party verification of a forestry operations adherence to these principles and criteria (Sample, 2000). Thus, following a successful third-party assessment and verification, a producer would be entitled to affix the FSC label directly to a product to inform consumers that it was produced from a forest managed in accordance with the FSC principles and criteria. Also in 1993, the International Standards Association (ISO) developed standards and guidelines for sustainable forest management that specified what "processes and procedures" a company would need to have in place to produce a quality product (systems certification) but not to certify actual performance or the quality of product produced. These "ISO 14001" standards thus document that a company has systems and processes in place that will ensure continuous improvement in management toward environmental improvement and sustainability over time.

But the FSC and ISO approaches did not fully meet the needs and concerns of the U.S. wood products industry. So, in 1995, the AFPA established the Sustainable Forestry Initiative (SFI[SM]) to develop a rigorous standard for sustainable forest management. The program requires member companies to follow a comprehensive system of sustainable environmental and forestry principles, practices, and performance measures as a condition of membership in AFPA, and a licensing program for nonmembers (www.afandpa.org.forestry/sfi; Berg & Cantrell, 1999). An independent, sustainable forestry board oversees the development and continuous improvement of SFI requirements (www.aboutsfb.org). Further, an external review panel of experts representing conservation, environmental, professional, academic, and public organizations provides a framework for independent review of the SFI program and its annual report, maintaining a Web site at www.sampsongroup.com.sfi/afandpa's.htm. So far, 16 companies have been expelled from AFPA membership for failure to meet the SFI standards. By 1999, more than 130 companies and 10 licensees representing 85 percent of paper production and 90 percent of industrial timberland in the United States participated in AFPA and the SFI programs (Berg & Cantrell, 1999), which today covers 105 million acres (42.5 million ha).

While most companies have embraced the SFI program, many have gone beyond applying principles and the standards internally and are seeking third-party verification that they are meeting the SFI standards in the field and management systems and have processes in place that meet the ISO 14001 requirements. Some companies pride themselves in going beyond these standards and requirements and are making an extra effort to provide visible verification. For example, Boise (formerly Boise Cascade) manages its 2.3 million acres (9,315 ha) in compliance with SFI program standards, and has gone beyond them to also include a comprehensive set of "Boise Forest Stewardship Values and Measures." Boise retains Pricewaterhouse Coopers LLP to provide a series of third-party forest management audits that assess compliance with both the SFI and its additional standards. Customers are also invited to accompany audit teams, and a special forest stewardship advisory council of nationally known forestry and conservation experts participates in reviewing audit results and recommending changes.

With the advent of green certification, forest management is responding to society and consumer demands for greater accountability, and the sustainability of forests into the future. Just as accountants certify that financial statements are accurate to certain standards, green certification can offer objective evaluations of whether sustainable, environmental standards are being met. This is a positive wave of the future for forestry.

SUMMARY

This chapter outlines forest management and research activities on private lands, which, in the United States, contain most of the nation's commercial timberland. These timberlands are owned by individual citizens, or by private companies, or corporations, as contrasted with *public* land owned in common by the people, and managed by a state, federal, county, or city agency.

Private forestlands owned by forest industry (13 percent of the nation's timberland) include land held by both small companies and large corporations that grow and process wood into products. *Nonindustrial, private forestland* (NIPFs) includes small ownerships held by persons from all walks of life, for whom forestry is at best an adjunct to their major vocation, but this is the largest ownership category of forest ownership and includes 58 percent of the nation's commercial timberland. Because NIPFs are so extensive, these lands are a focus of federal, state, and cooperative programs providing education, advice, technical assistance, and financial incentives to improve their management.

The forest industry generally contains the most productive forestland in the nation, and it is managed intensively utilizing a variety of practices over the life of forest stands. Some companies carry out their own research programs—testing and refining practices and developing basic information to enhance the productivity of their forestlands and improve efficiency of all their operations. An emerging focus of such research is on molecular biology aimed at enhanced understanding of life processes.

The forest industry is concerned about improving forestry's image and forest management on private lands. The American Forest and Paper Association (AFPA) is the public information organization of the industry. Through its American Forest Foundation, it sponsors the American Tree Farm System, Project Learning Tree, and other programs. During the past decade, the green certification movement has made a substantial impact on forestry on private lands. Many of the large companies are gaining certification for the sustainable management of most of their lands.

LITERATURE CITED

Alexander, Susan, and Brian Greber. 1991. Environmental ramifications of various materials used in construction and manufacture in the United States USDA Forest Service. Gen. Tech. Report PNW-277.

Argow, Keith A. 1993. "It's OK to Make Money on Our Woodlands." Forestry Commentary, *National Woodlands.*

Berg, Scott, and Rick Cantrell. 1999. Sustainable Forestry Initiative: Toward a Higher Standard *J. Forestry* 97(11):33–36.

Best, Constance. 2002. American Private Forests: Challenges for Conservation. *J. Forestry* 100(3):14–17.

De Coster, Lester, and Neil Sampson. 2000. Forest Fragmentation: Implications for Sustainable Private Forests. *J. Forestry* 98(3):4–8.

Journal of Forestry. 1998, March. Sustainable Forests: Sustainable Communities. Theme issue of eight articles. 96(3).

Lober, D., and M. Eisen. 1995. Retailing: Certification and the Home Improvement Industry. *J. Forestry* 93(4):39–41.

Sample, Alaric. 2000, Spring. Forest Management Certification: Factors Affecting Its Future Development in the United States. *The Pinchot Letter,* 8–13.

Simpson, Bob. 2001, November/December. Your Forests Are Certified. *Tree Farmer,* 5.

Smith, Brad, John Vissage, David Darr, and Raymond Sheffield. 2001. Forest Resources of the United States, 1997. Gen Tech Rpt. NC-219 USDA Forest Service, NC Research Station, St. Paul, Minn.

Society of American Foresters. 2002. Conservation Easements: Permanent Easements Against Sprawl. *J. Forestry* 100(3):8–12.

USDA Natural Resources Conservation Service (USDA-NRCS). 1999. Summary Report: 1997 Natural Resources Inventory (revised Dec. 2000). www.nhq. NRCS.USDA.gov/NRI/1997.

Vogt, Kristina, Bruce C. Larson, John C. Gordon, Daniel J. Vogt, and Anna Frangeres. 1999. Forest Certification: Roots, Issues, Challenges, and Benefits. CRC Press, Boca Raton, Fla.

Weyerhaeuser Company. 2002. Annual Report, 2000. Federal Way, Wash.

ADDITIONAL READINGS

American Forest Council. 1991. The American Forest: Facts and Figures 1991. Available from AFC, 1250 Connecticut Ave. N.W., Washington, D.C.

American Forest Foundation. 1990. Wise Use Brochure. 1250 Connecticut Ave. N.W., Washington, D.C.

Aplet, Gregory H., Nels Johnson, Jeffry T. Olson, and V. Aleric Sample, eds. 1993. Defining Sustainable Forestry. Island Press, Washington, D.C.

Ellefson, Paul V. 1992. *Forest Resources Policy: Process, Participants, and Programs.* McGraw-Hill, Inc., New York.

Hansen, Eric, Rick Fletcher, and James McAlexander. 1998. Sustainable Forestry, Swedish Style for Europes Greening Market. *J. Forestry* 96(3):38–43.

Jones, Stephen B., Glenn R. Glover, James C. Finley, Michael G. Jacobson, and A. Scott Reed. 2001. "Empowering Private Forest Landowners: Lessons from Pennsylvania, Alabama, and Oregon." *J. Forestry* 99(3):4–7.

National Research Council. 2002. Forested Landscapes in Perspective: Prospects & Opportunities for Sustainable Management of America's Nonfederal Forests. National Academy Press, Washington, D.C.

Schmidt, Ralph, Joyce K. Berry, and John C. Gordon. 1998. Forests to Fight Poverty: Creating National Strategies. Yale University Press, New Haven, Conn.

Sedjo, Roger. 1983. *Government Interventions. Social Needs and the Management of the U.S. Forests,* Resources for the Future, Washington, D.C.

Stanturf, John A., Stephen B. Jones, and William D. Ticknor. 1993. "Managing Industrial Forestland in a Changing Society." *J. Forestry* 91(11):6–11.

Zobel, Bruce, and John Talbert. 1991. *Applied Forest Tree Improvement.* Waveland Press, Prospect Heights, Ill.

STUDY QUESTIONS

1. Construct a pie chart showing the major divisions of forestland ownerships in the United States.
2. What do the acronyms *NIPF* and *AFPA* stand for?
3. Give three reasons why little or no forest management takes place on many small, private forestlands and woodlots.
4. Can you identify a small, private forest landowner within your family or among your acquaintances?
5. What factors led industrial forest owners to retain possession of their cutover lands?
6. How do certain companies insure their raw material supply without owning enough land to supply all of the timber themselves?
7. Discuss possible intensive management under the regeneration phase of a rotation.
8. How does the focus of molecular biology differ from the traditional research of forest biology?
9. What was the genesis of the American Tree Farm system? What does it claim to have done for American forests and lands?
10. Describe what green certification means for both producers and consumers.

CHAPTER 20

Urban Forestry

This chapter was prepared by R. Michael Bowman, city forester, Lewiston, Idaho, with revisions by the authors.

INTRODUCTION

Cities, towns, and urban and metropolitan areas (as opposed to rural and wild land areas) contain important natural resources, including trees and associated habitats for birds and other wildlife. Trees in various combinations with shrubs, ornamental plants, and lawns are the principal focal points of parks, greenbelts, cemeteries, arboretums, zoos, boulevards, landscaped buildings, residential yards, and planting strips along streets. In some places, ponds, lakes, streams, canals, and wetlands may augment urban environments. As economic and residential development spreads from the more highly developed urban centers to the suburbs, and into the urban-wild land interface beyond, the need for urban forestry grows (Ball, 1997).

Today, 80 percent of all Americans live in metropolitan areas, which average 33 percent tree cover (Dwyer et al., 2000; USDA Forest Service, 2001). Urban trees and forests play an important role in the livability of these urban environments and have important meanings for urban residents.

The many benefits of urban forests include moderation of climatic extremes, resulting in savings in heating and cooling costs; reduction of air pollution; storm water management; production of oxygen and some absorption of CO_2; habitats for birds, mammals, and fish, thereby providing a diversity of life; recreational places; beauty—a softening of the harsh lines of the built environment; and the considerable emotional benefits that arise from the more pleasing environment and the facilitation of deep psychological ties between people and urban trees (Dwyer et al., 1992). The value of urban trees and their associated habitats will grow as increasing numbers of urban residents become attached to these amenities and benefits.

The benefits of trees and associated natural attractions within cities have been recognized since ancient times. The historical evolution of urban beautification, and the deleterious effects of vegetative destruction, is well documented for the major cities of the world (Profaus et al., 1988). In the United States, the importance of urban trees is recognized by the designation *Tree City USA* for a growing number of communities proclaiming commitment to their urban forest resources.

Historical Use of Trees in Urban Design

Trees must have been an important part of earliest urban life, although their contribution, as far as we know, was primarily as a food source. One of the earliest accounts of trees prized for their beauty comes from the reign of Nebuchadnezzar in Babylon, ca. 550 B.C. According to legend, borne out by archaeological excavation, Nebuchadnezzar's wife yearned for the mountains of her homeland. Sparing no expense to simulate the mountain landscape, the king built roof-garden terraces that rose skyward, lavishly planted with trees of all kinds (Schneider, 1963). Planting of shade-giving trees in gardens to alleviate extreme temperatures was also stressed in Baghdad during the reign of Caliph Harun-al-Rashid, 763–809 B.C.

England contributed much to our knowledge concerning trees, including the first use of the term *arborist* in the late 1500s. In 1776, John Evelyn wrote a monumental work on trees, called *Sylva*, in which he discussed all known aspects of forestry and fruit trees, including arbor walks and the ornamental use of trees, and recommended species to plant (Chadwick, 1970). By the 1700s, the use of trees in urban settings, a respect for their importance in natural areas, and the knowledge of their growth and maintenance were well developed, at least among the educated classes. Interest in landscape design was growing. Visitors to botanical gardens and arboretums came away with new ideas for their own homes as diverse plant materials became available. Under Lancelot (Capability) Brown (1715–1783), the gardens of the great estates became less formal as the emphasis shifted from imposing highly stylized decoration to considering the capabilities of the landscape and creating lake-mirrored, tree-shaded places for relaxation and enjoyment.

In the early 1800s, open spaces in various London squares were laid out with planted trees and lawns. In fact, the image of London at that time was one of trees and grass. Many of these squares were residential green spaces surrounded by housing, resulting in quiet, comfortable residential neighborhoods (Zube, 1973). Tree-lined boulevards were introduced to Paris in the mid-1800s, mainly for defense in controlling the movement of troops but also providing beauty and shade (Grey & Deneke, 1986).

The art and science of forestry, including urban forestry, originated in Europe. Bernhard Fernow, a German-born and -trained forester and head of the U.S. Department of Interior's Division of Forestry, was an important force in the development of American forestry. Gifford Pinchot, who shaped U.S. forest policy and forwarded the establishment of the forest reserves, went to Europe near the turn of the century, after graduating from Yale University, to study for a year at the French forestry school. He also spent a month doing fieldwork for the Forstmeister in the city forest of Zurich, Switzerland (Watkins, 1992).

In the United States, the importance of urban green space and trees was recognized as early as 1682 by William Penn when he set aside public squares at regular intervals in the design of Philadelphia. In Savannah, Georgia, a visionary city father, General Oglethorpe, decreed in 1733 that old trees be left standing. Street and yard plantings have also been part of our heritage from the earliest years, sometimes started from nuts and seeds stuffed into the pockets of emigrants leaving Europe.

On a more formal basis, the first planned landscaping of a public area in the United States may have been Mount Auburn Cemetery, in what was then the outskirts of Boston. In 1831, this site attracted the attention of Dr. Jacob Bigelow and the Massachusetts Horticultural Society. By converting the site into the country's first "scenic cemetery," they not only beautified a burial ground, but unwittingly provided one of the most popular passive recreation areas in the Boston area. Today, with some 450 tree species, Mount Auburn is one of America's finest arboretums (Fig. 20-1).

Similarly, in the mid-1800s, Frederick Law Olmsted's Central Park plan pioneered the use of "green design" with trees and landscaping to bring "tranquility and rest to the mind" in New York City (Spirn, 1984). And on the frontier, more than a million trees were planted in Nebraska alone when J. Sterling Morton conceived the idea of Arbor Day in 1872. In the modern age, Ian McHarg's (1969) *Design With Nature* promoted the idea that cities and urban areas should be constructed in harmony with nature instead of with the idea that nature was to be conquered to meet human needs.

Today in North America, trees are an established and beloved part of many cities. These trees are there

FIGURE 20–1

"The View From Mount Auburn." An engraving by James Smillie, circa 1847. Founded in 1831 by the Massachusetts Horticultural Society, Mount Auburn Cemetery, in Cambridge, supports 450 native and exotic tree species on its 174 acres. *(Engraving from Mount Auburn Illustrated, 1847. Published by R. Martin, New York; courtesy Dave Barnett)*

as a result of citizen tree committees, community foresters, arborists, and dedicated citizens who have planted trees along streets, and of the enterprise of private owners who landscape their property. The reward has been enhancement of the city milieu, and the addition of the many other benefits trees bring to their surroundings.

Urban Forestry—A Definition

Although city dwellers and homeowners have had an interest in trees throughout history, that interest has been heightened by a growing environmental consciousness and the realization of the impacts of environmental degradation on health and welfare. *Urban forestry* is a specialized branch of forestry that has as its objective the cultivation and management of trees for their present and potential contribution to the physiological, sociological and economic well-being of urban society. Urban forestry in some areas goes beyond trees as sources of shade and beauty and emphasizes programs to revitalize cities. It seeks to operate in harmony with nature and citizen groups so people can benefit from trees and the habitats they engender. Achieving such goals requires thorough knowledge of ecological processes, which consider the interaction of flora, fauna, and human elements.

The term *urban forestry* was at first seen to be a contradiction in terms (Miller, 1996). Webster defines *urban* as "of, relating to, characteristic of, or constituting a city"; and forestry as "the science of developing, caring for, or cultivating forests." Webster defines an *urban forester* as "the manager of the trees and related ecosystems within a densely populated area" as contrasted with the term *arborist,* defined as "a person who deals with the health and form of individual landscape trees." Clearly, urban forestry must address many aspects absent or less pronounced in traditional forestry. Thus, diverse fields of expertise are involved.

Urban Foresters   What do these people do? When most people think of foresters, they have in mind managers of timberlands, or recreational and scenic areas. By contrast, the work of urban foresters is less known. The goal of urban foresters is to extend the life of urban trees, safeguard public health, and produce a reliable source of shade, beauty, and other benefits that result from healthy trees in parks and along streets, avenues, and in neighborhoods.

How, then, does an urban forester or municipal arborist differ from an arborist? In addition to the physiology and care of individual trees, the usual domain of the arborist, urban foresters are charged with the administration of the total urban forest (i.e., the cost-effective management of the human and material resources needed to sustain urban forest resources and integrate them with the community of people and surrounding development).

Importantly, urban foresters are educated to view trees collectively and to manage trees as an ecosystem, taking into consideration specific biological, social and economic conditions. Only such a broad, managerial view enables urban foresters to help municipal governments and taxpayers make wise decisions and get the most from their investment in trees. This does not preclude special treatment for individual trees, either those that are hazards or those that are highly valued and need help to survive (Fazio, n.d.).

Thus, urban foresters face challenges and opportunities similar to those on wildland areas—to improve diversity, increase outputs, and enhance the health of trees and their associated habitats—only in this instance, in densely populated areas with a high degree of human development.

URBAN FOREST VALUES AND BENEFITS

Financial and Economic Considerations

An urban tree is only infrequently appraised for the value of its timber. The dollar value of a street tree in good condition is often about 25 times that of an equivalent tree in a wildland setting (Moll, 1987).

The financial value of urban trees can be estimated based on replacement costs: removal and disposal of the former tree, and purchase, planting, and maintenance for a new, young tree with similar physical characteristics (species, size, and general condition) (Fig. 20-2). The International Society of Arboriculture (2000) has outlined the essential considerations in tree appraisal. Trees that are in some way unique, scarce, or in a special setting of historical or botanical importance should be evaluated by experienced tree appraisers using approaches that involve these considerations.

Dwyer & others (2000) estimate there are 3.8 billion trees in urban areas today, and nearly 20 times that number in metropolitan areas, defined as urban counties. Further, they estimate that urban areas have doubled in size in the past 25 years. Taken as a whole,

FIGURE 20–2
Street trees greatly enhance the livability of towns and cities. *(Seattle Engineering Department photo by Nicholas Cirelli)*

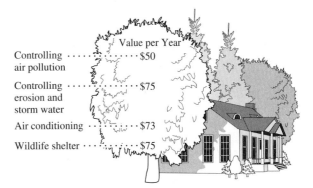

Value per Year

Controlling air pollution · · · · · · · · · · $50

Controlling erosion and storm water · · · · · · · $75

Air conditioning · · · · · · · · · · $73

Wildlife shelter · · · · · · · · · · $75

FIGURE 20–3
The aesthetic contributions of urban trees are only part of their value. Other values are illustrated here. *(Courtesy of American Forests. Thomas C. Wittmore drawing. From Mall and Young, 1992, Growing Greener Cities, Living Planet Press)*

the urban forest is a huge resource, and individually, urban trees can be very valuable.

Some estimates suggest the value of trees can represent 10–15 percent of a homeowner's property value. The usual valuation of urban trees ranges from a few hundred dollars to several thousand dollars for large trees in highly developed situations. Thus, by any estimate, the dollar value of urban street trees is considerable, and the total could be much more if the remaining available urban street tree spaces were filled. A key point is that individual trees in densely populated areas are valued more highly than individual trees in the forest, for a variety of reasons (Fig. 20-3).

But urban trees also have some negative aspects. According to Phillips (1993), tree roots can lift sidewalks and block pipes; branches can block street and commercial signs; branches and trees sometimes fall during storms, block streets, and cause power and telephone outages; leaves are an expense to clean up; and trees can interfere with utilities and people's views. To resolve these conflicts, municipalities need a street tree management plan.

Energy Conservation

Trees have been called a "low tech" solution to energy problems. Whether the goal is reducing the amount of money spent each month on utility bills or guiding the community and nation toward more efficient energy use, trees can provide part of the answer, because urban trees are important in reducing dependence on fossil fuels.

Trees save energy (1) by providing shade, thus reducing the need for air conditioning; (2) by breaking the force of winter winds and consequently lowering heating costs; and (3) by providing a renewable source of fuel that burns with less air pollution than many other fuels (when the right equipment is used). Trees also contribute to energy savings in more subtle ways. One is by reducing atmospheric CO_2, a key contributor to the greenhouse effect and the associated threat of global warming. While landscape trees in towns and cities may tie up relatively little CO_2 compared to large-scale forests, through their energy conservation effects they reduce the use of fossil fuels that release CO_2. By this reckoning, urban and community trees can be 10 to 20 times more effective than rural trees in reducing atmospheric CO_2 levels (Akbari et al., 1990; McPherson et al., 1989).

The properly planned introduction of trees throughout a city to create an adequate canopy over black-top can make a real difference. On the average, cities are 5° to 9° warmer than the rural areas that

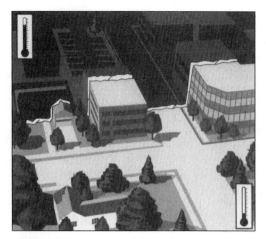

FIGURE 20–4
Trees not only add to the beauty and value of a home, but can help save energy. Planted in the right places around buildings, they can reduce energy used for heating and cooling by 10 to 15 percent. *(Courtesy of American Forests Cool Communities Program)*

surround them, a phenomenon known as the 'urban heat island' (Moll & Young, 1992). In the summer, the urban heat island is more than just uncomfortable. Nationally, it amounts to $1 million per hour in additional cooling costs while producing millions of tons of carbon dioxide, the major greenhouse gas; it also increases the likelihood of unhealthy smog levels. Three to 8 percent of current electricity demand is used to compensate for the heat island effect alone. Preliminary research suggests that, if current heat island trends continue, our cities will be 10° F (6° C) warmer in 50 years (U.S. Government Printing Office, n.d.).

Positive action can offset some of these trends. A minimum of three well-placed trees around a home (Fig. 20-4) can provide shade that will lower cooling costs by 10 to 50 percent (Moll & Young, 1992). By lightening the colors of parking lots and buildings so that heat-generating solar radiation is reflected, communities could reduce their peak cooling energy use by up to 50 percent. Tree planting and care are the least expensive ways to slow the buildup of carbon dioxide and reduce the threat of global warming (American Forests, n.d.).

These estimates are conservative. Nationwide energy savings could total billions of dollars annually if all cities planted trees to reduce heating and cooling

bills. Thus, urban trees can earn their keep. Moll and Young (1992) claim the money saved on energy in the United States during 1989 from urban trees exceeded the gross receipts from all national forest timber sales.

Aesthetic Considerations

Trees are attractive. For various reasons, those who could afford to, have usually surrounded themselves with trees. Urban trees can have a tremendous positive impact on both the physical attractiveness and social fabric of a community. Trees should be viewed as an integral part of the total urban structure. Trees frame and enhance views and add interest to a landscape. They heighten the sense of enclosure and perspective, thereby creating an impression of more or less space. They give definition, meaning, and utility to spaces between buildings and create a sense of unity in urban settings by reconciling widely divergent architectural styles. They provide diversity and life in a largely artificial environment.

Social and Psychological Issues

Trees not only enhance the visual quality of a community, they also make important contributions to its social qualities as well. In 1844, the New York Board of Health encouraged tree planting in urban areas to improve air quality and thus the liveability of the city and the health of residents. The close bonds between people and trees are enhanced by daily contact during all seasons of the year and by the distinct contrast between trees and the built environment. People love trees! Trees and forests convey serenity and beauty. A number of sensory dimensions may be involved, surrounding the individual with impressions of nature in an environment where natural things are at a minimum (Dwyer et al., 1991).

Recreational Values

The physical, social, and psychological benefits of urban vegetation enhance and expand recreation activities that take place in and around urban greenspace. Recreational use of urban forests and greenspace is extremely important to the life of any city and is compatible with and complements the other values and uses (Fig. 20-5). In Norway, Oslo's municipal forest provides an example of what can be achieved in urban planning. This multiple-use forest surrounding Norway's capital city provides the urban area's drinking water, some pulpwood and timber,

FIGURE 20–5
An urban forest or park, a consolidated area of trees and green-space in the city environment, provides pleasure and relaxation for residents. *Left:* A wooden boardwalk through a city park. *Upper:* People enjoying a passive form of recreation. *(Courtesy of Bellevue City Parks, Bellevue, Wa.)*

and serves as a recreation area for over a million visitors per year, who bike, hike, jog, ski, and picnic in the forest, most of which is closed to public vehicle traffic.

Even modest levels of urban vegetation can enhance recreation through the softening, shading, and buffering effects of trees and plantings. Walking, jogging, and running for health and pleasure are cases in point.

Ecological and Wildlife Considerations

Diverse communities of plants and animals in a city contribute to the overall health of an urban ecosystem and are also an important indicator of the city's livability for humans. Urban parks present many excellent opportunities for increasing awareness and knowledge about wildlife, thus enhancing pleasure from park visits.

Dick and Hendee (1986) found that 55 percent of the visitors to urban parks in Denver and Seattle made an observable response to wildlife they encountered. Such responses to wildlife increased with the numbers of wildlife, the variety of species pres-

ent, the physical and vocal activity of the animals, and their proximity to the visitors. Pigeons, squirrels, songbirds, and waterfowl are all popular attractions for visitors to parks and urban greenspace. In communities east and west, bird feeding and watching are popular activities (Fig. 20-6).

The general effects of urbanization on birds are fairly well known. Species favoring ecotones or forest edges fare quite well, while forest-dwelling species generally decline. Some species live at the urban farmland border and may depend on feeding by urban residents to survive during the winter, especially during extremely cold weather and heavy snowfall. As human impacts increase, bird populations tend to be dominated by a relatively few abundant, usually exotic, species. Insectivores, as well as cavity and ground nesters, decline under these circumstances. Knowledge of these general habitat relationships can be used in planning for urban residential and suburban environs, and undesirable impacts can be offset if bird habitat is considered in overall urban forest planning (DeGraff & Wentworth, 1986).

MAJOR ELEMENTS OF URBAN FORESTRY PROGRAMS

It must be noted first of all that many U.S. cities do not have a tree inventory or management program, and that initial management may well have to focus on identifying and dealing with immediate needs.

FIGURE 20–6
Bird watching and feeding are popular activities in urban forest areas. Here, visitors to a forested park observe friendly but wild Canada geese in southern Michigan's urban environment. *(Courtesy of Huron-Clinton Metropolitan Authority)*

For those that have inventories, urban forestry programs are in many ways quite similar to traditional forestry programs. Each requires an understanding, not only of the physical and biological resources, but of the social and institutional environments that define the purposes of management. In urban forestry, this includes establishing legal authority. Ideally, the resource is inventoried at some level, ranging from a quick "windshield" survey to a computerized system. This serves to identify its extent, qualities, and management needs, and a management plan is then prepared to establish priorities for actions necessary to meet objectives established by the legal authority.

Management of urban forest resources, after the inventory and planning phases, can be divided into several separate functions, such as workload planning and budgeting, insect and disease prevention and control, tree maintenance and replacement, and human resource management. Other functions vary with the extent and complexity of the urban area's forest resources, wildlife, and human populations.

Legal Authority

Ordinances enacted by state, county, municipal, city, neighborhood, or equivalent governmental units provide the legal authorization for urban forestry programs. They usually (a) define goals; (b) describe the authority, responsibility, and duties of urban foresters, arborists, or related positions; (c) create advisory committees or tree boards; (d) establish standards for the care of trees and other vegetation on public lands; and (e) specify the consequences for violations of specific sections of the ordinance.

Urban Forest Resources Inventory

In order to prescribe actions needed to maintain healthy urban trees and the related ecosystem, trees are inventoried by species, age, size, condition, spatial distribution, crown cover, and relationship to nearby features such as sidewalks and overhead wires (Fig. 20-7). These data can be assembled into reports and projections, usually by computer programs, and can then be used as a basis for systematic tree care and replacement and for increasing the species diversity and health of the ecosystem

TREE NUMBERING SYSTEMS
ASSUMPTIONS:
1. Assume lot line is halfway between houses, unless there is a fence.
2. Number the trees according to street address and in chronological order.
3. Corner lots continue with the house number.
4. Trees on vacant lots are given "fictitious" numbers followed by an X or other
letter to designate this arbitrary assignment of an address.
EXAMPLE:
Weston & Garrison – see numbers on street trees.

FIGURE 20–7
An example of a street tree inventory. The base map is a teledyne map, made from stereo pairs of aerial photos. The trees are numbered in chronological order according to house address numbers. The lot line is halfway between houses unless there is a fence. Considerable data on each tree is kept in a computerized file. Wellesley, Mass. *(From Phillips, Urban Trees)*

(Wagar & Smiley, 1990). Ideally, an inventory should describe the parts of the ecosystem, the structure and function of the ecosystem, and predict how it would change over time.

Inventory data, usually including information about individual trees and planting opportunities, are usually recorded in field logs. These data are then entered into a computer, using one of the many programs designed for this purpose. When a large number of trees must be inventoried, hand-held computers rather than data forms save money and reduce errors. Available programs range from inexpensive, user-friendly programs that can be maintained by a small town tree board, to expensive, sophisticated versions with maps, workload analysis functions, and other features designed for the needs of large cities. For example, in Paris, France, computer chips embedded in street trees are used to collect and monitor individual street tree data. Supplemental information about the urban forest ecosystem structure may be included, such as written descriptions of trends, photos, maps, and diagrams, in addition to some quantitative measures. The basic question for every piece of information is, "*How* does this apply?" That is, how does the information collected relate to the *purpose* of the inventory and management objectives? Specifically, how can it be used to make better management decisions?

Urban Forest Resource Planning

A management plan defines management goals and objectives, assumptions about the future, the characteristics of the resources being managed, and the management policies, programs, and actions to meet the goals and objectives of the plan. When a plan is developed, available material and fiscal resources must be considered. Successful planning includes public involvement to help establish the goals and objectives and often to develop strategies for action.

Many cities lack both management plans and inventories. With limited resources, they may be forced to focus on crises such as dying trees, hazardous trees, Dutch elm disease, or trees breaking up sidewalks. However, an inventory-based management plan is the best foundation for a scientific urban forest management plan.

Workload Planning and Budgeting

Analysis of the data gathered during the inventory will establish the existing condition of resources and suggest needs and priorities for cultural treatments such as tree planting, pruning, and removal. Meeting these needs by carrying out the necessary management tasks in a reasoned sequence requires careful workload planning and budgeting. Rarely will there be enough money and human resources to accomplish everything. Thus, urban foresters must be creative in devising a strategy and planning their work, not only to stay within the funds allocated by the administrators who control budgets, but also to justify funding requests.

Because appropriations of public funds for urban forestry are in competition with other needs and priorities, many urban and community forestry programs should seek alternative funding sources such as trusts. Trust funds are created from private donations, allocation of public funds, or even selling of bonds, and are often governed by a board of citizen directors from various walks of life. Spending from trusts is limited to the interest earnings they generate, but their continuity over time represents an advantage. Another source of financing may come from marketing urban forest products, such as leaves, firewood, or saw logs from trees removed because of age or hazardous condition (Fig. 20-8).

Volunteer Initiatives

Urban foresters often solicit volunteer workers and private contributions to help accomplish the many tasks that need to be done. Volunteers can inventory and plant trees in addition to helping with other parts of the enterprise (Fig. 20-9). Their involvement builds valuable public support and understanding, as well as helping to get the job done. However, management and coordination of volunteers is in itself a difficult and time-consuming activity. Again, a volunteer with people management experience might be found to handle this (Sharpe et al., 1994). People love trees and can be recruited to help protect them. Most states have an Urban Forestry Council, a grassroots citizen organization committed to expanding, improving, and preserving the state's urban tree resources. The primary mission is public education through promotion of exemplary urban forestry practices and policies. Many of these councils are nonprofit organizations funded by federal and state allocations, which partner with corporations and private businesses to generate community support for tree-planting initiatives. Urban Forestry Councils may also sponsor conferences and workshops on technical forestry and the benefits of trees, and provide hands-on training in conducting inventories and organizing local tree boards.

Local tree boards may consist of members appointed by the mayor or other city officials and are often charged with responsibility for guiding and/or developing an urban or community forestry program. This can include identifying goals, setting priorities, developing plans, and ensuring that they are carried out. Tree boards are typically comprised of five to seven citizens; tree board members may serve as the local experts on tree-related questions posed by private citizens and city administrators. They may also acquire training on technical aspects of urban forestry from State

FIGURE 20–8
Red oaks being removed after developing root rot in Cincinnati, Ohio. All wood
3 inches or less in diameter is chipped into mulch for new plantings. All larger wood
is either cut into firewood or into sawlogs. The sale of wood products provides funds
to plant new trees. *(Courtesy of Steve Sandfort)*

Forestry and Cooperative Extension Service program specialists (Fig. 20-10).

Insect and Disease Prevention and Control

One of the major factors to be considered in preventing serious insect and disease damage is tree species diversity. From the 1950s through the 1970s a Dutch elm disease (*Ophistoma ulmi*) epidemic destroyed majestic American elms in many cities. In the East, where elms were a predominant species, urban forests were severely impacted. Only in situations where technical know-how, public support, and an existing quick-moving organization could be brought to bear, such as in Minneapolis, were the trees saved (Price, 1993). Thus, inventory of urban trees and a plan guiding actions toward providing a balanced tree species mix is important to urban forest health.

Another consideration in the prevention of insect and disease damage is selection of tree species and cultivars that are insect and disease resistant and otherwise suited to the planting site. Drought-tolerant species and cultivars are especially important in the West because drought stress weakens trees, causing decline and making them an easy target for insect and disease attack. Species diversity, drought tolerance, and insect and disease resistance, along with a tree health-care program, are four key points in keeping urban forests and plantings healthy throughout their life cycle.

Despite the best efforts in planning, some problems will emerge. Thus, provision for regular inspections, selective spraying, pruning, repair, and tree removal will also be necessary (Fig. 20-11). Good communication with the public about the location of, and reasons for, these activities is essential.

Public Education and Involvement

Urban and community forestry programs require *citizen involvement,* just as do wildland forestry programs. In fact, urban forestry may elicit even stronger public response because of the large numbers of people in urban areas and the fact that the forest, as well as individual trees, may be literally in their front yards. Therefore, public education and citizen involvement are essential components of any urban forestry program.

FIGURE 20–9
Volunteers can assist urban forestry in many ways. *Left:* Volunteers in Cincinnati, Ohio place one of the 1,600 trees planted each year by volunteer help. *(Photo by Steve Sandfort) Right:* Neighbors pitch in to spruce up an urban trail in Seattle, Washington. *(Photo by Liz Ellis)*

Urban forestry public education programs should address a variety of topics and use diverse media such as short courses, news articles, signs and posters, workshops, and radio and television spots. Topics can include such things as: what is urban forestry, urban forest benefits, proper tree care and maintenance, and city tree ordinance requirements. Remember, the public can carry out many urban forestry functions. For example, proper tree pruning is a continuing challenge for urban forest managers, and incorrect pruning practices by poorly informed homeowners or even professional arborists can severely reduce the vitality of the trees along streets and in private yards. Making good information on proper pruning available can avert problems and promote tree health and appearance.

To be effective, urban forest managers must understand and communicate with the publics they serve. For example, a tree replacement project to reduce hazards near a high-voltage electrical power line may depend on removing large trees obstructing the power line. An uninformed public could stop the project because of resistance to cutting the large trees and misconceptions about the risk of not removing them. If benefits to individuals and the neighborhood are properly explained early on in the project, objections can be addressed and minimized.

Urban forestry offers a unique opportunity, not only to inform the public about the benefits of forest management in urban communities, but on wildland areas as well. The Growing Greener Cities Education Guide is an urban forest education program that has been teacher-tested and used in several communities. Programs such as Project Learning Tree and Project Wild, although focused on wildland areas, may use the urban forest as an outdoor classroom to teach stu-

FIGURE 20–10
Urban forest assistance. Most homeowners own at least one tree and sometimes need professional assistance. Here, a representative of a state forestry agency shares his knowledge on tree care with a homeowner in Georgia. *(Photograph courtesy of Georgia Forestry Commission)*

FIGURE 20–11
Tree removal is complicated in the urban environment. The technician must climb the tree using spurs and a safety belt, then remove it a section at a time in order to avoid utility wires, plantings, and houses. *(Photo by John C. Hendee)*

dents *how to think, not what to think* about natural resource issues. The goal of these public education programs is to cultivate understanding and support for stewardship of natural resources—including urban forest resources.

Utility Corridors

Utility corridors can greatly enhance or detract from the urban forest environment. Utility companies manage trees in utility corridors as part of providing safe and reliable electrical and other services to their customers. Some utility company practices regarding tree topping and pruning have created public conflicts that might have been avoided if the pruning had been carefully done and had preserved a natural appearance. City foresters should cooperate with utility companies so that appropriate and acceptable tree es-

tablishment, replacement, and pruning standards prevail. Planning for species diversity on utility corridors is important for production of wildlife habitat, ecosystem health, maintenance efficiency, and visual aesthetics. Planting the *right* tree in the *right* place is a management imperative. For example, planting trees with mature heights less than the height of power lines, or planning their gradual harvest and replacement within such confines, reduces impacts and conflicts along utility corridors (Fig. 20-12).

Utilization of Wood from Urban Forests

The biomass removed from urban forests can create an expensive nuisance and disposal problem that fills increasingly scarce landfill spaces. Yet it may also have potential to generate revenues. Many cities market residue from tree pruning and removal as fire

FIGURE 20–12
Trees planted beneath utility lines must be chosen carefully, as their heights at maturity should be less than the height of the power lines. *(Courtesy Puget Sound Power and Light)*

wood, mulch, boiler fuel, or bedding chips. In some unusual situations, wood from dead or dying mature trees can be made into attractive products. Cottage industries can manufacture items such as trophy bases and wooden pen and pencil sets from hardwoods removed from city streets. Secondary and vocational school wood shops may also utilize these materials (Fig. 20-13). Compost from urban forests can replenish and build soils. With the nation running out of landfill space and undergoing increased burning restrictions due to air-quality constraints, aggressive marketing of urban forest waste would seem a good strategy.

Examples of Urban Forestry in Action

Across the United States, there are many outstanding urban forestry programs. Cincinnati, Ohio, and Wellesley, Massachusetts are examples of cities with excellent programs organized to meet the demands of their respective communities.

Cincinnati, Ohio has approximately 380,000 people living along 1,000 miles (1,610 km) of streets within 88 sq mi (227 km²). As Cincinnati is bounded by the Ohio River and the state of Kentucky on the south, and surrounded by numerous townships, villages, and cities on other sides, new subdivisions and industrial parks are rare. Urban

foresters are thus challenged to change the patterns within a well-established landscape.

In 1976, concerned citizens spearheaded an effort to properly care for the tree assets of the city. In 1980, through the work of the mayoral committee on urban greenery, the city council adopted a professionally guided program based on the following principles:

- There would be maximum citizen involvement.
- Management would be by neighborhoods (management units) rather than by individual trees and street addresses.
- Funding would be based on a special front-footage assessment allowed by Chapter 727.01 of the Ohio Revised Code, with monies restricted to use on street trees.
- The program must be cost-effective, and so would have a small professional staff plan and administer contracts with private landscape and tree service companies, which would in turn have responsibility for almost 100 percent of the street tree work (Fig. 20-14).

The effective implementation of these principles has led to a 20-year management plan that guides the care of over 50,000 street trees by a small professional staff.

FIGURE 20–13

Upper: Seventh graders in a woodshop class learn about urban forest management from an urban forester. *Lower:* The project coordinator discusses how trees grow and ways they may contribute to the quality of urban life. *(Seattle Engineering Department photo by Nicholas Cirelli)*

FIGURE 20-14
The equipment used in street tree removal in Cincinnati, Ohio, is privately owned. All tree work, except voluntary planting, is contracted to tree service companies. *(Courtesy of Steve Sandfort)*

Wellesley, Massachusetts is a handsome residential town of 27,000 people located 11 miles west of Boston, Massachusetts. Its curving streets are lined with a variety of large and small native trees, introduced shade trees, and 75 new cultivar varieties.

To properly manage Wellesley's urban forest resource, the city chose to keep this responsibility within its Department of Public Works, Park and Tree Division. This innovative department has its street tree inventory updated every five years and a street tree master plan updated every eight years (see Fig. 20-7). With an urban forest comprised of 1,000 acres (450 ha) of parks and conservation land and 4,000 street trees, cost-effective implementation of the street tree master plan was necessary. The Division's labor force serves as the contractor hired to do pruning along the utility corridor. This provides income to the Division while reducing impacts of improper utility pruning. Almost all tree planting in Wellesley is done on private property. The trees are set back far enough to allow proper, healthy growth and are away from road salt spray and vehicle damage. The trees achieve a canopy just as quickly as those growing next to the street. All trees are maintained by the homeowner, not at taxpayer expense. The number of public trees is only half the number that it was 15 years ago, during which time the total number of trees in Wellesley's urban forest doubled.

Both Cincinnati and Wellesley have excellent programs, but they manage through different approaches. Cincinnati uses a small professional staff to administer private contracts for tree work and technical operations. Wellesley, by contrast, relies on its city department staff to maintain its parks and street trees (Fig. 20-15), but has privatized nearly half its trees by locating them on private lots adjacent to the street. As these examples show, creativity and novel approaches apply in urban forestry.

Cool Communities

Cool Communities is a national, action-oriented, environmental-improvement program of American Forests/Global ReLeaf and the U.S. Environmental Protection Agency. Cool Communities seeks to mobilize organizations, government agencies, and businesses, both locally and nationally, to create positive, measurable change in urban environmental conditions and energy consumption and to increase public awareness of these issues.

Cool Communities operates on three levels.

1. Partnership development: to unite the diverse resources in the community and nation.
2. Program implementation and monitoring: to provide scientific data on the energy-conserving benefits of trees and light-colored surfaces in various climates.
3. Public awareness and education: to raise people's awareness of these inexpensive, energy-conserving measures and to encourage their implementation.

Several model communities of various climates, sizes, and needs were selected by American Forests and the EPA to initiate Cool Communities including Tulsa, Oklahoma; Dade County, Florida; Tucson, Arizona; Frederick, Maryland; Springfield, Illinois; Austin, Texas; and Sacramento, California.

FIGURE 20–15
The staff and equipment employed to maintain the park and street trees of
Wellesley, Massachusetts. *(Photo by Leonard Phillips, Jr.)*

Using information from the EPA's guidebook *Cooling Our Communities,* and American Forests' tree-planting handbook *Growing Greener Trees,* local advisory groups in the communities plant trees, lighten surfaces, determine the resultant energy savings, and provide examples for other communities to develop similar community environmental improvement campaigns.

The Urban Tree House

The Atlanta Urban Tree House, an innovative community-based program, seeks to foster an interest in and understanding of natural resource concepts and careers for urban, at-risk youth. The program is set in a specially designed urban forest education structure in a local park where natural resource professionals from a variety of disciplines conduct lessons about ecology and natural resources in general, and forests specifically. The structure is an integral part of the program. It is built around a large oak tree, and is a 70-foot deck in the shape of the United States showing state boundaries, major rivers, long distance trails, and national forests. Environmental education programs are taught by program cooperators and include professionals from all levels of government, corporations, and private nonprofit organizations. Hands-on activities include learning first hand about urban forests as well as more distant na-

tional forests, as well as activities such as tree planting, landscaping with native plants, counting tree rings, wildlife identification, paper making, natural resource games, and theatrical productions presented by neighborhood youth (Fig. 20-16). The Atlanta Urban Tree House is the first in a nationwide network of Urban Tree Houses to link at-risk youth with national forests and other natural resources.

URBAN FORESTRY ORGANIZATIONS

Urban forestry enjoys support from many organizations ranging from professional and educational organizations such as the Society of American Foresters and International Society of Arboriculture, to voluntary citizen organizations such as garden clubs and environmental organizations. The organizations discussed here have a special focus on urban forestry.

Voluntary Citizen Organizations

These organizations are made up of citizens with an avocational interest in urban and community forestry. They work in a variety of ways to advance urban forestry concepts. A few examples follow.

The National Arbor Day Foundation   With over 1 million members, The National Arbor Day Foundation

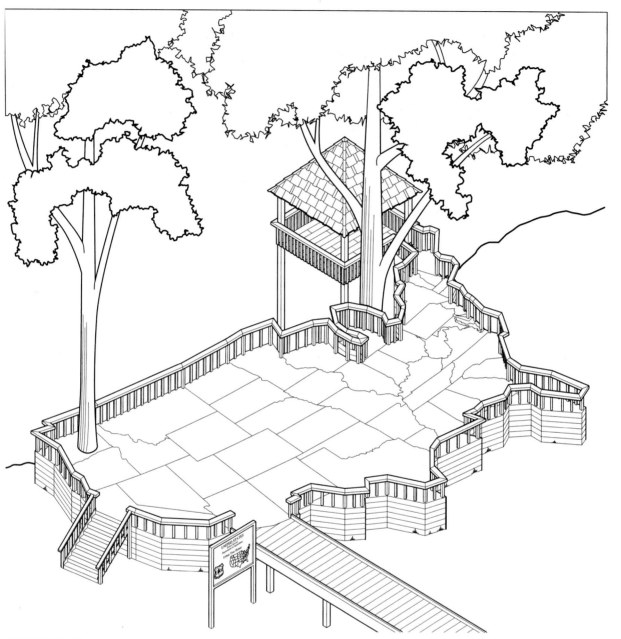

FIGURE 20–16
An artist's version of the Forest Service's Urban Tree House, shaped in the form of the United States, complete with the major rivers, trails, national forests, and forest experiment stations. The first Urban Tree House was built in Atlanta, Georgia; more will follow in other urban areas around the United States. (*Courtesy of 3-D Educational Media*)

is the world's largest tree-related conservation organization, with 61 percent of its budget funded from its $10 per year membership dues. Founded in 1972 during the centennial of Arbor Day to rekindle the spirit of tree planting in America, it has since developed other programs such as the Tree City USA recognition program and the Tree City USA technical bulletin series, which provides valuable, practical information. Other activities include public service messages for the mass media; children's programs; the promotion of trees for wildlife, windbreaks, erosion control, and other conservation purposes. The organization is also a sponsor of conferences and workshops at its Conference Center at the Arbor Day Farm in Nebraska City, Nebraska.

Tree People   Founded in 1973 by a 15-year-old boy with a love of planting trees, this group's mission is to help heal the environment by involving citizens of the Los Angeles area in the planting, care, and appreciation of trees and the urban forest. Through the Tree People's Citizen Forester training program, a combination of classroom presentations and field experiences, participants learn everything from basic tree physiology and care to fund-raising techniques. The training may include visits to Tree People's Coldwater Canyon Park, a 45-acre public park with 5 miles of nature trails, to recycling and composting displays, and to the Tree People's nursery, where drought and smog tolerant trees are grown.

Tree People has created a growing network of citizen foresters all across the Los Angeles area united in the goal of creating a new urban forest to nurture and sustain everyone. Since its beginning in 1973, Tree People has involved nearly a million children through its Environmental Leadership program, reaching 50,000 more children each year. Tree people's initiatives have grown to the point where the city of Los Angeles has embraced its initiative to forest the entire city (Tree People with Andy & Katie Lipkis, 1990).

American Forests   Formerly the American Forestry Association, this is the nation's oldest citizen conservation organization, established in 1875. It has been publishing articles on urban forestry in its magazine *American Forests* since 1895, and initiated urban forestry programs in the late 1970s. In 2002, it established *Forest Bytes,* a monthly Internet magazine with technical information, human- and tree-interest stories, and nationwide news of urban forestry.

In 1981, this organization established the National Urban Forest Council, a consensus-building organization designed to set priorities for the urban forest movement, formulate public policy, and stimulate positive action at the national level. Membership in the Council is open to everyone; a modest fee covers mailing costs. The minutes from the monthly meetings serve as an early-warning signal to members, alerting them to emerging issues and ideas that will impact the future of the urban forest movement. The Council serves as a support mechanism and clearinghouse for information relevant to state and local councils.

Stimulated by growing concern over *global warming, urban heat islands,* and *ozone depletion,* American Forests mounted new urban forestry initiatives in the late 1980s, including Global ReLeaf, a program emphasizing the importance of tree planting by individual and corporate citizens. This is one of several national efforts that have focused intense public attention on the need to plant and care for trees and forests, particularly in urban areas.

American Forests has partnerships with about 500 local citizen organizations and also provides education programs for teachers, promotes the values of urban forests through public service announcements and other means, and provides overall leadership to urban forestry through the National Urban Forestry Conference.

Urban Forestry Professional and Educational Societies

Dozens of professional and educational societies have an interest in urban and community forests. For example, the International Society of Arboriculture is a professional affiliation of 12,700 arborists who may work for city, municipal, educational, research, private, governmental, or utility organizations. The Society of American Foresters, the principal affiliation for 20,000 professional foresters, has an urban forestry working group for members interested in that field. The Society of Municipal Arborists provides accreditation to municipal and county forestry departments in a voluntary program of self regulation to establish high standards for municipal arboriculture. The American Society of Consulting Arborists and the National Institute of Municipal Law Officers are other professional affiliations that have members working in urban forestry. Some of these organizations and their Web sites are listed in Appendix G.

URBAN FORESTRY CAREERS

Urban forests offer many of the same managerial challenges as wildland forests, requiring the ability to visualize the forest over a long period of time and manage natural change to enhance desired forest uses and benefits. But the close interaction between people, development, and urban institutions imposes special challenges and expertise. Urban forestry programs are expanding in North America, and they offer promising career opportunities.

To become an urban forest professional requires formal education in forestry, horticulture, arboriculture, landscape design, or other disciplines having expertise pertinent to urban forestry. Many forestry schools have added urban forestry options to their programs (Hildebrandt et al., 1993). Forestry and the biological-, design-, or planning-based disciplines are important, but so are others including urban planning, landscape design, sociology, urban ecology, public relations, and marketing. These courses provide valuable supporting information for urban forestry careers.

Since urban forestry work is so diverse, a variety of work experiences in urban settings may be valuable preparation. For example, a job at a fast food restaurant may offer experience in marketing and public relations. Likewise, work in a park or arboretum, with a utility company, lawn and tree care service, private forestry consulting firm, environmental education center, or horticulture nursery can provide beneficial experience and skills.

SUMMARY

Trees and their associated habitats have long been recognized as valuable assets to metropolitan and urban areas, where 80 percent of the American population lives. The financial value of the nation's more than 3.5 billion urban trees on public and private property constitutes an enormous sum, and these trees and associated vegetation generate billions of dollars of value each year in reduced energy costs for heating and cooling. In addition, urban trees have a positive impact on both the physical attractiveness and social fabric of a community.

Urban forestry programs are implemented by urban foresters, municipal arborists, and others with diverse expertise. The best programs use planned approaches based on inventories of trees and associated resources, as well as policies and actions designed to meet the community's objectives. Public involvement is a key to the formulation and success of plans, and volunteer labor by citizens, together with their informed activity on public and private property, are important to urban forestry. Several cities in the United States have outstanding urban forestry programs that reflect their progressive character and the environmental concerns of their residents, who also enjoy the many benefits and values of well-managed urban trees and associated habitats.

The field of urban forestry is enlarging even as the area it covers expands into the urban wildland interface. Research into climate issues, and new attention to forest and energy issues, have resulted in a great expansion of research pertinent to urban and community forest issues.

LITERATURE CITED

Akbari, H., A. H. Rosenfeld, and H. Taha. 1990. Summer Heat Islands, Urban Trees, and White Surfaces. ASHRAE Transactions. 96(1):1381–1388.

American Forests. n.d. Cool Communities. Washington, D.C.

Ball, John. 1997. A New and Enhanced Role for Foresters. J. Forestry 95(10):6–10.

Chadwick, L. C. 1970. 3000 Years of Arboriculture: Past, Present and Future. Proceedings of the 46th International Shade Tree Conference, pp. 73a–78a.

DeGraff, Richard M., and J. Wentworth. 1986. "Avian Guild Structure and Habitat Associations in Suburban Bird Communities," *Urban Ecology* 12:399–412, Amsterdam, Netherlands.

Dick, Ronald E., and John C. Hendee. 1986. "Human Responses to Encounters with Wildlife in Urban Parks," *Leisure Sciences* 8(1):63–67.

Dwyer, John F., H. Schroeder, and P. Gobster. 1991. "The Significance of Urban Trees and Forests: Toward a Deeper Understanding of Values," *J. Arboriculture* 17(10):276–284.

Dwyer, J. F., E. G. McPherson, H. W. Schroeder, and R. A. Rountree. 1992. "Assessing the Benefits and Costs of the Urban Forest." *J. Arboriculture* 18(5): 227–234.

Dwyer, John F., et al. 2000. Connecting People with Ecosystems in the 21st Century: An Assessment of Our Nation's Urban Forests. Gen. Tech. Rep. PNW-GTR-490. USDA Forest Service, PNW Research Sta., Portland, Ore.

Fazio, James R. n.d. What is an Urban Forester? Tree City USA Bulletin No. 12. National Arbor Day Foundation.

Fazio, James R. n.d. How Trees Can Save Energy. Tree City USA Bulletin No. 21. National Arbor Day Foundation.

Grey, Gene W., and Frederick J. Deneke. 1986. *Urban Forestry* (2nd ed.), John Wiley and Sons, New York.

Hildebrandt, R. E., D. W. Floyd, and K. M. Koslowsky. 1993. "A Review of Urban Forestry Education in the 1990s." *J. Forestry* 91(3):40–42.

International Society of Arboriculture. 2000. *Guide for Plant Appraisal.* 9th ed. Savoy, Ill.

McHarg, Ian L. 1969. *Design with Nature.* The Natural History Press, Garden City, N.Y.

McPherson, E. Gregory, James R. Simpson, and Margaret Livingston. 1989. Effects of Three Landscape Treatments on Residential Energy and Water Use in Tucson, Arizona. Energy and Buildings. 13(1989):127–138.

Miller, Robert W. 1996. *Urban Forestry—Planning and Managing Urban Greenspaces.* 2nd ed. Prentice Hall, Englewood Cliffs, N.J.

Moll, Gary. 1987. The State of Our City Forests. *American Forests,* May/June.

Moll, Gary, and Stanley Young. 1992. *Growing Greener Cities.* Living Planet Press, Los Angeles.

Phillips, Leonard E. Jr. 1993. *Urban Trees.* McGraw-Hill, Inc., New York.

Price, Susan. 1993. "The Battle for the Elms." *Urban Forests,* 13(2):11–15.

Profaus, George V., R. Rowntree, and R. Loeb. 1988. "The Urban Forest Landscape of Athens, Greece: Aspects of Structure, Planning and Management," *J. Arboricultural* 12:83–107.

Schneider, Wolf. 1963. *Babylon is Everywhere.* McGraw-Hill Book Co., New York.

Sharpe, Grant W., Charles H. Odegaard, and Wenonah F. Sharpe. 1994. *A Comprehensive Introduction to Park Management.* Sagamore Publishing, Champaign, Ill.

Spirn, Anne Whiston. 1984. *Granite Garden: Urban Nature as Human Design. Basic Books.* Division of Harper Collins Publisher, Inc., New York.

Tree People with Andy and Katie Lipkis. 1990. The Simple Act of Planting a Tree: Healing Your Neighborhood, Your City and Your World. Tarcher Inc., Los Angeles, Ca.

USDA Forest Service. 2001. 2000 RFA Assessment of Forest and Range Lands Summary Report. Washington, D.C.

U.S. Government Printing Office. n.d. Cooling Our Communities. GPO document 055-000-00371-8.

Wagar, J. Alan, and E. Thomas Smiley. 1990. "Computer Assisted Management of Urban Trees." *J. Arboriculture* 16(8):209–215.

Watkins, T. H. 1992. "Father of the Forests." *J. Forestry* 90(1):12–15.

Zube, Ervin. 1973. The Natural History of Urban Trees in the Metro Forest, *Natural History* special supplement, 82(9).

ADDITIONAL READINGS AND INFORMATION

(See Appendix G. for Urban forestry-related organizations and their Web sites, which are excellent sources of information)

Bradley, Gordon A. (ed.), 1984. Land Use and Forest Resources in a Changing Environment: The Urban/Forest Interface. University of Washington Press, Seattle. (A valuable collection of expert papers on the subject.)

Ebenreck, Sara. 1989. "The Value of Trees," in G. Moll and S. Ebenbeck. (eds.), *Shading Our Cities, A Resource Guide for Urban and Community Forests,* Island Press, Washington, D.C.

Forestry. 1994, October. Theme issue on Community Forestry: More than Streets & Trees. 92(10).

Harris, R. W. 1992. Arboriculture: Integrated Management of Landscape Trees, Shrubs and Vines. 2nd ed. Prentice Hall, Upper Saddle River, N.J.

Lyle, John T. 1991. "Shaping Urban Ecosystems." *Renewable Resources J.* Summer 1991, pp. 17–20.

Merullo, Victor D., and Michael J. Valentine. 1992. *Arboriculture & The Law.* International Society of Arboriculture, Savoy, Ill.

Tree City USA Bulletin. The National Arbor Day Foundation, 100 Arbor Ave., Nebraska City, Nebr. 68410

Urban Forest Map—American Forests, P.O. Box 2000, Washington, D.C. 20013.

STUDY QUESTIONS

1. How would you define *urban forestry?*
2. Give two major reasons why urban and community forestry are increasing in importance.
3. Describe briefly the similarities and differences between urban and community forestry and wildland resource management.
4. How do urban forests conserve energy? Is a significant amount involved?
5. What are some of the nonfinancial values and benefits of urban forests? The financial benefits?

6. Is there an urban forest in your area?
7. How could you be involved in urban forestry? Does your nearest city have an urban forestry or tree organization described on its Web site?
8. Why is public involvement in the decisions of urban forest managers a necessity?
9. How do voluntary organizations and professional societies differ? Describe one of each involved in urban forestry.
10. Suggest five ways a person can prepare for a career in urban forestry.
11. Do you have a favorite tree you see every day? What species is it? Does it seem healthy? To what risks is it exposed?

CHAPTER 21

International Forestry

This chapter was substantially revised for the 6th edition by David A. Harcharik, Associate Deputy Chief for International Forestry, USDA Forest Service, (ret.) and updated for the 7th edition by the authors. Historical information was developed by the late Robert K. Winters, and the environmental education section by Dr. Sam Ham of the University of Idaho.

INTRODUCTION

This chapter discusses forests as a focus of international concern, looking at associated renewable resources where pertinent. The material traces the international roots of forestry, highlights key international forestry issues, and describes major international forestry programs and organizations that sponsor and carry out this work. International environmental education is presented as an activity essential to developing an appreciation for environmental values in developing countries so they will support sustainable forestry.

International Importance of Forestry

As described in Chapter 1, early people made the forest serve some of their basic needs and gradually made inroads upon it with fire, stone ax, and eventually the goat (Fig. 21-1). When the hunter–gatherers evolved to agriculture and domesticated livestock, they began a systematic attack on the forest. With this change, food became more plentiful, and population increased. This resulted in a need for more grazing land and cropland, beginning a process of deforestation that continues even today.

Forests and associated renewable resources always have been of local importance, but today forests are of international concern. Currently there exists growing appreciation for the value of forests to the global environment and world economies, in addition to their importance to local communities and ecosystems (Fig. 21-2). The health and status of the world's forests have become a top international priority as we understand more about their influences on global climate, the atmosphere, and the value of forest ecosystems in sustaining genetic diversity and contributing to human socioeconomic well-being. Emphasis worldwide centers on sustaining forests through sound management so they continue to provide a flow of goods and services, protecting the environment for present and future generations, and, on environmental education, so people will appreciate the importance of conservation and careful management of forests.

This chapter emphasizes forestry-related history, international forestry issues, and U.S. efforts to con-

FIGURE 21-1
A bas-relief, found on the tomb of the Pharaoh Akhouthtep of Egypt's fifth dynasty, demonstrates that the goat, even thousands of years ago in the Middle East, was well known for browsing on trees. *(Photo courtesy of FAO)*

tribute to solutions, including financial assistance to developing countries. But this should not overshadow the fact that other countries have growing forestry programs, and the United States also learns from them (see the *Journal of Forestry,* October 1997 for articles on forestry in India and South Korea; March 1998 from Sweden; November 1999 from British Columbia, Asia, and Europe). Several countries have forestry programs that predate those of the United States.

THE INTERNATIONAL ROOTS OF FORESTRY

While the European origins and early colonial applications of forestry seem far removed from today's international forestry, these had great influence on the evolution of forestry and renewable resource management in North America.

Europe

Although the Greeks and Romans were practicing tree management much earlier, forestry as a science has its roots in the shortage of ship timbers in northern Europe during the seventeenth century, when both France and Britain were building empires through sea power. During the 1660s and 1670s, Jean Baptiste Colbert, Prime Minister of France, attempted to increase naval power, but the existing French forests could not supply the high-quality naval timbers in the required quantities. Although initially much timber had to be purchased abroad, Colbert's ultimate aim was to make France self-sufficient in naval timber and other forest products.

To accomplish this goal of timber self-sufficiency, Colbert issued the Ordinance of Waters and Forests of 1669. It attempted to bring all French forests

FIGURE 21-2
Generalized distribution of the principal forest associations throughout the world. *(Courtesy of FAO)*

THE WORLD'S FOREST—MAIN VEGETATION ZONE

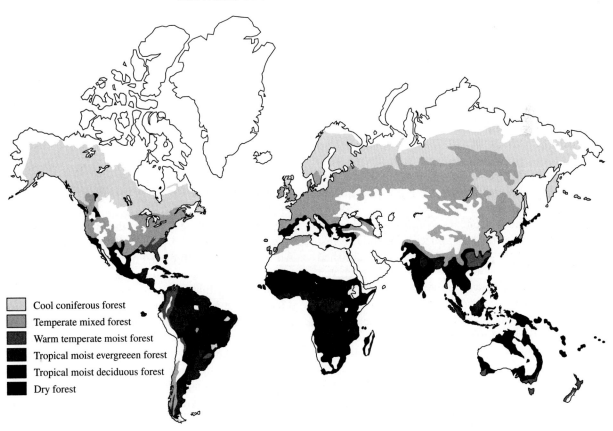

Cool coniferous forest
Temperate mixed forest
Warm temperate moist forest
Tropical moist evergreeen forest
Tropical moist deciduous forest
Dry forest

under systems of cutting that would provide a continuous yield of timber. At that time only the most rudimentary forestry principles, such as regulating the area harvested each year, were understood. A given forest ownership was simply divided into as many equal areas as the age in years of the trees at the anticipated harvest date. One such area was cut over each year. Reproduction was obtained from a few seed trees that were left, and some hardwood species that sprouted from the stump. Because of variations in the productivity and species composition of the annual cutting areas, there was no way the owner could forecast the quantity and character of the yield in any year. In fact, growth and yield information for individual species would be unavailable for more than a century, but the science of forestry was emerging.

Other events of the seventeenth century show the advance of the scientific method. Galileo invented the telescope and declared that the earth revolved around the sun. Anton van Leeuwenhoek invented the microscope, Robert Hooke identified living cells, and later Sir Isaac Newton established the laws of physical motion. German forest masters apparently were in tune with the new scientific spirit. Soon they were applying mathematical formulas to calculate forest growth, developing harvest schedules to spread the volume of forest yield over the number of years required to produce a new crop of timber, and preparing systematic forest operating plans. They began to establish experimental plots to test and demonstrate the utility of various silvicultural practices, such as the effects of thinning frequency and intensity on yields at various ages. Through such research they established a scientific basis for the practice of forestry.

Gradually these pioneers attracted pupils, and eventually some of them organized schools of forest practice. One of the earliest of these was established in 1768 at Ilsenburg, Germany, by Hans Dietrich von Zanthier. Others followed, among them the school headed by Johann Jacob Trunk at the University of Freiburg in Bresgau in 1795. Most of these schools depended entirely on the reputation of their founders and passed out of existence when these men died. Two, however, were exceptions—the forestry school established by Trunk in Freiburg, which continues to this day, and the one that is now a unit in the University of Dresden.

Although the French foresters had for some time held leadership in the practice of what might be called rule-of-thumb forestry, the Germans eventually changed all that by applying scientific methods. At first the French viewed the German theoretical forestry as impractical and clung to their old ways. After a time, however, the French leaders began to accept the new ideas and established a forestry school in Nancy in 1825. J. B. Lorentz, the first director, had been educated in the German school at Tharandt under the progressive leader Cotta. Lorentz faced opposition at first, but gradually the conservative elements in French forestry accepted the German concepts. As the nineteenth century advanced, the Germans and the French emerged as the undisputed leaders in the science of forestry (Fig. 21-3).

From these two countries the science and practice of forestry migrated, adjusting wherever necessary to fit local conditions in other parts of the world. German foresters were involved in forestry in Russia, Sweden, Spain, and other European countries, and had a strong influence on early forestry in the United States.

Beyond Europe

One of the more significant extensions of forestry was to India, where the European principles had to be adapted to subtropical forests. By the early 1800s, British officials were becoming alarmed about the diminishing supply of teak for ship timbers in India and Burma (now Myanmar). Thus, an Indian Forest Service was instituted, which ultimately grew to more than 600 foresters overseeing the work of more than 10,000 technicians, workers, and forest guards.

From India and Myanmar, the British carried the science of forestry to practically all their colonies. An officer of the Indian Forest Service would be temporarily detailed to study the forest situation in a British colony that was contemplating the creation of a local forest service. When his report had been studied and approved, he would usually be transferred to that colony to become the first head of the newly created forest service. In this manner, the science of forestry was introduced into practically all the British colonies. Other major European colonial powers generally were following a similar practice, and countries such as China and Japan brought in German or

FIGURE 21-3
Although there are no old growth forests remaining in Central Europe, some stands there have been under management for over 200 years, such as this mixed hardwood forest near Eberswald, Germany. *(USDA Forest Service photo by Dave Harcharik)*

French foresters to assist them in establishing forestry programs.

Scientific forestry was introduced into the United States when a young German forester, Bernhard Eduard Fernow, came to this country in 1876. He was first employed by various mining companies to manage their forests that were supplying wood for charcoal operations. Gradually he extended his influence to other fledgling U.S. forestry activities, and later was appointed chief of the Division of Forestry in the Department of Agriculture. In 1887, young Gifford Pinchot sailed for Europe to study at the Forest School in Nancy, France. Fernow and Pinchot were, therefore, the primary actors in the introduction of forestry to the United States. Pinchot went on to become the first chief of the Forest Service when it was created in 1905, and established in the Forest Service a sense of international commitment that he himself had acquired as a result of his earlier studies in Europe. In 1902, he traveled to Russia and China, and to the Philippines, where he advised on the creation of a forestry agency in that country.

MODERN INTERNATIONAL CONTRIBUTIONS TO FORESTRY

Many countries around the world continue to develop the science of forestry and share their findings with others. The world community, for example, has benefited substantially from German pioneering research on the effects of acid precipitation on forests; American, Australian, and Canadian advances in fire suppression technology; Scandinavian systems for harvesting small-dimension timber; French satellite imagery; and forest genetic research in central Europe and the United States. According to Gregersen et al. (1989), the United States has reaped considerable economic benefit from foreign research on structural particle board and the use of containerized forest tree seedlings in temperate and boreal zones. There also is much to learn from tropical regions on matters such as fast-growing plantations, game ranching, ecotourism, the use of medicinal plants, agroforestry systems, and social and extractive forestry aimed at meeting community needs in

FIGURE 21-4
Agroforestry, the combination of tree crops and agricultural crops, such as laurel
(Cordia alliodora) and coffee, shown here in Costa Rica. This is an important
alternative to shifting agriculture in many parts of the tropics. *(USDA Forest Service
photo by Dave Harcharik)*

developing countries (Fig. 21-4). For several current examples of forest science and management, see the special issue of the *Journal of Forestry* (2001) devoted to international forestry.

PRIORITY ISSUES

The international dimensions of forests and related renewable resources now command attention from political decision makers worldwide and even from heads of governments. For example, at the 1992 United Nations Conference on Environment and Development (UNCED), commonly referred to as the Earth Summit (and scheduled to be repeated in 2002 as the World Summit on Sustainable Development [WSSD]), 178 countries and 116 heads of state or government, the largest gathering of heads of state in history, convened to chart an international course of action for sustaining worldwide economic development while protecting the environment. Global partnership for sustainable development was the organizing theme (Ryan, 1992). Forestry figured prominently

on the agenda, and a statement on forest principles was agreed upon and signed. This outpouring of interest stemmed from apprehensions that worldwide forest loss and degradation will have far-reaching effects on the quality of human life. The Earth Summit also produced international agreements regarding climate change and biological diversity, and established an agenda on major international environmental issues, with provisions to monitor the effectiveness of these initiatives. Descriptions of some of the issues that give rise to the global concern over forests follow.

Tropical Deforestation

Deforestation, the conversion of forestland to other uses, has occurred throughout much of the world. However, deforestation in industrialized countries has essentially stopped, and forest cover there may even be stabilizing or increasing. Most attention today is focused on deforestation in the tropics. Concern stems from the accelerated rate of forest loss and degradation and the resulting impact on people and the environment, both locally and globally.

FIGURE 21-5
In this example of slash-and-burn agriculture a tropical rainforest undergoes conversion to row crops. Near a Meo village, Doi Suthep Pui, Thailand. (*Photo by Grant W. Sharpe*)

Estimates suggest that over half the tropical forests have already been lost, with the remainder undergoing conversion or harvesting at an alarming rate. The Food and Agriculture Organization of the United Nations (FAO) estimates that, from 1980 to 1990, tropical deforestation occurred at an average rate of 41.7 million acres (16.9 million ha) annually, an increase of 50 percent over the previous decade (Singh, 1993), as a result of slash-and-burn agriculture (Fig. 21-5), fuelwood gathering, conversion to pastures and row crops, poor logging practices, urbanization, and lack of provision for reforestation. Additional large areas are being seriously depleted and degraded, although technically they can still be classified as forests.

Clearing and degradation of tropical forests produces critical ecological and socioeconomic effects. Tropical soils, unprotected by vegetative cover, erode at an ever-increasing rate (Fig. 21-6). Sediment collects in reservoirs, reducing water storage capacity for irrigation, and hydroelectric output decreases, compromising multimillion dollar investments. Flooding becomes commonplace. In some semiarid regions, increasing desertification affects hundreds of millions of families. In Africa and throughout much of the tropics, the natural resource base is damaged and deteriorating, and suffers further from drought and overgrazing leading to slumping economics, short supplies of food and water habitat, and loss of wildlife (Fig. 21-7).

Tropical deforestation presents an extremely complex problem. It results from population pressures, poverty, inappropriate incentives for land use, inadequate government policies for land distribution and tenure, debt, low agricultural productivity, and weak government institutions. Better forestry practices alone, therefore, will not reverse the deforestation trend. However, forestry and natural resource professionals have much to contribute from current and future research results, environmental education, and training programs in resource management and planning. Perhaps the single, most important contribution will be to develop and demonstrate systems of tropical forest management that are more economical and beneficial than other ways of using the land. Until sustainable forestry is perceived to offer greater benefits (both monetary and nonmonetary) than competitive practices, exploitive deforestation will continue, except perhaps in a few small parks and reserves.

Loss of Biological Diversity

Biological diversity refers to the number of species of plants and animals, the genetic variation within species, and the range of landscapes or habitats

FIGURE 21-6
In many parts of the world, severe erosion is often the result of forest clearing, overgrazing, and poor road construction. This is especially true in dry, hilly terrain occasionally subject to heavy rains. Dominican Republic. *(USDA Forest Service photo by Dave Harcharik)*

available for life processes to occur within a given area. On a global scale, as much biological diversity as possible should be fostered so that a wide range of options are available for the use of all life forms, including humans. Worldwide, there is alarm over the daily loss of plant and animal species and other reductions in tropical biological diversity (Fig. 21-8). These losses are not limited to the tropics, but are especially acute there, because tropical forests may contain over half the species of life found on Earth. A loss of tropical forests and habitats may also directly affect diversity in temperate areas. Already there is concern over reported decline of several migratory bird species dependent on tropical forests for winter habitat. The ability of the tropical plant species to produce pharmaceuticals and a host of food and fiber products for future generations may be seriously limited without intensive management and protection of remaining forests (Fig. 21-9).

These losses involve a tradeoff between the potential benefits of preserving biological diversity and the present and future economic and social benefits of resource utilization. Effective management, and the creation of reserves over a species' range, often require a regional approach, international agreements, and financing, as well as training programs to ensure consistent management. Sparsely settled (frontier) areas are the focus of most multicountry reserve areas to date. Transboundary resource protection and management is diplomatically sensitive and complex, but some noteworthy efforts involving key wildlife species—"peace parks"—are emerging, such as for the snow-leopard in central Asia (Singh & Jackson, 1999) and the Kalahari Gemsbok Peace Park between South Africa and Botswana (www.peaceparks.org).

Biodiversity is preserved when intact ecosystems are protected. After assessing its first global forest survey using satellite data, the United Nations concluded that more than 80 percent of the world's remaining intact forests were located in just 15 countries, and that forest protection should focus on those countries (Kirby, 2001). The countries are: Australia, Bolivia, Brazil, Canada, China, Colombia,

FIGURE 21-7

Decades of removing fuel-wood for cooking and heating, in excess of growth, deplete the growing stock, creating ecological and socioeconomic problems. *Upper:* Fuel gatherers collect wood from a shrub forest in Turkey. *Middle:* Fires, fuel-wood cutting, and overgrazing have changed a former forest into a shrubland, in Turkey. *(Photos by Chad Oliver) Lower:* A charcoal kiln in South America. Most wood in the tropics is used for fuel, either directly as firewood or after being converted to charcoal. *(Courtesy of USDA Forest Service)*

FIGURE 21-8
The loss of forest cover, excessive hunting, and invasion of competitors nearly caused the extinction of the Puerto Rican parrot. The joint efforts of the USDA Forest Service, U.S. Fish and Wildlife Service, and Puerto Rican Department of Natural Resources in a captive breeding program and improved forest and wildlife management, have increased the breeding population. *(USDA Forest Service photo by Victor Cuevas)*

Democratic Republic of Congo, India, Indonesia, Mexico, Papua-New Guinea, Peru, Russia, the United States, and Venezuela.

Climate Change

Few environmental issues have generated as much international concern as global warming, often referred to as climate change. While some die-hard skeptics may still question the reality, it is generally accepted that global warming is occurring, and that forests are intimately involved. The potential impacts of global warming are so serious that research to further define its extent and how forests and forestry can help mitigate effects or reverse the trends is a high priority.

Confirmed increases in concentrations of carbon dioxide, methane, and other so-called greenhouse gases in the atmosphere have occurred since the beginning of the industrial age, and have escalated with continued industrialization. Atmospheric CO_2 concentration is increasing about 1.4 ppm per year due to fossil fuel burning and changing land uses

(Sanchez & Eaton, 2001). In theory, these gases will increase global temperature with disasterous consequences within a few decades. Scientists debate how great the temperature rise might be, but even a very small increase could produce major impacts. Changes in forest cover would occur; wildlife would thus be affected by the habitat changes, agricultural practices would need to be altered, and populations in low-lying coastal areas would need to relocate as sea levels rise in response to glacier and snow melt worldwide.

How are forests and forestry involved in global warming and climate change? Research suggests that about 80 percent of the carbon dioxide buildup stems from fossil fuel combustion by automobiles and home heating, and about 20 percent comes from forest clearing. But, since carbon is stored in wood and in forest soils, and growing trees consume carbon dioxide and produce oxygen, the presence of forests, and certain forest practices can do much to prevent or slow global warming and to mitigate its effects. This reality has been a force in shaping the direction and

FIGURE 21-9
Tropical forests, such as this one in Costa Rica, are the source of about one-half of the species of plants and animals currently known, and approximately one-third of the pharmaceuticals used around the world. *(USDA Forest Service photo by Dave Harcharik)*

importance of forestry for more than a decade. For example, preventing deforestation and expanding of forestation and reforestation efforts will help increase the consumption of CO_2 and the sequestration or storage of carbon in forest ecosystems (Sampson & Hair, 1992). Sedjo (1989) estimated that establishing tree plantations could postpone the carbon dioxide buildup by three to five decades, but that planting would need to be on the massive scale of 1.2 billion acres, dwarfing all previous efforts. But even on a lesser scale, such as in cities, woodlots, and small ownerships, reforestation has important symbolic, environmental, and economic benefits. Urban forestry programs also can play a major role by reducing energy consumed in heating and cooling (see Chapter 20, "Urban Forestry").

Adding relevance and urgency to forestry's role in reducing greenhouse gases to mitigate global warming is the international climate change agreement to the United Nations Framework Convention on Climate Change, called the Kyoto Protocol. Although not yet fully accepted by all nations, the Protocol requires industrialized nations to meet targets for reducing greenhouse gas emissions, but also provides possibilities for generating credits toward their targets by investing in emission-reduction projects (Rotter & Danish, 2000). These could include so-called carbon forestry projects aimed at enhancing the capacity of forests to consume and store carbon, or using wood to displace other "higher emission" products (Murray et al., 2000). Thus, the potential for forests, forest management practices, and wood products to help mitigate global climate change has become a major focus of forestry research, policy debate, and economic consideration (Harmon, 2001; Sanchez & Eaton, 2001; Skog & Nicholson, 1998). Other research is focusing on predicting changes in species composition, migration, and other vegetation response so that adaptive strategies can be developed. Such strategies may include breeding varieties of trees more resistant to stress, and forest and range management modifications to reduce loss to fires, insects, disease, and other disturbances. Concerns regarding global warming will continue to be a major influence on natural resource management and research programs in the future.

Air Pollution and Acid Rain

Air pollution and acid rain respect no political borders. A number of regional pollutants may damage forests, lakes, and streams, but the greatest threat comes from ozone and acidic deposition. Ozone damage to the forests has been demonstrated in Southern California and Mexico. Acidic deposition (fall-out from air pollution, including acid rain) is much more widespread, and is considered to be the cause of massive damage to forests in central Europe (Fig. 21-10). In North America, damage is less severe, but contributes to the death and reduced growth of high-elevation red spruce forests in the Northeast and in the southern Appalachians, and to the acidification of lakes in the northeastern United States and southeastern Canada.

Because of their complexity, potentially broad impacts, and obvious political implications, air pollution and acid rain are emerging as major forest and natural resource issues. Natural resource professionals have an important role to play in these issues, especially through providing data needed to predict and measure pollution impacts and by developing management approaches and strategies to reduce these effects. Of course, acid rain, air pollution, and global warming/climate change are related in that many of their causes lie in emissions from modern industrial processes and lifestyles that depend heavily on fossil fuel consumption.

Exotic Pests

The introduction of pests and diseases to regions where they do not occur naturally can cause staggering economic losses and grave ecological impacts. In the United States, considerable harm has been done to forests because of the introduction of, among others, Dutch elm disease (*Ophistoma ulmi*), chestnut

FIGURE 21-10

Acid precipitation and air pollution are thought to contribute to the decline and death of many coniferous forests in Central Europe. Thinning of the tree's crown is a symptom of damage. *(Courtesy of USDA Forest Service)*

blight (*Cryphonectria parasitica*), white pine blister rust (*Cronartium ribicola*), and the European gypsy moth (*Lymantria dispar*). Currently, there is major concern that the Asian gypsy moth (a close relative) may be transported on European and Asian cargo ships entering the United States, where it could cause immeasurable harm to coniferous forests. Similarly, other countries are sometimes hesitant to import wood products from the United States, fearing exotic pests could be introduced into their forests. A major issue in international forestry is anticipating, recognizing, and preventing spread of forest insects and diseases across national borders.

International Trade

Some nations enjoy a relative abundance of wood, while some must look to other parts of the world for their wood supplies. Yet other nations, such as the United States, have forests that could supply all their needs but because of economics, markets, and preference, choose to import a share of the wood products they consume, while also exporting some. Furthermore, some countries who most urgently need forest products can least afford them.

Thus, international trade in wood-based products has important and far-reaching consequences involving forestland use and management, timber harvesting, wood-based industrial development, and consumer demand. For example, Canada's lumber export policies affect the price of wood products in the United States, because a significant percentage of the wood products consumed in the United States comes from Canada. If such supplies are not available, prices will rise and wood products will flow in from elsewhere in the world, or log exports from the United States will be diverted to domestic use. The protection of ancient forests, and the declining national forest timber harvests in the western United States, help stimulate timber imports from other world regions, such as Siberia, where less-stringent environmental protection regulations exist. Under conditions where the protection of forests in the United States precipitates forest exploitation in another region, the import of wood products may be triggering environmental impacts elsewhere.

Research on international trade in forest products is needed to improve understanding—in an international setting—of wood and fiber supplies, comparative advantages in wood product production, solid wood and wood fiber consumption patterns, and to evaluate trade policy. Forest resource economics research directed to such areas could do much to improve the effectiveness with which the United States participates in the worldwide trade of forest products (Waggener, 1989).

INTERNATIONAL FORESTRY ASSISTANCE AND NATURAL RESOURCE ORGANIZATIONS

Coordinating and funding international forestry and natural resource work involves serious commitments by several organizations. For convenience, they can be grouped into (1) multilateral government organizations; (2) bilateral government organizations; and (3) nongovernmental organizations (NGOs). Together, donor governments and these organizations make available over $1 billion annually for international forestry work (Table 21-1). The United States is a major donor in terms of contributions but ranks far lower than other countries in terms of what we contribute in amount relative to gross national product (Fig. 21-11).

As summarized recently by Laarman (2001), foreign assistance for forestry is changing from aid to development of forest industries, as was common in the 1960s and 1970s by sending experts to underdeveloped countries, to capacity building and education so countries will have their own experts. Issues emphasized now—but ignored 20 years ago—include the importance of community participation in forest management, indigeneous land tenure rights, gender issues, and trees in small-scale agriculture—in all, a more social and people-focused orientation (Persson, 2000).

Considering all the agencies and organizations involved could create the impression of vast and growing government foreign aid; however, two key realities must be clarified. First, foreign assistance for forestry rose into the early 1990s but has declined since then. Second, while the United States is among the largest donor countries to forestry aid, it ranks 22nd compared to other countries as a percentage of GNP (see Table 21-1). Laarman (2001) estimates U.S. spending on international aid for forestry at about 40 cents per person or, expressed another way, about 1% of all U.S. humanitarian aid.

Multilateral Government Organizations

These are international bodies whose memberships include many individual governments. The United Nations system and the international development banks fall into this category. Some of the principal multilateral organizations active in forestry and natural resources are the Food and Agriculture Organization of the United Nations, the International Tropical Timber Organization, the World Food Program, and the World Bank and other multilateral banks. Laarman (2001) reports that eight multilateral organizations have provided 98 percent of the multilateral aid for forestry since 1992 but have reduced their funding by 29 percent during this period.

TABLE 21-1

FOREIGN AID IN RELATION TO NATIONAL INCOME, 1999.

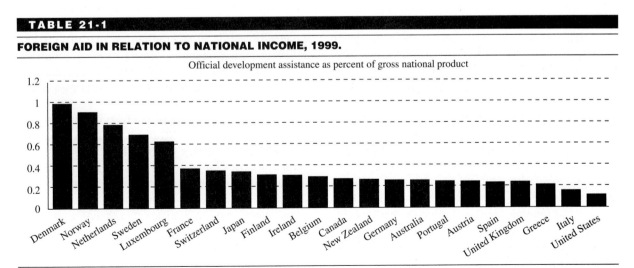

(*Source:* OECD, 2000c; reprinted from Laarman, 2001)

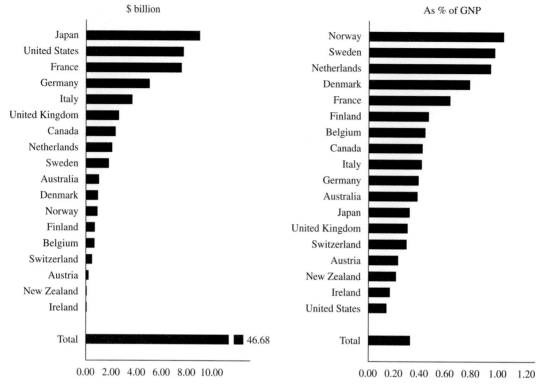

FIGURE 21-11

Net overseas development assistance from OECD countries in 1989. Graph at left represents total contributions. Graph at right represents contributions based on percentage of gross national product. *(Courtesy of Organization for Economic Cooperation and Development)*

The Food and Agriculture Organization (FAO) Founded in 1945 and headquartered in Rome, the FAO constitutes a leading international agency for exchange of technical information on world agriculture, fisheries and forestry and serves as a "world forestry organization." Four primary responsibilities shape its forestry activities: (1) the promotion and coordination of forestry research among its member nations; (2) the dissemination of statistics on forest area, timber volume and growth, and international trade in forest products; (3) direct technical assistance and advice to member nations, especially if one nation develops techniques of value to others; and (4) cooperation with other public international organizations. FAO carries out these responsibilities through information exchange, utilizing publications, technical meetings, and statistical assessments of forest resources and products, all financed by biennial assessments on member countries. Money from international bank trust funds helps pay for field proj-

ects. The United Nations Development Program also funds a large portion of the field program. FAO's field forestry projects include a broad array of forest management, training, and research activities generally aimed at strengthening institutions and building human resources in developing countries. Recipient nations are generally required to contribute a substantial portion of the project costs (see http://www.fao.org/forestry).

FAO is a good source of international forestry information through its Strategic Plan for Forestry, *Unasylva*, its quarterly journal of forestry and forest industries (available online), its on-line databases, and its report on the state of the world's forests.

The World Forestry Congress, a major international meeting often associated with FAO, convenes approximately every six years. Foresters from around the world meet at this congress to exchange technical information and to chart the future direction of forest policies.

FIGURE 21-12
The International Tropical Timber Organization works to see that tropical timber,
such as these dipterocarp logs in Sarawak, Indonesia, that are traded internationally,
will come from sustainable managed forests by the year 2000. *(USDA Forest
Service photo by Dave Harcharik)*

International Tropical Timber Organization (ITTO)   Created by treaty in 1983 out of concern for the sustainability of the tropical timber trade, ITTO was founded to promote sustainable management, marketing and use of tropical wood (Fig. 21-12). Headquartered in Yokohama, Japan, ITTO provides an international forum that seeks to engage both producing and consuming countries in dialogue. Certain aspects of the tropical timber economy, such as encouraging trading in timber from sustainable resources and the use of efficient processes of wood utilization, are stressed. A small ITTO staff receives guidance from the International Tropical Timber Council, which has representation from each of its 57 member countries, which together represent 95 percent of world trade in tropical timber and 75 percent of the world's tropical forests (http://www.itto.or.jp).

World Food Program (WFP)   Since its creation in 1963 to combat world hunger, the World Food Program, headquartered in Rome, Italy, has also become one of the world's largest providers of forestry assistance to developing countries, second only to the World Bank. Typical of other multilateral organizations, WFP's assistance to forestry projects grew from $27 million with 28 projects in 1974 to $183 million with 92 projects in 1992, but has since declined. Under its "Work for Food" program to promote self-reliance, WFP provides food aid to pay for reforestation, planting and maintaining trees, and related soil and water conservation activities (http://www.wfp.org). From 1963 to 2001, the World Food Program has invested $27.8 billion and 43 million metric tons of food to combat hunger, promote economic development and provide relief assistance in emergencies, worldwide.

The World Bank and Other Multilateral Banks
Founded in 1944, the World Bank Group (WBG) constitutes the largest of several international banks that support development work, including forestry. A multilateral lending agency owned by 180 member countries whose views are represented by a Board of Governors, the World Bank was created to help raise the standards of living in developing countries by channeling financial resources to them from developed countries. The World Bank Group has a large and complex structure, with a headquarters staff of 8,000 in Washington, D.C. and 2,000 field staff in 100 country offices worldwide. During 2001, the WBG provided 17.1 billion dollars to projects in

categories such as poverty, AIDS, the environment, and forestry. A database of their more than 9,500 projects funded from 1947 to the present is available on line, accessed from their excellent Website (http://www.worldbank.org).

The World Bank has the largest forestry lending program in the world.

A new program, launched in 1991 and headquartered at the World Bank, is the Global Environmental Facility (GEF) (http://www.gefweb.org). It is managed jointly by the World Bank, the United Nations Development Program (UNDP), and the United Nations Environment Program (UNEP). Recognizing that the responsibility for environmental problems is shared by industrialized and developing countries alike (171 countries are participating members), the GEF provides financial assistance to developing countries to help them undertake environmental work that will have global benefits. The GEF concentrates its grant monies on four areas: (1) protecting the ozone layer, (2) limiting greenhouse-gas emissions, (3) protecting biodiversity, and (4) protecting international waters. GEF projects give an environmental component to the overall World Bank portfolio of programs, which has been criticized in the past for being totally development-oriented.

Bilateral Governmental Organizations

Agencies of individual national governments engaging in direct, one-on-one work with other countries do so through bilateral organizations. A number of nations operate bilateral assistance programs in forestry. The U.S. government agencies most active in bilateral international forestry activities are the U.S. Agency for International Development (USAID); the U.S. Department of Agriculture, through its office of International Cooperation and Development (OICD) and through the Forest Service; and through the Peace Corps.

U.S. Agency for International Development (USAID) This bilateral program started as an outgrowth of the Marshall Plan (1948), which assisted in the recovery of European nations following World War II. In 1961, it became the U.S. Agency for International Development. AID's support for forestry varies depending on congressional funding levels. Strong in the 1950s and 1960s, it dropped to virtually zero in the late 1970s. Support rose again in the early 1980s and in 1992 reached the highest point ever, about \$205 million annually for forestry and related biodiversity projects. Since then, AID support for forestry has fallen by about half in real dollars (Laarman, 2001).

Funds for AID are appropriated by Congress and made available for development work through both loans and grants. Most of the AID programs are administered by AID offices (missions) overseas and are carried out by contractors. Forestry projects address a wide array of activities but focus on training, research, and on-site assistance for land management activities. These include reforestation, watershed protection, and agroforestry, and strengthening the recipient country's institutions through policy development. Encouraging involvement of local peoples forms an important element of AID projects. The host country carries out the project, usually with the advice of U.S. technical assistance teams provided by private firms, universities, or other government agencies.

USDA Forest Service The Forest Service has been active in international forestry since Gifford Pinchot, its first chief, visited the Philippines in the early 1900s. The mission of the USDA Forest Service international program is to promote sustainable forest management and biodiversity conservation worldwide (http://www.fs.fed.global). To implement this mission, international work by the Forest Service has four main dimensions: (1) strategic planning and policy development, (2) research and scientific exchange, (3) training, and (4) technical cooperation.

1. Strategic Planning and Policy Development. A major role of the Forest Service is to help determine and articulate U.S. policy on international forestry issues, in concert with the Department of State. Often the Forest Service represents the United States in international organizations and at conferences. For example, the Forest Service serves as the lead U.S. agency in dealing with the U.N. Food and Agricultural Organization (FAO) on forestry matters, is heavily involved in U.S. preparations for international conferences where forestry topics will be discussed, and provides advice on forestry issues to the multilateral banks and numerous other international organizations.

2. Research and Scientific Exchange. The Forest Service operates the largest forest research program in the world and shares and expands its research through international scientific exchanges. Many of its scientists and institutes participate in international work, especially the International Institute of Tropical Forestry in Puerto Rico, the Institute of Pacific Islands Forestry in Hawaii, and the Forest Products Laboratory in Madison, Wisconsin. The Forest Service has cooperative research projects with more than a dozen countries and has engaged in formal scientific exchange programs with more than 200 countries.

3. Training. The Forest Service attracts hundreds of international visitors each year, including land managers, scientists, and students, who wish to learn firsthand from U.S. forestry and resource conservation experience. Playing host to these international visitors and providing training through the organization of seminars, workshops, field trips, and short courses are important international activities of the Forest Service.

4. Technical Cooperation. Through its Forestry Support Program and the similar Disaster Assistance Support Program, the Forest Service helps AID, the Peace Corps, and other international organizations design forestry projects in developing countries, advises on technical aspects of project delivery, assists with recruitment of experts, and aids the response to emergencies and disasters overseas, such as fires, floods, hurricanes, medical problems, earthquakes, and civil strife.

The Peace Corps   Since its creation in 1961, 165,000 Americans have served as Peace Corps volunteers in 135 different countries to help third-world nations with a wide array of urban and rural development activities including forestry and natural resources and environmental management (http://www.peacecorps.gov) (Fig. 21-13). Peace Corps volunteers now represent the largest environmental grassroots force of any development agency. In 2002, Peace Corps had 7,000 volunteers in 70 countries with more than one-fourth of them working on environmental or agricultural projects, including forestry. Often volunteers work in close liaison with projects funded by FAO, AID, or other organizations.

In addition to its direct contribution to development, the Peace Corps allows Americans to gain practical experience overseas, as well as inspiration, which can form the foundation for international careers. Virtually every natural resource college in the nation now has Peace Corps alumni enrolled in graduate programs. Many of these men and women are planning to return to international conservation careers with the benefit of graduate training.

Nongovernmental Organizations (NGOs)

Nongovernmental organizations involved in international forestry and natural resources are numerous and diverse. Described here are the International Union of Forestry Research Organizations, the categories of conservation organizations, and private industries and firms.

International Union of Forestry Research Organizations (IUFRO)   This nonprofit, nongovernmental international network of forest scientists first organized in 1892. Its objectives focus on promotion of international cooperation in forestry and forest products research, and standardization of research techniques. Today, IUFRO includes more than 15,000 scientists from more than 700 member organizations in over 100 countries (http://iufro.boku.ac). Activities are organized primarily through its 240 specialized research groups in six technical divisions:

1. Forest Environment and Silviculture
2. Forest Plants and Forest Protection
3. Forest Operations and Techniques
4. Inventory, Growth, Yield, and Quantitative and Management Sciences
5. Forest Products
6. Social, Economic, Information, and Policy Sciences

IUFRO serves as a forum for the exchange of knowledge and experience in the field of forestry research at national and international workshops, seminars, and conferences (Fig. 21-14). Every five years, IUFRO organizes a World Forestry Congress, attended by as many as 2,000 participants. IUFRO maintains a small secretariat in Vienna, Austria, where records are maintained. Traditionally heavily dominated by scientists from industrialized countries, IUFRO also promotes the improvement of forestry research in less-developed countries through special projects and activities. IUFRO is a good source of forestry information derived from its technical meetings and proceedings.

FIGURE 21-13
Graduate foresters serve in the Peace Corps in a variety of forestry related projects. *Upper:* A U.S. Peace Corps volunteer discusses an introduced species of Casurina (beefwood) with his counterparts. The seedlings will be distributed for fuelwood production. *Lower:* A Peace Corps volunteer (in hat) poses with his forestry counterparts in the tree nursery he was responsible for starting. The seedlings will be used in school and community woodlots and in shelterbelt projects. Malkern Agriculture Research Station, Swaziland, Africa. *(Photos by Judy Lewis)*

Conservation Organizations These environmentally oriented, nongovernmental organizations (NGOs) are extremely diverse, and vary in size, expertise, base of support, and mode of operation. Some organizations illustrative of conservation NGOs in international work are the National Wildlife Federation, the Nature Conservancy, the Friends of the Earth, Rainforest Alliance, Greenpeace, the Pan American Development Foundation, and the World Wildlife Fund. There are hundreds more, including several large and power-

FIGURE 21-14
IUFRO organizes workshops, seminars, and study tours to promote the exchange of new developments in science and technology among scientists around the world. This scene is from an IUFRO meeting in eastern Germany. *(USDA Forest Service photo by Dave Harcharik)*

ful family foundations such as Ford, Rockefeller, Pew, MacArthur, Kellogg, and others. As well as providing money, conservation NGOs bring unique skills to environmental and development work by serving in private, people-to-people efforts. They range from local women's groups to national "umbrella" associations and internationally affiliated organizations. Some focus on grassroots, self-help forestry work, others on environmental activism and advocacy, and some on conservation policy. For example, one international conservation NGO, the Rainforest Alliance (http://www.rainforest-alliance.org), is a 17,000-member organization with 70 employees and a $6.8 million annual budget. Its mission is to protect people, ecosystems, and wildlife by implementing better business practices for biodiversity conservation and sustainability, including the certification of farms and forests for sustainable management. Conservation International (http://www.conservation.org) is dedicated to demonstrating that human societies are able to live harmoniously with nature. Headquartered in Washington, D.C., CI has projects in more than 30 countries on four continents.

The presence and influence of NGOs will also continue to grow, as citizen concerns for the world environmental situation grows. This was reflected at the 1992 Earth Summit in Brazil, where 20,000 NGOs gathered from around the world, and will be reflected again at the 2002 World Summit on Sustainable Development (WSSD) in Johannesburg, South Africa.

Private Industries and Firms

Some of the large wood products corporations are involved internationally through forestland ownership, wood products manufacturing in other countries, and the export and import of logs, lumber, and wood products (Laarman, 2001). For example, Weyerhauser is a large landowner in Canada and has opened plants in China and Mexico. International Paper has extensive holdings in South America. The paper industry is heavily involved in foreign trade for pulp and products. Thus, the financial and technical flows to forestry in other countries from the private sector are substantial, generating know-how, employment, and tax revenues in the countries involved. Increasingly, U.S. forest and wood products companies abroad are partners and leaders in certifying their operations for sustainability.

INTERNATIONAL ENVIRONMENTAL EDUCATION

Growing international awareness of the environment has focused worldwide attention on the degradation of natural resources and on the notion that some resources, such as biodiversity, the atmosphere, rainforests, and marine ecosystems, are "world resources," regardless of where they occur geographically. Concurrently, there is renewed interest and government support for the creation of public environmental awareness programs both in developed and developing countries. Government bilateral assistance agencies, multilateral development banks, and international conservation NGOs are earmarking unprecedented financial and human resources to assist developing countries in their efforts to heighten public awareness about relationships between the environment and quality of human life. This type of awareness raising is called *environmental education,* or *interpretation.*

According to Ham (2002), international environmental education can be divided into two broad categories—formal and nonformal. *Formal* environmental education programs take place within primary, secondary, and postsecondary school systems. Their purpose is to prepare the next generation of adults to be conscientious and informed resource users, consumers, and stewards. Elementary school children often are considered an important audience because their environmental beliefs are less rigid and they are more impressionable than adults. Secondary and postsecondary students are also important, because they are on the verge of entering the work force as both professionals and technicians.

Environmental education programs for school children represent a *long-term* educational solution to environmental problems (Ham & Sutherland, 1991). These programs are an "investment" in the sense that their goal is to nurture an informed future citizenry.

Nonformal environmental education programs typically are aimed at local landowners, consumers of print and broadcast media, community leaders, clubs and organizations, outdoor sports people, educators, and other people who use natural resource lands regularly or who are influential in the community. Nonformal programs are often directed at *adult* audiences—people whose decisions and behaviors affect *today's* environment, not just tomorrow's. Often whole communities are targeted by nonformal environmental education programs, but sometimes key audiences are singled out. Important among these audiences are farmers, teachers, business people, community opinion leaders, elected officials, mass media specialists, police, clergy, and, in some areas, even the military (see Table 21-2 for typology of audiences).

TABLE 21-2

TYPOLOGY OF AUDIENCES FOR ENVIRONMENTAL EDUCATION IN DEVELOPING COUNTRIES

Audience	Primary Mechanisms of Communication	Sphere of Influence
Local people	Staged events or organized tours; daily contact with agency or project personnel	Direct daily impact on condition of natural resources and on adjacent lands, significant political potential
National tourists	Outings or visits to protected areas	Intermittent impact on natural resources; direct political influence; local and national economic impact
Influential groups and citizens	Staged events or organized tours	Direct influence on policy and public opinion
Foreign tourists	Visits to protected areas (private or organized groups)	Intermittent impact on protected resource; indirect influence on policy; local and national economic impact; influence on international conservation NGOs in their country of origin

From Ham et al. (1989)

A major emphasis in nonformal environmental education has been the development of interpretive programs based in parks and protected areas, as well as through Elder Hostel programs and cruise ship lectures. In developed countries, these efforts typically represent the continuation of a historic tradition of interpretation, and are usually carried out by trained specialists working for organizations with an established environmental education function. By contrast, most developing countries lack this tradition, as well as the financial resources and technically trained staff necessary to establish one. Increasingly, they look to developed countries like the United States, not only as sources of economic and technical assistance, but also for training in interpretation, environmental education, and guiding methods. In many developing nations, ecotourism features learning about a country's environment, facilitated by trained guides and interpreters (Weiler & Ham, 2002).

SUMMARY AND OUTLOOK FOR CURRENT COOPERATION

As the new century begins, forests and related resources around the world command our attention. The importance of maintaining biodiversity and of fostering conscientious stewardship has brought new urgency to international aid and cooperation. The centrality of forestry and its related disciplines to international environmental and economic issues should be a source of pride, and of a sense of obligation. In the long run, if life on Earth is to be sustained at a level approaching our expectations, the natural environment and its resources will need to fuel sustainable development, and development in turn will need to protect environment and resources.

With world population now at 6 billion (UNFPA), some serious attempts to help reduce population growth or else have it reduced by disease, famine, or war—seem unavoidable. Although this issue has been ignored or delayed for political and religious reasons, sustainability in any world system will be impossible without success in this endeavor.

Unfortunately, the northern, mostly developed, countries arguing for better environmental protection and management find themselves pitted against the southern and less-developed tropical countries, which seek financial aid and technological assistance for economic development. This north–south debate is played out repeatedly in meetings sponsored by international organizations. The two sides are only beginning to realize that they are bound together by global ecosystems and that the quality of life in one hemisphere is inextricably linked to the quality of life in another. True progress in sustaining the world's forests will be made when the various countries and international organizations stop the north–south tug-of-war and begin to pull together to manage the global population increase, as well as environmental and economic systems for mutual benefit. Perhaps "mutual survival" would better express the importance of seriously addressing these issues.

LITERATURE CITED

Gregersen, H., J. Haygreen, S. Sindelar, and P. Jakes. 1989. "U.S. gains from foreign forestry research." *J. Forestry* 87(2):21–26.

Ham, Sam H. 2002. *Interpretation: A practical Guide for People with Big Ideas & Small Budgets.* North American Press/Fulcrum Publishing, Golden, Colo.

Ham, Sam H., and Dave S. Sutherland. 1991. Environmental Education and Extension on Fragile Lands. Chapter 11 in Young, P. (ed.). *Management of Fragile Lands in Latin America and the Caribbean: A Synthesis.* Development Alternatives, Inc., Bethesda, Md.

Ham, Sam H., Dave S. Sutherland, and James R. Barborak. 1989. Role of Protected Areas in Environmental Education in Central America. *J. Interpretation* 13(5):1–7.

Harmon, Mark. 2001. Carbon Sequestrian in Forests: Addressing the Scale Question. *J. Forestry* 99(4):24–29.

Journal of Forestry. 2001 July.: Special Issue on International Forestry. 99(7).

Kirby, Alex. 2001. 15 is Enough. Summary in Grist Magazine—Straight to the Source, (grist & gristmagazine.com) from BBC News, Alex Kirby, 20 August, 2001. http://news.bbc.co.uk/hi/english/sci/tech/newsid_1496000/1496833.stm.

Laarman, Jan G. 2001. Development Aid for Forestry: Unsustainable and in Need of Reform *J. Forestry* 99(7): 40–45.

Murray, Brian C., Stephen P. Prisley, Richard A. Birdsey, and R. Neil Sampson. 2000. Carbon Sinks in the Kyoto Protocol: Potential Relevance for U.S. Forests. *J. Forestry* 98(9):6–13.

Organization for Economic Cooperation and Development OECD. 2000. DAC Tables from 2000 Development Cooperation Report. www.oecd.org//dac/htm/dacstats/htm Last accessed by staff May 2001.

Persson, R. 2000. Assistance to Forestry: What Have We Learned? *International Forestry Review* 2(3):218–224.

Rotter, Jonathan, and Kyle Danish. 2000. Forest Carbon and the Kyoto Protocol's Clean Development Mechanism. *J. Forestry* 98(5):38–47.

Ryan, Robert J., Jr. 1992. "The Road From Rio." *Renewable Resources Jour.* Winter, 1992.

Sampson, R. N., and D. Hair (eds.). 1992. Forests and global change. Vol. 1: Opportunities for increasing forest cover. American Forests, Washington, D.C.

Sanchez, Felipe G., and Robert Eaton. 2001. Sequestering Carbon and Improving Soils: Benefits of Mulching and Incorporating Forest Slash. *J. Forestry* 99(1):32–36.

Sedjo, R. A. 1989. "Forests to affect the greenhouse effect." *J. Forestry* 87(7):12–15.

Singh, K. D. 1993. The 1990 Tropical Forest Resources Assessment. *Unasylva* (44)174.

Singh, Jaidev J., and Rodney Jackson. 1999. Transfrontier Conservation Areas: Creating Opportunities for Conservation, Peace and the Snow Leopard in Central Asia. *Intl. J. Wilderness* 5(3):7–12.

Skog, K. E., and G. A. Nicholson. 1998. Carbon Cycling through Wood Products: The role of wood and paper products in carbon sequestration. *Forest Products Journal* 48(7–8):75–83.

Waggener, Thomas R. 1989. "International Trade in Forest and Related Products," in Paul V. Ellefson (ed.), *Forest Resource Economics and Policy Research.* Westview Press, Inc., Boulder, Colo., Chap. 10.

Weiler, Betty, and Sam H. Ham. 2002. Tour Guide Training: A Model for Sustainable Capacity Building in Developing Countries *J. Sustainable Tourism* 10(1). (In Press)

ADDITIONAL READINGS

Brooks, David J. 1993. U.S. Forests in a Global Context. General Technical Report RM-228. Rocky Mountain Forest and Range Experiment Station. Fort Collins, Colo.

Ellefson, Paul V. (ed.). 1989. *Forest Resource Economics and Policy Research.* Westview Press, Inc., Boulder, Colo.

Evans, Julian. 1992. *Plantation Forestry in the Tropics: Tree Planting for Industrial, Social, Environmental, and Agroforestry Purposes.* (2nd ed.), Clarendon Press, Oxford.

FAO, 1993. The Challenge of Sustainable Forestry Management: What Future for the World's Forests? FAO of the United Nations. Rome, Italy.

Journal of Forestry. 2000, September. Special Issue on Forests, Carbon and Global Climate Change. 98(9).

Laarman, Jan, and Roger A. Sedjo. 1991. *Global Forests: Issues for Six Billion People.* McGraw-Hill, Inc., New York.

MacNeill, J., P. Winsemius, and T. Yakushiji. 1991. *Beyond Interdependence: The Meshing of the World's Economy and the Earth's Ecology.* Oxford University Press, New York.

Panayotou, Theodore, and Peter S. Ashton. 1992. *Not by Timber Alone: Economics and Ecology for Sustaining Tropical Forests.* Island Press, Washington, D.C.

Sedjo, R. A., R. N. Sampson, and J. Wisniewski. 1997. Economics of Carbon Sequestration in Forestry. CRC Press, New York.

Sharma, N. P. (ed.). 1992. *Managing the World's Forests: Looking for Balance Between Conservation and Development.* Kendall/Hunt Publishing Company, Dubuque, Iowa.

World Commission on Environment and Development. 1987. *Our Common Future.* Oxford University Press, Oxford.

Westoby, J. 1989. *Introduction to World Forestry.* Basil Blackwell, Oxford.

Whitmore, T. C. 1990. *An Introduction to Tropical Rain Forests.* Clarendon Press, Oxford.

STUDY QUESTIONS

1. Give three reasons why the health and status of forests in other countries have become important to the United States.

2. Discuss the connection between naval power and forestry in the 1600s.

3. Use the Internet to locate a volunteer international project involving forestry or the environment. Describe the project and why it is important.

4. List the international concerns found under "Priority Issues," and discuss one of these in a paragraph. Bring in material from your own reading, from TV viewing, or from the Internet to add to your comments.

5. How can forests, wood products, and forest management practices help mitigate global warning?

6. What is the FAO and what does it do?

7. What role does the World Bank (and other multilateral lenders) play in international forestry concerns?

8. What international roles does the U.S. Forest Service play?

9. Of those groups listed in Table 21-2, which would you judge to be the most important to reach, and why?

10. What is the trend in international forestry assistance in the United States? What is your view about this and why?

Major Legislation and Government Policy for Natural Resources in the United States, by Time Periods

Note: In the seventh edition, recent legislation is referred to in chapters where it is relevant. Interested readers are urged to use the Internet to seek current information on specific laws. Two good sites to start with are: The U.S. Senate Committee on Agriculture, Nutrition and Forestry (www.Senate.gov/~agriculture/Legislation/legislation.html) and the Society of American Foresters Legislative review (www.safnet.org/policy/review.htm).

Prior to 1770		Colonial development.
1770–1890		Period of both public land acquisition and disposal through settlement of the country, timber exploitation, development of widespread transportation facilities, and other subsidized internal improvements. Little progress toward the establishment of forest resource reserves or forestry policy.
1775–1783		American Revolution.
1820–1870		Industrial Revolution; westward expansion.
	1841	**Preemption Act:** Heads of families with certain qualifications could purchase 160 acres (64.7 ha) of public land (cheaply) for their own settlement, use, and benefit. Preemptor was supposed to improve the land and erect a dwelling.
1861–1865		Civil War.
	1862	**Homestead Act:** Broader than the Act of 1841. Free patent after 5 years of residence and cultivation. Claimant could elect to commute or pay up and secure patent at any time 6 months after time of filing.
	1862	**Morrill Act:** Several grants of public land for establishment of colleges of agriculture and mechanical arts. A second act in 1890 provided cash grants from public land sales. An amendment in 1903 provided that, in case these amounts were less than the amount appropriated, they could be paid from the Treasury.
	1872	**Yellowstone National Park Act:** Established Yellowstone National Park as "a public pleasuring ground" and ordered preservation of resources and wonders in natural condition.
	1873	**Timber Culture Act:** Offered to donate 160 acres (64.7 ha) of public land to any person who would plant 40 acres (16 ha) to trees and keep them growing for a period of 10 years. In 1878, an act reduced area and increased number of trees per acre to be planted. Repealed by the Forest Reserve Act in 1891.

1887		**Hatch Act:** Provided for financial assistance to states for agricultural experiment stations, including forestry research activities.
1891–1910		Forest Reserves established; later called National Forests; management policies of public forests created; state forestry programs initiated.
	1891	**Forest Reserve Act:** Authorized president to reserve any public timber or forest growth lands and proclaim them as public reservations. Also repealed Preemption Act of 1841 and the Timber Culture Act of 1873 and halted sales of public land.
	1897	**Forest Reserve Organic Act:** Provided for administration and use of existing and new forest reserves.
	1900	**The Lacey Act:** Outlawed interstate export or import of wildlife harvested or possessed against the laws of the state.
	1905	**Transfer Act:** Jurisdiction of the forest reserves transferred from Department of the Interior to Department of Agriculture.
	1906	**Antiquities Act:** Empowered the president to set aside, as national monuments, federally controlled areas containing prehistoric or historic structures, historic landmarks, or other objects of historic or scientific interest.
	1907	**Name Change Act:** Forest Reserves to National Forests.
	1908	**White House Conference.** Promoted national awareness of conservation.
1911–1932		National forests consolidated, new public forests established, and state forestry programs greatly improved.
	1911	**Weeks Act:** Promoted cooperative fire protection with states, provided for federal land acquisition on headwaters of streams and for land exchanges.
1914–1918		World War I.
	1914	**Smith-Lever Act:** Provided for cooperative agricultural extension work between U.S. Department of Agriculture and the land-grant colleges.
	1916	**National Park Service Act:** Created National Park Service in Department of the Interior, defined purposes for which national parks may be established, and authorized secretary of the interior to make such rules and regulations as he may deem necessary for their proper use and management.
	1918	**Migratory Bird Treaty Act:** Prohibited hunting or injury to wild birds moving among the United States, Canada, and Mexico.
	1922	**General Exchange Act:** When public interest will benefit, United States lands from the public domain in national forests may be exchanged for other lands if national forest lands are of equal or greater value—or an equal value of timber can be given in exchange.
	1924	**Clarke-McNary Act:** Authorized up to $20 million annually for cooperative fire control (authorization amended in May 1972 and increased to $40 million annually), authorized a study of tax laws, distributed tree seeds or planting stock, and provided for land acquisition of watershed lands under certain conditions.
	1928	**McSweeney-McNary Research Act:** Authorized and appropriated for investigations, experiments, and tests in order to promulgate, demonstrate, and determine the best methods of reforestation and of growing, managing, and utilizing timber forage and other forest products, maintaining watersheds, forest protection, and determining underlying economic considerations in forestry. Appropriated money and directed that a comprehensive forest survey be made and kept current with federal, state, and private cooperation.
	1929	**Migratory Bird Conservation Act:** Authorized purchase of new areas for waterfowl refuges.
1933–1942		Improved private forestry practices, established multiple-use forestry principle, and recognized forestry employment possibilities.

1933   **Civilian Conservation Corps (CCC) program:** Attacked the problem of unemployment. CCC enrollees were used in conservation work.

1933   **Lumber Code:** Broad federal legislation to invite and assist various industries to adopt codes of fair competition as a means of improving business. Committed forest industries to leave their lands in productive condition after logging, providing certain public cooperation was extended. Declared unconstitutional in 1934.

1934   **Taylor Grazing Act:** Authorized secretary of the interior to establish grazing districts on certain public lands, to dispose of isolated tracts, and to use 25 percent of receipts for certain range improvements and for other purposes. There were amendments to the 1934 act in 1936 to 1938 and in 1939 and later.

1934   **Migratory Bird Hunting Stamp Act:** Requires hunter over age 16 to buy a $1 federal waterfowl stamp before hunting migratory waterfowl.

1935   **Soil Conservation Service:** October 1933 Soil Erosion Service established in Department of the Interior. Transferred to Department of Agriculture in March 1935 by Executive Order. Soil Conservation Act of April 27, 1935, recognized need for cooperative soil conservation and directed secretary of agriculture to proceed on all phases of soil conservation.

1937   **Norris-Doxey Farm Forestry Act:** Authorized farm forestry in cooperation with the states.

1937   **Pittman-Robertson Act:** Provided federal aid for wildlife from excise taxes on sporting arms and ammunition. Funds used for wildlife research and development of game refuges and public hunting areas.

1941–1945   World War II.

1942–1964   Forestry research intensified, industrial forestry practices greatly improved, multiple-use and sustained-yield concept intensified, and woodland forestry practices increased.

1946   **Bureau of Land Management established:** Consolidated General Land Office and Grazing Service (Taylor Grazing Act lands) to form the Bureau of Land Management in the Department of the Interior.

1947   **Forest Pest Control Act:** Established government policy to protect all lands, regardless of ownership, from destructive forest insect pests and diseases. Authorized secretary of agriculture to cooperate with other federal, state, and local agencies and private concerns and individuals in detecting and controlling outbreaks.

1950   **Cooperative Forest Management Act:** Repealed Norris-Doxey Act. Appropriated $2,500,000 to enable secretary of agriculture to cooperate with state foresters in providing technical services to private landowners and processors. Raised to $5 million and then to $20 million in 1972. (See Cooperative Forestry Assistance Act of 1978.)

1950   **Dingell-Johnson Act:** Provided federal aid for fisheries by matching state excise tax on fishing rods and reels, lures, and other sport fishing equipment. Funds used for fish restoration and management projects.

1956   Bureau of Sports Fisheries and Wildlife and Bureau of Commercial Fisheries reorganized into U.S. Fish and Wildlife Service.

1958   **Outdoor Recreation Resources Review Act:** Established a commission to study recreation opportunities and to recommend policies and programs. Called ORRRC report.

1960   **Multiple-Use, Sustained-Yield Act:** Directed secretary of agriculture to develop and administer the renewable resources of the national forests for multiple use and sustained yield. Congress directed that the national forests should be managed for outdoor recreation, range, timber, watershed, and wildlife and fish purposes. Supplements Act of June 4, 1897, Organic Act. Defines multiple use and sustained yield.

1962   **Food and Agriculture Act:** Amended previous farm bills. Among other things, authorized payments for retiring cropland and continuing conservation reserve contracts under Soil Bank Act for conserving and developing soil, water, forest, wildlife, and recreation resources.

1962   **McIntire-Stennis Cooperation Research Act:** Authorized secretary of agriculture to cooperate with the states in carrying out a coordinated forestry, range, and related research program. Act specifies how financial and other assistance will be given. Provides research funds to forestry departments, schools, and colleges in land grant universities.

1963   **Bureau of Outdoor Recreation Act:** Organic act for BOR establishment, Department of Interior, April 2, 1962. Authorized several cooperative and coordinating functions that had been recommended by the Outdoor Recreation Resources Review Commission in its 1961 report. Renamed in 1978 the Heritage Conservation and Recreation Service. In 1981, its activities were transferred to the National Park Service.

1964–present   Public participation in policy management decisions of public domain, reexamined past legislation and management programs, and shifted emphasis toward environmental concerns including aesthetics. Wilderness designation an important issue.

1964   **Land and Water Conservation Act:** Provided funds for expanded federal land acquisition and for grants to states for planning, acquisition, and development of recreation areas under BOR leadership. Funding sources from surplus property sales, motorboat fuel taxes, and admission fees. Amended in 1968 to include appropriations from general revenues, and up to $200 million annually from outer continental shelf oil and gas leasing.

1964   **Wilderness Act:** Established the National Wilderness Preservation System. Directed secretaries of the interior and of agriculture to review additional lands within their administration within 10 years and to recommend to Congress new wilderness areas for inclusion into the system.

1965   **Clean Air Act:** An act to improve, strengthen, and accelerate programs for the prevention and abatement of air pollution.

1965   **White House Conference on Natural Beauty.**

1965   **Water Quality Act** of 1965 with 1972 Federal Water Pollution Control Act: Set standards for water quality to be implemented largely by the states with federal guidelines and backup enforcement authority.

1966   **National Wildlife Refuge System Act:** Provided for the conservation, protection, and propagation of native species of fish and wildlife, including migratory birds, that are threatened with extinction; to consolidate authorities; and for other purposes.

1968   **Wild and Scenic Rivers Act:** Declared a national policy to preserve certain rivers in their free-flowing condition.

1968   **North Cascades Complex Act:** Established the North Cascades National Park and Ross Lake and Lake Chelan National Recreation Areas.

1968   **Redwood National Park Act:** Preserved significant examples of coastal redwood, *Sequoia sempervirens.* 58,000-acre (23,472-ha) national park.

1968   **National Trails Act:** Declared Appalachian Trail in the east and Pacific Crest Trail in the west as initial components of the national system. Designated 14 other trails for study for possible addition to the system as national scenic trails. Also addressed trail opportunities in metropolitan and rural areas.

1969   **National Environmental Policy Act:** The purpose of the act was to declare a national policy that will encourage productive and enjoyable harmony between people and their environment, to promote efforts to eliminate or prevent damage to the environment, to promote understanding of the ecological systems and natural resources important to the nation, and to establish a council on environmental quality.

1970 **Environmental Quality Improvement Act:** Required all federal agencies to prepare an environmental impact statement (EIS) on major federal actions significantly affecting the quality of the human environment.

1970 **Environmental Protection Agency** was established with authority for monitoring water quality, air quality, pesticide, radiological, and solid waste standards.

1971 **RARE I (Roadless Area Review Evaluation)** by Forest Service to review remaining roadless areas in the National Forests and recommend those suitable and desirable for wilderness designation.

1971 **Wild and Free Roaming Horse and Burro Act:** Directed secretary of the interior to manage and protect such animals from capture, branding, harassment, or death, and to maintain for them specific sanctuaries on the public lands. Both BLM and Forest Service required to protect feral population on public lands.

1971 **Volunteers in the National Forest Act of 1972:** Provided for the recruitment and training of individuals to perform volunteer work supporting Forest Service programs and activities without compensation.

1971 **Alaska Native Claims Settlement:** Settled Indian and Eskimo claims resulting from Alaska purchase, and provided for federal and state public land selections.

1972 **Marine Mammal Protection Act:** Banned the killing and importing of whales and nearly all marine mammals. The moratorium can be waived after determining the current status of the species and the likely impact of proposed hunting.

1973 **Endangered Species Act:** Gave federal protection to species determined to be threatened with, or in danger of, extinction.

1973 **Monongahela controversy:** Federal District Court ruled that Forest Service harvesting practices violate 1897 organic act.

1973 **Agricultural and Consumer Protection Act:** Allowed for long-term contracts with landowners for installing conservation practices that will (1) improve fish, wildlife, and recreation resources; (2) enhance the level of management of nonindustrial private forest lands [forestry incentives contracts for owners of 500-acre (202 ha) or less by federal cost sharing for planting and timber-stand improvement]; and (3) develop long-term cover for wildlife.

1974 **Forest and Rangeland Renewable Resources Planning Act (RPA):** Called for long-range planning by the Forest Service to ensure that the United States has an adequate supply of forest resources in the future, while maintaining the quality of the environment. The two major requirements are that the Forest Service periodically submit to Congress a renewable resources assessment (every 10 years for all forest, range, and related lands in United States) and a long-range renewable resource program (updating every 5 years and always planning at least 45 years ahead).

1974 **Youth Conservation Act:** The YCC and the YACC made permanent organizations with a $60 million authorization. States can participate on a 50–50 basis. The Department of the Interior and the Department of Agriculture (Forest Service) are to administer the programs jointly.

1976 **Fish Conservation and Management Act:** Restricted foreign fishing in U.S. territorial waters. Established eight regional fishery management councils to determine which fisheries require conservation and management.

1976 **Federal Land Policy and Management Act:** Provided organic authority for Bureau of Land Management, as well as detailed management directions.

1976   **National Forest Management Act:** Amended the RPA (1974) planning process as well as requiring full public participation in the development and revision of land management plans for the National Forest System. Provides comprehensive new authorities for the management, sale, and harvesting of national forest timber; provides statutory protection for national forests created from the public domain; and provides direction for bidding on national forest timber, road building associated with timber harvesting, reforestation, salvage sales, and the handling of receipts from timber sales activities.

1977   **RARE II: Second Roadless Area Review** by Forest Service to correct perceived deficiencies in RARE I, and to speed up wilderness designation.

1977   **Soil and Water Resources Conservation Act (RCA):** Called for appraisal of all nonfederal land and water resources, analysis of program efficiency, and periodic recommendations to Congress by Soil Conservation Service.

1978   **Cooperative Forestry Assistance Act of 1978:** Authorized technical assistance, cost sharing, and resource protection programs on nonfederal forest lands, through cooperative agreements with state forestry agencies. Repeals and consolidates legislative authority for cooperative forestry assistance and forestry incentives enacted since 1924, specifically portions of Clark-McNary Act of 1924, Forest Pest Control Act of 1947, the Cooperative Forest Management Act of 1950, and sections of the Agriculture and Consumer Protection Act of 1973.

1978   **Forest and Rangeland Renewable Resources Research Act of 1978:** Authorized a forestry research program in resource management, environmental protection, forest products utilization, and resource assessment. Repeals, updates, and expands the authority for forestry research provided originally by the McSweeney-McNary Act of 1928.

1978   **Public Rangeland Improvement Act of 1978:** Amended the Federal Land Policy and Management Act of 1976 and the Wild Horse and Burros Protection Act of 1971 to provide additional direction and authorities for management of public rangelands. Establishes a statutory grazing fee formula in 16 western states for the years 1979–1985, excluding the National Grasslands.

1978   **Renewable Resources Extension Act of 1978:** Provided for an expanded and comprehensive extension services program for forest and rangeland renewable resources.

1978   **Heritage Conservation and Recreation Service (HCRS):** Created to identify, evaluate, and encourage protection of the nation's natural and historic resources and plan and fund recreation programs. Assumed most of the functions of the former Bureau of Outdoor Recreation, the office of Archaeology and Historic Preservation, and the Natural Landmarks program.

1978   **Endangered American Wilderness Act:** Designated 1.3 million acres (526,500 ha) in 10 western states as part of National Wilderness Preservation System and established by precedent some direction for future wilderness designations.

1978   **Comprehensive Employment Training Act (CETA) Amendments:** Provided funds for employees in recreation and other areas.

1978   **National Energy Conservation Policy Act:** Included studies of off-highway recreation vehicles and bicycle transportation facilities.

1979   **Archaeological Resources Protection Act:** Protected archaeological resources on public lands and Native American lands. Includes pottery, basketry, bottles, weapons, weapon projectiles, tools, structures, pit houses, rock paintings, rock carvings, intaglios, graves, human skeletal remains, or any piece of the foregoing items.

1980   **National Wild and Scenic River System:** Added 13 wild rivers to system.

1980   **Alaska National Interest Land Conservation Act (ANILCA):** Provided for the designation and conservation of certain public lands in the state of Alaska, including the designation of units of the National Park, National Wildlife Refuge, National Forest, National Wild and Scenic Rivers, and National Wilderness Preservation Systems, and for other purposes.

1980   **Wood Residue Utilization Act of December 19, 1980:** Purpose is to develop, demonstrate, and make available information on feasible methods that have potential for commercial applications to increase and improve utilization of all wood residues.

1981   **Heritage Conservation and Recreation Service (HCRS)** abolished. Most duties transferred to the National Park Service.

1982   **Designation of the Mount St. Helens National Volcanic Monument** in the state of Washington, to be administered by the USDA Forest Service.

1982   **National Wild and Scenic Rivers Act:** Set aside and protected special rivers.

1984   **Forest and Rangeland Renewable Resources Act:** Provided for an updated assessment of rangelands.

1985   **Presidential Commission on Outdoor Recreation Resource Review Commission:** Called for an update on the previous recreation findings of the 1958 ORRRC report.

1986   Congress initiated challenge grants to the USDA Forest Service and to Fish and Wildlife programs.

1986   **Emergency Wetlands Resource Act:** Promoted conservation of migratory waterfowl, including acquisition of wetlands and other essential habitat.

1990   **Americans With Disabilities Act:** Provided a clear and comprehensive mandate for the elimination of discrimination against individuals with disabilities in areas of employment, housing, public accommodations, transportation, communication, recreation, institutionalization, health services, voting, and access to public services.

1990   **Food, Agriculture, Conservation, and Trade Act:** Provided for wetland reserves, tree planting, watershed protection, and control of weeds and pests.

1990   **Amendment to the Clean Air Act:** Provided for attainment and maintenance of national ambient air quality standards.

1990   **Native American Graves Protection and Repatriation Act:** Provided for the protection of Native American graves.

1991   **Intermodal Surface Transportation Efficiency Act (ISTEA):** Provided for greater use of the national system of scenic byways.

1991   **Historic Preservation Act amendments:** Revised the relationship between the federal government and state preservation officers.

1992   **Energy Policy Act:** Prohibited the licensing by the Federal Regulatory Energy Commission of any new hydroelectrical power project located within units of the National Park System that would have adverse effect on federal lands within any such unit.

1997   **National Wildlife Refuge System Improvement Act:** Recognized hunting, fishing, wildlife viewing, environmental education, and interpretation as priority public uses.

2002   **Farm Security and Rural Investment Act of 2002:** Known as the "Farm Bill," this law includes a conservation title renewing and strengthening existing forestry and conservation cost-share programs for private landowners, and establishing new ones, including a new Forest Land Enhancement Program, and a new Sustainable Forestry Outreach Initiative. A strong statement of support for forestry and conservation on private lands, for the first time this Farm Bill includes mandatory funding levels for some cost-share programs, rather than leaving them to annual, congressional appropriations.

Metric Conversions

SPECIAL FORESTRY METRIC CONVERSIONS

Symbol	To convert	Multiply by	To find	Symbol
ft²/ac	Square feet per acre	0.2296	Square meters per hectare	m²/ha
ft³/ac	Cubic feet per acre	0.06997	Cubic meters per hectare	m³/ha
ft³/sec	Cubic feet per second	101.941	Cubic meters per hour	m³/h
ft/sec	Feet per second	1.097	Kilometers per hour	Km/h
gal/ac	Gallons per acre	11.2336	Liters per hectare	l/h
gal/min	Gallons per minute	0.0757	Liters per second	l/s
lb/ac	Pounds per acre	1.1208	Kilograms per hectare	Kg/ha
lb/ft³	Pounds per cubic feet	16.0185	Kilograms per cubic meter	Kg/m³
no./ac	Number (i.e., trees) per acre	2.471	Number per hectare	No/ha
ton/ac	Tons per acre	2.242	Tonnes per hectare	t/ha

METRIC CONVERSION FACTORS* (APPROXIMATE CONVERSIONS TO METRIC MEASURES)

Symbol	When you know	Multiply by	To find	Symbol
		Length		
in	inches	2.54	centimeters	cm
ft	feet	30.48	centimeters	cm
ft	feet	0.3048	meters	m
yd	yards	0.91	meters	m
ch	chains	20.1	meters	m
mi	miles	1.61	kilometers	km
		Area		
in^2	square inches	6.45	square centimeters	cm^2
ft^2	square feet	0.09	square meters	m^2
yd^2	square yards	0.83	square meters	m^2
ch^2	square chains	404.7	square meters	m^2
mi^2	square miles	2.58	square kilometers	km^2
ac	acres	0.405	hectares	ha
		Mass (weight)		
oz	ounces	28	grams	g
lb	pounds	0.45	kilograms	kg
	short tons (2000 lb)	0.9	tonnes	t
		Volume		
pt	pints	0.47	liters	l
qt	quarts	0.95	liters	l
gal	gallons	3.8	liters	l
ft^3	cubic feet	0.03	cubic meters	m^3
yd^3	cubic yards	0.76	cubic meters	m^3
		Temperature		
°F	Fahrenheit	5/9 (after subtracting 32)	Celsius	°C

*Courtesy National Bureau of Standards, U.S. Department of Commerce.

Common and Scientific Names of Trees Used in the Text

Common name	Scientific name
Alaska-cedar	*Chamaecyparis nootkatensis*
alder, red	*Alnus rubra*
ash, black	*Fraxinus nigra*
blue	*F. quadrangulata*
green	*F. pennsylvanica*
Oregon	*F. latifolia*
white	*F. americana*
aspen, bigtooth	*Populus grandidentata*
quaking	*P. tremuloides*
baldcypress	*Taxodium distichum*
basswood, American	*Tilia americana*
bayberry, southern (waxmyrtle)	*Myrica cerifera*
beech, American	*Fagus grandifolia*
birch, gray	*Betula populifolia*
paper (white)	*B. papyrifera*
river	*B. nigra*
sweet	*B. lenta*
western white	*B. papyrifera var. commutata*
yellow	*B. alleghaniensis*
blackgum (see also tupelo)	*Nyssa sylvatica*
buckeye, Ohio	*Aesculus glabra*
yellow	*A. octandra*
burningbush, eastern (wahoo)	*Euonymus atropurpureus*
butternut	*Juglans cinerea*
California-laurel	*Umbellularia californica*
cascara	*Rhamnus purshiana*
catalpa, northern	*Catalpa speciosa*
cherry, black	*Prunus serotina*
chestnut, American	*Castanea dentata*
chinkapin, golden	*Castanopsis chrysophylla*
cottonwood, black	*Populus trichocarpa*
eastern	*P. deltoides*
cucumbertree	*Magnolia acuminata*
dogwood, eastern	*Cornus florida*
Douglas-fir	*Pseudotsuga menziesii*

Common name	Scientific name
elm, American	*Ulmus americana*
slippery	*U. rubra*
eucalyptus	*Eucalyptus* spp.
fir, balsam	*Abies balsamea*
California red	*A. magnifica*
grand	*A. grandis*
noble	*A. procera*
Pacific silver	*A. amabilis*
Shasta red	*A. magnifica* var. *shastensis*
subalpine	*A. lasiocarpa*
white	*A. concolor*
hackberry	*Celtis occidentalis*
hemlock, eastern	*Tsuga canadensis*
mountain	*T. mertensiana*
western	*T. heterophylla*
hickory, bitternut	*Carya cordiformis*
mockernut	*C. tomentosa*
shagbark	*C. ovata*
water	*C. aquatica*
honeylocust	*Gleditsia triacanthos*
hophornbeam (ironwood)	*Ostrya virginiana*
incense-cedar	*Libocedrus decurrens*
juniper, Rocky Mountain	*Juniperus scopulorum*
kiawe	*Prosopis chilensis*
koa	*Acacia koa*
larch, eastern (see also tamarack)	*Larix laricina*
western	*L. occidentalis*
locust, black	*Robinia pseudoacacia*
macadamia	*Macadamia ternifolia*
madrone, Pacific	*Arbutus menziesii*
magnolia, southern	*Magnolia grandiflora*
mamani	*Sophora chrysophylla*
mangrove	*Rhizophora mangle*
maple, bigleaf	*Acer macrophyllum*
black	*A. nigrum*
red	*A. rubrum*
silver	*A. saccharinum*
sugar	*A. saccharum*
mulberry, red	*Morus rubra*
oak, black	*Quercus velutina*
bur	*Q. macrocarpa*
California black	*Q. kelloggii*
California live	*Q. agrifolia*
cherrybark	*Q. falcata* var. *pagodaefolia*
chestnut	*Q. prinus*
chinkapin	*Q. muehlenbergii*
live	*Q. virginiana*
northern red	*Q. rubra*
Nuttall	*Q. nuttallii*
Oregon white (garry)	*Q. garryana*
overcup	*Q. lyrata*
post	*Q. stellata*
scarlet	*Q. coccinea*
southern red	*Q. falcata*
swamp chestnut	*Q. michauxii*
swamp white	*Q. bicolor*
water	*Q. nigra*
white	*Q. alba*
willow	*Q. phellos*

Common name	Scientific name
osage-orange	*Maclura pomifera*
pecan	*Carya illinoensis*
persimmon, common	*Diospyros virginiana*
pine, bristlecone	*Pinus aristata*
eastern white	*P. strobus*
jack	*P. banksiana*
jeffrey	*P. jeffreyi*
limber	*P. flexilis*
loblolly	*P. taeda*
lodgepole	*P. contorta*
longleaf	*P. palustris*
Monterey	*P. radiata*
ocote	*P. teocote*
pitch	*P. rigida*
pond	*P. serotina*
ponderosa	*P. ponderosa*
red	*P. resinosa*
shortleaf	*P. echinata*
slash	*P. elliottii*
sugar	*P. lambertiana*
Virginia	*P. virginiana*
western white	*P. monticola*
whitebark	*P. albicaulis*
pinyon	*P. edulis*
singleleaf	*P. monophylla*
Mexican	*P. cembroides*
planertree (waterelm)	*Planera aquatica*
planetree	*Platanus occidentalis*
poplar, balsam	*Populus balsamifera*
swamp	*P. heterophylla*
Port-Orford-cedar	*Chamaecyparis lawsoniana*
redcedar, eastern	*Juniperus virginiana*
western	*Thuja plicata*
redwood	*Sequoia sempervirens*
sassafras	*Sassafras albidum*
sequoia, giant	*Sequoia gigantea*
silverbell	*Halesia* spp.
spruce, black	*Picea mariana*
blue	*P. pungens*
Engelmann	*P. engelmannii*
red	*P. rubens*
Sitka	*P. sitchensis*
white	*P. glauca*
sugarberry	*Celtis laevigata*
sweetgum	*Liquidambar styraciflua*
sycamore, American (planetree)	*Platanus occidentalis*
tamarack	*Larix laricina*
tanoak	*Lithocarpus densiflorus*
tupelo, black (blackgum)	*Nyssa sylvatica*
swamp	*N. s.* var. *biflora*
water	*N. aquatica*
walnut, black	*Juglans nigra*
white-cedar, Atlantic,	*Chamaecyparis thyoides*
northern	*Thuja occidentalis*
willow, black	*Salix nigra*
yellow-poplar (tuliptree)	*Liriodendron tulipifera*
yew, Pacific	*Taxus brevifolia*

From Elbert L. Little, 1979. "Checklist of United States Trees (Native and Naturalized)," Agriculture Handbook no. 541, U.S. Department of Agriculture Forest Service, Washington, D.C.

Common and Scientific Names of Mammals and Birds Used in the Text

Mammals	Scientific name
antelope, pronghorn	*Antilocapra americanus*
bat, little brown	*Myotis lucifugus*
bat, red	*Lasiurus borealis*
bear, black	*Ursus americanus*
bear, grizzly	*Ursus arctos*
beaver	*Castor canadensis*
buffalo (bison)	*Bison bison*
burro	*Equus asinus*
caribou, barren ground	*Rangifer caribou*
cattle, domestic	*Bos taurus*
cougar	*Felis concolor*
coyote	*Canus latrans*
deer, axis	*Axis axis*
deer, black-tailed	*Odocoileus hemionus columbianus*
deer, black-tailed Sitka	*Odocoileus hemionus sitkensis*
deer, mule	*Odocoileus hemionus*
deer, white-tailed	*Odocoileus virginianus*
elk, Rocky Mountain	*Cervus elaphus*
ermine	*Mustela erminea*
ferret, black-footed	*Mustela nigripes*
fisher	*Martes pennanti*
fox, red	*Vulpes fulva*
goat, mountain	*Oreamnos americanus*
martin, pine	*Martes martes*
mice, deer	*Peromyscus maniculatus*
mink	*Mustela vison*
mongoose	*Herpestes auropanetatus*
moose	*Alces alces*
muskrat	*Ondatra zibethicus*
nutria	*Myocaster coypus*
otter, river	*Lutra canadensis*
otter, sea	*Enhydra lutris*
porcupine	*Erethezon dorsatum*
rabbit, cottontail	*Sylvilagus* spp.

Mammal	Scientific name
raccoon	*Procyon lotor*
rat, Norway	*Rattus norvegicus*
sheep, barbary	*Ammotragus lervia*
sheep, bighorn	*Ovis canadensis*
sheep, desert bighorn	*Ovis canadensis nelsoni*
sheep, domestic	*Ovis musimon*
skunk, striped	*Mephitis mephitis*
squirrel, fox	*Sciurus niger*
squirrel, northern flying	*Glaucomys sabrinus*
vole, tree	*Arboremus longicaudus*
wolf, gray	*Canus lupis*
woodchuck (ground hog)	*Marmota monax*

Birds	Scientific name
chukar partridge	*Alectoris chukar*
coot, American	*Fulica americana*
condor, California	*Gymnogyps californianus*
cowbird	*Molothrus ater*
crow	*Corvus brachyrhychos*
eagle, bald	*Haliacetus leucocephalus*
falcon, peregrine	*Falco peregrinus*
goose, Canada	*Branta canadensis*
goose, snow	*Chen caerulescens*
grouse, ruffed	*Bonasa umbellus*
gull, herring	*Larus argentatus*
loon	*Gavia* spp.
murrelet, marbled	*Brachyramphus marmoratus*
owl, northern spotted	*Strix occidentalis*
pelican, white	*Pelecanus erythrorhynchos*
pheasant, ring-necked	*Phasianus colchicus*
ptarmigan, willow	*Lagopus lagopus*
quail, Mearns	*Cyrtonyx montezumae*
sparrow, English	*Passer domesticus*
starling, European	*Sturnus vulgaris*
tern, arctic	*Sterna paradisaea*
turkey, wild	*Meleagris gallopavo*
warbler, Kirtland's	*Dendroica kirtlandii*
woodcock, American	*Philohela minor*
woodpecker, red-cockaded	*Dendrocopos borealis*

Common and Scientific Names of Range Vegetation

Forbs	Scientific name
bur clover	*Medicago hispida*
filaree	*Erodium cicutarium*
knapweed	*Centaurea* spp.
leafy spurge	*Euphorbia esula*
Russian thistle	*Salsola kali*
starthistle	*Centaurea solstitialis*
tall larkspur	*Delphinium occidentale*

Grasses	Scientific name
big bluestem	*Andropogon gerardi*
black grama	*Bouteloua eriopoda*
bluebunch wheatgrass	*Agropyron spicatum*
blue grama	*Bouteloua gracilis*
bluejoint reedgrass	*Calamagrostis canadensis*
buffalo grass	*Buchloe dactyloides*
cheatgrass	*Bromus tectorum*
cottongrass	*Eriophorum* spp.
crested wheatgrass	*Agrophyron cristatum*
Idaho fescue	*Festuca idahoensis*
Indian grass	*Sorghastrom nutrans*
inland saltgrass	*Distichlis spicata*
intermediate wheatgrass	*Agropyron intermedium*
little bluestem	*Schizachyrium scoparium*
medusa head	*Taeniatherum asperum*
pineland threeawn	*Aristada stricta*
porcupine grass	*Stipa spartea*
Russian wildrye	*Elymus junceus*
squirreltail	*Sitanion hystrix*
soft chess	*Bromus mollis*
switchgrass	*Panicum virgatum*
tobosa grass	*Hilaria mutica*
wild oats	*Avena fatua*

Woody	Scientific name
alder	*Alnus* spp.
alligator juniper	*Juniperus deppeana*
American elm	*Ulmus americana*
big sagebrush	*Artemisia tridentata*
basin	spp. *tridentata*
mountain	spp. *vaseyana*
Wyoming	spp. *wyomingensis*
broom snakeweed	*Gutierrezia sarothrae*
ceanothus	*Ceanothus* spp.
chamise	*Adenostoma fasciculatum*
chinquapin oak	*Quercus muhlenbergia*
cranberry	*Vaccinium oxycoccos*
creosote bush	*Larrea tridentata*
dwarf willow	*Salix arctica*
eastern redbud	*Cercis canadensis*
Joshua tree	*Yucca brevifolia*
low sagebrush	*Artemisia arbuscula*
manzanita	*Arctostaphylos patula*
mesquite	*Prosopis juliflora*
ocotillo	*Fouquieria splendens*
palo verde	*Cercidium floridium*
plains cottonwood	*Populus deltoides*
Rocky Mountain juniper	*Juniperus monosperma*
saguaro	*Carnegiea gigantea*
salmonberry	*Rhus spectabilis*
shadscale	*Atriplex confertifolia*
silver sagebrush	*Artemisia cana*
single-leaf pinyon	*Pinus monophylla*
true pinyon	*Pinus edulis*
Utah juniper	*Juniperus osteosperma*
western juniper	*Juniperus occidentalis*
willow	*Salix* spp.
winterfat	*Ceratoides lanata*

Professional Societies for Natural Resource Personnel

Organization Address/Website	Membership	Publications
American Fisheries Society 5410 Grosvenor Lane #110 Bethesda, MD 20814-2199 (301) 897-8016 www.fisheries.org	9,000 4 geographic divisions, sections and chapters	*Transactions Am. Fish. Soc. Progressive Fish Culturist* *N. Am. Jour. Fisheries Mgt.* *Jour. Aquatic Animal Health* *Fisheries* *N. Am. Jour. Aquaculture*
American Water Resources Assoc. 4 West Federal Street P.O. Box 1626 Middleburg, VA 20118-1626 (540) 687-8390 www.awra.org	2,800 25 state chapters	*JAWRA* (bi-monthly) *Water Resources Impact* (bi-monthly)
Soil and Water Conservation Society 7515 NE Ankeny Road Ankeny, IA 50021 (515) 284-1227 www.swcs.org	10,000 103 chapters	*Journal of Soil and Water Cons.* *Conservagram* (newsletter)
Canadian Institute of Forestry 606 151 Slatter Street Ottawa, Ontario KIP 5H3 Canada (613) 234-2242 www.cif-ifc.org	2,500 23 sections	*The Forestry Chronicle* 6/yr *The Forestry Update* 12/yr *The Forestry Dialogue* (to mbrs parliament)
The Wildlife Society 5410 Grosvenor Lane Bethesda, MD 20814 (301) 897-9770 www.wildlife.org	8,900 7 geographic sections 53 chapters 62 student chapters	*Jour. Wildlife Mgt.* (quarterly) *Wildlife Soc. Bulletin* (quarterly) *Wildlife Monographs* (irreg.) *Wildlife Policy News* (bi-monthly newsletter)

Organization Address/Website	Membership	Publications
Society of American Foresters 5400 Grosvenor Lane Bethesda, MD 20814 (301) 897-8720 www.safnet.org	18,000 33 state societies 264 chapters 28 subject matter workgroups	*Journal Forestry* (monthly) *Forest Science* (quarterly) *Northern Jour. of Applied Forestry* (quarterly) *Southern Jour. of Applied Forestry* (quarterly) *Western Jour. of Applied Forestry* (quarterly) *The Forestry Source* *Annual Convention* proceedings
Society for Range Management 445 Union Blvd., Ste. 230 Lakewood, CO 80228 (303) 986-3309 www.srm.org	5,350 21 sections	*Jour. Range Mgt.* (monthly) *Rangelands* (monthly) *Range Monographs* (irreg.)
The National Association for Interpretation P.O. Box 2246 Fort Collins, CO 80522 (970) 484-8283 www.interpnet.com	3,600 10 regions 10 subject matter networks	*Legacy* 6/yr *Jour. Interpretation Res.*
The Natural Areas Association P.O. Box 1504 Bend, OR 97709		*Natural Areas Journal*
The International Association for Society & Natural Resources		*Society and Natural Resources*
The George Wright Society P.O. Box 65 Hancock, Michigan 49930-0065 www.georgewright.org/	Unknown	*George Wright Forum*

Urban and Community Forestry Professional and Educational Organizations

American Planning Association (APA)
122 S. Michigan Ave., Ste. 1600
Chicago, IL 60603
(312) 431-9100
www.planning.org

American Society of Consulting Arborists
15245 Shady Grove Rd., Ste. 130
Rockville, MD 20850
(301) 947-0483
asca-consultants.org

American Society of Landscape Architects (ASLA)
636 Eye St., NW
Washington, DC 20001-3736
(202) 898-2444
www.asla.org

Associated Landscape Contractors of America
150 Eldon St., Ste. 270
Herndon, VA 20170
(703) 736-9666
www.alca.org

Council of Tree & Landscape Appraisers
15245 Shady Grove Rd., Ste. 130
Rockville, MD 20850
(301) 947-0483

International Society of Arboriculture (ISA)
P.O. Box 3129
Champaign, IL 61826-3129
(217) 355-9411
www.isa-arbor.com

National Arborist Association (NAA)
3 Perimeter R., Unit 1
Manchester, NH 03103
(603) 314-5380
www.natlarb.com

National Association of State Foresters (NASF)
444 North Capitol Street, NW
Hall of States, Ste. 540
Washington, DC 20001
(202) 624-5415
www.stateforesters.org

National Association of Towns and Township
444 North Capitol Street, NW
Hall of States, Ste. 208
Washington, DC 20001-1202
(202) 624-3550
www.natat.org

National Institute of Municipal Law Officers
1000 Connecticut Ave., Ste. 902
Washington, DC 20036
(202) 466-5421

National League of Cities (NLC)
1301 Pennsylvania Ave., NW, Ste. 550
6th Floor
Washington, DC 20004
(202) 626-3010
www.nlc.org

National Recreation and Park Association
22377 Belmont Ridge Rd.
Ashburn, VA 20148-4501
(703) 858-0784
www.nrpa.org

Society of American Foresters (SAF)
5400 Grosvenor Lane
Bethesda, MD 20814
(301) 897-8720
www.safnet.org

Society of Municipal Arborists
P.O. Box 11521
St. Louis, MO 63105
(314) 862-3325
www.urban-forestry.com

American Forests
P.O. Box 2000
Washington, DC 20013
(202) 955-4500
www.americanforests.org

The National Arbor Day Foundation
100 Arbor Avenue
Nebraska City, NE 68410
(402) 474-5655
www.arborday.org

Tree People
12601 Mulholland Dr.
Beverly Hills, CA 90210
(818) 753-4600
www.treepeople.org

National Environmental Organizations

Organization/Website	Members in 2000	Activity	Founded
National Wildlife Federation www.nwf.org	4,000,000	Conservation	1936
Greenpeace U.S.A. www.greenpeaceusa.org	250,000	Environment	1971
Humane Society of the United States, The www.hsus.org	7,000,000	Animal Welfare	1954
National Arbor Day Foundation www.arborday.org	1,000,000	Urban Forestry	1972
World Wildlife Fund www.worldwildlife.org	1,000,000	Wildlife	1961
Clean Water Action www.cleanwateraction.org	700,000	Clean Water	1971
National Audubon Society www.audubon.org	550,000	Conservation	1905
Sierra Club www.sierraclub.org	700,000	Conservation	1892
Nature Conservancy, The www.tnc.org	550,000	Conservation	1951
Ducks Unlimited www.ducks.org	700,000	Waterfowl Conservation	1937
The Wilderness Society www.wilderness.org	200,000	Conservation	1935
Fund For Animals www.fundforanimals.org	200,000	Animal Welfare	1967
National Parks and Conservation Association www.npca.org	400,000	Park Conservation	1919
Natural Resources Defense Council www.nrdc.org	50,000	Environmental	1970

Organization/Website	Members in 2000	Activity	Founded
Environmental Defense www.edf.org	300,000	Environmental	1967
Defenders of Wildlife www.defenders.org	230,000	Wildlife Conservation	1947
Rails-to-Trails Conservancy www.railtrails.org	100,000	Recreation/Conservation	1985
Trout Unlimited www.tu.org	125,000	Wildlife Conservation	1959
National Wild Turkey Federation www.nwtf.org	315,000	Wildlife Conservation	1973
Izaak Walton League of America www.iwla.org	50,000	Conservation	1922
Rocky Mountain Elk Foundation www.rmef.org	245,000	Wildlife Conservation	1984
Friends of the Earth www.foe.org	50,000	Environmental	1969
Save the Redwoods League www.savetheredwoods.org	40,000	Conservation	1918
Quail Unlimited www.qu.org	45,000	Wildlife Conservation	1981
White Tails Unlimited www.whitetailsunlimited.com	55,000	Wildlife Conservation	1982
American Forests www.americanforests.org	34,818	Conservation, Forestry	1875
National Woodland Owners Association www.nationalwoodlands.org	33,500	Private Forests	1983
Earth Island Institute www.earthisland.org	33,000	Environmental	1982
Rainforest Action Network www.ran.org	30,000	Environmental	1985
Ruffled Grouse Society	23,000	Wildlife Conservation	1961
Tree People www.treepeople.org	20,000	Conservation	1976
Adirondack Council www.adirondackcouncil.org	18,000	Environmental	1975
Audubon Naturalist Society of the Central Atlantic States www.audubonnaturalist.org	13,000	Conservation	1897
American Rivers www.amrivers.org	12,000	Conservation	1973
Earth First www.earthfirstjournal.org	10,000	Environmental	1979
Friends of the River www.friendsoftheriver.org	7,300	California Conservation	1973
Renewable Natural Resources Foundation www.rnrf.org	17 organizations	Conservation	1972
Natural Resources Council of America www.naturalresourcescouncil.org	85 organizations	Conservation	1946

Glossary

Abiotic diseases: Diseases caused by environmental conditions or factors such as atmospheric deposition and pollution, nutrient imbalance, adverse temperatures, lighting, soil compaction, and flooding.

Adaptation: Any genetically determined alteration in structure or function of an organism, resulting from natural selection, that betters the organism's ability to survive and reproduce.

Adaptive forestry: See *ecosystem management*.

Adventitious: A root or bud appearing in an abnormal position or place, and developing from previously undifferentiated cells.

Allowable cut: The yield of timber that can be harvested annually from an area under a given plan, usually on a continuing basis; that is, without depleting the area as a whole. The concept implies that growth will balance the removals.

Angiosperms: Plants that have seeds enclosed in an ovary; includes the group of trees generally broadleaved and deciduous.

Animal unit month (AUM): The amount of forage needed to sustain one 1,000-lb. cow plus calf (an animal unit), or their equivalents, such as five to seven sheep, for 1 month.

Annual ring: The ring formed in the xylem each year by the contrast of the spring wood and the summer wood, seen in trunk cross sections in both the heartwood and the sapwood.

Autecology: The study of individual organisms in relation to their environment.

Autotrophs: Organisms capable of self nourishment; for example, through photosynthesis.

Biodiversity: Refers to the variations in plants and animals, their genetic makeup and biological processes, and the ecological nitches they occupy in structurally diverse forest landscapes.

Biological controls: Controlling organisms through favoring natural enemies of these undesirable species, or through inhibiting the target organism's reproductive and growth processes by use of fungi, bacteria, or viruses.

Biomass: The total organic material by volume or weight on a given site; in forestry usually only plant materials, and sometimes only certain species are considered.

Biotechnology: Applying the biological sciences to a problem; includes studies in gene splicing, micropropagation, tissue culture, and genetic engineering.

Biotic diseases: Diseases caused by pathogens.

Board feet: A unit of measure representing a board one foot square and one inch thick.

Boles: The trunks of trees.

Boreal forest: The northern coniferous forest; also called the *taiga*.

Broadcast burn: A controlled fire over a designated area to reduce fuel hazard or accomplish some other management objective.

Browse: Tender shoots and twigs of shrubs and trees used as food by animals; to eat on these.

Browse line: The uniformly trimmed line of tree branches above which animals have not been able to reach while browsing.

Bucking: Cutting trunks of trees into specified lengths after felling.

515

Cambium: The very thin growing layer of tissue between the bark and wood in woody plants.

Canopy: The roof of the forest formed by the crowns of the dominant trees and other vegetation.

Carrying capacity: The number of animals an area can adequately support without deterioration of the site. Also used to describe the number of recreationists who can use one site without deteriorating the physical or social conditions (social carrying capacity).

Chain: A unit of length equal to 66 feet (20.12 meters).

Chipper: A machine used for reducing logs and branches to chips.

Clearcutting: Removing most or all of the existing stand of trees, thus producing an environment for reproduction that is not influenced by the previous stand.

Climax forest: A stand in which a certain equilibrium appears to exist, in that the death of individual overstory trees allows replacement by younger individuals of the same species. Current thinking discounts this concept, emphasizing dynamic conditions rather than equilibrium.

Cold deck: Stacked logs awaiting transport to a mill.

Commerical forest (timberland): Forest lands that will grow at least 20 cubic feet per acre per year (1.4 cu meters per hectare).

Computer simulation: Use of a computer to simulate real-world conditions, such as a forest stand and how it would react to various treatments.

Conks: The sporophores or fruiting bodies of heart or root rot fungi; also known as wood mushrooms; appear on the trunks of infected trees.

Conservation: Husbanding and wise use of renewable natural resources such as forests, fisheries, and wildlife.

Continuous forest inventory (CFI): A means of obtaining projections of long-range growth and mortality of a forest through a system of permanent sample plots; data from these plots are collected periodically.

Cover type: The species or mix of tree species growing in a particular area.

Crown: The topmost portion of trees formed by leaves and branches.

Crown fire: A fast-spreading wildfire that advances from top to top of trees or shrubs, more or less independently of the surface (ground) fire.

Cruising timber: Examining forestland, usually by sampling, to determine species and estimate volumes and grades of standing timber.

Cubic foot: A unit for measuring wood equivalent to a cube with 12-inch sides.

Cultivar: A plant that originated and continues to grow only under cultivation.

DBH: Diameter at breast height; 4 1/2 ft (137 cm) above ground.

DDT: A powerful persisting insecticide whose use is now restricted by law due to damaging environmental effects.

Delimber: An excavator-based machine with an attachment that removes a tree's limbs and bucks the top off the tree at the landing. A log loader then places the processed logs on a truck.

Dendrology: The classification and identification of trees.

Diatom: Microscopic algae that form a large part of the plankton of both fresh water and salt water.

Ecology: The science dealing with relationships between plants and animals and their environments; forest ecology deals with how forest ecosystems interact and grow.

Ecosystem: A network of interacting organisms and their habitats.

Ecosystem management: An approach to forest management that, instead of concentrating on the forest stand, focuses on the forest landscapes and the integration of vegetation, human, wildlife, physical and biological sciences in natural resource management. Sometimes called *new forestry* or *adaptive forestry,* although others would argue these latter terms (approaches) are distinct.

Ecotone: A transitional zone between two plant communities, containing species characteristic of both communities.

Edge: The edge of the forest, whether natural or resulting from cutting. Because of the variety of cover and food sources, edge has been considered a productive area for wildlife.

Endemic stage: The presence, in normal (low) numbers, of a population of organisms in a given environment.

Entomologist: A professional or specialist working in that branch of zoology dealing with insects.

Environmental impact statement (EIS): A systematic, interdisciplinary report showing the social and physical impacts, including adverse effects, that proposed actions will have on the environment and offering alternatives for the proposed action. Required under the National Environmental Policy Act (NEPA) for proposed actions in which significant impacts are anticipated.

Epidemic stage: The presence, in abnormally high numbers, of a population of organisms in a given environment.

Erosion (geologic): The slow weathering of the earth's surface and the accompanying removal of material. (accelerated) A rapid movement of soil, and perhaps even rock and vegetation, caused by excess surface runoff.

Even-aged stand: A forest stand in which, due to previous fire or clearcutting, for instance, the majority of the trees have grown up essentially of the same age.

Exotic forest pests: The forest pests not native to a region.

Extensive forest management: Application of minimum management practices (protection from fire and little more) over a broad area of forest holdings (compare with *intensive forest management*).

Externalities: A term used in resource economics to cover such intangible resource values as recreation and wildlife; may also represent negative values, such as pollution. Values that are not fully registered in the marketplace.

Fee-simple: Ownership of real property with no restrictions attached. The clear title to land or property.

Feller-buncher: A machine that fells trees using a mechanical shear or a disc saw as an attachment. A feller-buncher can accumulate several trees before creating just the right size bunch for a grapple skider to take to the landing. A mechanical delimber or a whole tree chipper might be waiting at the landing to further process the tree.

Feral: Returned to a wild state from domestication (horses, goats, dogs, cats, hogs).

Fire triangle: Represents the three factors—heat, oxygen, and fuel—that are required in combination for a fire to burn.

FLIPS: Forest-level information-processing system; a Forest Service network of interconnected minicomputers to facilitate nationwide information transfer. Also called the "DG" for the company (Data General) that provided the system.

Forest: A community of trees, shrubs, herbs, and associated plants and organisms covering a considerable area, utilizing oxygen, water, and soil nutrients as it attains maturity and reproduces itself.

Forest development patterns: General trends in natural plant invasion, replacement, and dominance in a bare, or a disturbed, area.

Forest pests: Insects and related organisms and pathogens that damage trees and have the potential to be detrimental to achievement of resource management objectives.

Forest recreation: See *wildland recreation.*

Forestry: The science, the art, and the practice of managing and using for human benefits the natural resources that occur on, and in association with, forestlands.

Forwarder: A machine with a crane that can load logs onto its chasis and piggy-back them to a road where it can sort and pile them or load them directly onto a truck.

Fusee: A large-headed friction match, able to burn in the wind. Used in the ignition of prescribed fires such as burning-out and backfiring.

Geomorphology: The various land forms sculptured by surface forces.

Girdling: Cutting through a tree's bark, all around the tree, deeply enough to interrupt the flow of food to the roots and cause the death of the tree.

Ground fire: A fire that burns along the ground consuming forest litter and natural slash, but sometimes humus, peat, or other highly organic soil as well.

Gymnosperms: A large group of plants, including trees, producing seeds not enclosed in an ovary; in the instance of trees, generally cone-bearing evergreens.

Gyppo: A small, independent contract logger.

Hardwood: A milling classification, referring to wood produced from deciduous trees, such as oaks and maples.

Harvester: A purpose-built machine with a head (attachment) for felling and processing trees into log lengths. Sometimes the processing head itself is referred to as a harvester.

Harvesting: The cutting and removal of trees from the forest for product utilization.

Heartwood: The dark-colored, inactive area of a tree's trunk, (seen in cross section) its main function is to give the tree rigidity and support.

Heterotrophs: Organisms that require an external supply of energy contained in complex organic compounds to maintain their existence.

High-yield forestry: An intensive forest management system, first associated with the Weyerhaeuser Company but now used in variations by many companies, involving clean logging—removing all usable wood; preparing the ground for the new crop of genetically superior seedlings or seed; replanting by hand, machine, or by hydroseeding, usually within 12 months; fertilizing; and thinning until the trees are ready to harvest again and a new cycle commences.

Hydric soil: Soil that contains hydrogen.

Hydrologic cycle: The water cycle involving precipitation, evaporation, transpiration by plants, and surface and underground flow to the oceans.

Hydrophyte: A plant growing in water or very wet soil.

Industrial forest landowners: Forest landowners having extensive holdings of commercial timberland, as well as one or more types of processing plants, such as sawmills, pulp plants, or paper mills. Georgia Pacific, Smurfit-Stone Container Corporation, Weyerhaeuser, International Paper, and Mead-Westvaco are typical examples.

Integrated pest management: A strategy to control pests combining biological, silvicultural, and chemical treatments to effect stated management goals.

Integration (in manufacturing): The capability to utilize timber fully, such as a sawmill, a box factory, and a pressed-board plant all in one location, owned by the same company.

Intensive forest management: Management utilizing all or most of the up-to-date technology including fire management, and insect and disease protection; can also include thinning, fertilizing, and prompt reforestation after harvesting, often with specially selected seedlings. Because of the cost, usually limited to highly productive areas. See *high-yield forestry.*

Interpretive program: A service offered to visitors to parks, forests, and other outdoor recreation areas; includes talks, guided tours, and cultural interpretation, as well as exhibits, signs, self-guided tours, publications, and other information of use to the visitor. This information flow may also assist in solving management problems and in enhancing the visitors' understanding and appreciation of natural resources.

Introduced forest pests: Exotic pests that have become established in a region other than their native provenance.

Laissez-faire: "Allow to act," a French phrase that came to stand for the theory that the government should interfere as little as possible to the functioning of the economy; that the economy should be allowed to act so that the marketplace would control demand and supply, and thus determine the price and allocation of commodities. Leave alone. Hands off.

Land ethic: Code and standard for land management consistent with ecologically sound principles, and aiming for better harmony between people's needs for resources and nature's ability to sustain healthy and productive land.

Landing: A place where logs are collected while waiting for futher transport.

Linear programming: An analytical model that can be used to produce the mathematically best plan of action to meet some stated goal; in forestry mostly used for regulating forests and scheduling harvests.

Litter: (forest) A blanket of organic material, consisting of fallen branches, twigs, leaves, flowers, and fruits, as well as animal residues, found on the forest floor in various stages of decomposition.

Loessal soils: Fine-grained fertile loams deposited mainly by wind.

Log rule: A formula or table that gives the volume of a log on the basis of its diameter and length.

Mast: Food for wildlife consisting of the fruit of such trees as American beech, walnuts, oaks, and hickories.

Mechanized: Harvesting and processing timber with machines, reducing manual labor on the ground.

Mensuration: An adaptation of mathematics to the measurement of forested areas, of single trees and of logs, of total biomass, and of other units of forest products.

Metamorphosis: Changing form; complete metamorphosis in many insects encompasses four stages of development: egg, larvae, pupa, and adult; gradual metamorphosis goes from egg to nymph to adult.

Morphology: That branch of biology dealing with the form and structure of animals and plants.

Multiple-use management: Managing lands for many uses simultaneously in combination, such as providing outdoor recreation, forage, timber, watershed protection, and fish and wildlife. Seeking the best combination of uses and coordinated management without impairment of the productivity of the land, with consideration being given to relative resource values and not necessarily the greatest return on the dollar.

Muskeg: Northern bog, characterized by sphagnum moss, grasses, sedges, spruce, and tamarack trees.

Mycorrhiza: A symbiotic association of the mycelium of a fungus with the roots of certain plants, in which the hyphae form a closely woven mass around the rootlets of the plants. They serve as extensions of the plants' root systems, helping to absorb water and nutrients from the soil, and enhance resistance to pathogens.

Naval stores: Oleoresin extracted from pine forests, consisting of turpentine and rosin and their products; historically used in wooden ship maintenance to seal and repair seams and cracks. Although no longer used for those purposes, these silvachemicals, obtained chiefly from the pulping process as by-products, have many other uses in paints and manufacturing.

New forestry: See *ecoystem management.*

Nomadic agriculture: A primitive form of crop rotation. Also called *shifting agriculture,* describing the cutting and burning (slash and burn) of forests to obtain land to cultivate for crops. Although these areas are usually abandoned after a few years since better results can be obtained by moving on and clearing a new area, the farmers may return after the trees have regrown to begin the cycle over. They may also leave certain trees useful to their subsistence.

Nonconsumptive use: Viewing or appreciating natural resources without consuming or removing them for pleasure, commercial gain, or food; the term is usually used in connection with wildlife.

Nondeclining even flow: The stretching out of timber harvesting schedules, usually where there are large, existing old-growth timber volumes, to avoid a rapid decline in the sustainable harvest after all the old-growth has been removed. Used in the National Forest Management Act of 1976 as a criterion for timber harvest on the National Forests.

Nonindustrial private forest landowners: A term describing landowners who, as a group, own 57 percent of private commercial forestland, but who are not principally in the business of growing and processing trees. Sometimes referred to as *NIPF*.

Old growth: Mature or overmature stands of timber found in areas that have been undisturbed for many decades or centuries; usually trees of large size and large volume per acre.

Outdoor recreation: See *wildland recreation.*

Overstory: Dominant or tallest trees in a stand.

Pathogens: Biotic agents capable of causing disease; usually parasitic fungi, bacteria, viruses, and other microorganisms, as well as parasitic seed plants such as mistletoes. Do not include insects and related organisms.

Peavy: A hand tool for turning logs.

Peeler: A high-grade log from which rotary-cut veneer can be peeled in order to make plywood.

Perturbations: Natural or humanly caused disturbances, such as fire, erosion, windthrow, or avalanche.

Pheromone: A hormonal secretion stimulating a physiological or behavioral response from an individual of the same species.

Phloem: The inner bark of a tree where carbohydrates are transported downward through the tissue.

Playa: The sandy, salty, or mud-caked floor of a desert basin that temporarily becomes a shallow lake after a heavy rain.

Policy: A predetermined course of action seeking to forward general goals; forest policies are governing principles, plans, or courses of action designed to achieve goals through management and use of forest resources.

Prescribed fire: Controlled application of fire according to a carefully devised and closely monitored plan.

Preservationist: A person subscribing to a philosophy of saving or preserving renewable natural resources rather than using them for commodity production.

Presuppression: Preparations in advance of wildfire in order to better attack and suppress fires that do occur.

Provenance: Specific place of origin of plant or other materials; region wherein climatic and other growth conditions are similar.

Public domain: Originally, those national lands acquired through cession, acquisition, treaty, purchase, trade, or exchange; today the unreserved, unappropriated parts of the orginal public domain; now called *public lands.*

Rangeland: Mainly unforested lands dominating much of the western half of the continent, where precipitation is low and the principal vegetation is grass, forbs, and shrubs such as sagebrush.

RARE: Roadless Area Review and Evaulation; a Forest Service study to determine suitability of roadless areas on the National Forests for formal wilderness classification. RARE I took place in the early 1970s, and the second study called RARE II took place in the late 1970s.

Regeneration: Replanting of trees after harvest or fire, or reproduction that occurs naturally through seed from standing trees or other mechanisms.

Remanufacture: To make into a different form of wood product, such as lumber into furniture or window moldings.

Riparian: Streamside.

Roads: Generally classified as skid trails, spur roads, and main haul roads.

Rotation age: The age at which a given stand of trees will be harvested; harvesting maturity; completion of the growth–harvest cycle.

Rough and rotten trees: Poor-quality or decaying trees not suitable for milling for lumber. Sometimes called *cull* trees.

Roundwood: Timber in round form; a term used for pulp logs and firewood.

Sapwood: See *xylem.*

Scaling logs: Determining the volume of logs before they are converted into lumber or other wood products. Usually done at a log landing or when logs have been loaded on a truck or railroad car.

Second growth: The forest growth following harvest, fire, windthrow, or other removal of the preexisting stand.

Selection cutting: The removal of certain individual trees of an existing stand to provide space for reproduction, creating a seedling environment strongly influenced by the remaining stand and leading to a stand with several age classes.

Seral stages: The series of ecological stages occurring successively in plant or animal communities.

Shelterwood cutting: Partial removal of the existing stand, producing a variety of seedling environments; after regeneration is established the remaining overstory is usually removed.

Shifting agriculture: See *nomadic agriculture.* Sometimes called *slash-and-burn* agriculture.

Shovel logging: Relaying logs from a harvest area to roadside using a hydraulic log loader.

Shrub: Woody, perennial plant, usually with several stems branching from the ground; seldom more than 10 feet in height.

Side: A group of people and machines working together as a unit. A contractor might have a "skidder side" and a "cable side" working on the same "show."

Silviculture: The applied science of reproducing and manipulating a forest in order to fulfill stated management objectives.

Site index (SI): A numerical rating of a soil's productivity potential, expressed as the height a dominant tree has attained by a certain age, in that soil. A good Douglas-fir soil would have an SI of 150, 50-year index, meaning the dominant Douglas-fir trees would be 150 feet tall at age 50.

Skid: The bunch of logs pulled behind a track or wheel skidder. Also known as a pull, turn, drag, or twitch.

Skidder: A machine on tracks or wheels used to drag logs from the forest to a landing where they can be further processed and loaded on a truck. Skidders are equipped with cables or grapples.

Slash: Unused residues from harvesting, consisting mainly of limbs, leaves, bark, and undesirable specimens of commercial species, as well as commercially undesirable species of any size or condition.

Software: Programs or sets of instructions developed for use in a computer, to enable it to carry out certain functions.

Softwood: A milling classification, referring to the wood of conifers such as pine, fir, and cedar.

Springboards: Boards timber fallers stood on while felling large trees; wedged into cuts above the butt swell of the huge old-growth trees, they were used to support the fallers. Now seen only in old photographs, although the holes cut for them may be seen on old-growth conifer stumps in the West.

Stand density: A qualitative measure of tree stocking.

Stereoscopic pair: Two aerial photographs with enough overlap and consequent duplication of detail to allow three-dimensional examination of an object or area seen in both photographs.

Stocking: The number of trees in a stand compared with the ideal number for best commercial growth and management. As in a "fully stocked stand," or an "over-stocked stand," that would benefit from thinning.

Stratification: Moistening, controlled temperatures, and other practices to facilitate the germination of seeds when planting in a forest nursery.

Strip cropping: Alternate rows of crops, usually planted to follow the contour of the land to help control erosion.

Stumpage: The estimated volume and value of standing timber. The value received by the landowner for trees harvested.

Subclimax forest: Refers to a stand where one species is apparently being replaced by another through natural succession, or development patterns.

Succession: See *forest development patterns.*

Surface fire: Fire that burns surface vegetation and dead leaf, twig, and branch litter of the forest floor.

Sustained yield: The yield that, theoretically, a forest can produce continuously at a given intensity of management; reproduction and growth replacing removals.

Synecology: The ecological consideration of organisms in relation to their community.

Taxonomy: The science or technique of identification, naming, and classification.

Technology transfer: Promoting the use of modern technical innovations or practices through talks, demonstrations, exhibits, bulletins, and other means; getting the results of research out into the field. The diffusion, adoption, and implementation of new ways of doing things.

Terracing: Growing crops on hillsides by making a series of flat steps rather than planting on a slope.

Thinning: The removal of selected trees from an immature stand to stimulate growth on the remaining trees by allowing them access to more sunlight, moisture, and nutrients.

Tillage: Cultivation of the soil by turning it over with a plow, disk, or harrow.

Timber harvesting: The planned removal of trees from the forest for the purpose of meeting resource management objectives.

Transpiring: Giving off moisture as vapor through the surface of leaves.

Tree: Woody perennial plant, with a single, well-defined stem and generally a definite crown; usually 20 feet or more in height at maturity.

Tree farm: A formal classification of private forests having a certified management plan. Sponsored by the American Forest Foundation of the forest industry to promote good management and recognize accomplishments on the areas so designated, where reproduction and growth of trees for commercial purposes takes place.

Type conversion: Changing the species or mix of species growing in an area, through planting or other management techniques.

Veneer: A thin sheet of wood, sliced or peeled on a rotary lathe to a uniform thickness; used as a facing for other woods; also refers to the various layers in plywood.

Virgin forest: A mature forest that has not been humanly modified, or modified only by fire protection; an old-growth forest.

Visitor day: A measurement of visitor use at a recreation site; may be a 12-hour or 24-hour visit, depending on agency practice.

Volume table: Shows the average volume trees or logs will yield when measured by diameter and merchantable length.

Watershed: The total area contributing drainage to a stream or river.

Wilderness: A legal classification for predominately roadless land, where natural processes shape the landscape, opportunities for solitude and primitive forms of recreation exist, and human influence remains minimal.

Wildfire: A fire burning out of control in a forest or range area.

Wildland recreation: Also called *forest recreation, outdoor recreation,* or *natural resource recreation;* embraces those forms of recreation dependent on natural resources. Activity-oriented recreation, by contrast, is characterized by dependence on facilities.

Wildlife: Nondomesticated animals; usually referring to mammals, fish, and birds. If hunted by humans, referred to as "game."

Wolf tree: A vigorous dominant tree with a broad, spreading crown. Branches may extend all the way to the ground; usually grows in an open area. These characteristics make such a tree of limited value for timber, but may increase interpretive and wildlife habitat potential.

Xylem: Commonly known as *sapwood;* that part of the tree trunk that transports water and mineral from roots to leaves.

Yarding: Dragging or hauling of logs to a landing or collecting area for further transport.

Index

Note: Page numbers in **boldface** type refer to illustrations, photographs, or tables.